ADVANCED SEMICONDUCTOR AND ORGANIC NANO-TECHNIQUES

Part II

Tunable Bandgaps and Nanotubes

Related Publications

SEMICONDUCTORS AND SEMIMETALS
A Treatise

Edited by   *Robert K. Willardson*   *Eicke R. Weber*
Consulting Physicist   Department of Materials Science
12722 East 23rd Avenue   and Mineral Engineering
Spokane, WA 99216-0327   University of California at
                                      Berkeley
                                      Berkeley, CA 94720

Recent Series Titles

Volume 68   Isotope Effects in Solid State Physics
                       ISBN 0-12-752177-1

Volume 69   Recent Trends in Thermoelectric Materials Research I
                       ISBN 0-12-752178-X

Volume 70   Recent Trends in Thermoelectric Materials Research II
                       ISBN 0-12-752179-8

Volume 71   Recent Trends in Thermoelectric Materials Research III
                       ISBN 0-12-752180-1

Volume 72   Silicon Epitaxy
                       ISBN 0-12-752181-X

Volume 73   Processing and Properties of Compound Semiconductors
                       ISBN 0-12-752182-8

Forthcoming Series Titles

Volume 74   Silicon Germanium Strained Layers and Heterostructures
                       ISBN 0-12-752183-6

Thin Film Diamond—*Nebel and Ristein*

Laser Crystallization of Silicon—Fundamentals to Devices—*Nickel*

Industrial Applications of Chalcogenide Semiconductor Glass—*Fairman and Ushkov*

Electric Levels and Properties of Chalcogenide Semiconductor Glass—*Fairman and Ushkov*

Formation and Structure of Chalcogenide Semiconductor Glass—*Fairman and Ushkov*

Other Related Titles

Quantum Coherence, Correlation and Decoherence in Semiconductor Nanostructures
ISBN 0-12-682225-5

Interlayer Dielectrics for Semiconductor Technologies
ISBN 0-12-511221-1

Handbook of Nanostructured Materials and Nanotechnology—5 Volume Set
ISBN 0-12-513760-5

Nanostructured Materials and Nanotechnology—Concise Edition
ISBN 0-12-513920-9

ADVANCED SEMICONDUCTOR AND ORGANIC NANO-TECHNIQUES

Part II

Tunable Bandgaps and Nanotubes

Editor

HADIS MORKOÇ

DEPARTMENTS OF ELECTRICAL ENGINEERING AND PHYSICS
VIRGINIA COMMONWEALTH UNIVERSITY
RICHMOND, VIRGINIA

ACADEMIC PRESS

An imprint of Elsevier

Amsterdam • Boston • London • New York • Oxford • Paris
San Diego • San Francisco • Singapore • Sydney • Tokyo

ACADEMIC PRESS
An Imprint of Elsevier
84 Theobald's Road, London WC1X 8RR, UK
http://www.elsevier.com

ACADEMIC PRESS
An Imprint of Elsevier
525 B Street, Suite 1900 San Diego, California 92101-4495, USA
http://www.elsevier.com

Set   ISBN 0-12-507060-8
Part I   ISBN 0-12-507061-6 Nanoscale Electronics and Optoelectronics
Part II   ISBN 0-12-507062-4 Tunable Bandgaps and Nanotubes
Part III ISBN 0-12-507063-2 Physics and Technology of Molecular and Biotechnology Systems

A catalog record for this book is available from the Library of Congress

A catalogue record for this book is available from the British Library

Typeset by Newgen Imaging Systems (P) Ltd, Chennai, India
Printed and bound in Great Britain by MPG Books, Bodmin, Cornwall

03 04 05 06 07 MP 9 8 7 6 5 4 3 2 1

Contents

Chapter 1   Engineering the Electronic Structure and the Optical Properties of Semiconductor Quantum Dots   1

M. De Giorgi, R. Rinaldi, T. Johal, G. Pagliara, A. Passaseo,
M. De Vittorio, M. Lomascolo, R. Cingolani, A. Vasanelli,
R. Ferreira, and G. Bastard

Chapter 2   GaN-Based Modulation Doped FETs and Heterojunction Bipolar Transistors 51

H. Morkoç

Chapter 3   Ultraviolet Photodetectors based on GaN and AlGaN   147

H. Temkin

Chapter 4   Organic Field-Effect Transistors for Large-Area Electronics   191

C. D. Dimitrakopoulos

Chapter 5   Organic Optoelectronics: The Case of Oligothiophenes

G. Gigli, G. Barbarella, M. Anni, and R. Cingolani

Chapter 6   Single-Walled Carbon Nanotubes for Nanoelectronics

M. S. Fuhrer

Preface

Novel heterostructure devices have made many inroads in telecommunications and make it possible to amplify signals and detect negligible signals leading to technologies such as digital telephones and small direct satellite broadcast dish antennas. Indications are that the same technology is poised to bring about automobile guidance and collision avoidance systems. Key among these technologies is the modulation doped field-effect transistor (MODFET), which can loosely be considered the compound semiconductor analog of the ubiquitous metal oxide semiconductor field-effect transistor (MOSFET). The MODFET utilizes a pseudo-two-dimensional carrier gas, the concentration of which is modulated by a gate potential. In a MODFET, a larger-bandgap material with a high doping concentration (assumed to be n-type in this discussion) is grown on a lower-bandgap intrinsic material. Because the undoped low-bandgap material has no donor atoms cluttered about, impurity scattering no longer inhibits the carrier mobility and saturation velocity. Early versions of MODFETs relied on GaAs as the conducting medium and AlGaAs as the donor layer of larger bandgap in relation to GaAs. Shortly thereafter, the material system expanded to include the InGaAs/InAlAs structures on InP substrates, and SiGe strained layers on Si substrates. Although the InP-based heterostructures are capable of operating at higher frequencies, in the end, the structure that turned out to have the widest range of applications is the pseudomorphic AlGaAs/InGaAs/GaAs PMODFET. The material system for MODFETs has been extended to include GaN with total continuous power outputs approaching 30 W in devices with a few millimeter gate periphery, the realm of expensive, non-reproducible, power combining technologies. These have applications in portable radar, telecommunications, and broadcasting. The chapter by H. Morkoç delves into this.

Once conceived of having applications only in a small segment of telecommunications, semiconductor lasers have made great inroads in many aspects of technological wonder, with ever-evolving structures using new semiconductors for extending the wavelength of operation, both on the short- and long-wavelength ends. Compact optical emitters with very low power consumption coupled with detectors and fiber optics have improved long- and short-haul communication systems to the point where one can no longer tell the distance while conversing on the telephone. While wide-bandgap semiconductors were recognized as prime candidates for efficient compact blue/green light sources in the early 1960s, nearly three decades passed before the promise of blue/green LEDs and diode lasers finally saw the light at the end of the tunnel by the early 1990s. While early demonstrations were based on ZnSe, which did not make it because of reliability problems and its limitation to the green wavelengths only, another semiconductor, GaN, became the medium in which lasers with longevity acceptable to the industry were developed. It now appears increasingly likely that the 405–415-nm InGaN laser, or perhaps its even shorter wavelength AlGaInN cousins, will form the basis of the next generation of optical disk technology, to supercede the current DVD format. There have been recent feasibility demonstrations using the 405-nm laser in conjunction with a re-writable, dual-layer, phase-change standard 12-cm optical disk, yielding a storage capacity of 27 Gbytes per side with a user data transfer rate of 33 Mbps. This compares with approximately 5 Gbytes per side for the present DVD using a 670-nm red laser used commonly at the time of this writing. Details can be found in the chapter by A. V. Nurmikko.

The GaN material system did not leave the detector field untouched either. One of the driving forces is that GaN-based detectors operating in the ultraviolet part of the spectrum would have a multitude of applications, from astronomy and spectroscopy to terrestrial detection of flame and rocket exhaust. At altitudes below 20 000 ft, absorption by the ozone layer virtually eliminates solar radiation in the spectral region of 240–280 nm. Photodetectors operating in this wavelength range would not respond to solar illumination if designed properly and thus would not be limited by background radiation. There are several stringent requirements among which are (i) high photoresponse in the 240–280-nm range, (ii) low leakage currents resulting in noise equivalent power of $\sim10^{-15}$ W, and (iii) complete insensitivity to visible light, that is, out of band rejection on the order of 10^8–10^9, implying very rapid sensitivity roll-off. Progress made in GaN-based detectors where the active absorbing layer is AlGaN is very promising, as described in the chapter by H. Temkin.

Quest for better and more versatile structures for optoelectronic devices (lasers, far-infrared detectors, memories) and quantum computers fuelled an interest in reduced dimensional structures. However, semiconductor quantum dots (QDs) are also of particular interest for the study of basic quantum mechanical effects (QD-based lasers are discussed in Part I). Carriers in QDs are confined in three dimensions, which leads to energy quantization and to the formation of electronic shells that resemble those of natural atoms. For this reason, the QDs are often referred to as "artificial atoms". The atomic-like properties of such systems can be fully exploited only through the complete control of the geometry (shape, size) and composition of the dots.

The simultaneous knowledge of these parameters is necessary to engineer the wavefunctions and to exploit the dot properties in functional devices. Various techniques such as electron microscopy techniques, including transmission electron microscopy (TEM) and scanning electron microscopy (SEM), and scanning probe microscopy techniques, including scanning tunneling microscopy (STM) and atomic force microscopy (AFM), are necessary to probe the physical characteristics of QD structures. Compositional information at nanoscale is obtained by specific experiments such as spatially resolved electron energy loss spectroscopy (EELS). To probe the atomic-like optical properties of a single dot, local spectroscopy methods are necessary, involving local excitation and light collection from a single QD. Near Field Scanning Spectroscopy on nanomesas containing a few dots (possibly a single dot) is normally used for spatially resolved luminescence spectroscopy. However, a better resolution that allows direct wavefunction imaging can presently be accomplished only by tunneling current induced luminescence and scanning tunneling spectroscopy. Modeling is another important ingredient for the design of artificial atoms. The solution of a full three-dimensional Schrödinger equation including different contributions to the dot confining potentials, like the strain field and the existence of built-in electric fields induced by the piezoelectricity, indeed becomes crucial for the complete understanding of the artificial atom properties. The details of this field of study are discussed in the chapter by M. De Giorgi *et al.*

Parallel to development of electronic and optoelectronic devices based on inorganic semiconductors, the quest for reduced cost, disposability, and flexibility has fuelled a flurry of activity in exploring plastics for devices. Specifically, applications such as dielectric coatings, field-effect transistors, and light emission from plastics is just gaining steam. There are two types of devices that involve the use of organics. In one, an organic dye containing red and green (or simply yellow) dies are pumped with a InGaN-based blue LED, and the combination of blue from the InGaN LED and yellow from the organic medium leads to soft white light. In the all-organic case, an organic LED is prepared in full and biased to produce light. The color of light is dependent on the types of dies that are in the recombination medium. To distinguish this from other types of LEDs, the name OLED for organic LED has been coined.

These plastic LED materials are attractive in that they have applications in large-area displays for back lighting, active matrix displays, and even stimulated emission. Large area, physical flexibility, and low cost are the attractive features afforded by the organic technology. Until recently, light-emitting polymers (LEPs) were little more than a scientific curiosity. In the wake of rapid scientific progress, particularly in operation lifetime, a bright future is now seen for organic emitters for indoor displays, background panels, and nightlights built around relatively large organic molecules. There are also efforts to fabricate transistors based on polymers with the hope of constructing displays having built-in control circuitry in much the same way as liquid crystal displays. Until recently, the damper was the short longevity and, to some extent, brightness.

Organic emitters of recent vintage are in some ways similar to the semi-conductor varieties, taking advantage of multi-layers, with indium tin oxide (ITO) serving as a hole injector, contacts such as Mg and Ag (10:1 serving as electron injector), and a medium for recombination referred to as AlQ_3 was demonstrated for the first time at Kodak Research Laboratories. Doping AlQ_3 with various substances determines the emission wavelength and along the same lines dopant species can be customized for white light with a desired spectrum. This early report demonstrated an efficiency of $1.5 l/W$, brightness of $1000 \, cd/m^2$ coupled with a voltage of $10 \, V$. The aluminum complex transporting electrons, AlQ_3 [tris (8-hydroxyquinolin) aluminum], emits at $520 \, nm$ (green), and Nile Red doped AlQ_3 emits in the red at $600 \, nm$. The hole transporting layer (HTL) triphenyldiamine derivative (TPD), emitting around 410–$420 \, nm$, can be used for blue. All of these layers together culminate in the generation of the three primary colors that, when their concentrations are adjusted appropriately, produce white light. White light with luminance in the range of $10\,000 \, cd/m^2$ is possible. In general, the emission spectrum is wide, compared to semiconductor emitters. Therefore, in applications where the hue is important, the spectrum can be made narrower by placing the emitter region into a cavity. Though the concept has been around for quite a while, it was not applied to plastic LEDs until the last few years. In fact, in cases where the light emission from the active medium cover a wide range, as in the case where multiple dies are used, the cavity action can be used to tune the wavelength. OLEDs are discussed in the chapter by G. Gigli *et al.*

With the advent of organic semiconducting materials, efforts have been underway for quite some time to exploit these plastics that are easy to deposit on transparent substrates, at a potentially much lower cost. The generic term for FET where the channel is separated from the gate by an insulating material, other than SiO_2, is insulated gate FET, or IGFET for short, or organic thin-film field-effect transistors (OTFT). IGFETs based on conjugated polymers, oligomers, or other molecules have been fabricated and studied in the past. Initially, industrial applications of organic semiconductors exploited their photoconductive properties in xerography. However, the potential for applications of organic semiconductors with much broader impact became clear with the initial demonstrations of OTFTs based on either small organic molecules or conjugated polymers. The impressive improvements in performance and efficiency of organic devices during the last decade attracted the interest of the optoelectronics industry and paved the way to practical applications for organic semiconductors.

As in traditional inorganic semiconductors, organic semiconductors function either as p-type, in which the majority charge carriers are holes, or n-type, in which the majority charge carriers are electrons. The most widely studied organic semiconductors have been p-type. However, in the last decade, several new reports on OTFTs based on n-type organic semiconductors have appeared in the literature. Because of the relatively low mobility of organic semiconductors, OTFTs cannot rival the performance of field-effect transistors based on single-crystalline inorganic semiconductors, such as Si, Ge, and GaAs, which have charge carrier mobilities (μ) of three or more orders of magnitude higher than organics. Consequently, OTFTs can be competitive for applications requiring large areas, low-temperature processing, structural flexibility, and especially low cost. Among those applications are switching devices for active matrix flat panel displays (AMFPDs), organic light emitting diodes (OLEDs) and organic sensors, low-end smart cards, radio-frequency identification (RFID) tags, and electronic tickets sporting organic integrated circuits. The world of plastic electronics is discussed in the chapter by C. D. Dimitrakopoulos.

Switching gears somewhat, spawned by the synthesis of the C_{60} fullerene molecule, dubbed the buckyball, various tubular forms of carbon (carbon nanotubes) with diameters under 20 nm were discovered. Carbon nanotubes could be made conducting or semiconducting, and excellent electron transport properties were soon realized. Single-walled nanotubes (SWNTs) are naturally the simplest of these structures. They are nothing more than a single graphite plane rolled into a thin tube. Methods used in the synthesis of SWNTs do not produce nanotubes without large dispersion in their physical properties necessitating large-scale purification of SWNTs, for which methods have been developed. The cohesion of these molecular crystals occurs through van der Waals interactions and possibly other interactions involving electron correlations. It is commonly observed that the individual SWNTs coalesce into tube-like strands. Switches and simple circuits based on carbon nanotubes have already been demonstrated, and transistor operation has also been demonstrated. The details can be found in the chapter by M. S. Fuhrer.

HADIS MORKOÇ
Richmond, VA
October 2002

List of Contributors

Numbers in parentheses indicate the pages on which the authors' contributions begin.

M. ANNI (241), *National Nanotechnology Laboratory of INFM, University of Lecce, via Arnesano 73100 Lecce, Italy*

G. BARBARELLA (241), *ISOF, Area della Ricerca CNR, Via Gobotti 101, I-40129 Bologna, Italy*

G. BASTARD (1), *Laboratoire de Physique de la Matiere, Condensee, Ecole Normale Superieure, 75005 Paris, France*

R. CINGOLANI (1, 241), *National Nanotechnology Laboratory of INFM, University of Lecce, via Arnesano 73100 Lecce, Italy*

M. DE GIORGI (1), *National Nanotechnology Laboratory of INFM, C/O University of Lecce, Department of Innovation Engineering, Lecce, Italy*

M. DE VITTORIO (1), *National Nanotechnology Laboratory of INFM, C/O University of Lecce, Department of Innovation Engineering, Lecce, Italy*

C. D. DIMITRAKOPOULOS (191), *IBM T. J. Watson Research Center, Yorktown Heights, NY 10598*

R. FERREIRA (1), *Laboratoire de Physique de la Matiere, Condensee, Ecole Normale Superieure, 75005 Paris, France*

M. S. FUHRER (293), *Department of Physics, University of Maryland, MD 20742-4111*

G. GIGLI (241), *National Nanotechnology Laboratory of INFM, University of Lecce, via Arnesano 73100 Lecce, Italy*

T. JOHAL (1), *National Nanotechnology Laboratory of INFM, C/O University of Lecce, Department of Innovation Engineering, 7310 Lecce, Italy*

M. LOMASCOLO (1), *IME-CNR, Instituto per lo studio di nuovi Materiali per l'Elettronica National Nanotechnology Instititu of INFM, C/O University of Lecce, 73100 Lecce, Italy*

H. MORKOÇ (51), *Virginia Commonwealth University, Department of Electrical Engineering and Physics Department, Richmond, VA 23284-3072*

A. NURMIKKO (345), *Division of Engineering, Brown University, Providence, RI 0292*

G. PAGLIARA (1), *National Nanotechnology Laboratory of INFM, C/O University of Lecce, Department of Innovation Engineering, Lecce, Italy*

A. PASSASEO (1), *National Nanotechnology Laboratory of INFM, C/O University of Lecce, Department of Innovation Engineering, Lecce, Italy*

R. RINALDI (1), *National Nanotechnology Laboratory of INFM, C/O University of Lecce, Department of Innovation Engineering, 73100, Lecce, Italy*

H. TEMKIN (147), *Department of Electrical and Computer Engineering, Texas Tech University, Lubbock, TX 79409*

A. VASANELLI (1), *Laboratoire de Physique de la Matiere, Condensee, Ecole Normale Superieure, 75005 Paris, France*

Advanced Semiconductor and Organic Nano-Techniques (Part II)
H. Morkoç (Ed.)

CHAPTER 1

Engineering the Electronic Structure and the Optical Properties of Semiconductor Quantum Dots

M. De Giorgi, R. Rinaldi, T. Johal, G. Pagliara, A. Passaseo, M. De Vittorio, M. Lomascolo, and R. Cingolani

UNIVERSITY OF LECCE, LECCE

A. Vasanelli, R. Ferreira, and G. Bastard

ECOLE NORMALE SUPÉRIEURE, PARIS

The study of the optical and electronic properties of zero-dimensional (0D) semiconductor heterostructures has been the subject of intense investigation in the last few years. The interest for such nanostructures is mainly due to their application to opto-electronic devices (laser, far infrared detectors, memories) (Arakawa and Sakaki 1982; Weisbuch and Vinter 1991; Bimberg *et al.* 1999) and quantum computers (Kane 1998). However, semiconductor quantum dots (QDs) are also of particular interest for the study of basic quantum mechanical effects. Carriers in quantum dots are confined in three dimensions since their de Broglie wavelength is comparable to the dot size. This leads to energy quantization and to the formation of electronic shells which resemble those of natural atoms. For this reason, QDs are often referred to as "artificial atoms" (Kastner *et al.* 1993; Ashoori 1996; Gamman 2000). The atomic-like properties of such systems can be fully exploited only through the complete control of the geometry (shape, size) and composition of the dots. The simultaneous knowledge of these parameters is necessary to engineer the wavefunctions and to exploit the dot properties in functional devices. This is a tremendous task, as it requires the refinement of growth methods and nanoscale structural, optical, and electrical characterizations at the forefront of the present technologies. In this chapter, we discuss the latest developments of the field, with special attention to the methodologies developed for the control and the engineering of the electron states of InGaAs/GaAs QDs. Valuable information on the shape and size of the QDs can be obtained by a proper combination of electron microscopy techniques, like transmission electron microscopy (TEM) and scanning electron microscopy (SEM), and scanning probe microscopy techniques, such as scanning tunnelling microscopy (STM) and atomic force microscopy (AFM). Compositional information at nanoscale is obtained by specific experiments such as spatially resolved electron energy loss spectroscopy (EELS). To probe the atomic-like optical properties of a single dot, local spectroscopy methods are necessary, involving local excitation and light collection from a single QD. Near field scanning spectroscopy on nanomesas containing a few dots (possibly a single dot) is normally used for spatially resolved luminescence spectroscopy. However, a better resolution which allows a direct wavefunction imaging can presently be accomplished only by tunnelling current induced luminescence and scanning tunnelling spectroscopy. Modelling is another important ingredient for the design of artificial atoms. The solution of a full three dimensional Schrödinger equation including different contributions to the dot confining potentials, like the strain field and the existence of built-in electric fields induced by the piezoelectricity, becomes indeed crucial for the complete understanding of the artificial atom properties.

In what follows, we provide a brief review of these topics, in the attempt of describing the strategy for the engineering and fabrication of artificial atoms as a whole.

I. Fabrication and Structural Analysis of InGaAs/GaAs Quantum Dots

Different techniques are used for the fabrication of quantum dots, namely: lithographic patterning and etching of quantum well structures (Beaumount 1991; Snow *et al.* 1993; Fujita *et al.* 1995), growth on patterned substrates (Sugawara 1995; Hartmann *et al.* 1997) or by using stressors (Lipsanen *et al.* 1995). QDs have also been obtained from monolayer (ML) fluctuations of the well–barrier interface in quantum well structures (Gammon 1996). However, the most successful technique is the self-assembled growth by the Stranski–Krastanov (SK) mode (Goldstein *et al.* 1985; Ledentsov 1996; Rinaldi 1998; Passaseo *et al.* 2001c), which allows the realization of quasi-0D semiconductors of excellent structural quality. In such a growth mode, the formation of 3D islands (QDs) is driven by the strain field induced by the deposition of a few MLs of a highly strained material on a buffer layer. When the deposited layer (wetting layer, WL) exceeds a critical layer thickness (CLT), a transition from the 2D to the 3D growth occurs. At the CLT, the accumulated strain energy due to the lattice mismatch makes the island surfaces energetically more favourable than the flat surface, resulting in a uniformly islanded surface. The islanding process has been described using a strain-induced roughened growth front model, in which a roughened surface is stable for wavelengths above a minimum value (Srolovitz 1989). Since the islands are generally capped both the QDs, the WL and the barrier around the QDs are elastically strained (Xie *et al.* 1994) thus influencing the carrier confinement properties (Grundmann *et al.* 1995; Cusack *et al.* 1996). To increase the optical density of QD structures (Solomon *et al.* 1996), several QD layers should be grown in the structures. Since the local strain in the barrier above the dots is responsible for the vertical stacking of the dots, the top islands preferentially nucleate above the lower ones, forming vertical columns of dots (Xie *et al.* 1995; Solomon *et al.* 1996). The distance among the layers is a very important parameter. When this distance is comparable to the dot height, tunnel coupling occurs and a sort of dot-molecule is formed. On the other hand, for dot distances larger than ~ 10 nm, vertically stacked dots turn out to be uncoupled.

The prototype samples discussed in this chapter consist of InGaAs/GaAs single and six-fold stacked quantum dots, grown by metal–organic chemical vapour deposition (MOCVD) in the SK growth mode on (001) exactly oriented GaAs substrates. Details about the growth conditions can be found in Passaseo *et al.* (2001c).

The first step in the study of QD sample is the structural characterization of the samples, which provides the relevant information about the dot shape and size. AFM and STM (Wu *et al.* 1997; Xue *et al.* 1999; De Giorgi *et al.* 2000; Marquez *et al.* 2001) are techniques commonly used to investigate the structure of uncapped quantum dots. The first one gives important information about dot density and size dispersion. Figure 1 displays a typical plan view AFM image of a planar array of QDs. In this particular sample, the dot density is about 5×10^9 dots/cm^2 and the size dispersion is reduced to some $\pm 10\%$, resulting in a spectral broadening of the order of ± 15 meV. Detailed information about the shape of uncapped QDs can be obtained by high-resolution STM images which visualize the exact crystallographic faceting of the nanostructures. In Fig. 2, a typical topographic image [bias = 2.0 V, tunnelling current = 0.5 nA; Fig. 2(a)], together with the plan view STM image [Fig. 2(b)] and a line-scan across the centre of the single QD [Fig. 2(c)] are shown. In the image displayed in Fig. 2(a), the QD shape exhibits a complex arrangement of crystal facets resulting in a slightly asymmetric shape. The side-walls are a combination of {111} and {110} planes resulting in steep inclinations with respect to the [100] plane. The base is an octogon with elongated sides aligned along the [001] and [010] directions of the substrate crystal.

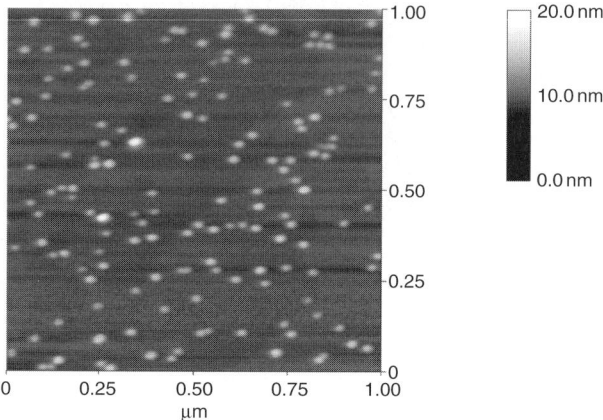

FIG. 1. Topographic plan-view AFM image on uncapped MOCVD InGaAs/GaAs QDs.

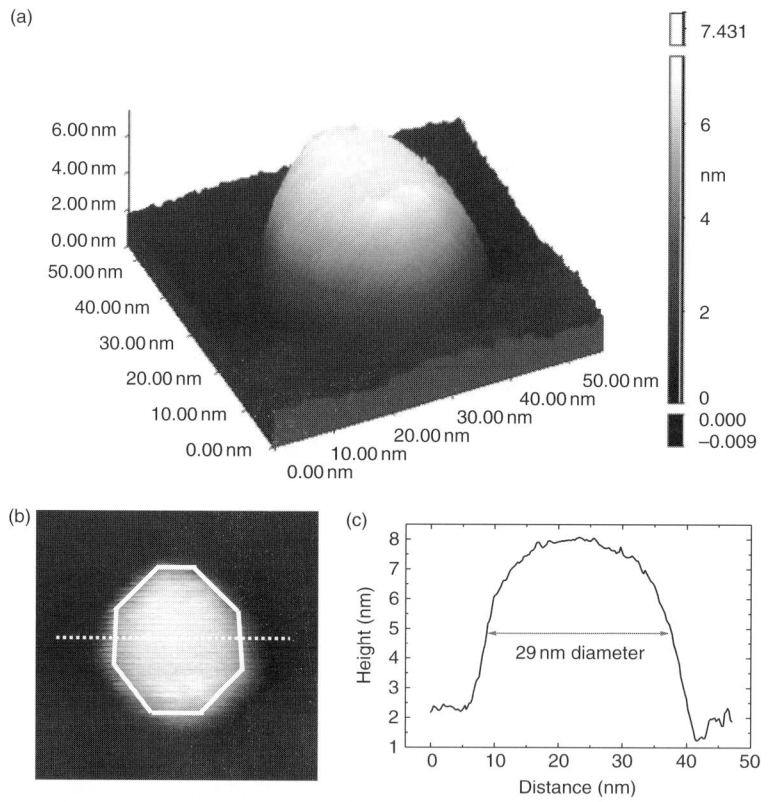

FIG. 2. Topographic STM image (a) on a single uncapped QD together with the plan-view of the STM image (b) and the line-scan profile across the QD (c).

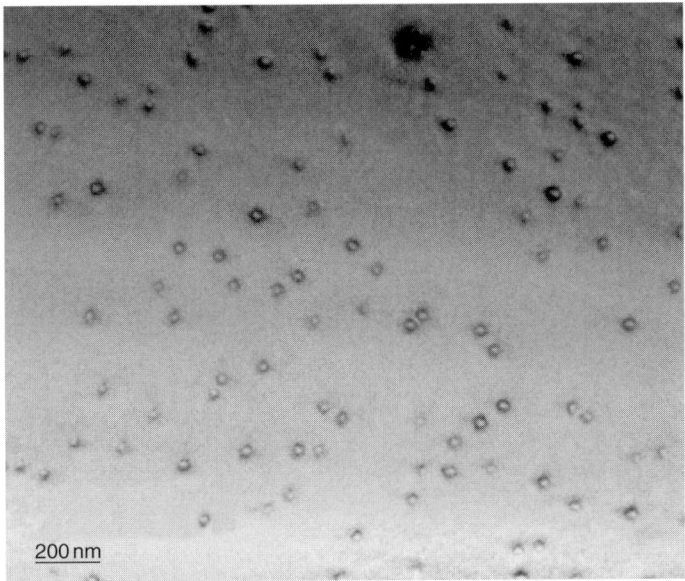

FIG. 3.   (001) Plan-view TEM image of capped InGaAs/GaAs QDs.

To obtain structural informations on capped QDs, TEM investigation is fundamental. In particular, two beam and on zone multi-beam plan-view TEM images are routinely used to investigate such nanostructures (Ruminov *et al.* 1995; Ledentsov *et al.* 1997; Liao *et al.* 1998; Zou *et al.* 1999). Figure 3 shows a representative (001) plan-view TEM image performed on a capped QD sample. In spite of the large contrast achieved by the technique, the interpretation of the diffraction image is very difficult, due to the convolution of strain contrast and composition dependent contrast. This makes the experimental determination of the capped QD shape still rather controversial (Georgsson *et al.* 1995; Ruminov *et al.* 1995; Lian *et al.* 1998; De Giorgi *et al.* 2000). To overcome this difficulty, high-resolution TEM (HRTEM) measurements are performed along different crystallographic directions. In Fig. 4(a), an HRTEM image of one dot in the ⟨001⟩ zone axis is shown. Due to the chemical sensitivity of such zone axis, it is possible to distinguish from the phase contrast features the regions where In is located and, consequently, to get information about the real dot shape (De Giorgi *et al.* 2001). In fact, the diffraction pattern of the GaAs lattice exhibits spots arranged in squares with a bright spot at the centre [Fig. 4(b)], whereas in the InGaAs lattice the central diffraction spot disappears [Fig. 4(c)]. A detailed analysis of the HRTEM images shows that the capped MOCVD InGaAs/GaAs QDs have a cross-section similar to a truncated cone, with an angle between the dot base and the dot side equal to 54°. The sizes of the dots strongly depend on the growth conditions.

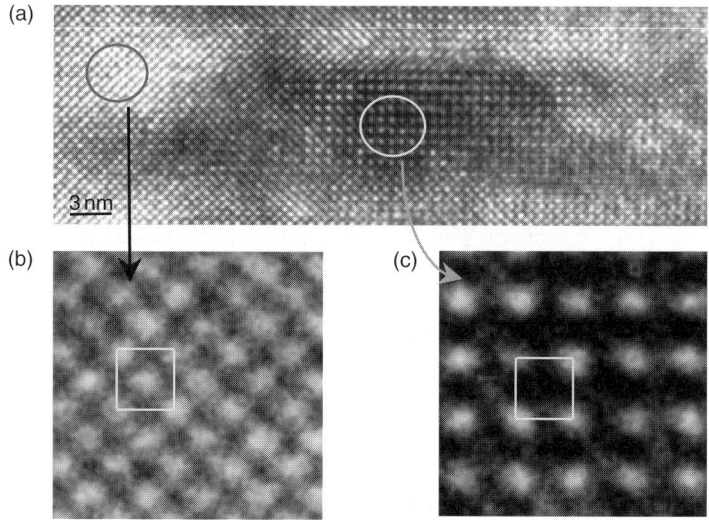

FIG. 4.   (a) High-resolution cross-section TEM image of a single dot along the ⟨001⟩ zone axis. Diffraction pattern in the GaAs barrier region (b) and in the InGaAs dot region (c).

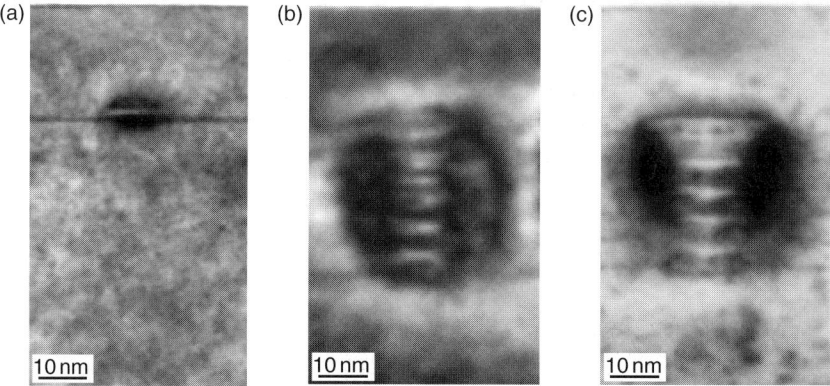

FIG. 5.   ⟨001⟩ cross-sectional low magnification TEM images of a single dot sample (a), and of a uniform (b) and non-uniform (c) vertically stacked dot heterostructure.

In vertically stacked QD samples, TEM measurements give additional information about the uniformity of the islands along the stack. Figure 5 shows the cross-sectional low-magnification TEM images obtained from three samples, consisting, respectively, of a single [Fig. 5(a)] and six-fold stacked dot layers [Fig. 5(b) and (c)]. The comparison between the images of the vertically stacked QD samples [Fig. 5(b) and (c)] shows that, although the dimensions of the bottom layers are nearly the same, the vertical size dispersion can be quite different. In particular, the sample in Fig. 5(b) shows dots which are vertically aligned without extended defects and with a rather uniform size (only the topmost dots exhibit a small enlargement). Conversely, the sample in Fig. 5(c) exhibits an increase of the dot size. These important differences are related to the different growth conditions that influence the quality of the stacked structures in terms of the vertical size uniformity. The QDs are formed after the deposition of only 4 ML of InGaAs in the sample showed in Fig. 5(b) and 6 ML in the sample of Fig. 5(c) resulting in a different total strain in the structure. The occurrence of size dispersion along the stack [like in Fig. 5(c)], in fact, causes considerable inhomogeneous broadening of the optical spectra thus preventing the use of these materials for QD lasers. The achievement of uniform stacked dots is thus a crucial prerequisite to increase the optical density of the active material without introducing large inhomogeneous broadening.

II. Correlation between Strain Field and TEM Contrast

Information about the uniformity of vertically aligned dots can be obtained directly from plan-view TEM images once the correlation between TEM diffraction contrast and localized strain field of single and vertically stacked dots is known. Compared to cross-section TEM measurements, this kind of characterization is easier and faster because it does not need sample preparation. The study of this correlation is discussed in this section.

Figure 6(a)–(c) shows the on-zone plan-view bright field (BF) corresponding to the same sequence of samples reported in Fig. 5. These images

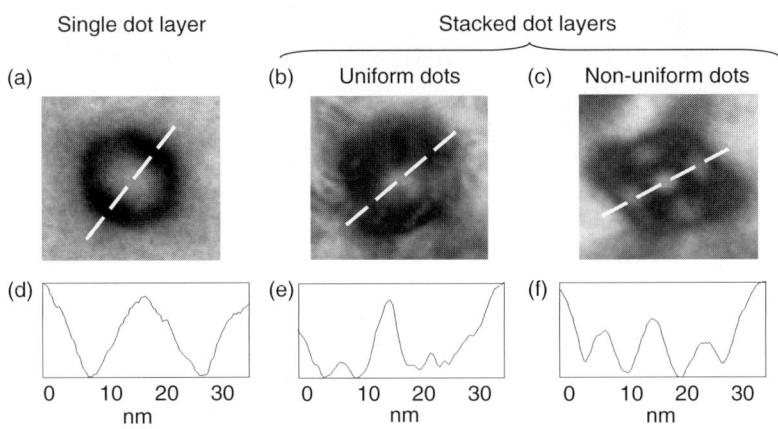

FIG. 6. [100] Plan-view images obtained in the on-zone BF imaging conditions from the single (a), uniform (b), and non-uniform (c) vertically stacked dot sample showed in Fig. 5. The contrast line-scans, performed along the ⟨001⟩ directions on both images, are also displayed in (d) and (f).

have been obtained in the [001] zone axis, that is, with the electron beam propagating along the growth direction. They show striking differences in the contrast pattern. In particular, for the single dot [Fig. 6(a)], and the uniform stack of dots, [Fig. 6(b)] the contrast is characterized by an external dark region, of nearly circular shape, with a bright spot at the centre. The stacked dots with a non-uniform size along the stack [Fig. 6(c)] show a completely different feature, that is, an intensity modulation, resulting in a flower-like pattern. Figure 6(d)–(f) displays the contrast line-scan performed along the ⟨010⟩ directions, that is, the intensity modulation along the dashed lines in Fig. 6(a)–(c). A main central maximum is observed in the line-scan profiles of Fig. 6(d) and (e) whereas Fig. 6(f) shows three maxima of comparable intensity, the external ones being lightly weaker and symmetric with respect to the central one.

It is known that the white/black diffraction contrast in the TEM images is due to the inhomogeneous lattice strain associated to the 3D islands (Yao *et al.* 1991). This induces a local variation of the lattice planes orientation, resulting in a local modification of the electron diffraction conditions. The different contrast observed in the single and stacked dot samples is a quite general feature observed in many other samples, regardless of the growth conditions. This suggests that the main parameter that affects the contrast pattern of plan-view images is the dot size uniformity along the stacking direction. The strain field associated with dot families of different sizes should overlap, inducing a modulation of the strain along the stacking direction, which results in a modification of the electron diffraction condition. This hypothesis has been checked by a study of the correlation between the TEM diffraction contrast and the localized strain field in the single and vertically stacked dots, and, in particular, the effect of the linear combination of strain fields associated with dots of different dimensions. The exact strain field in capped QDs is generally calculated by the finite element method (Nishi *et al.* 1994; Benabbas *et al.* 1996). However, at first approximation, it can be assumed that it is equal to the strain field felt by QDs induced by a stressor (Tulkki and Heinämäki 1995) of simple parallelepiped shape. In this approximation, the QD strain field is calculated by using the analytical method developed in Mazzer *et al.* (1998), which provides results very similar to those obtained by the finite element method. The functions that describe the surface profile (the stressor pattern) are developed in Fourier series: the stress tensor, σ, is calculated by superimposing the stress field associated with each cosine component. Since the amplitude of the relevant Fourier components are much smaller than the corresponding wavelength, one looks for a solution of the elasticity equations:

$$(1 + \nu)\nabla^2 \boldsymbol{\sigma} + \nabla^{\mathrm{T}}\nabla \mathrm{Trace}(\boldsymbol{\sigma}) = 0 \tag{1}$$

$$\nabla \cdot \boldsymbol{\sigma} = 0 \tag{2}$$

having the form of a series expansion:

$$\boldsymbol{\sigma}(x, y, z) = \sum_{\alpha=0}^{\infty} t^{\alpha} \boldsymbol{\sigma}^{\alpha}(x, y, z) \tag{3}$$

where ∇^{T} is the transpose of the gradient vector ∇ and ν is Poisson's ratio. The boundary conditions are given by the requirement that no net force acts on the free surface. In the case of stressors having a simple parallelepiped shape, the first-order expansion of the hydrostatic component of the stress field is given by $\sigma_{\mathrm{hy}} = t\sigma_{\mathrm{hy}}^{(1)} + \ldots$, where

$$\sigma_{\mathrm{hy}}^{(1)} = -\frac{\pi}{6} \sigma_0 (1 + \nu)$$

$$
\begin{aligned}
\times \{ & \chi_y(z)[\phi_y(-x, z) + \phi_y(x, z)] + \chi_y(-z)[\phi_y(-x, -z) \\
& + \phi_y(x, -z)] + \chi_y(x)[\phi_y(-z, x) + \phi_y(z, x)] \\
& + \chi_y(-x)[\phi_y(-z, -x) + \phi_y(z, -x)] \}
\end{aligned} \tag{4}
$$

with

$$\phi_y(u, v) = \frac{L_u + u}{\sqrt{(L_u + u)^2 + (L_v + v)^2 + y^2}}$$

$$\chi_y(u) = \frac{L_u + u}{\sqrt{(L_u + u)^2 + y^2}} \tag{5}$$

L_x and L_z are the dimensions of the dot. The strain tensor $\boldsymbol{\varepsilon}$ is then obtained by applying Hooke's law. By this method, it is found that the hydrostatic strain component is constant inside the dot region and zero outside. This is consistent with the results obtained by other methods in capped QDs (Pryor *et al.* 1998; Andreev *et al.* 1999). A small strain modulation is found at the edges of the dots.

It is worth noting that the displacement vector $\mathbf{u}$ is related to the strain tensor $\boldsymbol{\varepsilon}$ by the well-known equation

$$\varepsilon_{ij} = \frac{1}{2} \left(\frac{\partial u_i}{\partial x_j} + \frac{\partial u_j}{\partial x_i} \right) \tag{6}$$

Only the strain components leading to a local variation of lattice planes which are parallel to the electron beam (parallel to the z axis) give rise to contrast in the TEM plan-view images, namely ε_{xx}, ε_{zz}, ε_{xz}, and ε_{zx}. Since the off-diagonal components are smaller than the diagonal terms, only the term $\varepsilon_{xx} + \varepsilon_{zz}$ is considered in this discussion.

First, let us consider a simple model with only two stacked QDs of different sizes. We label the bigger dot as "dot A" and the smaller one as "dot B". In

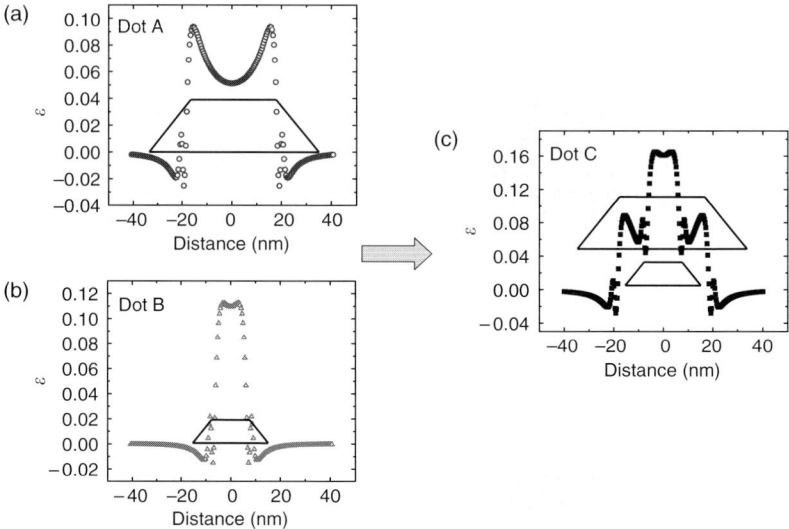

FIG. 7.   Plot of the calculated strain components $\varepsilon_{xx} + \varepsilon_{zz}$ along the diagonal direction for two dots having different sizes. Circles (a) and triangles (b) indicate the strain field of the big (dot A) and small dot (dot B), respectively. Squares (c) represent the coherent superposition of the strain field associated with dots A and B.

Fig. 7, the calculated strain component $\varepsilon_{xx} + \varepsilon_{zz}$ along the cross-section of the dots is plotted for two nanostructures. In the single dot A [Fig. 7(a)] the dilated region is larger than for the single dot B [Fig. 7(b)]. As a consequence, the electrons transmitted at the centre of the stacked QDs (around $x = z = 0$) feel a strain dilation in both QDs, whereas the electrons at the edges of the dots (at about 15 nm away from the centre) feel a strain dilatation only in dot A and a strain compression in dot B. The resulting strain field experienced by the electrons is the coherent superposition of the strain fields associated with the two dots [Fig. 7(c)]. An absolute maximum occurs at the centre of the structure, due to the combination of the strain field related to expanded regions [Fig. 7(a) and (b)], whereas the secondary maxima are generated by the superposition of regions where the structure is expanded (in the bigger dot A) and compressed (in the smaller dot B). It follows therefore that the combination of strain fields associated with dots of different dimensions (the size of dot A is approximately twice that of dot B) induces a modulation of the strain [Fig. 7(c)], which in turn causes a modulation of the electron diffraction.

It is interesting to analyse how such modulation changes as a function of the size of the two dots (L_A and L_B). Figure 8(a) and (b) displays the calculated contour plot of the $\varepsilon_{xx} + \varepsilon_{zz}$ component for single and two stacked dots of same dimension. The bright and dark zones correspond to the

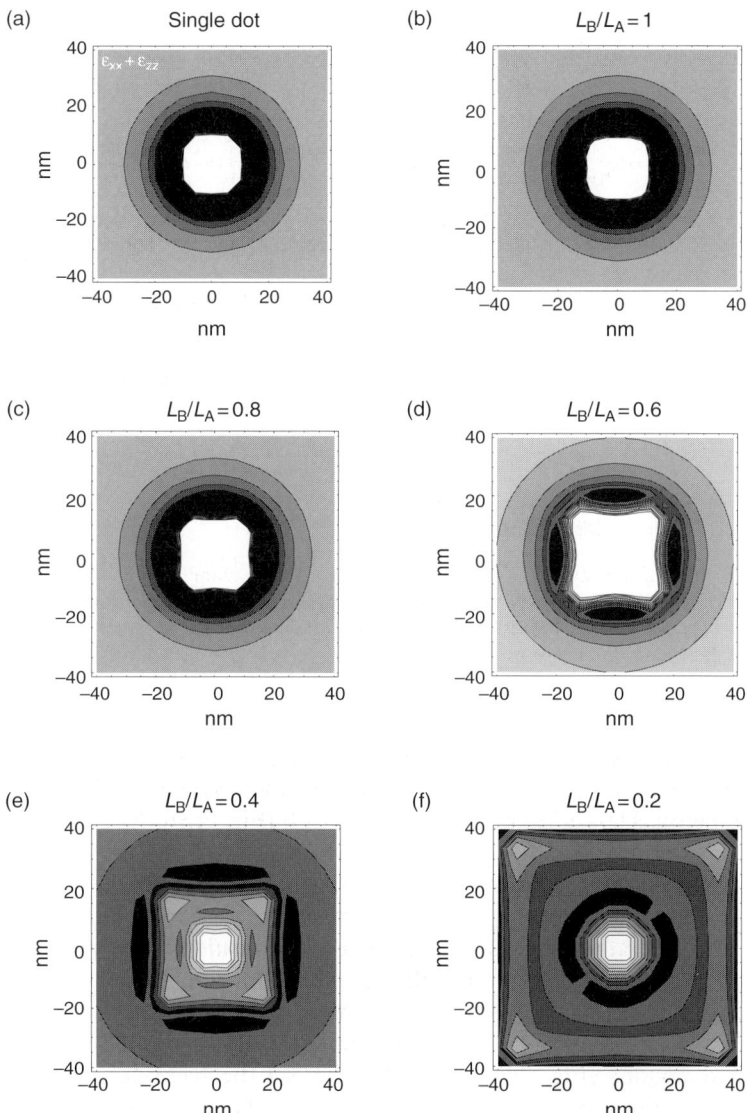

FIG. 8.   Contour plot of the calculated strain field ($\varepsilon_{xx} + \varepsilon_{zz}$) for single (a) and vertically stacked QDs (b)–(f) as a function of the ratio between the sizes (L_A and L_B) of the two dot families.

expanded and compressed regions respectively, whereas the grey zones display the intermediate conditions. As expected, in the single dot of Fig. 8(a), strain dilatation is observed at the centre of the island, whereas strain compression occurs at the edges. As a consequence, the transmitted electron beam is diffracted in a different way at the centre and at the edges of the QDs resulting in the diffraction contrast of Fig. 6(a). In the two vertically stacked layers, for $L_A = L_B$, the contour plot coincides with that of the single dot [Fig. 8(a)]. This is expected since in the case of coherent superposition of strain fields associated to QDs having the same size, no superposition between the expanded and compressed regions occurs. When the ratio L_B/L_A decreases [Fig. 8(c)–(f)], that is, when the difference between the dot dimensions of the two families increases, the modulation of the strain field becomes evident. A flower-like pattern is found only for $L_B/L_A \leq 0.5$.

The model can be readily extended to the more complicated case of Fig. 5(c), in which the sample shows six dot layers whose sizes change continuously from the bottom to the top layers. A realistic calculation was performed by assuming three dot families A, B, and C with $L_B = 0.8L_A$ and $L_C = 0.4L_A$ (see scheme of Fig. 9). Again, we find a strain modulation, which induces a central maximum, and four weaker secondary maxima (Fig. 9). Even though a quantitative analysis of the contrast pattern and the strain field would need advanced simulations of the TEM images, in which the modelled strain field is used as an input for the dynamical electron scattering (Janssens et al. 1995; Benabbas et al. 1996; Ledentsov et al. 1997; Liao et al. 1998), these results show that the different diffraction contrast arises from the coherent superposition of the strain fields associated with dots of different sizes. This induces a modulation of the total strain field which, in turn, results in a modification of the electron diffraction conditions. Therefore, from the plan-view diffraction contrast, it is possible to get information about the uniformity of the dots in the stack and about the variations of their relative sizes.

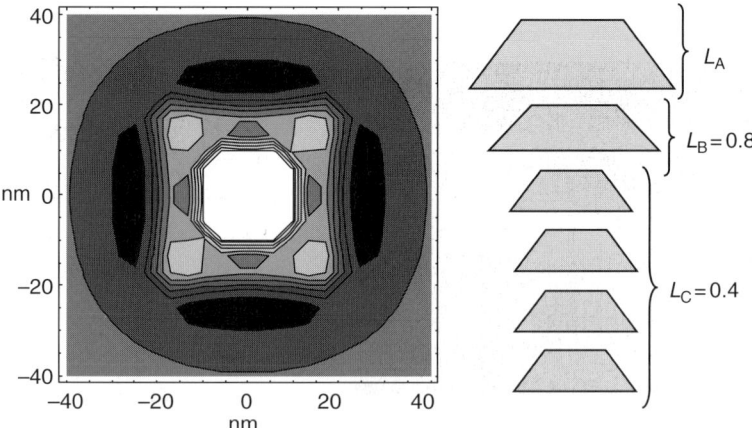

FIG. 9.   Contour plot of the calculated total strain field ($\varepsilon_{xx} + \varepsilon_{zz}$) for a six-layer model. The dot dimensions are considered such that $L_B = 0.8L_A$ and $L_C = 0.4L_A$. On the right side: scheme of the six-layer dot structure.

III.   Chemical Analysis

The chemical composition is another important parameter that influences the properties of QDs. Compositional fluctuations and the occurrence of chemical gradients within the dots causes strong variation of the confinement potential and, in turn, of the carrier wavefunctions. A nanoscale assessment of composition and stoichiometric disorder is a quite difficult task. TEM and energy dispersive X-ray (EDX) spectroscopy or EELS measurements can be performed in order to have structural and chemical information on quantum dots. The EDX and EELS measurements provide information about the relative In/As concentration ratio along the line-scan direction, with a sub-nanometer spatial resolution.

The composition profile of a real $In_xGa_{1-x}As$ QD is not necessarily similar to the uniform composition of an $In_xGa_{1-x}As$ alloy layer. In fact, the composition profile may depend on many parameters of the growth process, including: (i) the nominal composition of the dot layer; (ii) the growth temperature (Joyce *et al.* 1998); (iii) strain-driven In enrichment of the dots (Krost *et al.* 1999; Liu *et al.* 2000); and (iv) compositional change during capping (Garcia *et al.* 1997), and post-growth annealing (Leon *et al.* 1996; Lobo *et al.* 1998). In addition, intermixing may further change the average In concentration of the dots, possibly enriching the In content in the core of the dot. In order to determine quantitatively the compositional disorder in QDs, high spatial resolution EDX and EELS experiments are necessary (Sato *et al.* 1998; Siverns *et al.* 1998).

Figure 10(a) shows a low-resolution scanning transmission electron microscopy (STEM) image in the $[1\bar{1}0]$ direction on the sample. The position-resolved EDX and EELS experiments were performed by scanning an

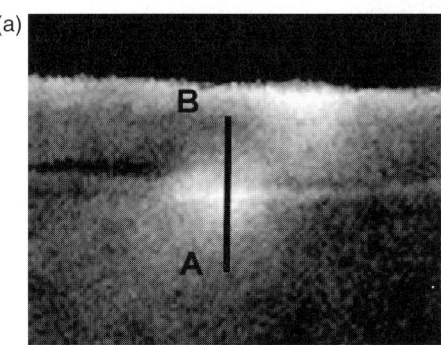
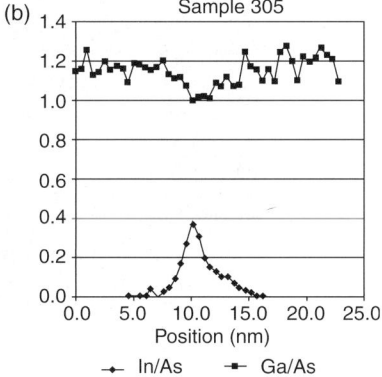

FIG. 10.   (a) Low-resolution scanning TEM image of a single dot. (b) EDX spectra obtained by scanning an electron probe across the quantum dot (line-scan A–B) showing the Ga/As and In/As atomic ratio.

electron probe (with a spot size of 2 Å) across the dot. The In/As and Ga/As concentration ratios deduced by EELS are plotted as a function of position in Fig. 10(b). The In/As ratio increases as the Ga/As ratio decreases. The quantitative determination of the compositional In profile from such experimental data is very difficult because the energy loss signal comes not only from the dot or WL, but also from the GaAs barrier surrounding the dot. In order to have a simple and reliable calibration of the absolute In content obtained by the EELS, we scale the composition of the WL measured far from the dot with the actual composition deduced from the photoluminescence spectra of the excitonic gap of the WL. As expected, the In content found for the WL corresponds to the nominal In concentration ($x \sim 0.5$), whereas the In profile across the dot shows a clear fine structure.

The In content as a function of position obtained by scanning the electron probe across the dot (line-scan 1) and the WL (line-scan 2) (Fig. 11) are

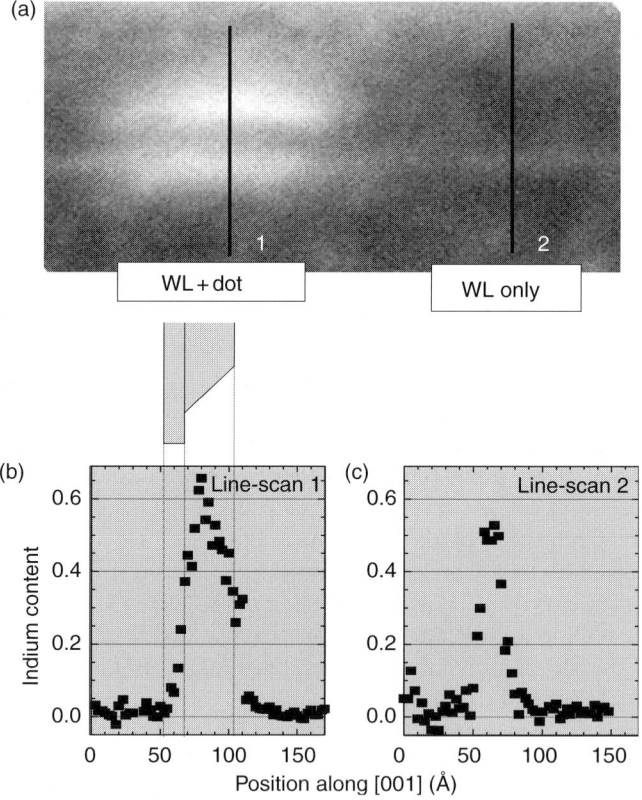

FIG. 11.   Scanning TEM image along the [1$\bar{1}$0] direction (a) and EELS scan [(b) and (c)] showing the compositional profile across the dot (line-scan 1) and the WL (line-scan 2). The measured In content is shown in (b) for the dot and in (c) for the WL, far from the dot.

compared in Fig. 11(b) and (c), respectively. The spatial resolution is sub-nanometric. Looking at the WL [Fig. 11(c)], the In concentration shows a gaussian profile with a full width at half maximum (FWHM) equal to 1.4 ± 0.3 nm. This is in excellent agreement with the WL thickness obtained by HRTEM measurements and with the nominal growth parameters. Due to the high spatial resolution of the experiments, the broadening of the compositional profile indicates that the WL has sharp interfaces with an inter-diffusion limited to about 1–2 ML. Unlike the case of the WL, the scan across the dot [Fig. 11(b)] shows a considerable broadening and an asymmetry in the In compositional profile. The In/As concentration in the dot is larger than that of the WL far from the dot. On the contrary, the WL underneath the dot seems to be In depleted. Such depletion is probably due to In diffusion into the dot during the growth. The In concentration in the WL underneath the dot is found to be $x = 30\%$, whereas in the dot a non-uniform compositional profile ranging from 30% at the base, to 65% at the core, and to 30% at the top is observed. In Fig. 11(b), it is also possible to see a smooth In tail extending for about 3 nm in the GaAs barrier. This confirms that there is a considerable In diffusion outside the dot boundaries. These results clearly indicate the existence of a compositional variation across the dot. This induces a modification of the band offset, of the strain field and effective masses and of the energy levels of QDs [see, e.g., Shumway *et al.* (2001)].

IV.   Modelling of the Wavefunctions

For a meaningful interpretation of the experimental results obtained in QDs, electronic structure calculations must be correlated to the precise information about the structural properties of the QDs discussed previously. Considerable effort has been made on the theoretical modelling of QD systems using different approaches: effective masses (Cusack *et al.* 1996; Fonseca *et al.* 1998), eight-band $\mathbf{k} \cdot \mathbf{p}$ theory (Jiang and Singh 1997; Pryor 1998a; Stier *et al.* 1999) and empirical pseudo-potential theory (Zunger 1998). The calculations of the electronic structure of QDs have been performed at various levels of approximation, assuming different QD shapes such as lens- (Lelong and Bastard 1996) or disk-shape (Marzin and Bastard 1994), pyramids with different facets (Stier *et al.* 1999), and truncated pyramids. The differences in the predicted QD ground- and excited-state emission lines, as well as in the intersublevel energies, reflect the different dependencies on the geometry (shape, aspect ratio, etc.) and material parameters assumed in the calculations. In the following, we shall use effective mass calculations which do not account for atomistic-like features (e.g., the details of the bonds at the interfaces) that are known to cause small effects in GaAs/Ga(In)As dots. Nevertheless, if needed, these properties can be obtained by appropriate methods (pseudo-potentials or tight-binding) and,

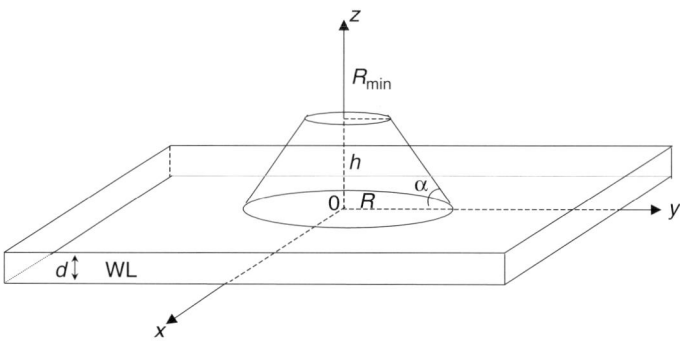

FIG. 12.   Scheme of the QD nanostructure and WL.

when expressed into the effective mass framework, calculated by perturbation theory. The single band effective mass-like calculation has already been used to describe several experiments such as far infrared magneto-transmission (Hameau *et al.* 1999), and capacitance–voltages curves (Lelong *et al.* 1998). It has also been used to describe the electronic levels of both single and stacked (Vasanelli *et al.* 2001a) dot layers and satisfactorily employed to interpret the photoluminescence spectra of QD samples.

In the model, non-interacting electrons (e) and holes (h) are subjected to 3D confinement potentials $V_e(\mathbf{r}_e)$ and $V_h(\mathbf{r}_h)$. The structural analysis, performed on MOCVD InGaAs/GaAs QD samples and discussed in the previous section, shows that the dots have a truncated cone shape (Fig. 12) of height h, base radius R, and basis angle $\alpha = 54°$. The dot lies on a InGaAs WL of thickness d. Since the dot density is approximately 10^9–$10^{10}\,\mathrm{cm}^{-2} \ll 1/R^2$ for $R \sim 10\,\mathrm{nm}$, one can neglect lateral coupling among the dots. The detailed analysis of the compositional profile shows that the In composition is non-uniform. However, let us assume initially a uniform In composition and discuss later the effects of grading on the dot levels. The one-particle Hamiltonian is given, in cylindrical coordinates, by

$$H_{e(h)} = \underbrace{E^z_{ce(h)} + E^{\rho,\theta}_{ce(h)}}_{\text{Kinetic energy}} + \underbrace{V_{e(h)}(z, \rho)}_{\text{Confining potential}} \tag{7}$$

with

$$E^z_{ce(h)} = -\frac{\hbar^2}{2m^*_{e(h)}} \frac{\partial^2}{\partial z^2} \tag{8}$$

$$E^{\rho,\theta}_{ce(h)} = -\frac{\hbar^2}{2m^*_{e(h)}} \left(\frac{1}{\rho} \frac{\partial}{\partial \rho} + \frac{\partial^2}{\partial \rho^2} + \frac{1}{\rho^2} \frac{\partial^2}{\partial \theta^2} \right) \tag{9}$$

$V_{e(h)}(z, \rho)$ is the axially symmetric confining potential for electrons (e) and holes (h) in the nanostructure, which is assumed piecewise constant, that is zero in the barrier region (GaAs) and constant $V_0^{e(h)}$ in the dot and WL regions (InGaAs). Thus, the confining potential can be written as

$$V_{e(h)}(z, \rho) = V_{WL}(z, \rho) + V_{dot}(z, \rho) \tag{10}$$

where V_{WL} and V_{dot}, potentials in the WL and in the dot, respectively, are equal to

$$\begin{cases} V_{WL}(z, \rho) = -V_0^{e(h)} \, Y[z] \, Y[d-z] \\ V_{dot}(z, \rho) = -V_0^{e(h)} \{ Y[z-d] \, Y[h+d-z] \, Y[R+(d-z)/\tan(\alpha) - \rho] \} \end{cases} \tag{11}$$

$Y[z]$ is the Heaviside function. Note that the origin of the z axis is at the bottom of the wetting layer. The height of the confining potential, $V_0^{e(h)}$, is obtained by taking into account the contribution of the strain, which modifies the energy gap of the crystal

$$\begin{cases} V_0^e = Q_0 \Delta E_{gap} - Q_\varepsilon E_0 \\ V_0^h = (1 - Q_0) \Delta E_{gap} - (1 - Q_\varepsilon) E_0 \pm E_1 \end{cases} \tag{12}$$

where ΔE_{gap} is the difference between the energy gap of the GaAs barrier and the InGaAs quantum well, Q_0 is the offset ratio for the unstrained bands [$Q_0 \approx 65\%$ (Di Dio *et al.* 1996)] whereas $Q_\varepsilon = a_c/a$, a_c being the hydrostatic deformation potential for the conduction band and a the total hydrostatic potential. E_0 and E_1 are the shift of the fundamental gap (which is proportional to the isotropic part of the strain tensor) and the valence-band splitting (given by the anisotropic parts of the strain tensor), respectively (Pollak 1990). Finally, the effective masses for electrons (e) and holes (h), $m_{e(h)}^*$ have been linearly interpolated between those of GaAs and InAs. Due to the cylindrical symmetry, the Hamiltonian $[H_{e(h)}, L_z] = 0$, with $L_z = -i\hbar(\partial/\partial\theta)$, and the solutions can be written as

$$\psi_{nl}(z, \rho, \theta) = e^{il\theta}\varphi_{nl}(\rho, z) \tag{13}$$

with $l = 0$ for S-like levels, $l = \pm 1$ for P-like levels, $l = \pm 2$ for D-like levels, and so on. n is an integer that labels different states that have the same l.

It is clear that the Hamiltonian $H_{e(h)}$ [Eqs. (7)–(11)] is not separable in z and ρ. This is due to the non-separability of the potential, which couples the motions along z (growth axis) and ρ (in-plane radial coordinate). However, the aspect ratio h/R is small in Ga(In)As dots and it is sensible to use separable wavefunctions $\psi_{nl}(z, \rho, \theta)$ such as

$$\psi_{nlj}(z, \rho, \theta) = e^{il\theta} \, F_{nl}(\rho) \, \varphi_j(z) \tag{14}$$

where $j = 1, 2, \ldots$ for different z-motions associated with the same radial solution. Thus, the states of dot nanostructure are characterized by three indexes:

$$|n, l, j\rangle$$

where $j = 1, 2, 3, \ldots$, and describes the motion along z; $l = 0, \pm 1, \pm 2, \ldots$ is due to the cylindrical symmetry; and $n - 1$ give the knots number of the radial wavefunction. 1S, 1P, ..., no knot; 2S, 2P, ..., one knot; and so on.

The radial wavefunction is chosen as

$$F_{nl}(\rho) = N_{nl}\,\rho^{|l|}\,\exp\left\{\frac{-\rho^2}{2\beta_{nl}^2}\right\}\,P_{nl}(\rho^2) \tag{15}$$

where N_{nl} is the normalization constant, β_{nl} the variational parameter, and $P_{nl}(\rho^2)$ a polynomial of degree $n - 1$ in ρ^2.

$$P_{nl}(\rho^2) = \sum_{i=0}^{n-1} C_{nl,i}\,\rho^{2i} \tag{16}$$

and imposed $C_{nl,0} = 1$. Note that the radial function $F_{n,l}(\rho)$ is dependent upon a single variational parameter, β_{nl}, the others ($\beta_{n' < nl}$) being known from orthogonalization with lower states. Inserting the total solution, $\psi_{nlj}(z, \rho, \theta)$, in the Schrödinger equation and integrating with respect to the radial variables, one gets the effective 1D problem:

$$\left[-\frac{\hbar^2}{2m_{e(h)}^*}\frac{d^2}{dz^2} + \langle T_\perp \rangle_{nl} + V_{nl}^{\text{eff}}(z)\right]\varphi_j^{nl}(z) = E_{nlj}\,\varphi_j^{nl}(z) \tag{17}$$

where

$$\langle T_\perp \rangle_{nl} = -\frac{\hbar^2}{2m_{e(h)}^*}\int_0^\infty \rho\,d\rho\,F_{nl}(\rho)\left(\frac{d^2}{d\rho^2} + \frac{1}{\rho}\frac{d}{d\rho} - \frac{l^2}{\rho^2}\right)F_{nl}(\rho) \tag{18}$$

$$V_{nl}^{\text{eff}}(z) = \int_0^\infty \rho\,d\rho\,F_{nl}(\rho)V_{e(h)}(z, \rho)F_{nl}(\rho) \tag{19}$$

Figure 13 displays the effective potentials for electrons $V_{1S}^{\text{eff}}(z)$ (solid line) and $V_{1P}^{\text{eff}}(z)$ (dashed line) which give rise to two 1S-like states (1S1 and 1S2) and one 1P-like state (1P1), respectively, for a single dot. For such effective potentials, $\varphi_j^{nl}(z)$ can be found analytically in the regions where $V_{nl}^{\text{eff}}(z)$ is constant (that is everywhere outside the dot region). The squared envelopes of the two 1S-like bound levels are also shown in Fig. 13. This approach gives the best ensemble of separable bound solutions with gaussian-like profile for

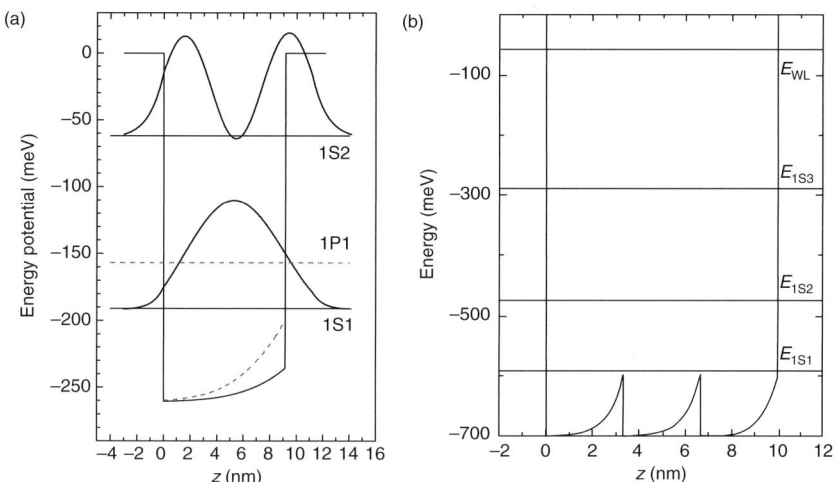

FIG. 13. Typical shape of the effective potential $V^{\text{eff}}_{1S}(z)$ (solid line) and $V^{\text{eff}}_{1P}(z)$ (dashed line) for a single QD (a), and three coupled QDs (b). The energy levels E_{nlj} and the corresponding z functions are also shown.

the radial motion. The usefulness of separable trials has been pointed out previously for the case of single InAs dots (Marzin and Bastard 1994). Note finally that this approach allows to recover automatically the WL solution $[\varphi(z) = \chi_{\text{WL}}(z)$ and $E = E_{\text{WL}}]$ when $\beta_{nl} \to \infty$.

Compared to other methods (Marzin and Bastard 1994; Grandmann *et al.* 1995; Lelong *et al.* 1998; Ferreira and Bastard 1999; Hameau *et al.* 1999; Stier *et al.* 1999), the present method is numerically lighter. A similar variational procedure was also proposed to describe the first few bound states of one dot using a separable solution with three variational parameters and a gaussian shape for both the in-plane and z-motions (Ferreira and Bastard 1999). However, the method developed in this section is expected to give a better solution than in Ferreira and Bastard (1999), since for the radial motion we seek for the best $\varphi(z)$. Another advantage of this method is to be very efficient for a system composed of a stack of coupled dots (Vasanelli *et al.* 2001a).

Figure 14 shows the calculated energies of the first few electron levels of a single dot as a function of its radius, R ($h/R = 1/4$ and $\alpha = 54°$). At small R values (not shown in the figure) the ground level saturates towards the energy of the WL. With increasing R, the number of the bound electron states increases, whereas the energy separation decreases. This is more evident in Fig. 15, where we plot the energy separation between the 1P1 state and 1S1 state as a function of the dot radius.

It is clear that only the states of energy below the WL edge are localized in the dot. In addition to the truly bound states, the calculation also show a few levels energetically located between the WL edge and the GaAs barrier (Fig. 14). These states resemble dot resonances in the WL continuum.

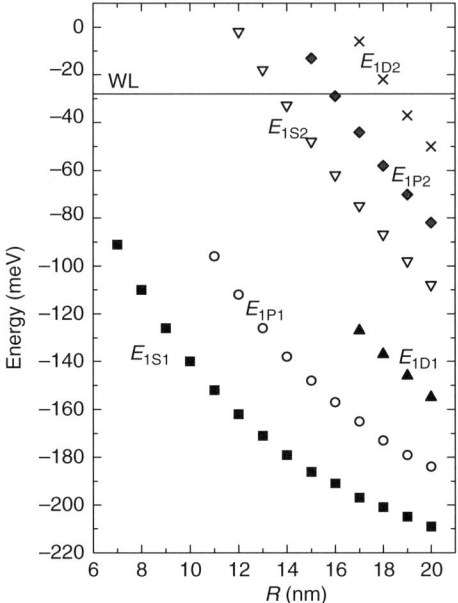

FIG. 14.   Bound electron states of a single InGaAs/GaAs QD as a function of the dot base radius, R. The aspect ratio is $h/R = 1/4$ whereas the angle $\alpha = 54°$. The solid line shows the WL bound state.

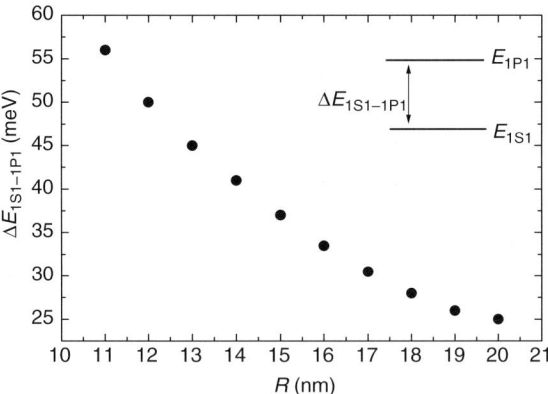

FIG. 15.   Energy separation between the ground state 1S1 and the first excited state 1P1 as a function of the dot radius, R.

The electron density contour plots of the bound levels of dots are shown in Fig. 16. The ground state (1S1) [Fig. 16(a)] has a peak at the centre of the dot and decays steeply towards the barrier regions. Quite generally, the ground state of actual Ga(In)As dots was found to be very well localized inside the

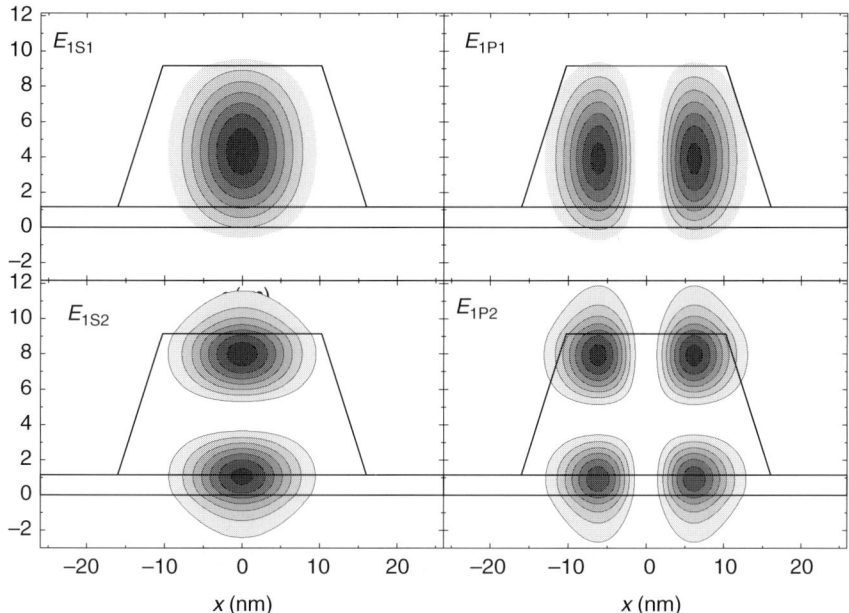

FIG. 16.   Electron density contour plots of a $In_{0.5}Ga_{0.5}As/GaAs$ QD of dimensions $R = 16$ nm, $h = 8$ nm, $\alpha = 54°$, $d = 4$ ML. E_{1S1} is the ground state of the electrons, whereas E_{1P1}, E_{1S2}, and E_{1P2} are the first, second, and third excited states, respectively.

dot. The excited states exhibits node lines and/or node plane: the 1P1 state [Fig. 16(b)] has a node line along the z direction, the 1S2 state [Fig. 16(c)] has a node plane and, finally, the 1P2 state [Fig. 16(d)] presents both a node line and a node plane. Moreover, the corresponding wavefunctions are more and more delocalized in the (x, y) plane with increasing l, due to the $\rho^{|l|}$ term (Eq. 15). Note that the bound electron wavefunctions extend very little into the WL. For a given dot size, the binding energy of electrons and holes increases nearly linearly with increasing In content. This results in a red-shift of the ground interband transition (Fig. 17) since the electrons and hole become more tightly bound to the dot.

The accuracy of this approach has been checked by comparison to the model of Stier *et al.* (1999) for pyramidal dots. In our case, the dots are approximated by cones of radius such that their basal area is the same of the pyramid and with the same heights. Using the same material parameters, a good quantitative agreement with Stier *et al.* (1999) has been found for the ground state: the two calculations differ by less than 5 meV over the entire R range. Degenerate 1P levels differ at most by 15 meV. Such a good agreement with a multi-band formalism witnesses that the one band problem describes the essential features of the physics of electron states in QDs.

M. DE GIORGI *ET AL.*

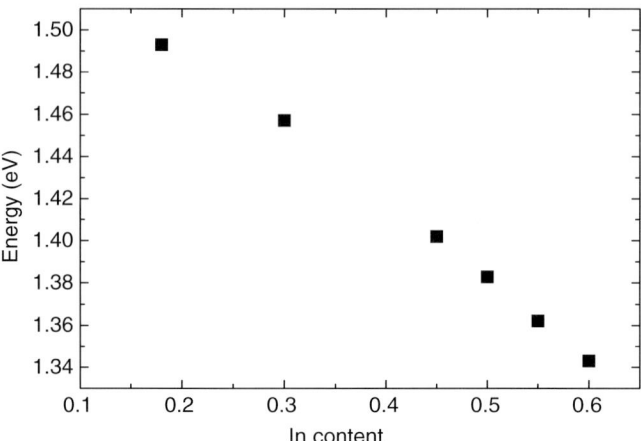

FIG. 17.   Ground level transition energy as a function of the In content in a dot.

Let us now describe the electron structure of vertically stacked QDs. For stacked dots that are well separated along z, a molecular-like description applies. In particular, the levels of the structure can easily be assigned to the tunnel coupling of states belonging to different dots. For tightly coupled structures, on the contrary, a description in terms of an effective cylinder-like confinement profile is more adequate, since the envelopes are well delocalized along the stacking direction and exhibit only small lateral oscillations nearby the WL regions (this regime for QDs is the equivalent of the pseudo-alloy regime of short period superlattices).

Figure 18 shows the 1S and 1P electron energies of a two dot structure (barrier potential equal to 697 meV, $R = 3$ nm, and $\alpha = 30°$) versus the distance D separating the two WLs. The WL edge is also plotted (solid curve) and corresponds to the ground solution of a two thin coupled quantum wells problem. For $D \leq 3$ nm, the two dots are attached. The height of the first dot is kept constant ($h = 3$ nm) while the height of the lower one varies from 0 to 3 nm. For $D = 0$, one has the solutions of a single dot floating on a WL of thickness $2d$, which binds only one 1S and one 1P levels (like the single dot–single WL situation). For $D \geq 3$ nm, the two identical dots are separated by an increasingly thick barrier. The energy curves for the 1S1 and 1P1 states exhibit a pronounced kink vs D (at $D = 3$ nm). This might seem surprising since, according to the Feynman–Hellmann theorem, one would expect the energy and its derivative to be continuous vs the dot separation. Actually, there is no contradiction since the two dots Hamiltonian changes drastically when the dots become detached: there is an extra step-like barrier, which widens with increasing D. In contrast, the 1S2, 1P2 vs D curves do not exhibit appreciable kinks. This is associated with the fact that a node in the z dependence of their envelope functions is expected in the vicinity of the

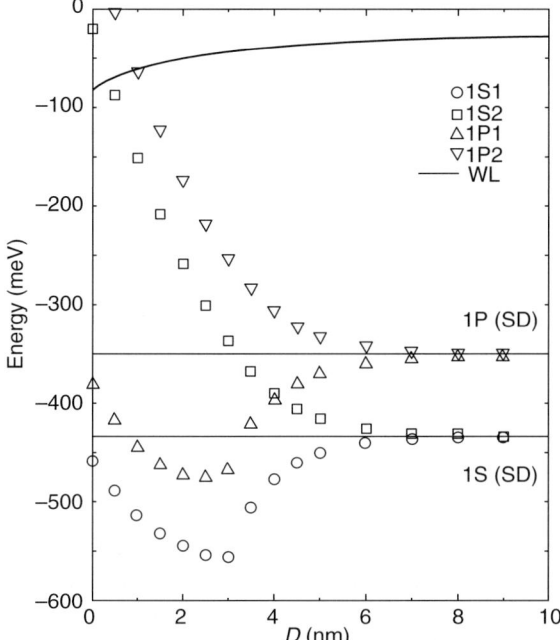

FIG. 18.   Energies of the 1S- and 1P-like states for the two coupled dots ($R = 10$ nm) as a function of the separation D between the two WLs. The height of the second dot is $h_2 = 3$ nm. The height h_1 of the first dot varies from 0 to 3 nm.

middle of the structure, where the effect of the dots detachment should be important. In the $D > 3$ nm region the envelopes of the two 1S1 and 1S2 (or 1P1 and 1P2) levels increasingly localize inside the dot regions, and resemble more and more the symmetric and anti-symmetric mixture of the two single dot solutions (energies $E_{1S1(SD)} = -433$ meV or $E_{1P1(SD)} = -351$ meV). The large dot-to-dot separation ($D > 3$ nm) corresponds to the molecular regime in which atomic orbitals localized in a given dot start to hybridize at large D to give rise to bonding and anti-bonding levels, of increasingly large separation vs $1/D$. The $D < 3$ nm region has, on the other hand, no counterpart in molecular physics, since in the latter case the nuclei are point-like particles (while dots have a non-zero size) whose Coulomb (or filled shell) repulsion overshadows the transfer integrals at small distance (while dots are neutral).

Structures with a large number of tightly stacked dots have been used to manufacture lasing devices (Sugawara *et al.* 1999). Pryor (1998b) has discussed the case of an infinite 1D array of vertically coupled InAs dots separated by GaAs barriers and calculated the miniband width as a function of the interdot separation. In the case of multi-dot structures in which there is no barrier the picture of a cylinder-like 3D profile for the confining potential gives a good first estimate for the 1Sj electron eigenstates. This is essentially due to the fact that the dots touch each other and the InAs region form a corridor inside the GaAs barrier with an effective radial thickness controlled by β_{1S}. The small departures from this result come mainly from the corrections to the cylindrical profile [or to the departures of $V_{1S}(z)$ from a square well shape $V_{SW}(z)$] nearby the WL regions between the dots. Note that such a "cylindrical" picture becomes less valid for the excited (P, D, ...) dot levels, for which $V_{nl}(z)$ and $V_{SW}(z)$ become rather different inside the dot regions. Actually, the calculated $\varphi(z)$ for more excited states differ significantly from the square well solution: they display a localization inside different dot regions, as due to the higher value of $V_{n,l>0}(z)$ inside the dots, as compared to $V_{1S}(z)$ (cf. Fig. 13 for a single dot). Figure 19 shows the contour plot of the first two bound electron states in a double-dot structure separated by $D = 3.5 \text{ nm}$.

The calculated energies of the bound 1S and 1P states as a function of the number of dots N_{dot} are shown in Fig. 20. A single dot ($N_{dot} = 1$) binds only one 1S and one 1P levels. For two dots, the symmetric-like and antisymmetric-like solutions discussed above are obtained. For large N_{dot}, the energies of the N_{dot} 1S-like and N_{dot} 1P-like levels pack within a finite energy range with increasing N_{dot}. The energies of the 1S- and 1P-like ground states become N_{dot} independent after a few dot periods, as shown in Fig. 20, but the convergence is slower for the energies of the topmost 1S- and 1P-like states (it is reached only if $N_{dot} \geq 15$). For $N_{dot} \to \infty$, the effective potential $V(z)$ becomes periodic (identical dots are assumed) and the eigenstates are Bloch-like. The calculations of 1S and 1P miniband widths have been performed: $\Delta_{1S} = 492.4 \text{ meV}$ and $\Delta_{1P} = 473.9 \text{ meV}$. It is worth noting that for $N_{dot} \geq 4$, the parameter related to the radial degrees of freedom become N_{dot} independent and equal to $\beta_{1S} = 3.43 \text{ nm}$ and $\beta_{1P} = 3.17 \text{ nm}$. Δ_{1S} and Δ_{1P} were

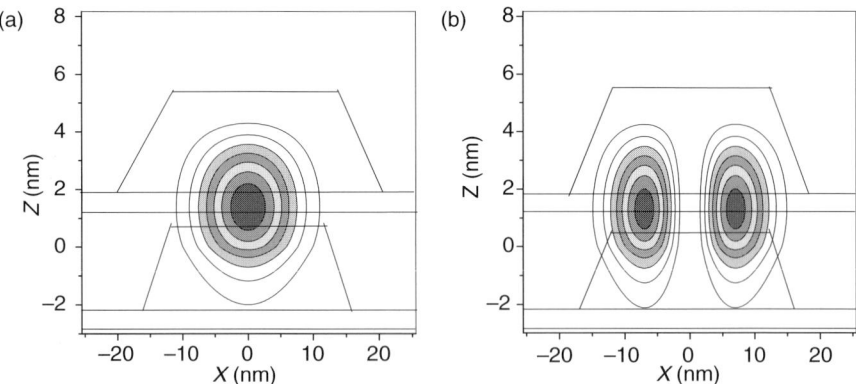

FIG. 19. Electron density contour plots of the first two bound states in a double-dot structure with a distance D between the two WLs equal to 3.5 nm.

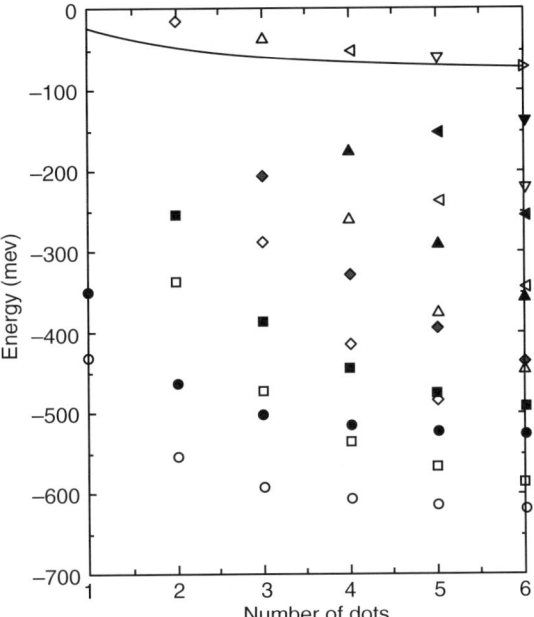

FIG. 20. 1S-like (open symbols) and 1P-like (bold symbols) levels of a multi-dot structure as a function of the number of (identical) dots. $R = 10$ nm, $h = 3$ nm, and $\alpha = 30°$.

calculated for $V(z)$ obtained with these values. We found that: (i) the mini-bands are very broad, a direct consequence of the strong vertical coupling between the dots; and (ii) $\Delta_{1S} > \Delta_{1P}$, due to the differences between $V_{1S}(z)$ and $V_{1P}(z)$ (see Fig. 13). The calculations also predict, in agreement with Pryor, that for $D > 4$ nm the gap between the first and second miniband is greater than 30 meV while the width of the first miniband is less than 20 meV.

As pointed out in the previous section, the actual InGaAs/GaAs dots are far from being compositionally homogeneous. The In diffusion inside and outside the dots has to be taken into account in the energy level calculations. The inclusion of a non-uniform In profile results in the inversion of the electron–hole dipole in the ground state of graded dot compared with the ungraded situation. The calculations of the ground 1S electron and 1S hole states lead to an expectation value of the electron (hole) position z_e (z_h) along the growth axis such that $\langle z_e \rangle \geq \langle z_h \rangle$. Hence, there exists a permanent dipole when a dot contains an electron–hole pair in the ground state that is non-zero, lined up along z and such that $\langle d_{eh} \rangle = -|e|(\langle z_e \rangle - \langle z_h \rangle) \leq 0$. Thus, in the presence of a not too large electric field $\mathbf{F}$ applied along z, the ground transition energy $E_{T(2)}$ varies like:

$$E_{T(2)}(\mathbf{F}) = E_{T(0)} - d_{eh}\mathbf{F} - \beta \mathbf{F}^2 \tag{20}$$

where the first term represents the response of the permanent dipole, while the second term comes from the induced dipole. $\beta = \beta_e + \beta_h$ depends on the polarizability of the electron and hole wavefunctions via the usual second-order term:

$$\beta_\gamma = \sum_{j=1} |\langle \phi^\gamma_{1Sj} | ez | \phi^\gamma_{1S1} \rangle|^2 / [E^\gamma_{1Sj} - E^\gamma_{1S1}] \qquad (21)$$

and is always positive for the ground state. Hence, for a homogeneous dot, one expects $E_{T(2)}(\mathbf{F})$ to be represented by a shifted parabola whose maximum occurs at a positive electric field ($-d_{eh}/2\beta > 0$ if $d_{eh} < 0$). The experiments performed by Fry *et al.* (2000) on dots inserted into the intrinsic part of reversed biased p–i–n diodes showed the opposite behaviour. Since there was no way with uniform dots to interpret the data either by effective mass type of calculations or by more elaborate computation schemes, one was forced to give up the assumption of homogeneous dots.

Let us now discuss the grading effects on a specific example, which is the graded dots shown in Fig. 11. In the calculations, the dot is approximated by a truncated cone of basis radius R, top radius R_{top}, and height h ($R = 10$ nm, $R_{top} = 8$ nm, $h = 3.5$ nm) floating on a WL of thickness $d = 1.85$ nm. The effective masses is taken: $0.067m_0$ for electrons and $0.112m_0$ in the plane and $0.337m_0$ along z for holes. The total dot potential V_{dot} is the superposition of two contributions. The first one is piecewise constant and equal to $V_e = -233$ meV ($V_h = -162$ meV) inside the truncated cone, $V_{Wle} = -150$ meV ($V_{Wlh} = -105$ meV) in the WL region and vanishes in the GaAs barrier. The second is given by the product of a gaussian function of ρ, whose thickness follows the dot radius for different z values, and of a function of z, which reproduces the In concentration profile. Because of the cylindrical symmetry, and using the separable form for bound states discussed previously, the solutions are written in the form of Eq. (14) leading to an effective potential $V^{eff}_{nl}(z)$ along the z direction [see Eq. (19)]. In Fig. 21, we show this effective potential for the 1S electron states without (dashed line) and with (solid line) the In grading contribution for a single dot. The horizontal lines indicate the ground (1S1) electron levels in the two cases. A consequence of this non-uniform In profile is that the mean positions of the electron and of the hole in their respective ground levels are now inverted: we find $p_{eh} = \langle z_e \rangle - \langle z_h \rangle = -0.17$ nm, whereas without grading the same quantity is $+0.05$ nm, with $\langle z_\gamma \rangle = \langle \phi^\gamma 1S1 | z | \phi^\phi 1S1 \rangle$ for $\gamma = e$ or h. This implies that the QD possesses an inverted permanent dipole moment, that is, $d_{eh} = -|e|P_{eh} > 0$. This positive dipole would in turn bring the maximum of $E_{T(2)}(\mathbf{F})$ in a field region compatible with experiments.

To study the effect of a uniform electric field $\mathbf{F}$, we diagonalize the electrostatic perturbation within the basis of the zero field bound states. In Fig. 22, the two upper curves are the fundamental transition energies as a

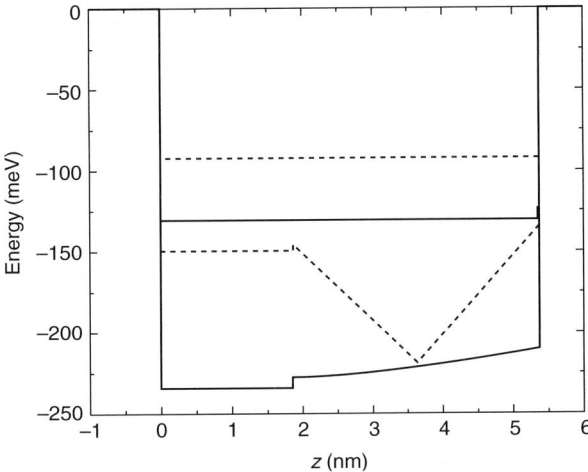

FIG. 21. 1S effective potential along the z direction for electrons in a single dot structure with the uniform potential (solid line) and the non-uniform one (dashed line). Horizontal lines: calculated 1S levels.

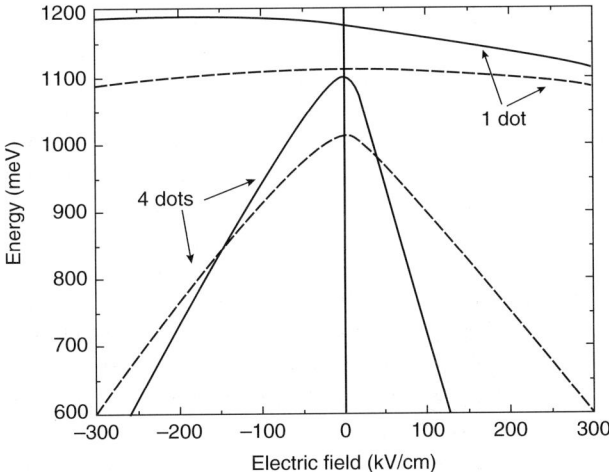

FIG. 22. Electric field dependence of the fundamental interband transitions for a single dot ($N_{dot} = 1$, two upper curves) and for a structure with four attached dots ($N_{dot} = 4$, two lower curves) with the uniform effective potential (dashed lines) and the non-uniform one (solid lines).

function of **F** for a single QD when considering (solid line) or not (dashed line) the grading effects. Thus, $E_{T(2)}(\mathbf{F})$ has a symmetric profile around a maximum value at $\mathbf{F_M} = -d_{eh}/(2\beta)$. In the absence of grading, the Stark shift in Fig. 22 is roughly parabolic, with a maximum at approximately $+10\,\mathrm{kV/cm}$, due to the small (positive) value of p_{eh}. When the grading is taken into account, the **F** dependence of the transition energy is asymmetrical, with a maximum around $-190\,\text{kV/cm}$, due to the large (negative) value of p_{eh}.

Large electric fields ($|\mathbf{F}| > 100\,\text{kV/cm}$) have to be applied to the structure to obtain a non-negligible Stark shift in the case of a single dot. This is a general feature of all single dot structures which arises from the flatness of the dots (2–3 nm in height) and recalls the very small Stark shift of narrow 1D quantum well structures. More polarizable structures are obtained by vertically stacking QDs. Strongly stacked QD structures (i.e., with no barrier between the dots) have already been used to manufacture lasing devices (Fry *et al.* 2000). It has been shown above that the electron moves in these structures rather freely along a vertical InGaAs corridor (Vasanelli *et al.* 2001b). This behaviour allows the creation of artificial quantum wells while, at the same time, avoiding the formation of a 2D continuum as is found in the case of 1D quantum wells. A large Stark tunability of the interband transitions in the absence of grading effects is calculated, in fact as large as in single quantum well structures. To deal with grading effects in stacked dots, we empirically include the effect of In diffusion. As experimentally shown by Zhang *et al.* (2001), during the growth process of several layers of QDs, the In content increases and the Ga content decreases in the QDs as subsequent layers are grown. For this reason, we have considered a confining potential that is deeper in the topmost dot of the stack. The two lower curves in Fig. 22 show the Stark shift for the fundamental transition in a structure composed of four dots when considering (solid line) or not (dashed line) the grading effects. Figure 22 shows that, even with grading, multi-stacked dots display a wide Stark tunability of their interband transitions. We can firstly note a red-shift of the zero field transition energy with increasing number of dots. This arises from the increase of the total height of the structure and hence from the lowering of the electron and hole energies. Secondly, the maximum field $\mathbf{F}_{\text{M}}$ is smaller and the electron–hole mean separation is larger for multi-dot structures, as compared to the single dot case (for instance, we calculate $\mathbf{F}_{\text{M}} \approx -30\,\text{kV/cm}$ and $p_{\text{eh}} = -0.76\,\text{nm}$ for two stacked dots). Finally, the energy vs field curve for four dots is strongly asymmetric: it is roughly linear for high positive and high negative fields but with rather different slopes. These two features are associated with the competition between the strong "internal" (due to the non-uniform In profile) and applied fields. In particular, the non-parabolic profiles in Fig. 22 point out that a simple second-order perturbative treatment is unable to describe the huge Stark shifts in single and multi-dot structures, contrary to the usual situation in 1D quantum wells and for ungraded dot structures. Once the electron and hole eigenstates are known, the absorption spectrum related to interband transitions may be obtained Sajal *et al.* (1991). Due to size fluctuations, the absorption spectrum of an in-plane distribution of dots is a broad line. In order to reproduce qualitatively an actual statistical ensemble, we assume

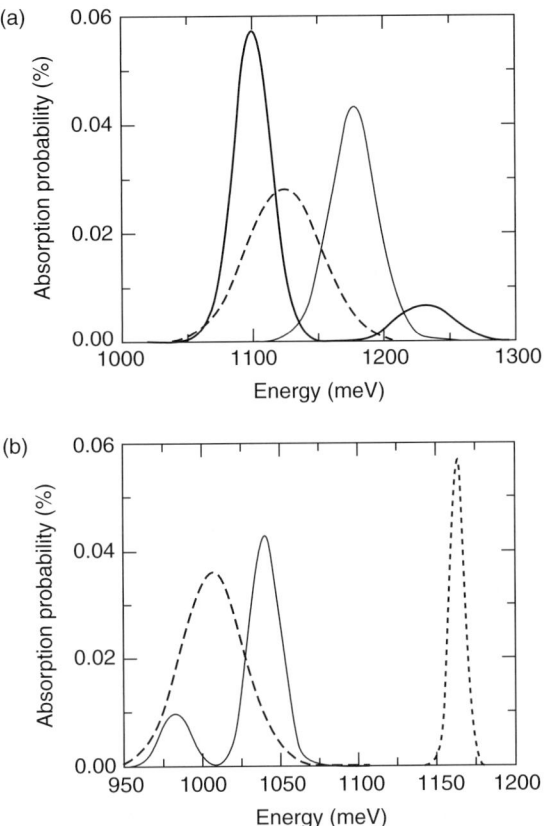

FIG. 23.   Absorption probability for a single dot (a) and for a structure of two dots (b). The thick lines represent the fundamental transitions, whereas the thin lines represent the transitions among the excited states in a uniform In distribution structure (dashed lines) and in a graded one (solid lines).

that R and h are distributed following gaussian functions centered respectively in $R_0 = 10$ nm and $h_0 = 3.5$ nm, with standard deviations $\sigma_R = 0.3$ nm and $\sigma_h = 0.6$ nm. For simplicity, we also assume that a given height fluctuation is equally shared between the dots of a given stack (R_{top} is the same for all the dots of a stack and such that for each dot the basis angle $\alpha = \tan^{-1}[h/(R - R_{\text{top}})] = 60°$). Figure 23 shows the light absorption probability spectrum for the fundamental (thick lines) and for the first excited (thin lines) transitions of a single dot [Fig. 23(a)] and of a structure with two attached dots [Fig. 23(b)], both with a uniform In distribution (dashed lines) and with a graded profile (solid lines) at $\mathbf{F} = +200$ kV/cm. For the single dot case, the first excited transition involves the $1S1_e$ and $1S2_h$ levels and has a non-negligible absorption probability only in the presence of the electric field. In the presence of the grading, both the ground and the first excited transitions are red shifted and their energy separation increases. Also, the transition lines are broader, since the electron and hole energies become more sensitive to the height variations. The broadening of both fundamental and excited transitions decreases with increasing number of dots, as previously shown (Sajal *et al.* 1991). In addition, the excited transition becomes more probable than the fundamental one in both kinds of structures (with and without grading effects). It is worth noting that fundamental and excited transitions are much more separated in the In graded structure: this may represent a significant advantage for electro-modulation and for quantum computation.

V. Photoluminescence Experiments

As discussed so far, the calculation of the dot electronic structure is very useful not only to study dependence of the dot properties on size, shape, and composition but also to interpret the photoluminescence (PL) experiments. Representative PL spectra measured at 10 K for different excitation power intensities are shown in Fig. 24. With increasing the excitation intensity, the PL spectra show a clear band filling dynamics, which is characteristic of QDs with uniform size distribution. This monomodal size distribution is typical of samples having a low dot density ($\approx 10^9 \, \mathrm{cm}^{-2}$). On the contrary, samples with

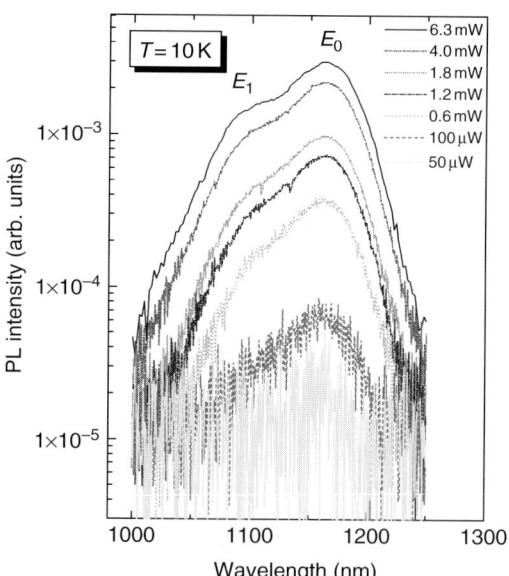

FIG. 24. PL spectra performed at 10 K for different excitation power intensity (from 50 μW to 6.3 mW). E_0 and E_1 label the ground state and the first excited state peak transitions, respectively.

dot density in excess of 10^{10} cm^{-2} often show a bimodal size distribution resulting in emission spectra which exhibit structures of constant intensity ratio, independent of the excitation intensity. The ground level emission (E_0) peak is at 1167 nm with an FWHM of about 46 meV, whereas the first excited states (E_1) is observed at 1104 nm (FWHM equal to 66 meV) as obtained by the Gaussian deconvolution of the PL spectra. The spectra show a inhomogeneous broadening due to the non-uniformity of the QD size as well as the composition and shape. Even very high quality samples display this type of broadening. The energy splitting between the ground level and the first excited level is about 60 meV. The theoretical results for this sample confirm the existence of only two bound levels in the conduction band (CB), (E_{1s1}^e, E_{1p1}^e) and of more than two bound levels in the valence band (VB), (E_{1s1}^h, E_{1p1}^h, E_{1s2}^h, and so on). The optical transitions involving levels of different symmetry are very weak, whereas those involving levels of same symmetry are more intense. Among these, the main transitions are those between the ground levels and the first excited levels in CB and VB, respectively, that is $E_0 = E_{1s1}^e \rightarrow E_{1s1}^h$ and $E_1 = E_{1p1}^e \rightarrow E_{1p1}^h$, the others having much smaller oscillator strengths. The theoretical energy splitting ($E_1 - E_0$) is estimated to be 65 meV, in good agreement with the experimental data. The PL spectra of the QD structure for different temperatures are shown in Fig. 25. With increasing temperature, it is possible to observe two main effects: (i) a redshift of the peaks; and (ii) a reduction of the luminescence intensity. The redshift of the peaks is due to the energy-gap reduction with temperature (Sajal *et al.* 1991) whereas the intensity reduction is due to the thermal escape of the carriers from the dot levels and to the thermal activation of non-radiative

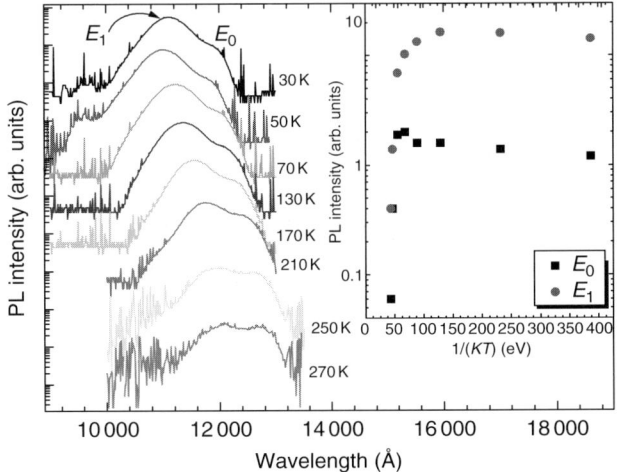

FIG. 25.   PL spectra performed at different temperatures. Inset: Arrhenius plot for the ground level and the first excited transition.

recombination channels. Plotting the integrated emission intensity of the luminescence peaks (obtained by the Gaussian deconvolution) vs $1/(KT)$ (see Arrhenius plot in the inset of Fig. 25), we found that the depletion of the first excited state corresponds to a small increase of the integrated intensity of the ground level. This is due to the capture in the ground level of carriers escaped from the excited state. Above 180 K, the ground state also begins to empty due to the thermal escape effect.

Since for the realization of a real dot device, it is necessary to grow the nanostructures on doped substrates, the study of the optical properties of the dots on different substrates, namely semi-insulating (SI) and n^+-type (Si doped), is very important. In Fig. 26, we display the PL spectra at different excitation intensities of two dot samples, grown in the same growth conditions, on an SI [Fig. 26(a)] and n^+-type (Si doped) [Fig. 26(b)] substrate. The emission efficiency with respect to the WL is very different in the two samples. This is due to a small change of the strain field experienced by the InGaAs layer grown on the different substrates, which results in a slightly different critical layer thickness (CLT). In fact, in the heavily doped substrates, the strain field is increased by a very small quantity with respect to the SI

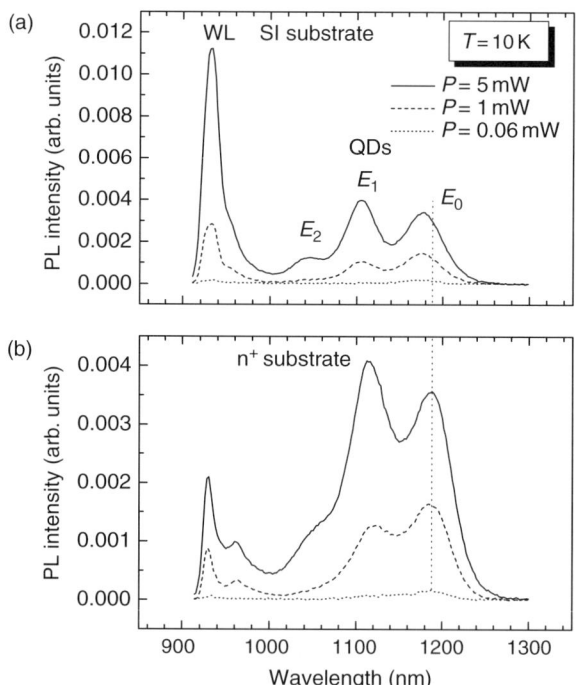

FIG. 26.   PL spectra of two QD samples grown under the same growth conditions on SI, (a) and n^+-type (b) substrate. The spectra have been performed at 10 K for different excitation power intensities, ranging between 0.06 and 5 mW.

substrate, due to the smaller covalent radius of the Si atoms (preferentially incorporated on Ga sites in the GaAs lattice) with respect to Ga atoms. This results in a slightly lower lattice parameter (by a factor 10^{-5}), which was measured by X-ray diffraction in the doped substrate with respect to the undoped GaAs. The difference of the strain field also induces a small variation of the dot size, which in turn causes a small energy shift of the luminescence peaks (see Fig. 26).

In QD device structures realized by MOCVD, another interesting effect occurs. The careful control of the growth parameters allows the tuning of the photoluminescence emission wavelength in the range from 1 to 1.5 μm (Tatebayashi *et al.* 2001). However, the electroluminescence (EL) and the photocurrent (PC) spectra of laser structures grown on n-type GaAs substrate (p–i–n structure) are found to systematically shift to a shorter wavelength [Fig. 27(a)] as compared to identical samples grown on p-substrate. In p–i–n structures, grown on n-substrate, the PC spectra display no ground

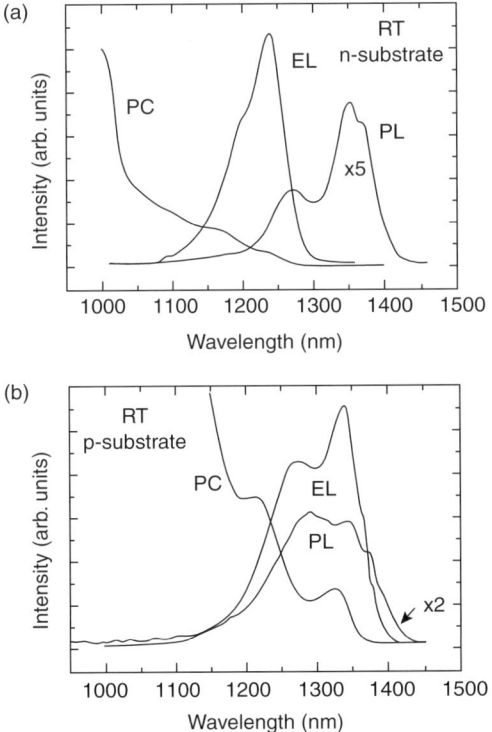

FIG. 27.   Small room temperature (RT) PL emission of the QD sample with an $In_xGa_{1-x}As$ barrier of $x = 0.1$, compared with and EL and PC spectra of the corresponding laser structure grown (a) on n^+-doped substrates (p–i–n structure) and (b) on p^+-doped substrates (n–i–p structure).

state resonance [Fig. 27(a)]. As a consequence EL emission is obtained only from the excited states [Fig. 27(a)]. By growing the same layer sequence on p-type GaAs substrate (n–i–p structure), that is, by reversing the direction of the built-in electric field of the device, one obtains EL emission also from the ground state [Fig. 27(b)]. The comparison between PL, PC, and EL spectra was done by systematically varying the composition of the $In_xGa_{1-x}As$ barrier (Passaseo *et al.* 2001b). The In content in the barrier is changed to tune the emission wavelength of the ground level from 1.28 μm up to near 1.4 μm (Passaseo *et al.* 2001b) at room temperature. Figure 28 shows the comparison of the emission wavelength of the ground state and of the first excited state, taken from the PL and from the EL spectra, for the samples grown on n^+- and p^+-substrate, as a function of the barrier composition. In all the p–i–n samples [Fig. 28(a)] (grown on the n-type substrate), the EL emission occurs near the first excited state, whereas EL emission from the ground state is observed in all the structures grown on p-type substrate [Fig. 28(b)], confirming the non-commutative nature of the QD devices. This

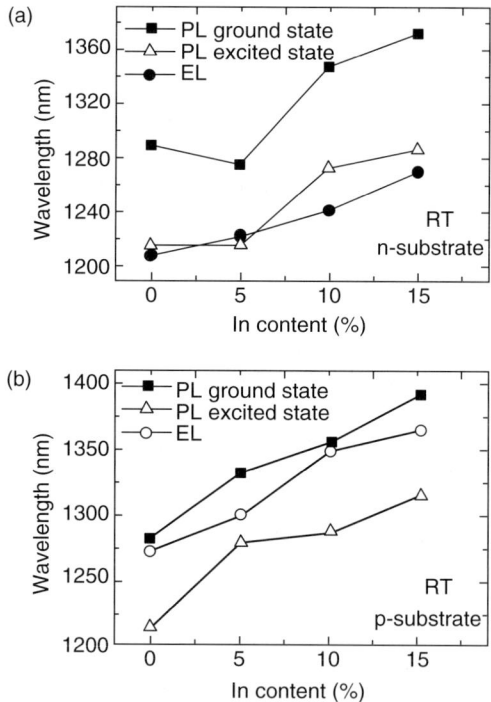

FIG. 28.   Room temperature emission wavelength of the ground state and the first excited state of the PL test structures as a function of the $In_xGa_{1-x}As$ barrier composition for x ranging between 0 and 0.15, grown on (a) n-type and (b) p-type substrates, together with the EL emission wavelength of the corresponding p–i–n and n–i–p laser structures.

effect is probably related to the existence of the permanent electron–hole dipole moment (Passaseo *et al.* 2001a) (discussed in the previous section), which results parallel to the electric field of the device junction in p–i–n structures and anti-parallel in the n–i–p one. Therefore, the study of this effect is important for the design of efficient laser structures emitting at 1.3 μm at room temperature (Passaseo *et al.* 2001a; Tatebayashi *et al.* 2001), which are very interesting for the second window of the telecommunications. Note that the PL spectra measured on samples without topmost contact layer, do show the same emission wavelengths, regardless of the substrate.

VI.   Single Dot Spectroscopy

1.   WAVEFUNCTION SPECTROSCOPY

As discussed in the previous section, the study of the electronic and optical properties of an "ensemble" of QDs has been hindered by the inhomo-geneous broadening of the spectral features (usually >30 meV in the spatially averaging techniques such as photoluminescence), due to the non-uniformity in the dot size (usually ≈10%), shape, and composition. This broadening obscures Coulomb effects involving energy shifts at the approximately milli-electron Volt level. More recently, the development of spatially resolved techniques has circumvented these problems, enabling the investigation of the optical properties of individual dots for which the broadening is purely homogeneous (<50 μeV) (Marzin *et al.* 1994; Gammon *et al.* 1995; Empedocles *et al.* 1996; Bacher *et al.* 1999; Landin *et al.* 1999; Dekel *et al.* 2000b; Thompson *et al.* 2001). Sharp, narrow lines, as in gas-phase spec-troscopy, are observed in the single dot spectra where energy separations and intensities reflect the shell-like structure and population of the QD states. However, there is a departure from the analogy with atomic physics due to the appearance of multi-peaked structures which arise as consequence of the charging (Hartmann *et al.* 2000; Finley *et al.* 2001a,b; Karlsson *et al.* 2001; Regelman *et al.* 2001) and the multi-particle capture characteristics of the QDs (Dekel *et al.* 1998; Bayer *et al.* 2000). In this section, recent reports of single dot scanning tunnelling spectroscopy (STS), scanning near-field optical microscopy (SNOM), and tunnelling current induced luminescence (TCIL) are presented.

STS of isolated QDs, produced by colloidal synthesis (Alperson *et al.* 1995, 1999; Banin *et al.* 1999; Millo *et al.* 2000, 2001) as well as cross-sectional studies of epitaxially self-assembled QDs have been reported in recent years (Legrand *et al.* 1998; Grandidier *et al.* 2000). In these studies, the discrete nature of the electronic structure and the spatial variation of the confined electronic wavefunctions of single QDs was demonstrated. In the following, we present the STS imaging of intact uncapped $In_{0.5}Ga_{0.5}As$ QDs grown by strain-driven self-assembled epitaxy on a GaAs matrix. The origins and nature of the measured tunnelling current is investigated by comparing with calculated tunnelling current spectra.

tunnelling spectra have been measured simultaneously with topographic images of single QDs. During the spectroscopy imaging measurements, that is, the spatial variation of the tunnelling current, as a function of the sample bias, the tunnelling current feedback loop is switched off and the set point which established the tip–sample distance was that of the tunnelling conditions for the topographic images. In this way, all purely topographic dependencies of the tunnelling current were removed.

Typical *I–V* curves measured on the QD and in the WL region are shown in Fig. 29. Since in this quantum dots the chemical composition of the WL and QD is nominally the same, the differences in the tunnelling current spectra, such as the smaller zero current gap of the QD with respect to the WL can be indicative of the smaller electronic gap which results in quantum confinement and the formation of QD states (Franceschetti and Zunger 2000). At negative sample biases [Fig. 29(a)–(d)], which correspond to tunnelling out of occupied states, there is a clear contrast in the current signal between the region of the QD and the WL. This current contrast persists and exhibits a spatial variation within the structure of the QD over the voltage range −1.50 to −0.51 V [Fig. 29(a)–(d)] and then disappears abruptly. In the voltage range between −0.51 and 0.60 V [Fig. 29(d)–(f)], there is an absence of contrast to within the noise levels of these measurements. This voltage range extends over the region of the QD energy gap and the Fermi level resides at the middle of this gap. Very little or no current contrast is observed in the positive

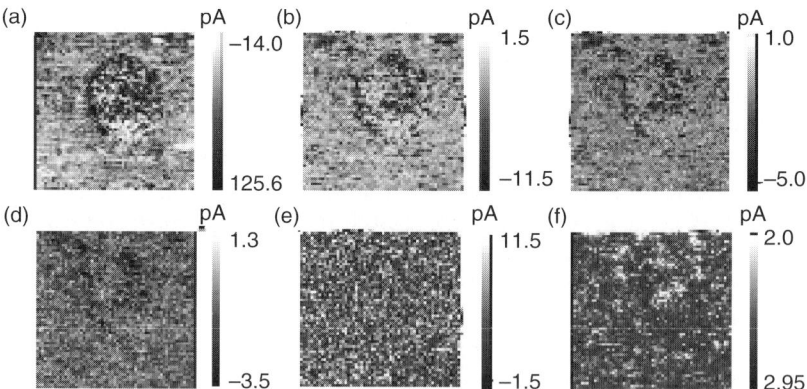

FIG. 29.   Current images, $80 \times 80\,nm^2$, of a single, isolated QD as a function of the bias voltage (a) −1.50 V, (b) −0.79 V, (c) −0.65 V, (d) −0.54 V, (e) 0.00 V, and (f) 0.61 V. The tunneling set-point, defining the tip–sample separation, is given by a bias voltage of 2.0 V and tunnelling current of 0.5 nA.

voltage range corresponding to tunnelling into the unoccupied states of the QD. The spatially resolved spectroscopic features at negative sample biases (such as those shown in Fig. 29) were observed from QDs of different samples as well as using different tips (both W and Pt–Ir tips). Therefore, it can be assumed that these features are independent of the tip density of states and tip effects. Since current contrast was observed only in the negative bias range, when tunnelling out of the occupied states, the proceeding discussion shall be confined to the occupied states of the QD and WL, that is, the "hole" states.

In seeking to interpret the measured tunnelling current images the spatially resolved tunnelling current of an isolated InAs QD situated above a WL in vacuum has been calculated. Following Tersoff and Hamann (1985), the tunnelling current is given by

$$
I \approx \sum_{n,l \geq 0,j} g_l \left[\frac{1}{1 + e^{-\beta E_{nlj}}} - \frac{1}{1 + e^{-\beta(E_{nlj}+\phi)}} \right] |F_{nlj}(\mathbf{r}_t)|^2
$$
$$
+ \frac{S}{4\pi^2} \int d^2k \left[\frac{1}{1 + e^{-\beta E_k}} - \frac{1}{1 + e^{-\beta(E_k+\phi)}} \right] |\psi_{\mathbf{k}}(\mathbf{r}_t)|^2 \tag{22}
$$

where the QD bound states, $|F_{nlj}(\mathbf{r}_t)|^2$, are described by quantum numbers, n, l, j, and the WL eigenstates, $|\psi_{\mathbf{k}}(\mathbf{r}_t)|^2$, by the 2D wavevector $\mathbf{k}$, $\beta = 1/kT$ and the applied bias, ϕ, is defined as $\phi = \phi_t - \phi_s$ where ϕ_t the tip chemical potential and ϕ_s is the sample chemical potential. The QD bound states can be calculated following the envelope function approach developed in the previous section. Since the QDs are uncapped, the influence of the vacuum is treated by using a high (2 eV) potential barrier.

The tunnelling junction comprises both the vacuum barrier and the surface of the QD and the WL, which will invariably be oxidized. To a first approximation, this oxide can be treated as an insulating layer. There exists a large body of work on tunnelling spectroscopy of such junctions, where when treating the insulator barrier as a vacuum barrier the gross features of the calculated tunnelling current spectra have been qualitatively correct (Duke 1969; Wiesendanger 1998).

For the calculation of the spatially defined tunnelling current the tip–sample distance has been defined for a constant tunnelling current for an applied bias of $\phi = 2$ eV. In these conditions, the tip position variation simulates the spatial extent of the charge density associated with the localized, confined states of the QD and is accordingly shown in Fig. 30(a) and compared with the geometric shape of the QD. The lateral extent of the charge density is accentuated in these calculations which do not accurately account for the decay of the wavefunctions outside the QD.

Calculated tunnelling current spectra are shown in Fig. 30(b) together with the calculated tunnelling current intensity profiles (Fig. 31) for a series of biases for different locations in the QD and WL. At a low bias of -0.53 eV [Fig. 31(f)], the contribution to the tunnelling current is due solely to the

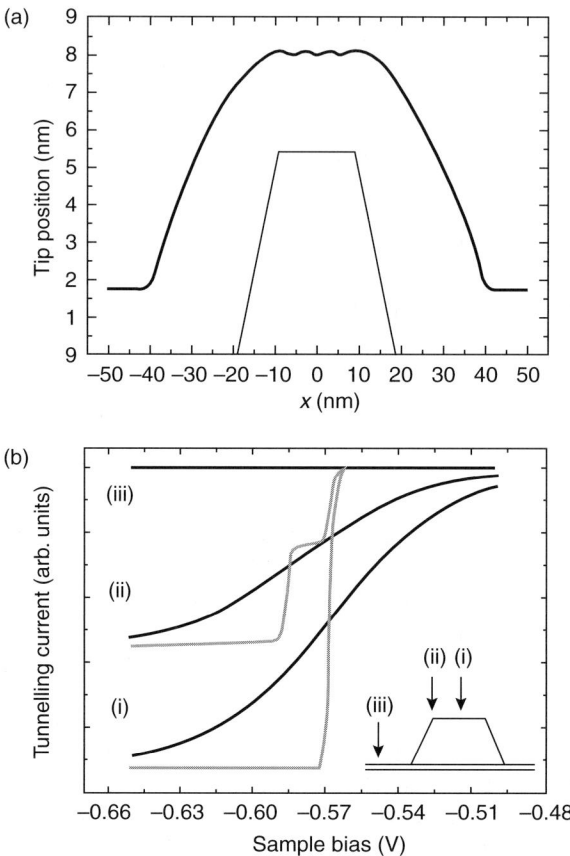

FIG. 30. (a) The calculated tip variation, $z(x)$, across the QD, and (b) calculated tunnelling spectra, $I(\phi)$, for (i) the centre, (ii) periphery of the QD, and (iii) the WL region at 300 K (black) and 10 K (grey).

1S-like state which is spherically symmetric and gives rise to a sharp, localized contribution to the tunnelling current at the center of the QD. With increasing bias to -0.54 eV [Fig. 31(e)], there is the contribution due to the 1P-like states which results in an additional tunnelling current signal encircling the centred 1S signal. Further increasing the bias leads to the onset of tunnelling current originating from the higher level eigenstates and then due to the delocalized WL states. The influence of finite temperature results in the mixing of these contributions, the consequence of which, at room temperature, is the inability to distinguish the onset of tunnelling current due to the different eigenstates. However, it is the spatial resolution afforded by the STM which allows these states, to some degree, to be identified.

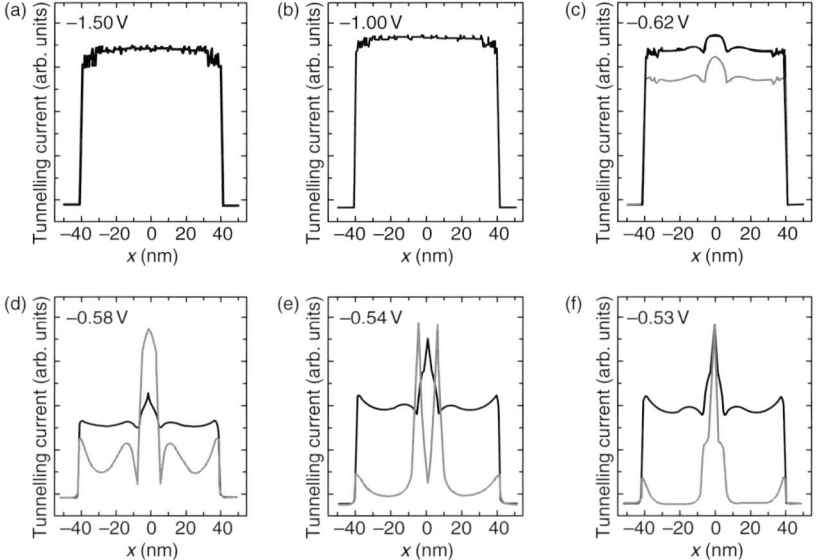

FIG. 31.   The calculated tunnelling current profiles, $I(x)$, for bias voltages of (a) -1.50 V, (b) -1.00 V, (c) -0.62 V, (d) -0.58 V, (e) -0.54 V, and (f) -0.53 V at 300 K (black) and 10 K (grey).

In comparing the measured tunnelling current images with the calculated tunnelling current profiles there are similarities in the spatial extent and symmetry. The highly localized structure at the centre of the QD observable at a voltage of -0.54 V compares well with the calculated contribution due to the 1S- and 1P-like eigenstate of the single QD. The increase in the spread of the intensity at the centre of the QD for the -0.65 V bias together with the appearance of the current intensity at the periphery of the QD can be iden-tified with the higher eigenstates of the QD and the contribution due to the delocalized WL states. The absence of current contrast in the positive bias range together with the very measurement of a sizeable, permanent tunnel-ling current when tunnelling into or out of the localized, confined states of the QD requires further discussion. The electronic structure of the single QD is best described by occupied and unoccupied, highly localized, atomic-like states. The maximum number of electrons or holes that the single QDs can host when excited is, in the case of the QDs reported here, 16 and this charge density is delocalized over the volume of the QD. This would imply, when considering a QD in complete isolation, that tunnelling into these states, will result in a tunnelling current density well below the detection limit of the STM. The clear observation of an increase in the tunnelling current at the QD implies that, as in the case of point defects at semiconductor surfaces (De la Broise *et al.* 2000), to achieve a permanent tunnelling current measurable by STM the QD needs to exchange carriers with the bands of the host GaAs crystal and the WL continuum by the emission and capture of electrons and holes. When tunnelling out of the occupied states of the QD (the negative bias) the rate of capture of electrons has to be efficient and vice versa when tunnelling into the unoccupied states (the positive bias), the evacuation of electrons should be efficient. Carrier relaxation in QD by phonon scattering (Heitz *et al.* 1997), Auger scattering (Ferreira and Bastard 1999) and electron–phonon coupling leading to the formation of "ever-lasting" polarons (Hameau *et al.* 1999) have been reported as well as escape of electrons and holes by thermal activation and tunnelling into the WL (Kapteyn *et al.* 2000). From these STS measurements alone it is not possible to distinguish the dominant mechanism for the carrier relaxation in the QD. However, it is clear that hole escape from the QD is sufficiently fast. And since below the dot layer there is a reservoir of free carriers due to the n-type doping of the GaAs layer, the hole states are filled in a time-scale faster than the tunnelling out of these states [estimated to be of the order of picoseconds (Fertig 1990)]. So there is an *effective* higher density of occupied states resulting in a measurable tunnelling current from the occupied QD state.

In the case of the unoccupied states, there are a number of conspiring factors, which may give rise to a real, and an apparent occupation of the unoccupied states thus reducing the contrast between the QD and the WL in the current images in positive bias. At room temperature, there is a finite probability that these states are partially occupied due to intrinsic and extrinsic carriers. Alternatively, defects at the surface of the dot act as trapping centres, which prevent the population of the unoccupied states of the QD. Although this process will reduce the tunnelling current, it should not completely prevent tunnelling into the unoccupied states of the dot. A further explanation for the absence of a strong contrast in the positive bias is the absence of an efficient evacuation process of the QD electrons. The time-scale of electron escape from charged, completely isolated QDs has been predicted to be of the order of hours (Martorell *et al.* 2001). However in the case of QDs embedded in a GaAs matrix there is the possibility of radiative recombination, de-excitation by coupling to the lattice phonons (Heitz *et al.* 1997) and/or tunnelling into the GaAs barrier (Kapteyn *et al.* 2000). For radiative recombination to be the major de-excitation path, there is a need for the availability of hole states into which the electron can fall. For this argument doping becomes an important issue. To date, the relaxation processes in the QD are not completely understood, therefore at this point conclusive explanations are not forthcoming and further work is required, for example STS measurements of QDs grown on p-type substrates are desirable.[1]

---

[1] Current contrast in only the positive bias was reported by Shumway *et al.* (2001), since their samples were p-type doped a similar explanation is probable.

In conclusion, in the voltage-dependent current images, the charge density associated with the discrete occupied states of the QD, the hole wavefunctions, can be identified by comparison with theoretical calculations of the tunnelling current. The analogy with point defects is instructive in understanding the origins of the measured tunnelling current from the discrete, localized states of the QDs. The important consideration is that an exchange of carriers with the hosting material, the bulk GaAs and the InGaAs WL, is required to form a measurable, permanent current. This suggest that STS imaging of 0D structures, such as QDs, can only be interpreted correctly when not treating the QD as an isolated entity but very much part of the complete system which involves the hosting crystal.

2.   OPTICAL SPECTROSCOPY

Near-field scanning optical microscopy (NSOM) is another powerful technique to study single dot optical properties with a sub-wavelength spatial resolution (Betzing et al. 1987; Berzig et al. 1991). There are two distinctly ways of operating an NSOM: (i) the sample is excited in near field by using a tip of sub-wavelength aperture and the luminescence is collected in far field (emission mode), as it will be shown in the following; and (ii) the luminescence is collected in near field through the tip (collection mode). In collection mode the spatial resolution is determined by the dimension of the tip, whereas in emission mode, the diffusion of the generated carriers and excitons dominates the resolution of the technique. This effect can be removed by isolating the single dot in a nanomesa, of typical dimensions $200 \times 200 \times 30 \, \text{nm}^3$, realized by electron beam lithography (EBL) at the centre of a typically $6 \times 6 \, \mu\text{m}^2$ window. The single dot luminescence is excited in near field condition by the continuous wave 514.5-nm line of an Argon laser and it is collected in far field and dispersed by a 0.3 m of focal length spectrograph equipped by a cooled low-noise CCD detector (Andor Technologies). Time integration of the order of several minutes (5–10 min) is used in most experiments.

In Fig. 32, the excitation intensity dependence of the single dot PL spectra is shown. Sharp lines associated with single dot recombination processes can be observed. The main features of the spectra are: (i) At very low excitation intensity (0.26 µW) a single sharp line, with a FWHM of 1.5 meV, appears at 1.223 eV (Fig. 32, bottom line). (ii) With increasing excitation intensity, five spectral features appear in the energy range of 10 meV below the fundamental line. All these lines are related to the s shell of the QD. (iii) A further increase of pumping intensity causes the saturation of the s-shell lines and the appearance of two p-shell bands at 1.258 eV, which are split by about 3 meV. When the excitation intensity is varied between 2.9 and 9.0 µW, the p lines are dominated by the emergence of components which are red-shifted by 3.5 meV with respect to the original p-shell signal. (iv) At very high excitation intensity (9 µW), the s shell disappears, the p shell becomes more structured,

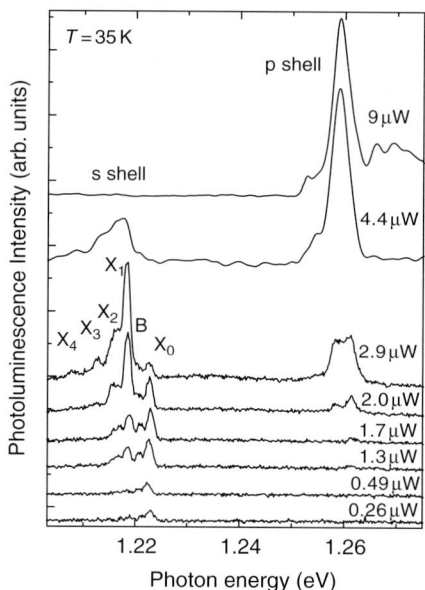

Fig. 32. PL spectra of a single InGaAs QD, excited by the 514.5-nm line of an Argon laser at 35 K, as a function of the excitation intensity, with increasing intensity from the bottom upwards (0.26, 0.40, 1.3, 1.7, 2.0, 2.89, 4.4 and 9.0 μW).

with a shoulder in the low energy tail of the high intensity feature at 1.258 eV. The appearance of the spectral features as function of the excitation density is a clear signature of the progressive shell filling and of the few-particle effects occurring in the dot with increasing carrier occupation (Lomascolo *et al.* 2002).

The identification of the different peaks is possible by means of detailed theoretical calculations (Rontari *et al.* 1999; Bayer *et al.* 2000; Dekel *et al.* 2000a; Brasken *et al.* 2001; Lomascolo (Lomascolo *et al.* 2002)). Taking into account the Coulomb interaction and correlation between few particles, the optical spectra can be determined as a function of the exact number of electrons and holes in a single QD. Nevertheless, due to the continuous wave nature of the optical excitation in many experiments, a steady state condition in the creation–recombination mechanisms of electron–hole pairs in the dot is achieved. It is therefore reasonable to expect that the measured optical spectra by temporal averaging, encompass contributions due to transitions involving different numbers of carrier captured by QD. With this assumption, and analyzing the variation of the integrated intensities of the salient components of the spectra (determined by fitting the spectra peaks with lorenztian and gaussian line-shapes) with increasing power intensity, we determined the recombination nature of the peaks. The single peak X at 1.223 eV (Fig. 32) arises from the recombination of the exciton ground state, which corresponds to the configuration where the carriers occupy the respective S shells. With increasing photoexcitation intensity, recombination from the multi-exciton states are expected to show up in the optical spectra. The peak B_0 at 1.221 eV (Fig. 32) corresponds to the transition between the neutral bi-exciton and the exciton ground state, whereas the peak x_1 at 1.218 eV to the isoenergetic transition between the charged bi-exciton and the exciton ground state, and between the charged exciton and the neutral exciton ground state. As the number of carriers is further increased, the interaction and correlation between the carriers induce new peaks in the s and the p shell is progressively filled.

A further enhancement of the spatial resolution can be achieved by tunnelling current-induced luminescence (TCIL). The high spatial resolution is afforded by the use of a scanning tunnelling microscope tip to inject carriers, either electrons or hole, into capped QDs. A planar geometry is employed where carriers are injected into the cap layer directly above the plane of QDs. In unprocessed samples, the multitude of QDs are accessed, so that spectral diffusion, excitation transfer and interdot tunnelling can occur. The TCIL was measured by using an STM operating in ultra-high vacuum (UHV) equipped with a liquid helium flow cryostat[2] and Pt–Ir tips. The luminescence signal was collected by a lens and dispersed by a 0.3-m monochromator and detected by a cooled Si CCD. The spectral resolution in the region of 1 μm was measured to be 0.25 meV and typical integration times for individual spectra was 1000 s. In Fig. 33(a)–(c), TCIL spectra collected when the tip is in tunnelling contact at different points of the cap surface, with tunnelling conditions of a gap voltage of −1.5 V and a tunnelling current of 5 nA, are shown. No TCIL signal was detected for gap voltages less than −1.5 V, or for positive biases. The first spectra, Fig. 33(a), exhibits a single well-defined peak at 1.294 eV with an FWHM of 2.0 meV. Figure 33(b), at another point of tunnelling contact on the sample surface, shows a doublet at 1.303 eV with a 4.2 meV splitting where the two components have FWHMs of 2.1 meV. These coupled peaks are accompanied by broader (FWHM ≈ 3.5 meV), less intense peaks at separations of 26–31 meV on either side. At a different point on the sample surface, the spectrum, shown in Fig. 33(c), is dominated by a complex structure with a main peak at 1.257 eV with an FWHM of 2 meV accompanied by a low-energy structure as well as small, broad peaks on the high-energy side, at 1.300 and 1.356 eV. A striking aspect of this spectrum with respect to previous ones in Fig. 33(a) and (b) is the four-fold increase in photon intensity.

It is instructive to compare the TCIL spectra with the PL spectrum of the sample. The PL spectra, measured by exciting by a He–Ne laser, 30 mW power, is shown in Fig. 34. The QD PL emission is centred at 1.22 eV, with an

---

[2]By the use of liquid Helium, the nominal temperature on the sample surface should be 25 K. We expect a higher temperature due to the larger sample size with respect to that quoted in the Omicron VT STM manual.

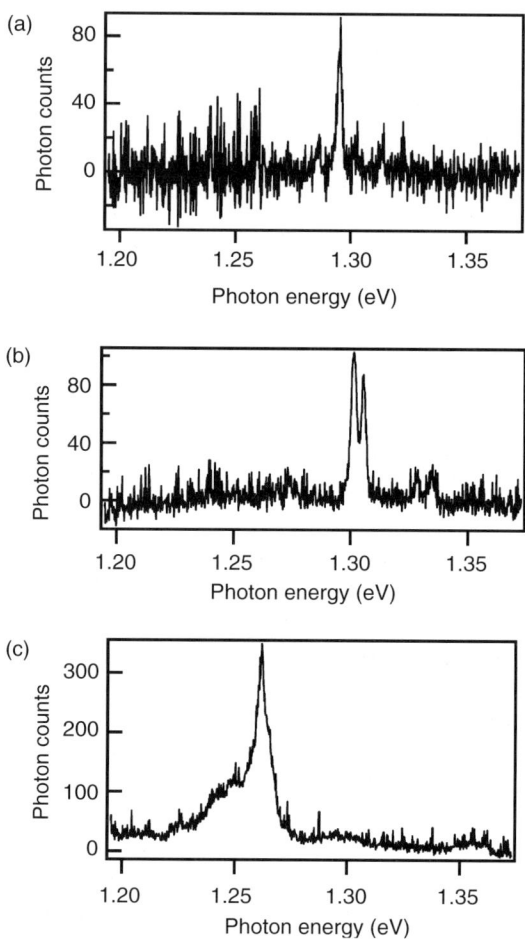

FIG. 33. TCIL spectra, gap voltage of −1.5 V and tunnelling current of 5 nA, where tunnelling contact has been made at different locations on the GaAs cap surface, (a) from a single QD, (b) from a pair of coupled QDs and (c) at a different point on the sample surface from ensemble QDs.

FWHM of 32 meV, the 1.29–1.34 eV region corresponds to the emission from the smaller QDs and the peak centred at 1.37 eV is due to quantum well-like emission from the WL. By this comparison, it is observed that the TCIL peaks observed in Fig. 33(a) and (b) do not lie strictly in the energy region of the main QD PL emission. Such a result can be easily explained.

Following Renaud and Alvarado (1991), the TCIL consists of a series of processes: (i) tunnelling out of (and therefore injecting holes into) the GaAs cap region; (ii) transport of ballistic holes across cap layer which are captured by WL and QD states; and followed by (iii) radiative recombination.

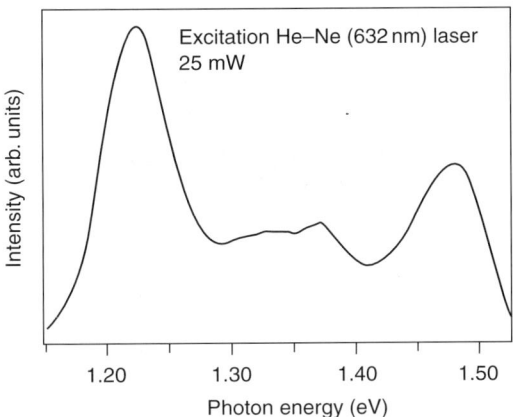

FIG. 34.   Ensemble PL spectrum of the $In_{0.5}Ga_{0.05}As$ QD (main peak 1.22 eV), the WL (1.37 eV), and GaAs ($\approx$1.51 eV).

The hole injection is characterized by a cone of diffusion, which defines the spatial resolution of the technique. Such a cone angle is not well defined and it is dependent on various parameters such as the tip dimensions, shape, the surface morphology, and composition (Dahne-Prietsch and Kalka 2000). The influence of these extraneous factors was observed in the TCIL spectra measured using different tips (not reported here) where, due to the poor quality of the tip, broad TCIL spectra, which are averaging over a wide area, were commonly observed. The spectra reported here exhibit the narrowest TCIL peaks observed, $\approx$1.5–2.0 meV and thus, it is proposed, exhibit high spatial resolution.

Since the gap voltage is -1.5 V, energetic holes in the barrier region are created. Prior to recombination at the QDs, the injected holes must therefore undergo relaxation processes to dissipate their excess kinetic energy. In the voltage range used in these experiments, it is reasonable to assume that the dominant scattering mechanism is that of longitudinal optical (LO) phonon scattering. By virtue of the high density of available states into which holes can de-excite, the probability of multiple scattering events and their further dispersion is high (Sprinzak *et al.* 1997).

Upon hole injection in the vicinity of the smaller dots (which are of the higher proportion), capture occurs at one of these smaller dots, which have higher confinement energies and therefore lower residual barriers, and a single peak is observed at 1.294 eV [Fig. 33(a)]. The appearance of this high-energy peak, which is not in the range of the main ensemble PL signal, is counter-intuitive, and would imply that injected carriers are more inclined to relax into, and then radiatively recombine from states which are not the lowest potential states of the system. However, the spatial resolving ability of TCIL allows the almost random selection of different points of the sample, and thus the capture of holes by the smaller QDs with higher confinement energy.

At a different point on the sample [Fig. 33(b)], coupled small QDs are filled, giving rise to ground state excitonic emission of the symmetric and anti-symmetric bonding states of the coupled QD states (the broader peaks are probably due to Raman scattering of LO phonons). Finally, Fig. 33(c) corresponds to the filling of hole states of the larger QDs (the main emission at 1.257 eV), and their neighbouring dots by interdot tunnelling processes. The initial relaxation of the injected holes into the smaller QD states is evident by the emission at 1.30 eV. The calculated differences in the exciton ground state energies for the two families of dots is ≈ 45 meV. This correlates well with the observed separation between the TCIL signals, which originates from the smaller and larger dots. To conclude, the tunnelling current induced luminescence spectra of single, coupled and an ensemble of InGaAs QDs can be selectively analysed, following the complex energy dependent relaxation of the holes, which results in hole capture by QDs of different sizes depending on their spatial location.

VII. Conclusions

Recent developments about self-assembled QDs have been presented, with special attention to the methodologies developed for the control and the engineering of the electron states of MOCVD-grown InGaAs/GaAs QDs. We showed that the optical and electronic properties of such nanostructures strongly depend on parameters, such as size, shape, composition, and strain of the dots. To have a high control of these parameters, different techniques have been developed, such as high-resolution TEM and STM, EDX and EELS measurements. In addition, local spectroscopy techniques have been used to study important effects in individual quantum dot. For instance, NSOM and STS measurements permit to observe the Coulomb interaction between few carriers inside a single QD, whereas a direct wavefunction imaging can be obtain by scanning tunnelling luminescence measurements. Finally, we showed that electronic structure calculations are fundamental for a meaningful interpretation of the experimental results obtained in QDs.

ACKNOWLEDEGMENTS

The authors would like to thank Dr Massimo Catalano and Dr Antonietta Taurino for the TEM measurements performed on quantum dots. Iolena Tarantini and Angelo Melcarne for the technical support to the growth and characterization.

Funding from INFM, European Community, and Agilent Technologies is gratefully acknowledged.

REFERENCES

Alperson, B., S. Cohen, I. Rubinstein, and G. Hodes, *Phys. Rev. B* **52**, R17017 (1995).
Alperson, B., I. Rubinstein, G. Hodes, D. Porath, and O. Millo, *Appl. Phys. Lett.* **75**, 1751 (1999).
Andreev, A. D., J. R. Downes, D. A. Faux, and E. P. O'Reilly, *J. Appl. Phys.* **86**, 297 (1999).
Arakawa, Y., and H. Sakaki, *Appl. Phys. Lett.* **40**, 939 (1982).
Ashoori, R. C., *Nature* **379**, 413 (1996).
Bacher, G., R. Weigand, J. Seufert, V. D. Kulakovskii, N. A. Gippius, A. Forchel, K. Leonardi, and D. Hommel, *Phys. Rev. Lett.* **83**, 4417 (1999).
Banin, U., Y. W. Cao, D. Katz, and O. Millo, *Nature* **400**, 542 (1999).
Bayer, M., O. Stern, P. Hawrylak, S. Fafard, and A. Forchel, *Nature* **405**, 923 (2000).
Beaumount, S. P., in *Low-Dimensional Structures in Semiconductors*, NATO ASI Series. Series B, Physics, Vol. 281 (1991).
Benabbas, T., P. François, Y. Androussi, and A. Lefebvre, *J. Appl. Phys.* **80**, 2763 (1996).
Berzig, E., J. K. Trautman, T. D. Harris, J. S. Weiner, and R. L. Kostelak, *Science* **251**, 1468 (1991).
Betzing, E., M. Isaacson, and A. Lewis, *Appl. Phys. Lett.* **51**, 2088 (1987).
Bimberg, D., M. Grundmann, and N. D. Ledentsov, in *Quantum Dot Heterostructures*, Wiley, New York (1999).
Brasken, M., M. Lindberg, D. Sundholm, and J. Olsen, *Phys. Rev. B* **64**, 353121 (2001).
Cusack, M. A., P. R. Briddon, and M. Jaros, *Phys. Rev. B* **54**, R2300 (1996).
Dahne-Prietsch, M., and T. Kalka, *J. Electron. Spect. Related Phenom.* **109**, 211 (2000).
De Giorgi, M., A. Vasanelli, R. Rinaldi, M. Anni, M. Lomascolo, S. Antonaci, A. Passaseo, R. Cingolani, A. Taurino, M. Catalano, and E. Di Fabrizio, *Micron* **31**, 245 (2000).
De Giorgi, M., A. Taurino, A. Passaseo, M. Catalano, and R. Cingolati, *Phys. Rev. B* **63**, 245302 (2001).
Dekel, E., D. Gershoni, E. Ehrenfreund, D. Spektor, J. M. Garcia, and P. M. Petroff, *Phys. Rev. Lett.* **80**, 4991 (1998).
Dekel, E., D. Ghersoni, E. Ehrenfreund, J. M. Garcia, and P. M. Petroff, *Phys. Rev B* **61**, 11009 (2000a).
Dekel, E., D. V. Regelman, D. Gershoni, E. Ehrenfreund, W. V. Schoenfeld, and P. M. Petroff, *Phys. Rev. B* **62**, 11038 (2000b).
De la Broise, X., C. Delerue, M. Lannoo, B. Grandidier, and D. Stievenard, *Phys. Rev. B* **61**, 2138 (2000).
Di Dio, M., M. Lomascolo, A. Passaseo, C. Gerardi, C. Giannini, A. Quirini, L. Taffer, P. V. Giugno, M. De Vittorio, D. Greco, A. Convertino, R. Rinaldi, L. Vasanelli, and R. Cingolani, *J. Appl. Phys.* **80**, 482 (1996).
Duke, C. B., *Tunneling in Solids*, Academic Press, New York (1969).
Empedocles, S. A., D. J. Norris, and M. G. Bawendi, *Phys. Rev. Lett.* **77**, 3873 (1996).
Ferreira, R., and G. Bastard, *Appl. Phys. Lett.* **74**, 2818 (1999).
Fertig, H. A., *Phys. Rev. B* **65**, 2321 (1990).
Finley, J. J., A. D. Ashmore A. Lemaitre, D. J. Mowbray, M. S. Skolnick, I. E. Itskevich, P. A. Maksym, M. Hopkinson, and T. F. Krauss, *Phys. Rev. B* **63**, 073307 (2001a).
Finley, J. J., P. W. Fry, A. D. Ashmore, A. Lemaitre, A. I. Tartakovskii, R. Oulton, D. J. Mowbray, M. S. Skolnick, M. Hopkinson, P. D. Buckle, and P. A. Maksym, *Phys. Rev. B* **63**, 161305 (2001b).

Fonseca, L. R. C., J. L. Jimenez, J. P. Leburton, and R. M. Martin, *Phys. Rev. B* **57**, 4017 (1998).

Franceschetti, A., and A. Zunger, *Phys. Rev. B* **62**, 2614 (2000).

Fry, P. W., I. E. Itskevich, D. J. Mowbray, M. S. Skolnick, J. J. Finley, J. A. Barker, E. P. O'Reilly, L. R. Wilson, I. A. Larkin, P. A. Maksym, M. Hopkinson, M. Al-Khafaji, J. P. R. David, A. G. Cullis, G. Hill, and J. C. Clark, *Phys. Rev. Lett.* **84**, 733 (2000).

Fujita, S., S. Maruno, H. Watanabe, Y. Kusumi, and M. Ichikawa, *Appl. Phys. Lett.* **66**, 2754 (1995).

Gammon, D., *Phys. Rev. Lett.* **76**, 3005 (1996).

Gammon, D., *Nature* **405**, 899 (2000).

Gammon, D., E. S. Snow, and D. S. Katzer, *Appl. Phys. Lett.* **67**, 2391 (1995).

Garcia, J. M., G. Medeiros-Ribeiro, K. Schmidt, T. Ngo, J. L. Feng, A. Lorke, J. Kotthaus, and P. M. Petroff, *Appl. Phys. Lett.* **71**, 2014 (1997).

Georgsson, K., N. Carlsson, L. Samuelson, W. Seifert, and L. R. Wallenberg, *Appl. Phys. Lett.* **67**, 2981 (1995).

Goldstein, L., F. Glas, J. Y. Marzin, M. N. Charasse, and G. Le Roux, *Appl. Phys. Lett.* **47**, 1099 (1985).

Grandidier, B., Y. M. Niquet, B. Legrand, J. P. Nys, C. Priester, D. Stiévenard, J. M. Gerard, and V. Thierry-Mieg, *Phys. Rev. Lett.* **85**, 1068 (2000).

Grundmann, M., O. Stier, and D. Bimberg, *Phys. Rev. B* **52**, 11969 (1995).

Hameau, S., Y. Guldner, O. Verzelen, R. Ferreira, G. Bastard, J. Zeman, A. Lemaître, and J. M. Gérard, *Phys. Rev. Lett.* **83**, 4152 (1999).

Hartmann, A., L. Loubies, F. Reinhardt, and E. Kapon, *Appl. Phys. Lett.* **71**, 1314 (1997).

Hartmann, A., Y. Ducommun, E. Kapon, U. Hohenester, and E. Molinari, *Phys. Rev. Lett.* **84**, 5648 (2000).

Heitz, R., M. Veit, N. N. Ledenstov, A. Hoffmann, D. Bimberg, V. M. Ustinov, P. S. Kopev, and Zh. I. Alferov, *Phys. Rev. B* **56**, 10435 (1997).

Janssens, K. G. F., O. Van der Biest, J. Vanhellemont, H. E. Maes, and R. Hull, *Appl. Phys. Lett.* **67**, 1530 (1995).

Jiang, H., and J. Singh, *Phys. Rev. B* **56**, 4696 (1997).

Joyce, P. B., T. J. Krzyzewski, C. G. Bell, B. A. Joyce, and T. S. Jones, *Phys. Rev B* **58**, R15981 (1998).

Kane, B. E., *Nature* **393**, 133 (1998).

Kapteyn, C. M., M. Lion, R. Heitz, D. Bimberg, P. N. Brunkov, B. V. Volovik, S. G. Konnikov, A. R. Kovsh, and V. M. Ustinov, *Appl. Phys. Lett.* **76**, 1573 (2000).

Karlsson, K. F., E. S. Moskalenko, P. O. Holtz, B. Monemar, W. V. Schoenfld, J. M. Garcia, and P. M. Petroff, *Appl. Phys. Lett.* **78**, 2952 (2001).

Kastner, Mark A., *et al.*, *Phys. Today* **46**, 24 (1993).

Krost, A., J. Bläsing, F. Heinrichsdorff, and D. Bimberg, *Appl. Phys. Lett.* **75**, 2957 (1999).

Landin, L., M. E. Pistol, C. Pryor, M. Persson, L. Samuelson, and M. Miller, *Phys. Rev. B* **60**, 16640 (1999).

Ledentsov, N. N., *Solid State Electron.* **40**, 785 (1996).

Ledentsov, N. N., *et al.*, *Appl. Phys. Lett.* **70**, 2888 (1997).

Legrand, B., B. Grandidier, J. P. Nys, D. Stievenard, J. M. Gerard, and V. Thierry-Mieg, *Appl. Phys. Lett.* **73**, 97 (1998).

Lelong, Ph., and G. Bastard, *Solid State Commun.* **98**, 819 (1996).

Lelong, Ph., O. Heller, and G. Bastard, *Solid State Electron.* **42**, 1251 (1998).

Leon, R., Y. Kim, C. Jagadish, M. Gal, J. Zou, and D. J. H. Cockayne, *Appl. Phys. Lett.* **86**, 1888 (1996).

Lian, G. D., J. Yuan, L. M. Brown, G. H. Kim, and D. A. Ritchie, *Appl. Phys. Lett.* **73**, 49 (1998).

Liao, X. Z., J. Zou, X. F. Duan, D. J. H. Cockayne, R. Leon, and C. Lobo, *Phys. Rev. B* **58**, R4235 (1998).

Lipsanen, H., M. Sopanen, and J. Ahopelto, *Phys. Rev. B* **51**, 13868 (1995).

Liu, N., J. Tersoff, O. Baklenov, A. L. Holmes, and C. K. Shih, *Phys. Rev. Lett.* **84**, 334 (2000).

Lobo, C., R. Leon, S. Fafard, and P. G. Piva, *Appl. Phys. Lett.* **72**, 2850 (1998).

Lomascolo, *et al.*, *Phys. Rev. B* **66**, R041302 (2002).

Marquez, J., L. Geelhaar, and K. Jacobi, *Appl. Phys. Lett.* **78**, 2309 (2001).

Martorell, J., D. W. L. Sprung, P. A. Machado, and C. G. Smith, *Phys. Rev. B* **63**, 45325 (2001).

Marzin, J. Y., and G. Bastard, *Solid State Commun.* **92**, 437 (1994).

Marzin, J. Y., J. M. Gerad, A. Izrael, D. Barrier, and G. Bastard, *Phys. Rev. Lett.* **73**, 716 (1994).

Mazzer, M., M. De Giorgi, R. Cingolani, G. Porcllo, F. Rossi, and E. Molinari, *J. Appl. Phys.* **84**, 1 (1998).

Millo, O., D. Katz, Y. W. Cao, and U. Banin, *Phys. Rev. B.* **61**, 16773 (2000).

Millo, O., D. Katz, Y. W. Cao, and U. Banin, *Phys. Rev. Lett.* **86**, 5751 (2001).

Nishi, K., A. A. Yamaguchi, J. Ahopelto, A. Usui, and H. Sakaki, *J. Appl. Phys.* **76**, 7437 (1994).

Passaseo, A., G. Maruccio, M. De Vittorio, S. De Rinaldis, T. Todaro, R. Rinaldi, and R. Cingolani, *Appl. Phys. Lett.* **79**, 1435 (2001a).

Passaseo, A., G. Maruccio, M. De Vittorio, R. Rinaldi, R. Cingolani, and M. Lomascolo, *Appl. Phys. Lett.* **78**, 1382 (2001b).

Passaseo, A., R. Rinaldi, M. Longo, S. Antonaci, A. L. Convertino, R. Cingolani, A. Taurino, and M. Catalano, *J. Appl. Phys.* **89**, 4341 (2001c).

Pollak, F. H., *Semicond. Semimet.* **32**, 17 (1990).

Pryor, C., *Phys. Rev. B* **57**, 7190 (1998a).

Pryor, C., *Phys. Rev. Lett.* **80**, 3579 (1998b).

Pryor, C., J. Kim, L. W. Wang, A. J. Williamson, and A. Zunger, *J. Appl. Phys.* **83**, 2548 (1998).

Regelman, D. V., E. Dekel, D. Gershoni, E. Ehrenfreund, A. J. Williamson, J. Shumway, A. Zunger, W. V. Schoenfeld, and P. M. Petroff, *Phys. Rev. B* **64**, 165301 (2001).

Renaud, P., and S. F. Alvarado, *Phys. Rev. B* **44**, 6340 (1991).

Rinaldi, R., *Int. J. Mod. Phys.* **12**, 471 (1998).

Rontari, M., F. Rossi, F. Manghi, and E. Molinari, *Phys. Rev. B* **59**, 10165 (1999).

Ruminov, S., P. Werner, K. Scheerschmidt, U. Gösele, J. Heydenreich, U. Richter, N. N. Ledentsov, M. Grundmann, D. Bimberg, V. M. Ustinov, A. Yu. Egorov, P. S. Kop'ev, and Zh. I. Alferov, *Phys. Rev. B* **51**, 14766 (1995).

Sajal, P., J. B. Roy, and P. K. Basu, *J. Appl. Phys.* **69**, 827 (1991).

Sato, H., T. Sugahara, Y. Naoi, and S. Sakai, *Jpn. J. Appl. Phys., Part 1* **37**, 2013 (1998).

Shumway, J., A. J. Williamson, A. Zunger, A. Passaseo, M. De Giorgi, R. Cingolani, M. Catalano, and P. Crozier, *Phys. Rev. B* **64**, 125302 (2001).

Siverns, P. D., S. Malik, G. McPherson, D. Childs, C. Roberts, R. Murray, B. A. Joyce, and H. Davock, *Phys. Rev. B* **58**, R10127 (1998).

Snow, E. S., P. M. Campbell, and B. V. Shanabrook, *Appl. Phys. Lett.* **63**, 3488 (1993).

Solomon, G. S., J. A. Trezza, A. F. Marshall, and J. S. Harris, *Phys. Rev. Lett.* **76**, 952 (1996).

Sprinzak, D., M. Heiblum, Y. Levinson, and H. Shtrikman, *Phys. Rev. B* **55**, R10185 (1997).

Srolovitz, D. J., *Acta Metall.* **37**, 621 (1989).

Stier, O., M. Grundmann, and D. Bimberg, *Phys. Rev. B* **59**, 5688 (1999).

Sugawara, M., *Phys. Rev. B* **51**, 10743 (1995).

Sugawara, M., K. Mukai, and Y. Nakata, *Appl. Phys. Lett.* **74**, 1561 (1999).

Tatebayashi, J., M. Nishioka, and Y. Arakawa, *Appl. Phys. Lett.* **78**, 3469 (2001).

Tersoff, J., and D. R. Hamann, *Phys. Rev. B* **31**, 805 (1985).

Thompson, R. M., R. M. Stevenson, A. J. Shields, I. Farrer, C. J. Lobo, D. A. Ritchie, M. L. Leadbeater, and M. Pepper, *Phys. Rev. B* **64**, R201302 (2001).

Tulkki, J., and A. Heinämäki, *Phys. Rev. B* **52**, 8239 (1995).

Vasanelli, A., M. De Giorgi, R. Ferreira, R. Cingolani, and G. Bastard, *Jpn. J. Appl. Phys., Part 1* **40**, 1955 (2001a).

Vasanelli, A., R. Ferriera, H. Sakaki, and G. Bastard, *Solid State Commun.* **118**, 459 (2001b).

Weisbuch, C., and G. Vinter, in *Quantum Semiconductor Structures*, Academic press, Boston, MA (1991).

Wiesendanger, R., *Scanning Probe Microscopy and Spectroscopy*, Cambridge University Press, Cambridge (1998).

Wu, W., J. R Tucker, and G. S. Solomon, *Appl. Phys. Lett.* **71**, 1083 (1997).

Xie, Q., P. Chen, and A. Madhukar, *Appl. Phys. Lett.* **65**, 2051 (1994).

Xie, Q., A. Madhukar, P. Chen, and N. P. Kobayashi, *Phys. Rev. Lett.* **75**, 2542 (1995).

Xue, Q.-K., Y. Hasegawa, and H. Kiyama, *Jpn. J. Appl. Phys., Part 1* **38**, 500 (1999).

Yao, J. Y., T. G. Andersson, and G. L. Dunlop, *J. Appl. Phys.* **69**, 2224 (1991).

Zhang, Q., J. Zhu, X. Ren, H. Li, and T. Wang, *Appl. Phys. Lett.* **78**, 3830 (2001).

Zou, J., X. Z. Liao, D. J. H. Cockayne, and R. Leon, *Phys. Rev. B* **59**, 12279 (1999).

Zunger, A., *MRS Bull.* **23**, 35 (1998).

Advanced Semiconductor and Organic Nano-Techniques (Part II)
H. Morkoç (Ed.)

CHAPTER 2

GaN-Based Modulation Doped FETs and Heterojunction Bipolar Transistors

H. Morkoç

VIRGINIA COMMONWEALTH UNIVERSITY, RICHMOND, VIRGINIA

I. Introduction

Semiconductor nitrides such as aluminum nitride (AlN), gallium nitride (GaN), and indium nitride (InN) are very promising materials for their potential use in optoelectronic devices (both emitters and detectors), and high-power/temperature electronic devices, as have been treated in length and reviewed recently (Strite and Morkoç 1992; Morkoç *et al.* 1994; Mohammad *et al.* 1995; Mohammad and Morkoç 1996; Ambacher 1998; Morkoç 1999a; Pearton *et al.* 1999). These materials and their ternary and quaternary alloys cover an energy bandgap range of 1.9–6.2 eV, suitable for band-to-band light generation with colors ranging from red (potentially) to ultraviolet (UV) wavelengths. Specifically, nitrides are suitable for applications such as surface acoustic wave devices (Duffy *et al.* 1973), UV detectors (Razeghi and Rogalski 1996; Xu *et al.* 1997), Bragg reflectors (Fritz and Drummond 1995), waveguides, UV and visible light emitting diodes (LEDs) (Nakamura *et al.* 1994; Morkoç and Mohammad 1995, 1999), and laser diodes (LDs) (Nakamura *et al.* 1997) for digital data read–write applications. During the last several decades, lasers and LEDs have expanded remarkably both in terms of the range of emission wavelengths available and brightness. The nitride semiconductor-based LEDs have proven to be reliable in applications such as displays, lighting, indicator lights, advertisement, and traffic signs/signals. Additional possible applications include use in agriculture as light sources for accelerated photosynthesis, and in health care for diagnosis and treatment. Lasers, as coherent sources, are crucial for high-density optical read and write technologies because the diffraction-limited optical storage density increases approximately quadratically in the ideal case as the probe laser wavelength is reduced. Nitride-based coherent UV sources are attracting a good deal of attention for optical storage devices. Optical storage would enable the storage and retrieval of inordinate number of images and vast quantities of text with untold efficiency. Other equally attractive applications envisioned include printing and surgery.

When used as UV sensors in jet engines, automobiles, and furnaces (boilers), the devices would allow optimal fuel efficiency and control of effluents for a cleaner environment. Moreover, UV sensors that operate in the solar-blind region (260–290 nm) would have high detectivity because the ozone layer absorbs solar radiation at those wavelengths, thus virtually eliminating the radiation noise. Consequently, these detectors are expected to play a pivotal role in threat recognition aimed against aircraft and other vehicles (Razeghi and Rogalski 1996; Xu *et al.* 1997; Morkoç 1998a). GaN photodiodes (Brown *et al.*) exhibited zero-bias responsivities of about 0.21 A/W at 356 nm which decreased by more than three orders of magnitude for wavelengths longer than 390 nm. The noise equivalent power (NEP) at a reverse bias of 10 V is $(f > 100 \, \text{Hz})$ $6.6 \times 10^{-15} \, \text{W/Hz}^{1/2}$, which is extremely small (H. Temkin, Texas Tech University, private communication). Detector speed, while affected in terms of uniformity by the sheet resistance of the p-layer, which suffers from the notoriously low doping levels, is in the picosecond range (J. C. Campbell, University of Texas at Austin, private communication). Finally, the GaN-based detectors with AlN mole fractions approaching the solar-blind region of the spectrum have been fabricated into arrays for imaging. Detector arrays with pixel sizes of 32×32 have been fabricated and tested already (Schetzina *et al.* 1999). For a detailed treatise of GaN-based detectors, the reader is referred to Chapter 3 (Part II).

GaN's large bandgap, large dielectric breakdown field, fortuitously good electron transport properties (Kolnik *et al.* 1995; Bhapkar and Shur 1997; Ridley 1998) (an electron mobility possibly in excess of 2000 cm^2/V s and a peak velocity approaching 3×10^7 cm^{-1} at room temperature), and good thermal conductivity are trademarks of high-power/temperature electronic devices (Morkoç 1998b). Sheppard *et al.* (1999) have reported that 0.45-μm gate, high-power modulation doped FETs (MODFETs) on SiC substrates exhibited a power density of 6.8 W/mm in a 125-μm-wide device and a total power of 4 W (with a power density of 2 W/mm) at 10 GHz. Other groups have also reported on the superior performance of GaN-based MODFETs on SiC and sapphire substrates with respect to competing materials, particularly at X band and higher frequencies (Binari *et al.* 1997b; Ping *et al.* 1998; Sullivan *et al.* 1998; Wu *et al.* 1998). What is astounding is that researchers at HRL Laboratories have recently demonstrated GaN/AlGaN MODFETs prepared by MBE on SiC substrates, which exhibited a total power level of 6.3 W at 10 GHz from a 1-mm-wide device. What is more astounding is that the power level is not really thermally limited as the power density extrapolated from a 0.1-mm device is 6.5 W. When four of these devices are power combined in a single-stage amplifier, an output power of 22.9 W with a power added efficiency of 37% was obtained at 9 GHz (Micovic *et al.* 2001). Equally impressive is the noise figure of 0.85 dB at 10 GHz with an associated gain of 11 dB. The drain breakdown voltages in these quarter-micrometer gate devices are about 60 V, which are, in part, responsible for such a record performance (N. Nguyen and C. Nguyen, HRL Laboratories, private communication).

Applications of high-power GaN-based MODFETs include amplifiers operative at high power levels, high temperatures, and in unfriendly environments such as radar, missiles, satellites as well as in low-cost compact amplifiers for wireless base stations. Much of these applications are currently met by pseudomorphic modulation doped FETs (Henderson *et al.* 1986).

Though in its infancy, efforts are underway to exploit nitrides for bipolar transistors. However, the materials quality needs to be improved more before performance expected from GaN can be obtained. Difficulties include the notoriously low p-type doping and low diffusion length in epitaxial layers.

Nitride semiconductors have been deposited by vapor phase epitaxy [i.e., both hydride VPE (HVPE), Molnar *et al.* (1997), which has been developed for thick GaN layers and organometallic VPE (OMVPE), Yamaguchi *et al.* (1999), which has been developed for heterostructures], and in vacuum by a slew of variants of molecular beam epitaxy (MBE) (Morkoç 1998a). All the high-performance light emitters, which require high-quality InGaN, have been produced by OMVPE. On the other hand, MBE has been very successful in producing structures that do not require InGaN for optical emitters. Some examples are FETs and detectors. With its innate refined control of growth parameters, *in situ* monitoring capability, and uniformity, MBE is well suited for depositing heterostructures and gaining insight into the deposition/incorporation mechanisms. MBE's control over growth parameters is such that any structure can be grown in any sequence. The structures based on conventional compound semiconductors such as IR lasers for CD players, surface-emitting vertical cavity lasers, and high-performance pseudomorphic MODFETs have all been produced very successfully, most of them commercially, by MBE. Nitride growth, however, requires much higher temperatures than those used in producing conventional Group III–V semiconductors for which the MBE systems were designed. In addition, it has proved difficult to provide active N species at sufficiently high rates for nitride growth. Despite these mechanical/engineering limitations and its relatively late entry, MBE, with appropriate modifications, has already played a key role on a number of fronts, such as high-performance GaN-based MODFETs and fast solar-blind detectors. Interestingly, highest mobility two-dimensional electron gas (2DEG) systems were grown with MBE on templates prepared by HVPE, MOCVD, and bulk-GaN. Moreover, MBE-grown films on such templates produce very clean luminescence with only the excitonic transitions and their excited states observable.

In this chapter, the properties of nitride semiconductors, electron transport in bulk and two-dimensional systems based on nitrides, polarization issues, MODFET simulations, technology, and performance are discussed. A succinct discussion of nitride-based HBTs close the discussion.

II.   Crystal Structure of Nitrides

The Group-III nitrides crystallize in the wurtzite, zincblende, and rocksalt structures; the thermodynamically stable phase at ambient conditions has the wurtzite structure. The wurtzite structure has a hexagonal unit cell and, thus, two lattice constants, c and a. It contains six atoms of each type. The space grouping for the wurtzite structure is $P6_3mc$ in Hermann–Mauguin notation (C_{6v}^4 in Schoenflies notation). The wurtzite structure consists of two interpenetrating hexagonal close packed (HCP) sublattices, each with one type of atom, offset along the c-axis by 5/8 of the cell height ($5/8c$). The zincblende structure has a cubic unit cell. The space grouping for the zincblende structure is $F\bar{4}3m$ in Hermann–Mauguin notation (T_d^2 in Schoenflies notation).

The zincblende and wurtzite structures are similar. In both cases, each Group-III atom is coordinated by four nitrogen atoms. Conversely, each nitrogen atom is coordinated by four Group-III atoms. The main difference between these two structures lies in the stacking sequence of the closest packed diatomic planes. For the wurtzite structure, the stacking sequence of

(0001) planes is AaBbAa, whereas for the zincblende structure, the stacking sequence of {111} planes is AaBbCc.

1. GALLIUM NITRIDE

Although GaN is the most extensively studied of all the other Group-III nitrides, it is still in dire need of further investigations. Though improving, GaN has a relatively large n-type background carrier concentration due to native defects and impurities. The lack of commercially available native substrates exacerbates the situation. Combined with the difficulties in obtaining p-type doping and the somewhat arcane fabrication processes, stymied timely progress. Information available in the literature regarding many of the physical properties of GaN is still evolving and naturally controversial. This is, in part, a result of measurements being made on samples of widely varying quality. Nevertheless, what follows is believed to represent the state of knowledge regarding the physical properties of GaN.

a. Chemical Properties of GaN

Since Johnson *et al.* first synthesized GaN in 1932 (for details, see Morkoç 1999a), a great body of data has repeatedly indicated that GaN is an exceedingly stable compound and exhibits significant hardness, which is coveted. Moreover, owing to its wide energy bandgap, it is also an excellent candidate for device operation in high temperature and harsh environments. As a matter of fact, the majority of GaN researchers are currently interested in semiconductor device applications. While the thermal stability of GaN allows freedom for high-temperature processing, the chemical stability of GaN presents a technological challenge. Conventional wet-etching techniques used in semiconductor processing are not very successful for GaN device fabrication. Well-established chemical-etching processes are required for device technology development. Promising possibilities are various dry-etching processes under development, reviewed by Mohammad and Morkoç (1996) and more recently by Pearton *et al.* (1999).

The thermal stability and dissociation of GaN have also been examined. While some experimental studies of the stability of GaN conducted at high temperature suggested that significant weight losses occur at temperatures as low as 750°C, others contradicted it, and suggested that no significant weight loss should occur even at a temperature of 1000°C. GaN is less stable in an HCl or H_2 atmosphere than in N_2. Using mass spectroscopy, it was determined that $(GaN)_2$ dimers are the primary components of decomposition. Others observed only N_2^+ and Ga^+ to be the primary components in the vapor over GaN. Based on the measurement of apparent vapor pressure, the heat of sublimation of GaN was calculated to be 72.4 ± 0.5 kcal/mol.

The equilibrium N_2 pressure of GaN as a function of temperature was also calculated (Morkoç 1999a).

b. Mechanical and Thermal Properties of GaN

Wurtzite GaN has a molecular weight of 83.728 g/mol. At room temperature, its lattice parameters are $a_0 = 3.1892 \pm 0.0009$ Å and $c_0 = 5.1850 \pm 0.0005$ Å. However, for the zincblende polytype the calculated lattice constant based on measured Ga–N bond distance in wurtzite-GaN (WZ-GaN) is $a = 4.503$ Å. The measured value for this polytype varies between 4.49 and 4.55 Å, indicating that the calculated result is within acceptable limits.

Measurements over the temperature range of 300–900 K indicate a mean coefficient of thermal expansion of GaN in the c-plane of $\Delta a/a = 5.59 \times 10^{-6} \, \mathrm{K}^{-1}$. Similarly, in the temperature ranges of 300–700 and 700–900 K, the mean coefficient of thermal expansion in the c-direction is $\Delta c/c = 3.17 \times 10^{-6}$ and $7.75 \times 10^{-6} \, \mathrm{K}^{-1}$, respectively (Mohammad and Morkoç 1996).

The room-temperature value of the thermal conductivity measured, at $\kappa = 1.3 \, \mathrm{W/cm \, K}$, many years ago is a little smaller than the predicted (some time ago) value of 1.7 W/cm K. The recent value of 1.8 W/cm K measured by a new method, dubbed the scanning thermal microscopy (SThM), which measures the thermal conductivity near the surface, is closer to predictions (Asnin et al. 1999). Further, very recent calculations (Witek 1998) indicate the thermal conductivity of GaN to be near 4 W/cm K. Recent measurements, with long-standing credibility, made in very high quality GaN templates produced at Samsung Laboratories by Drs Y.-J. Park and S.S. Park indicated the thermal conductivity to be 2.3 W/cm K at room temperature (D. Morelli, G. Slack, L. Schowalter, Y. Park, and M. Morkoç, unpublished results). The Debye temperature (Θ_D) of GaN at 0 K was calculated to be $\Theta_D \sim 600$ K. Other thermal properties of WZ-GaN have been studied by a number of authors. The equilibrium vapor pressure of N_2 over solid GaN has been found to be 10 MPa at 1368 K and 1 GPa at 1803 K.

Experimental investigation of the elastic constants of WZ-GaN has been carried out by using X-ray diffraction. The estimates from elastic coefficients $-2C_{13}/C_{33}$ and measured values of Poisson's ratio $\nu_{\langle 0001 \rangle} = (\Delta a/a_0)/(\Delta c/c_0)$ of 0.372 and 0.378, respectively, are in good agreement. However, there is a wide spread in the reported values of elastic stiffness coefficients.

Young's modulus and Poisson's ratio for GaN films have also been measured. From the elastic stiffness coefficients, Young's modulus $E_{\langle 0001 \rangle}$ is estimated to be 150 GPa. The elastic properties of zincblende-GaN (ZB-GaN), using the values of WZ-GaN, were reported as an estimate. The bulk modulus for WZ-GaN was calculated by first principle and first principle orthogonalized linear combination of atomic orbitals (LCAO) calculations with the resulting values of 195 and 203 GPa, respectively. More recent calculations indicate the bulk modulus to be 207 GPa, which is good agreement with previous values (Wagner *et al.* 2000).

A group theory approach predicts that the optical phonon modes of WZ-GaN contain one A_1 mode, one E_1 mode, two E_2 modes, and two B_1 modes whose wavenumbers have been measured by Raman scattering with the assignment of E_1-TO, A_1-TO, E_1-LO, A_1-LO, and two E_2 phonons. The transverse optical (TO) and longitudinal optical (LO) phonon wavenumbers of ZB-GaN have also been determined. Much of the lattice dynamics in GaN and related materials have been revisited recently.

2.   ALUMINUM NITRIDE

AlN exhibits many useful mechanical and electronic properties. For example, high hardness, high thermal conductivity, resistance to high temperature and caustic chemicals combined with, in non-crystalline form, a reasonable thermal match to Si and GaAs, make it an attractive material for electronic packaging applications. However, the main interest in this semiconductor stems from its ability to form alloys with GaN producing AlGaN and allowing fabrication of AlGaN/GaN-based electronic and optical devices, the latter of which could be active from the green wavelengths well into the UV region. There is also some interest in exploiting AlN as substrate because the nitrogen vapor pressure over Al is several orders of magnitude smaller than on Ga. Contamination-free deposition environment, coupled with advanced procedures, has allowed researchers to grow improved-quality AlN. Consequently, many of the physical properties of AlN have been reliably measured and bulk-AlN synthesized.

When crystallized in the hexagonal wurtzite structure, the AlN crystal has a molar mass of 20.495 g/mol. The cubic form is hard to obtain and thus will be ignored. The space group symmetry is C_{6v}^4 ($P6_3mc$) and the point group symmetry is $C6v$ ($6mm$). The c/a ratio for this is $(8/3)^{1/2} = 1.633$. The reported lattice parameters range from 3.110 to 3.113 Å for a, and from 4.978 to 4.982 Å for c. The c/a ratio, thus, varies between 1.000 and 1.602. The deviation of the c/a ratio from that of the ideal wurtzite crystal is probably due to lattice stability and ionicity. While for the metastable zincblende polytype AlN has a value of $a = 4.38$ Å, the rocksalt structure retained at room temperature has a value of $a = 4.043$–4.045 Å.

a.   *Thermal and Chemical Properties of AlN*

AlN is an extremely hard ceramic material with a melting point higher than 2000°C. Thermal conductivity, κ, of AlN at room temperature has been predicted at ≈ 3.2 W/cm K. Measured values of κ at 300 K are 2.5 and 2.85 W/cm K.

Using X-ray techniques across a broad temperature range (77–1269 K), it was noted that the thermal expansion of AlN is isotropic with a room-temperature value of 2.56×10^{-6} K^{-1}. The thermal expansion coefficients of AlN measured have mean values of $\Delta a/a = 4.2 \times 10^{-6}$ K^{-1} and $\Delta c/c = 5.3 \times 10^{-6}$ K^{-1}.

The equilibrium N_2 vapor pressure on AlN is relatively low compared to that on GaN, which makes AlN easier to synthesize. The calculated temperatures at which the equilibrium N_2 pressure reaches 1, 10, and 100 atm. are 2836, 3088, and 3390 K, respectively.

Similar to GaN but even more so, AlN exhibits inertness to many wet etches. A number of AlN etches have been reported in the literature. However, none of these etches have been performed on high-quality single-crystal AlN. The surface chemistry of AlN has been investigated by numerous techniques, including Auger electron spectroscopy, X-ray and ultraviolet photoemission spectroscopy (XPS), ultraviolet photoelectron spectroscopy, and electron spectroscopy.

b. Mechanical Properties of AlN

Early investigations of the elastic properties of AlN were carried out on sintered polycrystalline specimens, due to the unavailability of large single crystals. This, however, paved the way to more refined measurements as single crystalline AlN became available. Somewhat related to the mechanical properties is the phonon structure of AlN, which has been the subject of numerous investigations. The phonon dispersion spectrum of AlN has 12 branches, three acoustic and nine optical. TO and LO phonon energies have been obtained from fits to IR reflectivity measurements, the results of which can be found in Morkoç (1999a). Raman-active optical phonon modes belong to the A_1, E_1, and E_2 group representations. Several Raman scattering studies on AlN have been conducted and the measured phonon energies can be found in Morkoç (1999a).

Using a Knoop diamond indenter, the hardness of AlN has been measured to be ~ 12 GPa on the basal plane (0001). Some anisotropy in Knoop hardness has been observed with the indent direction perpendicular to the c-axis, with measured values in the range of 10–14 GPa.

c. Electrical Properties of AlN

Due to low intrinsic carrier concentration and deep native defect and impurity energy levels, the electrical characterization of AlN has usually been limited to resistivity measurements. One such measurement on transparent

AlN single crystals yielded resistivities $\rho = 10^{11}–10^{13}\,\Omega\,cm$, a value consistent with other reports. However, it was found that impure crystals, which exhibited a bluish color, possibly due to the presence of Al_2OC, had much lower resistivities, $\rho = 10^3–10^5\,\Omega\,cm$.

The insulating nature of these early films hindered meaningful studies of their electrical transport properties. With the availability of refined growth techniques, AlN is presently grown with much improved crystal quality and shows both n- and p-type conduction. This has rejuvenated efforts to measure both electron and hole Hall mobility. Hall measurements in p-type AlN produced a very rough estimate of the hole mobility $\mu_p = 14\,cm^2/V\,s$ at 290 K.

d.   Optical Properties of AlN

Since an AlN lattice has a very high affinity for oxygen, it is almost impossible to eliminate oxygen contamination in AlN. Currently, commercially available AlN contains about 1–1.5 at.% oxygen. Some oxygen is dissolved in the AlN lattice while the remainder forms an oxide coating on the surface of each powder grain. After irradiation with UV light, AlN doped with oxygen is found to emit a series of broad luminescence bands in the near-UV frequencies at room temperature, no matter whether the sample was powdered, single crystal, or sintered ceramic. Two broad luminescence lines centered in the vicinity of 3.0 and 4.2 eV and greater than 0.5 eV wide, for samples contaminated with about 1–1.5 at.% oxygen, have been observed. Luminescence characteristics of polycrystalline sintered AlN samples have been investigated and a continuous shift of peak position in the UV luminescence line as a function of oxygen content up to a critical concentration of about 0.75 at.% has been observed. The luminescence lines beyond this limit of oxygen concentration remained stationary. Recently, however, AlN layers on both Si and sapphire substrates with sharp bandedge luminescence have been prepared.

High-quality AlN has been characterized by optical absorption and the room-temperature bandgap was determined to be direct with a value of about 6.2 eV. A broad emission spectrum range of 2–3 eV with a peak at 2.8 eV has also been observed. Other investigations confirmed the presence of a 2.8-eV peak, which had previously been attributed to oxygen impurity. Temperature-dependent optical absorption was performed, which led to a bandgap of 6.28 eV at 5 K compared to the room-temperature value of 6.2 ± 0.1 eV. An optical study of AlN reported on the luminescence of Mg and rare earth centers.

Measurements of the refractive index of AlN have been performed in amorphous, polycrystalline, and single epitaxial thin films. The values of refractive index are in the $n = 1.99–2.25$ range with several groups reporting $n = 2.15 \pm 0.05$. These values of refractive index are found to increase with increasing structural order, varying in the range of 1.8–1.9 for amorphous films, 1.9–2.1 for polycrystalline films, and 2.1–2.2 for single-crystal epitaxial films. The spectral dependence and the polarization dependence of the index of refraction have been measured and showed near constant refractive index in the wavelength range of 400–600 nm. Some of these measurements also show that, in the long-wavelength range, the dielectric constant of AlN, ε_0, lies in the range of 8.3–11.5, and that most of the measurements fall within $\varepsilon_0 = 8.5 \pm 0.2$. Other measurements in the high-frequency range showed the dielectric constant to be $\varepsilon_0 = 4.68$ and $\varepsilon_\infty = 4.84$. AlN has also been examined for its potential for second harmonic generation.

3. Indium Nitride

Pure InN has not received the experimental attention given to GaN and AlN. This is probably due to difficulties in growing high-quality crystalline InN samples and because of the existence of alternative well-characterized semiconductors such as AlGaAs and InGaAsP, which have similar energy bandgaps to that of InN (1.89 eV). Consequently, the practical applications of InN are restricted to its alloys with GaN and AlN.

InN remains the least understood of the nitrides. Morkoç (1999a) contains the measured physical properties of InN. Given the fact that the growth of bulk-single-crystal InN films using equilibrium techniques was unlikely, attention turned to deposition of thin films using non-equilibrium techniques. All of the data reported below, unless otherwise specified, were obtained from highly conductive n-type polycrystalline InN grown by non-equilibrium techniques.

a.   Crystal Structure of InN

InN normally crystallizes in the wurtzite (hexagonal) structure with C_{6v}^4 in Schoenflies notation ($P6_3mc$ in Hermann–Mauguin notation) space group and $C6v$ ($6mm$) point group symmetries. The zincblende (cubic) form has been reported to occur in films containing both polytypes. Although InN normally crystallizes in the wurtzite (hexagonal) structure, occasionally it crystallizes in the zincblende (cubic) polytype.

The average value of c_0/a_0 data from various measurements is about 1.615 ± 0.008, which is close to the more optimistic value of 1.633 determined from layers specially grown under significant precautions, best possible growth conditions, and presumably with reduced nitrogen vacancies. An examination of the tabulated data would indicate that, although a_0 values obtained from various sources are relatively close, c_0 values obtained from these sources are rather scattered. This may possibly be due to nitrogen deficiency, since nitrogen atoms are closely packed in (0001) planes. While the cubic polytype of InN yields a molecular cell volume of 30.9 Å, the hexagonal polytype yields a molecular cell volume of 31.2 ± 0.2 Å.

b. Mechanical and Thermal Properties of InN

The experimental density of InN obtained from Archimedean displacement measurement is 6.89 g/cm^3 at 25°C. This is comparable to 6.81 g/cm^3 estimated from X-ray data. Bulk modulus obtained from first principles calculations by a local-density approximation and by a linear muffin-tin orbital method is $B = 165$ GPa. When in a hexagonal structure, the second-order elastic moduli are C_{11}, C_{12}, C_{13}, C_{33}, and C_{44}, but there are no reports of these parameters yet. Since these figures depend on lattice constants, which are within some 10%, values of other nitrides can be used as a first approximation when absolutely warranted. Bulk modulus has been calculated from first principles by a local-density approximation and by a linear muffin-tin orbital method, suggesting a value $B = 165$ GPa.

InN has 12 phonon modes at the zone center (symmetry group $C6v$), three acoustic and nine optical with the acoustic branches near zero at $k = 0$. The IR-active modes are of $E_1(LO)$, $E_1(TO)$, $A_1(LO)$, and $A_1(TO)$ types. Moreover, a TO mode has been observed at 478 cm^{-1} (59.3 meV) by optical reflectance and 460 cm^{-1} (57.1 meV) by transmission.

The linear thermal expansion coefficients measured at five different temperatures between 190 and 560 K indicate that both along the parallel and perpendicular directions to the c-axis of InN these coefficients increase with increasing temperature. Deriving thermal conductivity data from the Liebfried–Schloman scaling parameter and assuming that thermal conductivity is limited by intrinsic phonon–phonon scattering, the thermal conductivity is about 0.80 ± 0.20 W/cm K. While the heat capacity for InN is $(9.1 \pm 2.9) \times 10^{-3}$ T (cal/mol K) at temperatures between 298 and 1273 K, the entropy is 10.4 cal/mol K at 298.15 K. The equilibrium partial pressure of N$_2$ over InN is about 1 atm. at 800 K, but it increases exponentially with temperature to 10^5 atm. at 1100 K.

c. Electrical Properties of InN

It is fair to state that no reliable experimental data for electron mobility in InN have yet been obtained. Notoriously large density of nitrogen vacancies are thought to lead to large background electron concentrations in InN. Because of all these factors, the electron mobilities obtained from various films have varied widely. The electron mobility in InN can be as high as 3000 cm^2/V s at room temperature. A recent study of the electron mobility of InN as a function of growth temperature indicates that the mobility of

UHV-ECR-RMS grown InN can be as much as four times the mobility of conventionally grown InN.

d. Optical Properties of InN

Values of the room-temperature InN direct bandgap reported by various groups range from 1.7 to 2.07 eV with the value of 1.89 eV measured by optical absorption. Reflection and transmission measurements have been performed, which led to an estimate of the effective mass $m_e^* = 0.11m_0$ and an index of refraction $n = 3.05 \pm 0.05$. The long wavelength limit of the refractive index was reported to be 2.88 ± 0.15. The temperature dependence of the bandgap of InN indicates a bandgap temperature coefficient of

$$(dE_g/dT) = -1.8 \times 10^{-4}\,eV/K$$

From reflectance data, the existence of a TO phonon mode at 478 cm^{-1} and an LO mode at 694 cm^{-1} were also deduced.

4. DILUTE GaAs(N)

When small amounts of N and As are incorporated into GaAs and GaN lattices, respectively, a large negative bandgap bowing parameter results. Consequently, with very small amounts of N in the GaAs lattice, its bandgap can be made very small, to a point where 1.3- and 1.5-μm lasers can all be built with GaAs technology. Anomalously large bandgap bowing parameters exhibited by GaAsN and GaNAs are caused by large chemical and size difference between As and N (Bellaiche et al. 1997; Bellaiche 1999; Kent and Zunger 2001; Wang 2001). Dependence of the bangap energy in GaAsN on nitrogen content originating from both the GaN point (in which case As is added) and the GaAs point (in which case N added) is shown in Fig. 1. Also shown is the bandgap variation with composition for other commonly used ternaries. The bold vertical line through GaAs represents the bandgap attainable with GaInAsN, at least in theory, while maintaining lattice matching to GaAs. The decrease in the lattice constant caused by N can be compensated with In added to the lattice. The potential of covering a large range of bandgap energies on GaAs substrates has attracted a great deal of interest in this material system. In fact, the first laser containing N was an InGaAs(N) active layer one. Due, in part, to extreme non-equilibrium conditions employed for growth, MBE is the dominant growth approach for dilute arsenides with nitrogen. The critical issues are compositional control, incorporation of more than a few percent of N, doping inefficiency, and layer quality. The situation is exacerbated on all fronts when the N concentration is increased for achieving 1.5-μm wavelength of emission. Post-growth

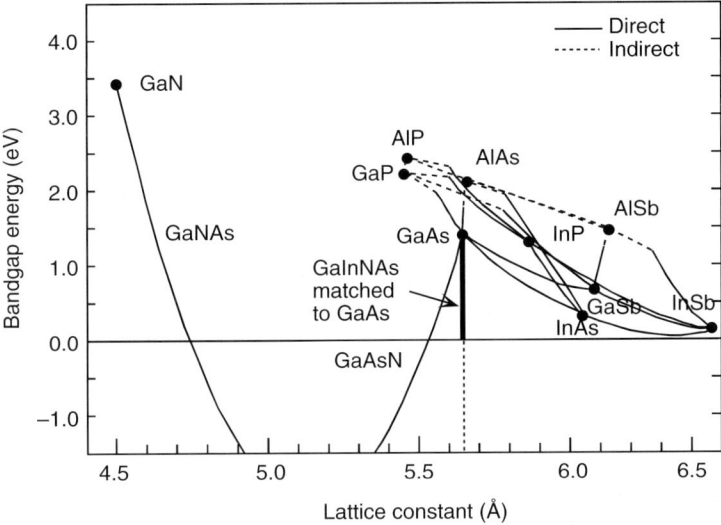

FIG. 1. Bandgap energy as a function of material composition for dilute GaAs(N) and GaN(As).

annealing is often employed to improve the crystal quality and/or increase Si dopant incorporation, alas at the expense of blue shift in the bandgap!

As alluded to earlier, the chemical and size difference between the N and As atoms are the challenges facing the experimentalist. In addition, generation of atomic nitrogen, not different from the technology required for hexagonal GaN growth (Morkoç 1999a, 2002), presents some attention to detail. While basic mismatch between N and As can be dealt with by growing the layers under non-equilibrium conditions, the issue of atomic nitrogen can be handled by compact RF sources that have seen a good degree of improvement of late. By adjusting the RF power and pressure in the cell, one can tailor the source to produce mostly the atomic species by optimizing the emission at 745 nm of wavelength. It should be mentioned that the substrate and most of the structure are zincblende and, consequently, the dilute material assimilates and assumes the same crystalline structure. The desired nitrogen concentrations are in the range of 1–10% for red-shifting the transitions out to as long as 1.55 μm. Larger growth rates lead to reduced incorporation of N in the lattice. Similarly, higher growth temperatures lead to the same. Consequently, when 1.55-μm wavelength material is desired, lower growth rates must be employed as well as lower growth temperatures. At substrate temperatures of 500°C or below, if very large As overpressure are employed, incorporation of N is limited because the fluence of atomic nitrogen is small. However, atomic nitrogen is very reactive and, therefore, compositional control should be much better as compared to quaternaries relying on P and As (InGaAsP).

As expected, due to dissimilarities of N and As, the luminescence properties of GaInNAs degrade rapidly with increasing nitrogen concentration. Remedies such as post-growth annealing are employed to enhance the luminescence efficiency of GaInNAs. However, this enhancement is accompanied with a blue shift in the transition in bulk and quantum well materials. Nitrogen and possibly In diffusion out of GaInAsN are thought to be responsible for the observed luminescence shift to shorter wavelengths.

Device applications of the dilute nitride materials as well as the pertinent issues are discussed in Chapter 7 (Part I). However, for completeness, a short reference will be made to edge emitting and vertical cavity lasers operating at 1.3 and 1.5-μm portions of the optical spectrum. Several groups have reported lasers operating at the 1.3-μm region (Kitatani et al. 1998, 2000; Nakatsuka et al. 1998; Sato and Satoh 1998; Kondow et al. 1999; Li et al. 1999; Ougazzaden et al. 1999; Fischer et al. 2000a; Yang et al. 2000), where the silica-based fiber dispersion is zero, and at the 1.5-μm region (Fischer et al. 2000b), where the loss is low, again for the silica-based fibers. Both are intended for telecommunication purposes. High-speed testing of these lasers has also been performed (Reinhardt et al. 2000) with data transmission rates as high as Gbit/s having been achieved already (Steinle et al. 2001). For interconnects and high-speed data links, vertical cavity surface emitting lasers (VCSELs) have received a great deal of attention. Now that dilute nitrides are becoming potential candidates for long-wavelength lasers, efforts are under way to explore VCSELs in this material system as well (Fischer et al. 2000c; Schneider et al. 2001).

III.   Electron Transport Properties in GaN and GaN/AlGaN Heterostructures

Electron mobility is a key parameter in the operation of n-channel FETs in that it affects the access resistances as well as the rate with which the carrier velocity increases with field. Consequently, we will treat the low-field mobility in GaN and its dependence on various scattering events first. This will be followed by treating the 2DEG mobility in MODFETs of direct relevance to technology importance. Ultimately, electron mobility is limited by the interaction of electrons with phonons, and, in particular, with optical phonons. This holds for bulk mobility as well as for that in AlGaN/GaN MODFETs.

The room-temperature electron mobility values in bulk-GaN grown with HVPE to a thickness of 60 μm was reported to be 950 cm^2/V s (Look et al. 1997). Freestanding GaN templates grown by HVPE exhibited room-temperature electron mobilities approaching 1400 cm^2/V s (Huang et al. 2001). Recent HVPE layers exhibit much higher mobilities on the surface of the layers, which approach that of free-standing templates. The values reported for metal organic chemical vapor deposition (MOCVD) grown layers were also in excess of $900 \, cm^2/V \, s$ (Nakamura *et al.* 1992), though the temperature dependence of mobility in this particular sample was rather unique. Early MBE layers exhibited mobilities as high as $580 \, cm^2/V \, s$ on SiC substrates, which at that time were not as commonly used as in recent times (Lin *et al.* 1993a). Typically, however, the MBE-grown films produce much lower mobility values of $100–300 \, cm^2/V \, s$ (Ng *et al.* 1998). The lower mobilities have been attributed to both high dislocation densities (Ng *et al.* 1998; Weimann *et al.* 1998; Look and Sizelove 1999) and elevated levels of point defects (Fang *et al.* 1998; Wook *et al.* 1998).

Dislocations are considered by some to be an important scattering mechanism in films having dislocation densities above $1 \times 10^8 \, cm^{-2}$ (Ng *et al.* 1998; Look and Sizelove 1999). One should keep in mind that these are preliminary attributes and more detailed experiments coupled with detailed analyses are needed to confirm the proposed models. Depending on the particulars of the growth and substrate preparation, GaN films grown by MBE typically have dislocation densities in the range of $5 \times 10^9–10^{10} \, cm^{-2}$ (Ng *et al.* 1998). With refined procedures, however, dislocation densities in the range of $8 \times 10^8–2 \times 10^9 \, cm^{-2}$ can be obtained when grown directly on sapphire substrates with AlN or GaN buffer layers. Reduction of dislocation density, and other scattering centers that are inherently related to dislocations, is really the key to achieving high mobility GaN resulting from the heart of the buffer layer and or early stages of growth. Based on the premise that the [002] X-ray diffraction is affected by screw dislocations and the [104] peak by edge dislocations and the fact that RF-nitrogen grown MBE layers produce excellent [002] peaks (in the 40–120 arcsec range) while the [104] peaks are wider and weaker (180–300 arcsec.), one can conclude that majority of dislocations in MBE layers is that of the propagating edge type.

The strength of MBE, that is, 2D growth, does not bode well for dislocation reduction as the edge dislocations propagate along the *c*-axis going right through the sample. The detailed picture is somewhat dependent on the particulars of the V/III ratio resulting in pitted or pit-free growth. Some sort of 3D growth at the early stages of growth, as in the case of growth from vapor, followed by a smoothing layer would help reduce dislocations. The other option is to use HVPE or MOCVD buffer layers for MBE growth. This approach led to record or near-record bulk ($1150 \, cm^2/V \, s$ at room temperature, even higher of late) (Heying *et al.* 2000b) and 2DEG ($53 \, 500 \, cm^2/V \, s$ at 4.2 K) mobilities (Manfra *et al.* 2000). It is clear that the buffer layers grown by the vapor-phase epitaxy method help eliminate the main problem associated with MBE: the poor quality of the buffer layer. The other long-standing obstacle for MBE, difficulties associated with sapphire and SiC substrate preparation, has been eliminated. In the case of sapphire, a high-temperature anneal in O_2 environment produces an atomically smooth and damage free surface (Cui *et al.*). In the case of SiC, some form of H_2 etching at elevated temperatures removes the surface damage caused by polishing (Powell *et al.* 1998), as in the case of sapphire. Controlling the Ga/N ratio and substrate temperature causes the dislocation density across the homoepitaxial interface to remain constant (Tarsa *et al.* 1997; Heying *et al.* 2000a). While the above two works are related to RF MBE, ammonia MBE has also produced GaN with very high electron mobilities (as high as $70\,000\,cm^2/V\,s$ at 4 K) when grown on bulk GaN wafers, which, in turn, were grown under high pressure and temperature conditions (N. Grandjean, CHREA-CNRS, Valbonne, France, private communication).

Electron mobility is one of the most important parameters associated with the material with great impact on devices. The temperature dependence of mobility and the carrier concentration can be used to extract fundamental information regarding scattering mechanisms (Rode 1975; Seeger 1982). As compared to the other III–V semiconductors, such as GaAs, GaN possesses many unique material and physical properties (Morkoç 1999a). However, the lack of high-quality material, until very recently, prevented detailed investigations of carrier transport. The earlier transport investigations had to cope with poor crystal quality and low carrier mobility, well below predictions (Chin *et al.* 1994; Rode and Gaskill 1995; Jain *et al.* 2000; Ridley *et al.* 2000).

We should point out that unintentionally doped GaN exhibits n-type conduction with a typical electron concentration of $\sim 10^{17}\,cm^{-3}$, with heavy compensation. Typical compensation ratios observed for MOCVD- and MBE-grown films are about 0.3, though a lower ratio of ~ 0.24 was reported for HVPE-grown crystals (Chin *et al.* 1994; Look *et al.* 1997; Dhar and Ghosh 1999). Compensation reduces the electron mobility in GaN for a given electron concentration. Another point that should be kept in mind is that GaN layers are often grown on foreign substrates with very different properties. The degenerate layer at the interface (caused by extended defects and impurities), spontaneous polarization at heterointerfaces, and piezoelectric effects all should be considered. Experiments show that, even for thick GaN grown by HVPE, the degenerate interfacial layer has an important contribution to the Hall conductivity, especially at low temperatures where freeze-out occurs for the donors in bulk, leading to domination by the interfacial layer (Joshi 1994; Look *et al.* 1997; Götz *et al.* 1998; Cheong *et al.* 2000). In these cases, the measured data must be corrected to extract meaningful numbers (Dhar and Ghosh 1999). The typical extended defect density of GaN grown by various techniques is $\sim 10^9\,cm^{-2}$ (Visconti *et al.* 2000b). In many cases, the dislocation and defect scattering may also limit the carrier mobility, especially at low temperatures (Ng *et al.* 1998; Zhu and Sawaki 2000). Finally, many material and physical parameters of GaN were not available for some of the previous simulations where those parameters were treated as adjustable parameters. Needless to say, reliable parameters are required in the calculation of the electron mobility and in the interpretation of experimental results accurately.

IV. Bulk Mobility in GaN

To treat bulk mobility, we made use of free-standing high-quality GaN templates grown by HVPE at Samsung where the questionable interface layer has been chemically removed. A quantitative comparison with theoretical calculations demonstrates that the one-layer and one-donor conductance model is sufficient to account for the measured data in the entire temperature range without considering any dislocation scattering and any adjustable parameter other than the acceptor concentration (Huang *et al.* 2001). The sample shows a low impurity concentration, a low compensation ratio, negligible dislocation scattering; the high electron mobilities derived from the Hall measurements show the high-quality transport properties associated with the sample.

A thick ~ 300-μm GaN film was first deposited by HVPE on the c-plane of sapphire. It was then thermally decomposed at the film/substrate interface and lifted off by scanning a laser beam with a photon energy larger than that corresponding to the GaN bandgap. Both sides of the free-standing GaN crystal were mechanically polished, and the Ga-face was dry etched to yield a smooth surface. Wet chemical etching was then applied to the N-face to remove some 30 μm of material, the region containing the high conductance degenerate layer. The final thickness of the GaN template is about 200 μm. The sample has also been characterized by X-ray diffraction, atomic force microscopy, and photoluminescence (Visconti *et al.* 2000a).

The Raman spectra were measured for this sample to derive the Debye temperature and the optical dielectric constant as literature values of these parameters vary depending on sample quality. The phonon energies measured in this sample are: $A_1(LO) = 737.0 \, cm^{-1}$; $A_1(TO) = 532.5 \, cm^{-1}$; $E_1(LO) = 745.0 \, cm^{-1}$; $E_1(TO) = 558.5 \, cm^{-1}$ with an accuracy of $\pm 0.5 \, cm^{-1}$. These phonon energies are very close to those measured by Azuhata *et al.* (1995) for a thick GaN film grown on sapphire. The optical phonon modes were measured in a variety of scattering configurations as indicated. As usual, $X = (100)$, $Y = (010)$, and $Z = (001)$. Here, Z was taken to be parallel to the growth direction. Figure 2(a) shows the $A_1(LO)$ and $A_1(TO)$ phonon spectra where the half-width of the phonon modes is only $8 \, cm^{-1}$. $E_1(TO)$ and E_2 phonon modes are also shown in Fig. 2(b).

Variable temperature Hall effect measurements were performed in the dark in a temperature range of 26.5–273 K using the van der Pauw geometry. Ohmic contacts were formed on the Ga-face with Ti/Al/Ti/Au metallization followed by rapid thermal annealing at 900°C for 30 s (Lin *et al.* 1994). Good

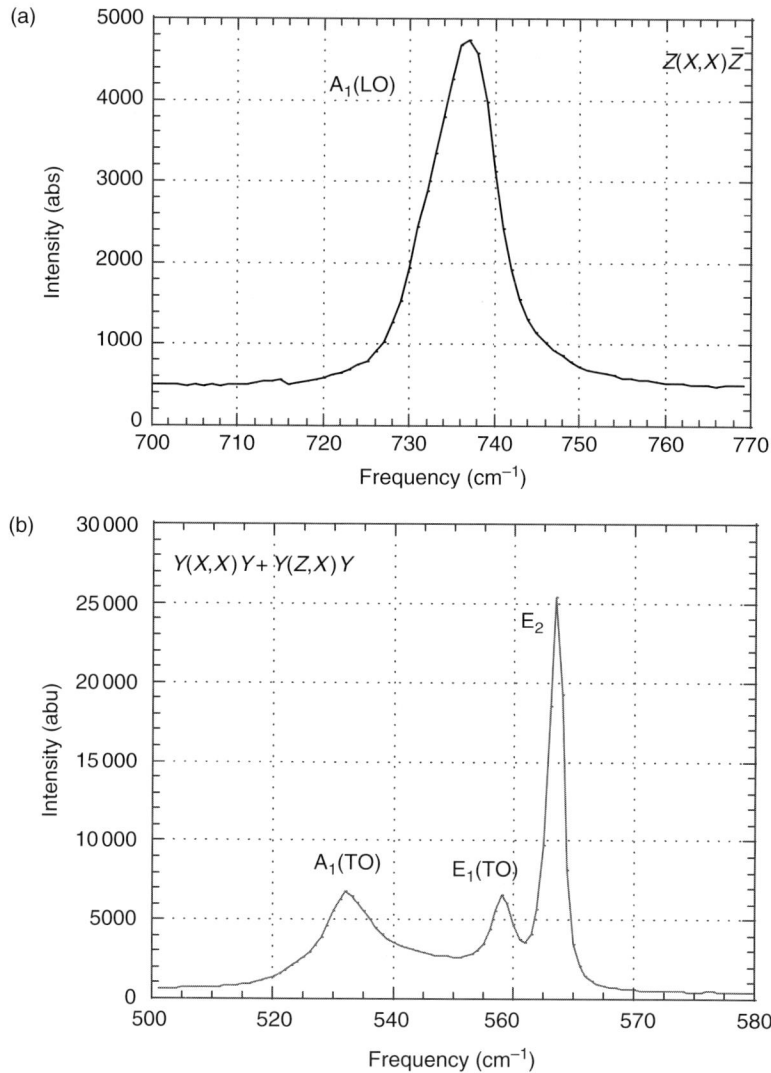

FIG. 2. (a) $A_1(LO)$ phonon spectrum of the sample investigated. The scattering configuration is shown in the upper right corner. (b) $A_1(TO)$ phonon spectrum of the sample investigated. The E phonon spectra are also shown. The scattering configuration is shown in the upper left corner.

ohmic contacts were verified over the measured temperature range inside the cryostat. Extreme care was taken for the accuracy of the experimental conditions such as the magnetic field, electrical current, and the sample temperature.

The measured Hall mobility and carrier concentration are shown in Figs. 3 and 4, respectively, as a function of temperature. A mobility of 1425 cm^2/V s was observed near room temperature (273 K) with the peak mobility being 7386 cm^2/V s at 48 K. The temperature dependence of both the Hall mobility

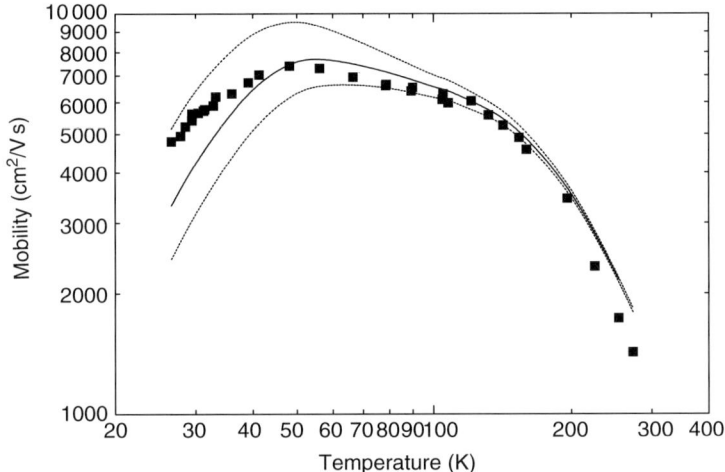

FIG. 3.   The measured Hall mobility data (solid squares) from the GaN template grown by HVPE as a function of temperature. The solid line is the calculated result using $N_a = 2.4 \times 10^{15}$ cm^{-3}, representing the best fit to the measured results. The upper and lower dotted lines are the calculated results using $N_a = 1.4 \times 10^{15}$ and 3.4×10^{15} cm^{-3}, respectively.

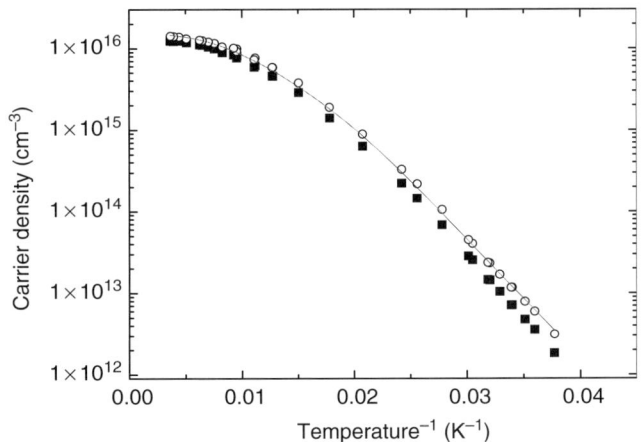

FIG. 4.   The measured Hall densities n_H (solid squares) as a function of reciprocal temperature from the GaN template grown by HVPE. The open circles represent the carrier density corrected by the Hall factor, $n = n_H r_H$. The solid line is the fit to the theoretical expression of charge balance with hole and neutral acceptor densities neglected.

and electron concentration (lack of saturation) point to the absence of a degenerate interfacial layer after etching. Therefore, only the single-layer model with a single donor was successfully applied to the analysis of the data.

The temperature dependence of the electron mobility was calculated by solving the Boltzmann transport equation (BTE) iteratively (Rode 1975). The following scattering processes were included: acoustic deformation potential scattering, piezoelectric scattering, polar optical phonon scattering, and ionized impurity scattering. Only one donor and one acceptor were assumed. The dislocation scattering was not considered because the dislocation density in our sample ($<10^6 \, \text{cm}^{-2}$) (Visconti *et al.* 2000a) is much less than that where dislocation scattering affects the transport properties ($10^8 \, \text{cm}^{-2}$) (Ng *et al.* 1998; Zhu and Sawaki 2000). Likewise, neutral-impurity scattering was not included since it is insignificant. The material parameters used in the calculations are listed in Table I.

The high-frequency dielectric constant for this sample was calculated using the Lyddane–Sachs–Teller relation using the phonon frequencies of A modes, which should represent a benchmark value due to the high quality of the sample. Importantly, instead of allowing the acoustic deformation potential to be an adjustable fitting parameter, the recent unscreened acoustic deformation potential ($E_{\text{ds}} = 8.54 \, \text{eV}$) was used. This value is deduced from recent high-mobility 2DEG system mobilities at low temperatures where the acoustic phonon scattering is important (Hsu and Walukiewicz 2001). If screening is included, a deformation potential of 12 eV describes experimental 2DEG mobilities better (W. Walukiewicz, private communication). However, this is not necessarily applicable here; in part, because the sample under investigation has a very low electron concentration and, in part, due to freeze-out at low temperatures. Therefore, the deformation potential without screening was used in our simulations as a fixed parameter. The other parameters were kept the same as those reported previously (Rode 1975; Morkoç 1999a) and shown in Table I.

TABLE I

MATERIAL PARAMETERS OF GaN USED IN THE CALCULATIONS

Parameter	Symbol (units)	Value
High-frequency dielectric constant	ε_∞	5.43
Low-frequency dielectric constant	ε_0	10.4
Polar phonon Debye temperature	θ_{LO} (K)	1060
Mass density	ρ (kg/m^3)	6.10×10^3
Sound velocity	v_{s} (m/s)	6.59×10^3
Piezoelectric coefficient	P	0.118
Acoustic deformation potential	E_{ds} (eV)	8.54
Effective mass	m^* (kg)	$0.22m_0$

The theoretical calculation was fitted to the measured data in concert with the temperature dependence of the electron concentration and charge neutrality. The donor and acceptor concentrations are used as the fitting parameters. The solid line in Fig. 3 represents the best fit to the measured data. The dotted lines give the estimated upper and lower bounds for the acceptor concentration. As shown, quantitative agreement with the measured mobility in the entire temperature range is obtained to within about 30%. The acceptor concentration obtained from the fitting is $2.4 \times 10^{15} \, \text{cm}^{-3}$.

The carrier concentration as a function of temperature is also calculated using the charge neutrality condition with hole and neutral acceptor density being neglected. The solid line shown in Fig. 4 represents the fit to the measured electron concentration corrected by the Hall factor. The impurity concentrations and the activation energy used in the fitting are $N_D = 1.76 \times 10^{16} \, \text{cm}^{-3}$, $N_A = 2.40 \times 10^{15} \, \text{cm}^{-3}$, and $E_D = 25.2 \, \text{meV}$. The agreement between the N_D and N_A values derived from the mobility and the carrier concentration data demonstrates the self-consistency of the results with the assumed model. The actual value and the temperature dependence of the Hall factor are implicit in calculations and are shown in Fig. 5, where a degeneracy factor of 2 was assumed.

Considering the screening effect, the donor binding energy E_{D0} can be calculated from $E_{D0} = E_D + \alpha N_D^{1/3} = 30.7 \, \text{meV}$, where the screening factor $\alpha = 2.1 \times 10^{-5} \, \text{meV cm}$. This result is close to the measured value from IR absorption (29.0 meV) (Wang *et al.* 1996). We note that the donor binding energy obtained here is higher than the 27.5-meV value derived for the HVPE sample with higher donor concentration of $1.25 \times 10^{17} \, \text{cm}^{-3}$ reported in Look *et al.* (1997). The discrepancy may be introduced from the two-layer conductance in the latter case. As shown earlier (Joshi 1994; Look *et al.* 1997;

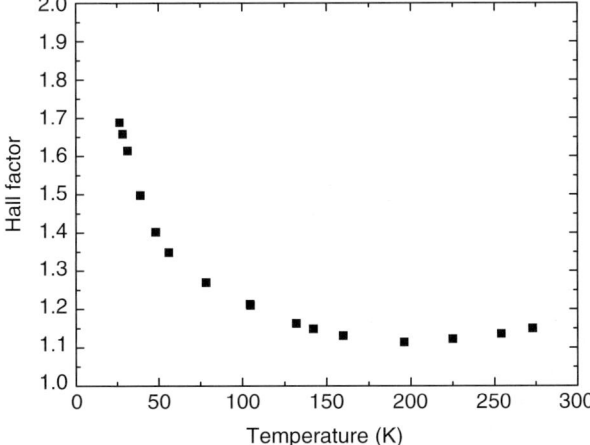

FIG. 5.   Temperature dependence of the calculated Hall factor, r_H.

Götz et al. 1998; Cheong et al. 2000), the interface conductance mainly affects the low-temperature data. Since the donor activation energy can only be accurately determined from the low-temperature region in which the ionized-impurity scattering is dominant, any high-conductivity interface contribution would make it difficult to deduce an accurate donor binding energy.

The measured impurity concentrations N_D and N_A as well as the compensation ratio (N_A/N_D) of 0.14 are among the lowest values reported for bulk-GaN grown by any technique. The peak mobility of 7386 cm^2/V s and the value near room temperature are also among the highest for bulk-GaN. The quantitative agreement between the calculations and the measured data demonstrates that the single-layer and single-donor model amply describes the conduction process. The results also show that dislocation scattering is not important for the sample under investigation in the entire measured temperature range, demonstrating good crystal quality.

V. Polarization Effects, Mobility, and Electron Concentration in 2DEG Systems

As mentioned in Section I, AlGaN/GaN heterostructures have been the subject of many recent investigations because of their potential for use in high-temperature, high-power devices (Aktas et al. 1995; Mohammad et al. 1995; Binari et al. 1997; Morkoç 1998b), due to the large band discontinuities and polarization-induced screening charge. While polarization effects cause a redistribution of weakly bound and free charges, they cannot directly produce free electrons to form a 2DEG (Hsu and Walukiewicz 1997; Morkoç et al. 1999, 2002a,b). In the GaN-based system, issues dealing with hetero-interfaces must include a discussion of polarization. Polarization induces a field that affects the interface charge through screening, causing the mobile carriers move to where the opposing fixed polarization charge resides (lies). Since nitrides are large bandgap materials that tend to be n-type with very low hole concentrations, mobile carriers are most likely donated by intentional donors or donor-like defects.

The AlGaN barrier (Hsu and Walukiewicz 1998) and the AlGaN surface (Smorchkova et al. 1999) have been suggested as the source of electrons. Positive surface charge has also been suggested to account for the experimental observations in the form of dependence of the 2DEG density on the thickness and/or alloy composition of the AlGaN barrier (Asbeck et al. 2000; Shur et al. 2000). Typically, the AlGaN barrier is grown on a relatively thick GaN layer to form the basis for the 2DEG system in a Ga-polarity sample. The inherent lattice mismatch causes a biaxial tensile strain, and the thermal mismatch causes a biaxial compressive strain in the growth plane. The resultant strain induces a macroscopic electric field in the polar material. In addition, due to the particular crystalline structure of the wurtzite lattice, a spontaneous polarization field is also found in both AlGaN and GaN even in the absence of strain. In heterostructures where the growth takes place along the (0001) direction, both the spontaneous and induced polarizations are directed opposite to the growth direction. The effect of polarization field on the position of the band edges has been calculated by several groups (Hsu and Walukiewicz 1998; Maeda *et al.* 1998; Oberhuber *et al.* 1998; Shur *et al.* 2000; Morkoç 2002a,b). Polarization-induced fields in Ga-polarity samples, just as any negative gate voltage induced field in FETs, increase the conduction bandedge in AlGaN barrier with distance from the interface. In the presence of free, weakly bound, and surface charge, the internal polarization field is screened by a redistribution of these charges. The surface states may be in the form of donor-like states, which donate their electrons to the lowest unoccupied energy states at the AlGaN/GaN interface.

The holy grail of GaN/AlGaN heterostructures is the debate on the origin of the carriers, which end up at the interface. The observed dependence of the 2DEG density on the thickness and composition of the AlGaN barrier has been linked to surface donor states, the binding energy of which is roughly equal to the Schottky barrier height in n-type GaN (Smorchkova *et al.* 1999). Though this may point to the same surface states being responsible for a possible and weak Fermi level pinning, it is not noticeable on the surface of GaN. In addition, pinning of the surface Fermi level in n-type GaN requires electrons to be transferred from bulk donors to surface acceptor states, whereas an excess of surface donors is required to form 2DEG in an AlGaN/GaN heterostructure. This inconsistency can be resolved, however, by assuming that the surface defects are amphoteric (i.e., they can act as either acceptors or donors depending on the circumstances) (Walukiewicz 1989). The polarization fields present in nitride heterostructures are strong enough to shift the Fermi energy at the surface of the AlGaN barrier below the charge transition level of the surface defects for Ga-face growth. This causes the surface defects on the barrier to transform from being acceptor-like to donor-like surface defects, which can provide the electrons for the 2DEG at the AlGaN/GaN interface (Walukiewicz 1989). Hsu and Walukiewicz (2001) elaborated on the surface donor-like defect that is likely to form at growth temperature and its manifestation as a source of carriers confined at the underlying interface between the AlGaN top layer and GaN below it. The model calculations appeared to be somewhat insensitive to parameters such as donor formation energy and surface Fermi level. One sensitive parameter is the strength of the polarization field, which will be discussed below.

Spontaneous polarization has only recently been fully understood. As pointed out earlier (Bykhovski *et al.* 1996), nitrides lack inversion symmetry and exhibit piezoelectric effects when strained along the [0001] direction. Piezoelectric coefficients in nitrides are almost an order of magnitude larger than in many of the traditional Group III–V semiconductors

TABLE II

PIEZOELECTRIC CONSTANTS AND SPONTANEOUS POLARIZATION CHARGE IN
NITRIDE SEMICONDUCTORS

	AlN	GaN	InN
$e_{33}{}^a$ (C/m^{2b})	1.46	0.73	0.97
$e_{31}{}^a$ (C/m^2)	−0.60	−0.49	−0.57
P_0 (C/m^2)	−0.081	−0.029	−0.032
$[e_{31}−(C_{31}/C_{33}^c)e_{33}]$	−0.86	−0.68	−0.90

[a] e_{31} and e_{33} are piezoelectric constants.
[b] C/m^2 is coulombs per square meter.
[c] C_{31} and C_{33} are elastic constants.

(Bernardini *et al.* 1997). In addition, WZ-GaN has a unique axis, thus allowing spontaneous polarization (P_0 whose values are given in Table II) to be present even in the absence of any strain. This can manifest itself as polarization charge at heterointerfaces. The magnitude of the polarization charge, converted to number of electrons can be in mid-10^{13} cm^{-2} level for AlN/GaN heterointerfaces, which is huge by any standards. For comparison, the interface charge in the GaAs/AlGaAs system used for MODFETs in less than 10% of this figure. An excellent review of the polarization effects can be found in Resta (1994) and references therein.

Let us compare the relative importance of spontaneous polarization to piezoelectric polarization. For a biaxially strained layer, the effective piezoelectric polarization is given by

$$P_z^{\text{piezo}} = [e_{31} − (C_{31}/C_{33})e_{33}]\varepsilon_{\perp}, \tag{1}$$

where $\varepsilon_{\perp} = \varepsilon_{xx} + \varepsilon_{yy}$ is the in-plane strain and C_{31} and C_{33} are elastic constants.

For Al$_x$Ga$_{1-x}$N coherently strained on a relaxed GaN substrate, the strain $\varepsilon_{\perp}$ is expected to be proportional to x and is given by $\varepsilon_{\perp} = 2(a_{\text{GaN}} - a_{\text{AlGaN}})/a_{\text{AlGaN}}$, which is $0.051x$ and is tensile. The piezoelectric polarization is then $P^{\text{piezo}} = -0.044x$, that is, pointing in the $[000\bar{1}]$ direction. The corresponding difference in spontaneous polarization between Al$_x$Ga$_{1-x}$N and GaN is also expected to be proportional to x, the AlN mole fraction, and is given by $\Delta P^{\text{spon}} = -0.052x$. Consequently, the two are in the same direction for this particular orientation, and are comparable in magnitude. The total polarization for AlN/GaN interface, which is defined in this case as the sum of the piezoelectric polarization and the differential polarization-charge is $-0.096x$. Note that these are all in C/m^2 and that 1 C/m$^2 = 0.624 \times 10^{15}$ electrons/cm^2. Thus, for x of the order of 0.1, we are dealing with total polarization charge of the order of mid-10^{12} cm^{-2}.

For a coherently strained $In_xGa_{1-x}N$ layer on relaxed GaN, the difference in spontaneous polarization is much smaller, $\Delta P^{spon} = -0.003x$. Furthermore, the $In_xGa_{1-x}N$ layer on GaN would be under compressive strain $\varepsilon_\perp = -0.203x$ and $P^{piezo} -+0.183x$. Here, the piezoelectric polarization dominates and is opposite in direction to the spontaneous polarization charge, but even larger in absolute magnitude. In the case of a coherently strained $Al_xIn_{1-x}N$ layer on a relaxed GaN layer, the situation is unique in that for $x = 0$ we revert to the InN on GaN case, and for $x = 1$ we revert to the AlN on GaN case.

Numerical figures can be generated for the total polarization charge following the expressions outlined above by using a linear extrapolation of the strain and differential spontaneous polarization. The total polarization at the interface is the sum of the piezoelectric and differential spontaneous polarization, $P_{total} = \Delta P_{sp} + P_{pe}$. Taking the normal modulation doped structures where the GaN buffer layer is assumed completely relaxed and the AlGaN barrier layer is assumed coherently strained, one arrives at the plot shown in Fig. 6 for the total polarization charge at the interface. Additionally, data for InGaN and AlInN on GaN are shown. For tensile-strained $Al_xGa_{1-x}N$ or $Al_xIn_{1-x}N$ (for large x values) on GaN layer is under tensile strain and the piezoelectric and the spontaneous polarization are negative and point in the same direction, thus they add up. The spontaneous and piezoelectric polarizations oppose one another for compressively strained $In_xGa_{1-x}N$ or $Al_xIn_{1-x}N$ (for small x values) layers. To calculate the differential spontaneous and piezoelectric polarization associated with alloys, one can employ a linear interpolation for the spontaneous polarization, piezoelectric, and elastic constants from the binary compounds (Ambacher *et al.* 1999). As discussed above, the piezoelectric polarization in coherently strained $Al_xGa_{1-x}N$ layers grown on GaN increases to $-0.044\,C/m^2$ for $x = 1$. For InGaN layers, the piezoelectric polarization increases up to $+0.183\,C/m^2$ for $x = 1$. The ternary $Al_{0.82}In_{0.18}N$ can be grown lattice-matched to GaN and the piezoelectric polarization vanishes. For lower Al concentrations, that is, $x < 0.82$, the piezoelectric polarization increases due to increasing biaxial compressive strain. For higher Al concentrations, that is, $x > 0.82$, the layer is under tensile strain and the piezoelectric polarization becomes negative (Fig. 6). Opposing strain and spontaneous polarization charge in $Al_xIn_{1-x}N$ cancel one another for a mole fraction of about $x = 0.7$ with the underlying assumption that the $Al_xIn_{1-x}N$ is coherently strained and the GaN layer on which it is grown is completely relaxed (Ambacher *et al.* 2000). The discussion of more accurate effective mass calculations can be found in the MODFET modeling section.

Some further words of caution about the above estimates are needed. If the AlGaN layers are not pseudomorphic but partially relaxed (by misfit dislocations, for example), then the piezoelectric effect would be reduced but the spontaneous polarization would still be present. If the interfaces are not

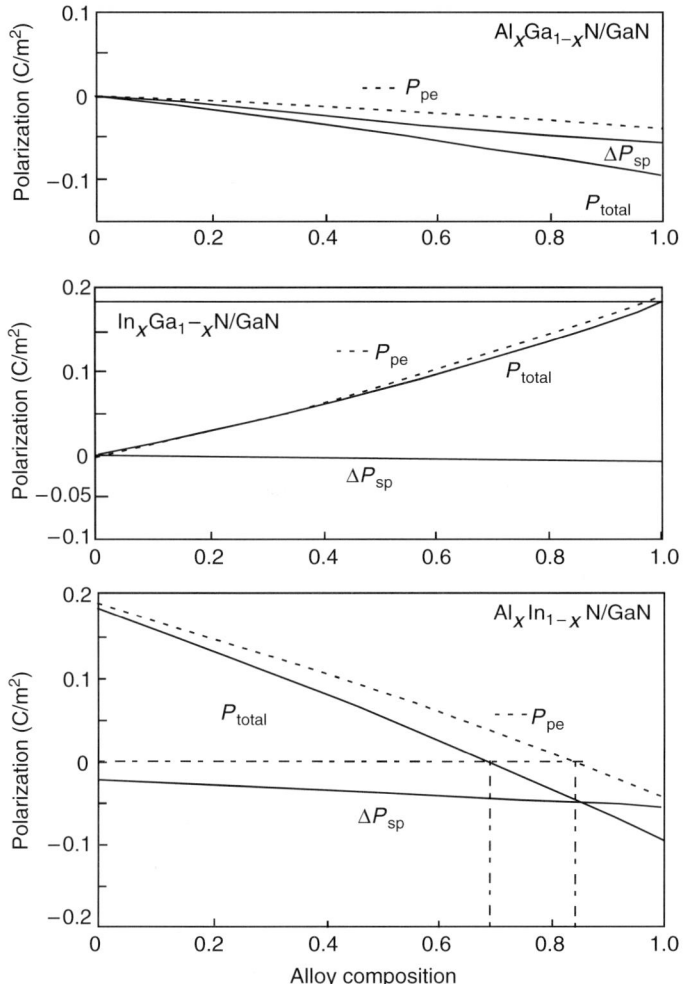

FIG. 6. Piezoelectric, spontaneous, and total polarization charge of coherently strained AlGaN, InGaN, and AlInN alloys grown on completely unstrained Ga polarity GaN buffer layers vs the alloy composition. The polarization values were determined by linear extrapolation of the physicals properties from the binary compounds. The figure is similar to that reported by Ambacher *et al.* (2000), but the parameter used are consistent with our previous publications.

atomically sharp but exhibit a certain degree of interdiffusion, the differences in spontaneous polarization would be reduced as well. Finally, if domains with inverted polarity exist, the overall polarization effects may be washed out. Also note that in an inverted structure with nitrogen (N) polarity towards the surface, it may be possible to create a two-dimensional hole-gas (2DHG) at the AlGaN/ GaN interface, provided that free holes are available.

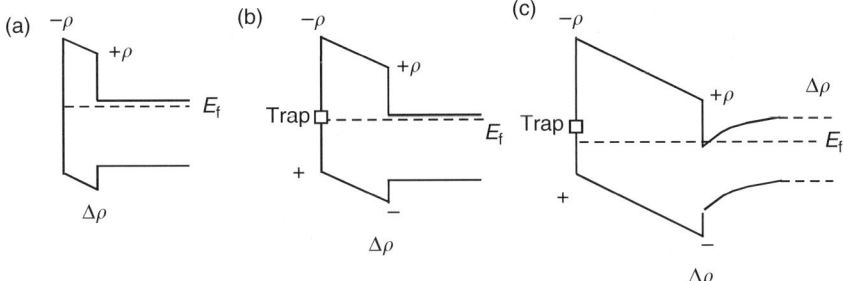

FIG. 7. Schematic representation of a very simple band structure of an AlGaN/GaN structure with varying AlGaN thickness, as one goes from (a) to (c), which demonstrates how the surface charge could participate in the screening of the polarization-induced charge or the field. Patterned after Smorchkova *et al.* (1999).

However, if an n-type GaN layer is placed on top, a 2DEG may form on top of the AlGaN layer.

Let us now consider a hypothetical case in which no free carriers exist. In this case, the polarization charge causes a linear band bending in $Al_xGa_{1-x}N$ with no change in the band of the underlying GaN as shown in Fig. 7. Let us also assume that there are surface states below within the bandgap as shown. As the AlGaN thickness is increased, the band bending would be such that the surface states would become ionized. The released electrons in the process would wind up at the interface, which tend to screen the polarization charge. If the surface-state concentrations were not sufficiently high, the surface Fermi level comes very close to the valence band. This implies that holes should be present within the AlGaN layer near the surface. This positive charge needs to be balanced by a negative charge at the interface, the source of which could be surface states. In reality, the semiconductor system contains defects and free carriers that would make the picture somewhat more complicated. For one thing, the Fermi level may not be able to come as close to the valence band as that shown in Fig. 7. The surface defect supposition has recently been forwarded with experimental backing (Smotchkova *et al.* 1999).

With Ga polarity, the conduction bandedge of $Al_xGa_{1-x}N$ will slope up towards the surface where the $Al_xGa_{1-x}N$ layer is. The band diagram shown would hold in this hypothetical case only if the $Al_xGa_{1-x}N$ layer thickness is such that the Fermi level at the surface is still a few kTs away from the valence band edge. If the $Al_xGa_{1-x}N$ thickness was made larger, the Fermi level near the surface would be very close to the valence band, causing a positive hole-charge at the surface, as shown in Fig. 7. A negative image charge would then form at the heterointerface tending to screen the polarization charge. This charge is different from the polarization charge in that it is mobile and can partake in, current flow in addition to altering the band diagram. In short, the polarization charge is a fixed charge. However, it would induce mobile charges to screen it if they are present. The recent device literature is confusing as some authors are implying as if the polarization-charge alone is capable of doping a semiconductor. Misleading nomenclature with no physical basis such as "piezo-doping" has already been coined. To make the point in another way, polarization charge causes band bending and any free carriers, if present in the system, would tend to screen the polarization charge. The screening charge represents the mobile carriers whose source could be native defects, impurity dopants (intentional or unintentional) and surface states.

Returning to modulation doped structures with $Al_xGa_{1-x}N$ barriers, the sign of the polarization is to produce a potential energy for electrons sloping down from the Ga face towards the N face. Thus, for a structure in which the Ga face is turned towards the surface, the potential will slope down from the $Al_xGa_{1-x}N$ surface towards the AlGaN/GaN interface and helps to drive carriers towards the 2DEG forming at this interface. For example, if there is an ohmic metal contact on the $Al_xGa_{1-x}N$ surface, electrons will flow towards the 2DEG below that layer.

The most favorable situation for enhancing sheet carrier concentration would occur for an InGaN quantum well on top of relaxed n-GaN and below an AlGaN barrier with the whole structure having cation polarity towards the surface. In that case, the field will slope down towards the InGaN/AlGaN interface in the quantum well and will help localize the carriers in the 2DEG. Note that the piezoelectric polarizations estimated here are based on the theoretical values for a perfectly insulating material. The free carriers that are present in each layer screen the field. For example, if free carriers are provided from metal contacts and they flow towards the 2DEG, this process sets up a screening field, which counters the polarization-induced field. Under equilibrium conditions, if they are reached, the net field is determined by the condition that the chemical potential for electrons (the Fermi level) must be constant throughout the structure. This depends on the doping and band bending in the substrate and, possibly, in each of the layers. At the very least, one may expect these fields to be reduced by a factor corresponding to the macroscopic dielectric constant, that is, a factor of order 10 but possibly larger if the conductivity of the layers increase by the free carriers in the system. Consequently, a more realistic expectation for the effect on sheet carrier concentration is on the order of 10^{11}–10^{12} electrons/cm^2.

The difference between these and traditional device structures without polarization effects is that for uniform dopant concentrations one obtains parabolically varying potentials with distance, whereas here the linear terms come from polarization on top of the parabolic terms. These linear terms lead to variations of the potential over a shorter distance scale determined by the thickness of the layers, whereas the parabolic terms correspond to the space-charge regions. Thus, the linear terms may help to localize carriers if the polarity of the structure is chosen properly.

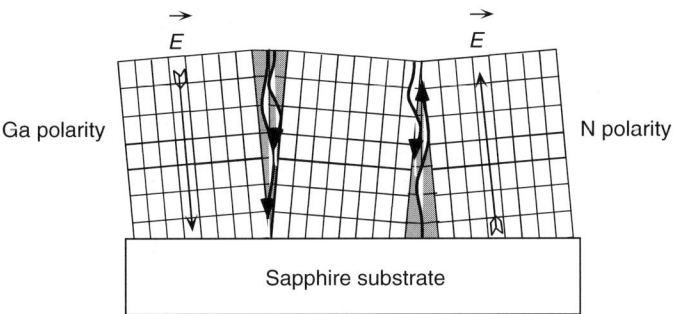

FIG. 8.   Artistic view of twisted and tilted columnar growth in GaN along with Ga and N polarity regions.

The immediate impact of this polarization is that the field generated by this process must be considered together with that induced by the applied voltage and charge redistribution. Moreover, as alluded to earlier, free carriers can also be drawn from any shallow and weakly bound impurities and metal in contact with the semiconductor. In any case, the free carriers would tend to screen the piezoelectric-induced polarization field. An additional complicating factor in nitrides in relation to polarization is that the semiconductor tends to twist and tilt in a columnar mode, in an effort to minimize strain as shown in Fig. 8. Multiple polarity has been confirmed in epitaxial GaN layers on sapphire substrates by convergent beam electron diffraction (Daudin *et al.* 1996; Potin *et al.* 1997; Vermaut *et al.* 1997). These columns do not necessarily have the same cation/anion ordering polarity as shown in Fig. 8. In addition, stacking faults would also lead to mixed polarity samples. In the presence of strain, Ga polarity domains and N polarity domains would have opposite polarization, causing increased scattering.

Figure 9 is a schematic representation of an ideal inversion domain boundary formed in growth along the [0001] direction. On the left of the boundary, the growth initiates with N, and on the right it begins with Ga. On the left-hand side, the bond along the [0001] direction is from Ga to N; this is called Ga polarity. On the right-hand side, the [0001] bonds are from N to Ga; this is called N polarity. In N polarity and under tensile strain, the PE field generated points towards the surface, whereas that for the Ga polarity region points in from the surface. When the strain is compressive, the direction of the field changes. Yet, an additional complicating factor is the asymmetry in the apparent (measured) barrier discontinuities between GaN and its binary and ternaries caused by polarization (Martin *et al.* 1994, 1996; Martin 1996), which we have not really discussed here.

To reiterate, the spontaneous polarization arises simply because of the ionicity of the bonds and the low symmetry in wurtzite. In fact, Bernardini and Fiorentini (1998) showed that the field that occurs in quantum wells is

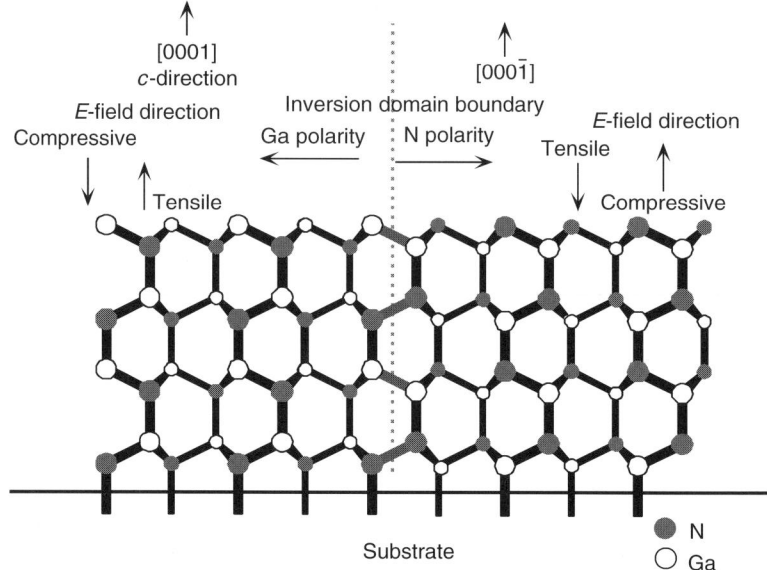

FIG. 9. Domains in GaN, with N polarity (nitrogen surface layer) on the right and Ga polarity (with Ga on surface) on the left-hand side under compressive residual strain. The arrows show the direction of the piezoelectric field in each of the domains. Dotted line indicates the schematic representation of an ideal inversion domain boundary formed in along the *c*-axis.

determined by the difference in spontaneous polarization between the two bulks and the PE contribution. The field (i.e., the slope of the potential) is quite independent of the offset (i.e., the dipole discontinuity that occurs at the interface between the two materials).

In considering a normal MODFET (N-MODFET) structure where the larger bandgap AlGaN donor layer is deposited on top of a GaN channel layer, see Fig. 10, both the spontaneous polarization and piezoelectric polarization must be accounted for. For an N-MODFET structure with Ga polarity, the potential will slope down from the surface of the AlGaN layer towards the AlGaN/GaN interface and will help to drive free electrons towards the interface forming a 2DEG as shown in Fig. 10. For example, if there is an ohmic metal contact on the AlGaN surface, electrons will flow towards the 2DEG below the AlGaN layer from contacts. Since nitride semiconductors in question have large bandgaps, thermal generation rates are minuscule and the role played by thermally generated carriers can be ignored.

A case of importance in GaN-based semiconductors is the polarity of the epitaxial layer, that is, whether the Ga or the N plane forms the surface. It is, therefore, essential that the N-face case also be considered. The case of an AlGaN (tensile strained)/GaN (relaxed) heterostructure with nitrogen

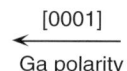

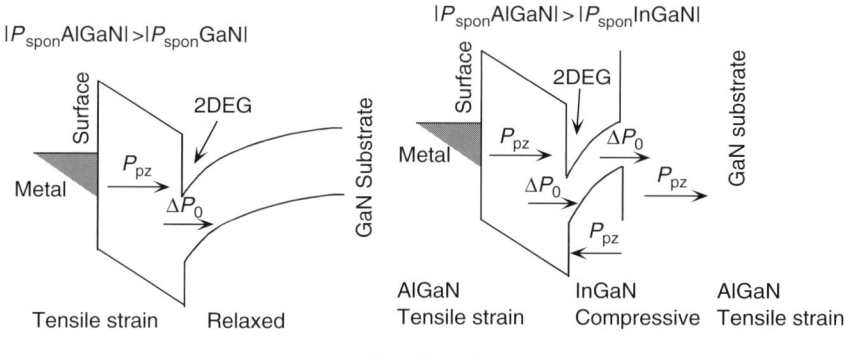

FIG. 10.   GaN-based normal modulation doped structures with Ga polarity and GaN and InGaN active layers. If the sign of strain was to change, to compressive from tensile, then the direction of the piezoelectric polarization would change. In that case the spontaneous and piezoelectric polarization charges would oppose one another, and the larger one would determine whether hole or electron accumulation is favored at the interface.

polarity for an n-type GaN and for a p-type GaN buffer layer is shown in Fig. 11, as in the cases depicted in Fig. 10, where the piezoelectric polarization and spontaneous polarization charges support one another. Unlike the Ga-polarity case, the polarization charge is such that screening charge will be made of holes if they are present in the film. If holes constitute the minority charge in the film, then the thermal process is the means by which they would be created. However, this process in a wide bandgap semi-conductor such as GaN is very slow and equilibrium condition may not be attained. If the strain in AlGaN were compressive, the direction of the piezo-electric polarization vector would change causing the piezoelectric polari-zation to counter the spontaneous polarization. This would actually represent the case when the epitaxial films are relaxed at the growth tem-perature and upon cooling to room temperature, the film would be under compressive strain if on sapphire substrates. This is due to the expansion coefficient of sapphire being larger than that of GaN. In such a case, the larger of the two would dominate and determines whether hole or electron accumulation would be favored. If, on the other hand, the film is grown on a SiC substrate, the strain due to thermal expansion would be tensile. This would lead to the case where the piezoelectric polarization and spontaneous polarization would support one another.

Inverted modulation doped structures can also be used to interrogate the picture in effect and perhaps take advantage of the unique features presented.

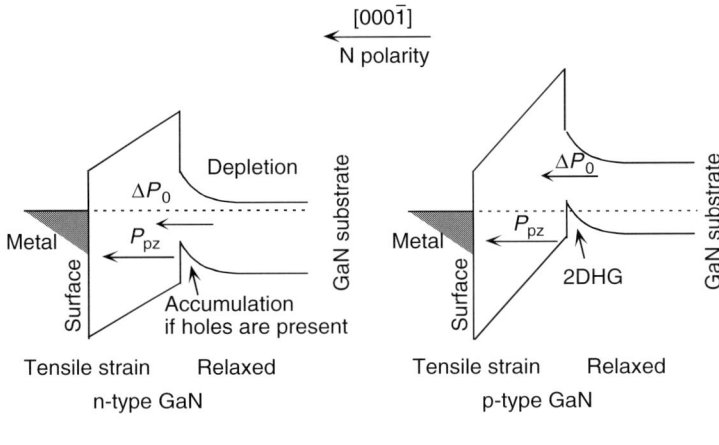

FIG. 11. AlGaN/GaN-based normal modulation doped structures with N polarity for two cases, one for n-type and the other for p-type buffer layer. If the sign of strain were to change, to compressive from tensile, then the direction of the piezoelectric polarization would change. In that case the spontaneous and piezoelectric polarizations would oppose one another, and the larger one would determine whether hole or electron accumulation is favored at the interface.

In such a case, the AlGaN layer precedes the GaN top layer where the charge accumulation would occur. The interface between the AlGaN layer and the bottom GaN layer, which is referred to as the buffer layer, would be graded to avoid a normal interface.

Figure 12 shows, schematically, an AlGaN/GaN-based inverted modulation doped structure with Ga and N polarities. As can be seen, in the case of Ga polarity and tensile strain in AlGaN, both the piezoelectric and spontaneous polarization vectors support each other leading to a hole accumulation at the interface if holes are present in the system. Since the thermal generation rate is very small, the semiconductor structure cannot be expected to reach equilibrium by this means at room temperature in a reasonable period of time. On the other hand, with N polarity and tensile strain in AlGaN, the structure favors electron accumulation at the interface. If the sign of strain was to change, to compressive from tensile, then the direction of the piezoelectric polarization would change. In that case the spontaneous and piezoelectric polarization charges would oppose one another, and the larger one would determine whether hole or electron accumulation is favored at the interface.

Ambacher *et al.* (2000), employing Hall effect and capacitance–voltage profiling measurements, measured the sheet carrier concentration and its profile in modulation doped structures at room temperature, the results of which are shown in Fig. 13. As seen in the figure, the sheet carrier concentrations are consistently larger than those expected from piezoelectric

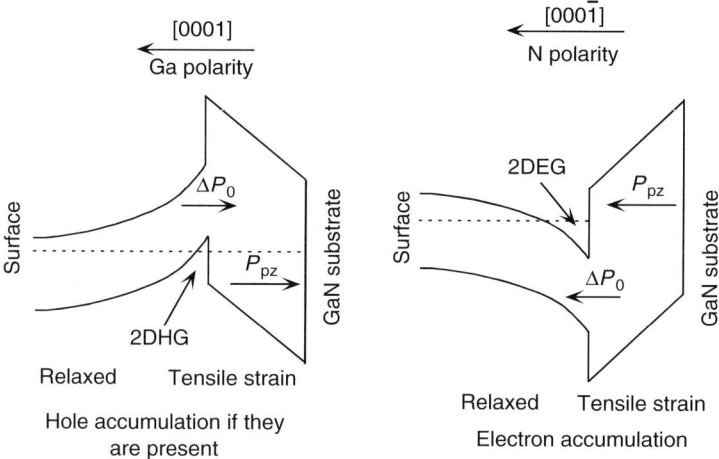

FIG. 12. AlGaN/GaN-based inverted modulation doped structures with Ga and N polarities. As can be seen, in the case of Ga polarity and tensile strain in AlGaN, both the piezoelectric and spontaneous polarization vectors support each other leading to hole accumulation at the interface if holes are present in the system. In the other source, thermal generation rate is very small and the semiconductor structure cannot be expected to reach equilibrium by this means at room temperature in a reasonable period of time. On the other hand, with N polarity and tensile strain in AlGaN, the structure favors electron accumulation at the interface. If the sign of strain were to change, to compressive from tensile, then the direction of the piezoelectric polarization would change. In that case the spontaneous polarization and piezoelectric polarization would oppose one another, and the larger one would determine whether hole or electron accumulation is favored at the interface.

polarization alone. Since the GaN buffer contribution was verified to be negligible, an additional source, namely spontaneous polarization, was invoked to explain the observations. The contribution from unintentionally doped barrier layer also was found to be negligible as the measured 2DEG concentration with increasing barrier thickness up to 650 Å ($x = 0.25$) did not appear to affect the results. The highest measured sheet carrier concentrations for $0.2 < x < 0.45$ are in good agreement with the calculated values of total polarization charge taking into account the spontaneous polarization and the measured strain relaxation of AlGaN. The maximum sheet carrier concentration for undoped AlGaN/GaN (and also for InGaN/GaN) heterostructures is limited to about $2 \times 10^{13}\,\text{cm}^{-2}$ due to strain relaxation of the top alloy layer. The calculated and measured sheet carrier concentrations of undoped MODFET structures with alloy compositions $x < 0.2$ are not in good agreement. The measured sheet carrier concentrations are up to $5 \times 10^{12}\,\text{cm}^{-2}$ smaller than the predicted ones. In addition, the large scattering of the measured sheet carrier concentrations for heterostructures with similar barrier thicknesses and alloy compositions is unexpected and much larger than the error in experiments. For a more accurate calculation of the

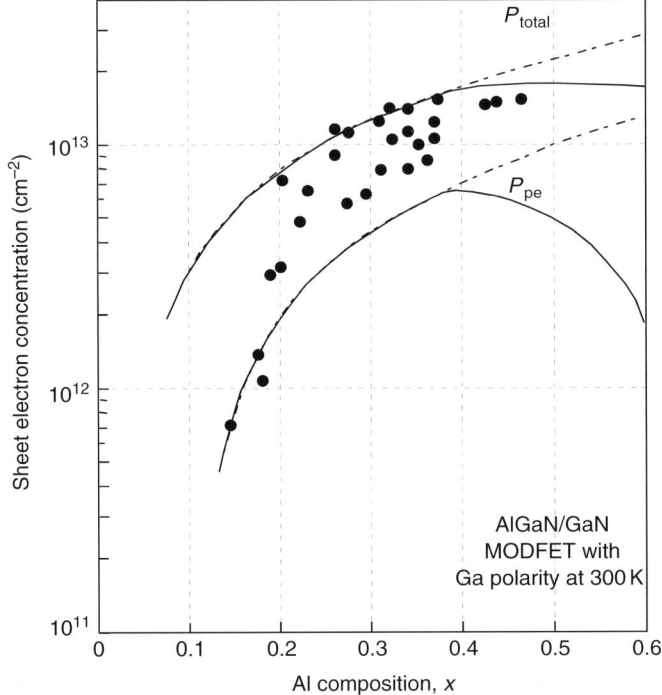

FIG. 13. Comparison of calculated and measured electron sheet carrier concentrations in Ga-face AlGaN/GaN heterostructures. The dashed lines are calculated results for pseudomorphic structures. The solid lines are indicative of the calculated results, which take the measured partial strain relaxation into account (for more details, see Ambacher *et al.* 1999, 2000).

polarization charge in a Ga-polarity modulation doped structure, see Section IX.1.

The physics literature appears to be fairly clear in that the polarization charge is a bound charge and that any free carriers act only to screen it. However, the device reports on GaN FETs have gone so far as to suggest that piezoelectric effect in and of itself is sufficient to provide the free carriers needed for devices. Polarization effects, particularly spontaneous polarization, have immense impact on measured band discontinuities. For example, the dependence of measured band discontinuities on the order in which the larger and the smaller band gap semiconductors are grown is one that can be attributed to polarization effects (Martin *et al.* 1996; Bernardini and Fiorentini 1998).

To reiterate, as a result of polarization, the static potential at the GaN/AlN interface is different from that at the AlN/GaN interface that gives rise to interface charge larger than the charge densities used in devices. A substantial level of effort has been expended toward determining band discontinuities, but the field is in desperate need of more in-depth investigations in improved structures. The observed asymmetry in AlN/GaN and GaN/AlN interfaces caused by spontaneous polarization is within the experimental errors of Martin *et al.* (1994, 1996) and Martin (1996). Inversion domains, see Fig. 8, combined with any strain in semiconductor nitrides lead to flipping PE fields, see Fig. 9, with adverse effects on our ability to characterize the films, let alone exploit this phenomenon for devices. Such flipping fields would also cause increased scattering of the carrier as they traverse in the c-plane. Simply put, identical device structures with different polarity layers would have widely differing performance underscoring the importance that these issues will have to be investigated and reconciled. The polarity mixing causes the PE-induced electric field to flip from one domain to the next, causing a variation in the sheet carrier concentration along the channel of a FET-like device. The same polarity mixing would have deleterious effects in the base of an HBT as well and depending on the polarity the induced field would either aid or impede minority carrier transit.

VI.   Partial Strain Relaxation

We, heretofore, dealt only with the coherently strained AlGaN top layer and the relaxed GaN buffer layer in the case of normal modulation doped structures and relaxed AlGaN and coherently strained GaN top layers in the case of inverted modulation doped structures. This is, of course, an assumption whose validity has not been confirmed. In reality, a mostly likely scenario is that a partial relaxation would take place coupled with a residual thermal strain. Ambacher *et al.* (2000) treated this issue of partial relaxation. As pointed out by Ambacher *et al.*, calculations of the critical thickness of InGaN and AlGaN grown on GaN show that for a typical AlGaN barrier thickness of about 20 nm, strain relaxation should occur for alloy compositions above $x = 0.2$. At the outset, we should mention that the critical thickness, the thickness above which the film begins to relax, is difficult to determine because of the large density of dislocations in this material system. Nevertheless, proceeding as if it can be done, does indeed give an estimate, which is very useful. In order to determine the alloy composition and relaxation of the barrier, the lattice constants $a(x)$ and $c(x)$ were measured by high-resolution X-ray diffraction (HRXRD) (Ambacher *et al.* 1999). For barrier thicknesses of about 25 nm, macroscopic strain relaxation was observed for alloy compositions of $x_c = 0.20$ and 0.38 for InGaN and AlGaN, respectively. In both cases, the degree of relaxation increased nearly linearly with increasing alloy composition for $x > x_c$. Naturally, for a barrier with a fixed alloy composition, the piezoelectric polarization and the sheet charge induced by that polarization decrease linearly with increasing level of

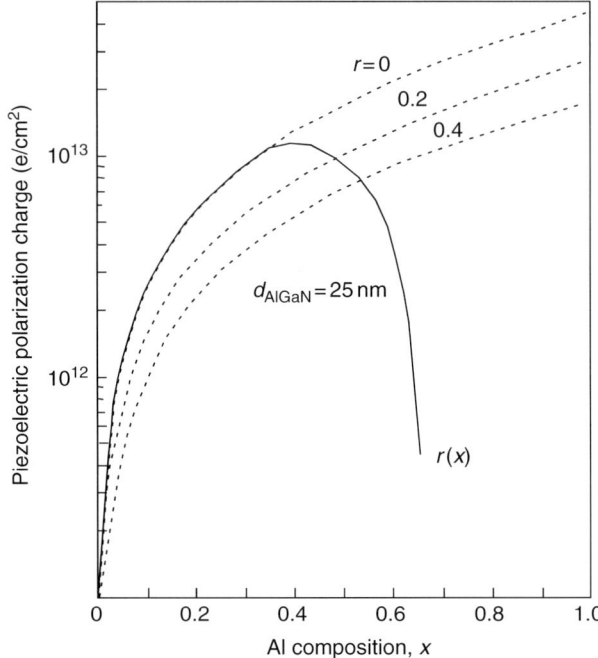

FIG. 14. Piezoelectric polarization-induced sheet charge at the interface of a Ga-face AlGaN/GaN heterostructure for different degrees of relaxation of the barrier layer (dashed lines). The solid line was calculated taking into account the measured degree of relaxation for AlGaN top layers with thicknesses of 25 ± 5 nm. The figures have been modified to be consistent with the piezoelectric and polarization charge figures used in our previous publications.

relaxation r. The polarization-induced charge at the interface for a Ga-face AlGaN/GaN heterostructure vs Al-concentration is shown for different degrees of strain relaxation in Fig. 14. The maximum sheet charge caused by piezoelectric polarization of a strained 30-nm-thick barrier was calculated using the measured degree of relaxation. As can be seen, for a barrier with $x = 0.4$, the total sheet charge induced by spontaneous and piezoelectric polarization decreases if AlGaN becomes partially or completely relaxed.

Hsu and Walukiewicz (2001) incorporated issues related to polarization into a normal modulation doped structure, calculating the dependence of parameters such as the sheet carrier concentration on structural parameters. While this issue will be treated in detail Section IX, we will discuss the treatment proposed by Hsu and Walukiewicz since it relates the sheet concentration to structural parameters and discusses the electron mobility.

Figure 15(a) shows the dependence of the 2DEG density as a function of Al content of an $Al_xGa_{1-x}N$ barrier with a thickness of 31 nm. Values of $A = 7.9 \times 10^4$ V/cm and $B = 1.15 \times 10^7$ V/cm in $P(x) = A + Bx$, describing the dependence of polarization on AlN mole fraction, led to good quantitative

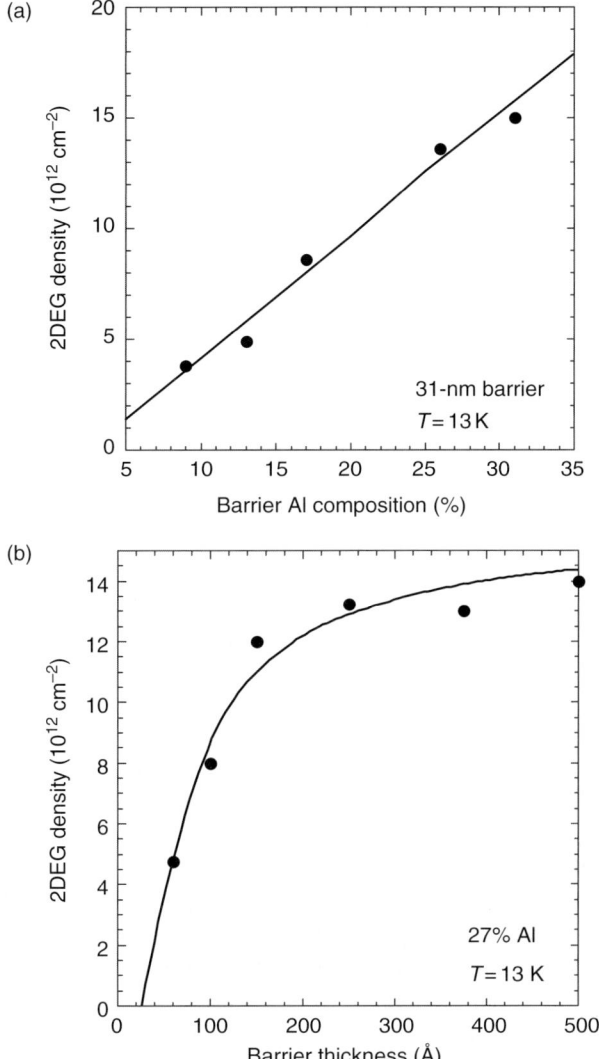

FIG. 15.   (a) 2DEG densities as functions of AlGaN barrier Al composition. (b) 2DEG densities as a function of thickness of an $Al_{0.27}Ga_{0.73}N$ barrier. After Hsu and Walukiewicz (2001).

agreement with experiments. In principle, parameter A is related to the uncompensated, spontaneous polarization field in GaN. Its relatively small value indicates that either this spontaneous polarization is very small or it is well compensated by charges from unintentional dopants within the AlGaN layer. Figure 15(b) shows the dependence of the 2DEG density on the thickness of an $Al_{0.27}Ga_{0.73}N$ barrier.

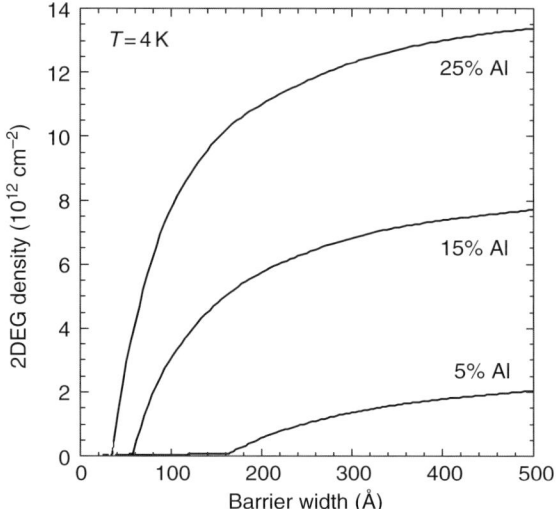

FIG. 16.   Calculated 2DEG densities in AlGaN/GaN heterostructures as a function of barrier width for three different Al compositions. After Hsu and Walukiewicz (2001).

Figure 16 shows the calculated dependence of the 2DEG density on the barrier thickness for several alloy compositions at 4 K, formed by surface donors in an unintentionally doped structure. For very thin barriers ($< 30 \, \text{Å}$ for a barrier with $x = 25\%$ and $< 160 \, \text{Å}$ for $x = 5\%$), the defect level is below the Fermi energy and so the surface defects are neutral. In this case, only the electrons from any dopant in the barrier are transferred to the GaN well and if one assumes a donor concentration of $5 \times 10^{16} \, \text{cm}^{-3}$, the density remains below $10^{11} \, \text{cm}^{-2}$. Once the barrier thickness increases to the point of bringing the defect level above the Fermi energy, electrons are transferred from these surface donors and the 2DEG density increases rapidly.

At very low alloy compositions (5%), the 2DEG density rises continuously at an approximately constant rate throughout the entire range of barrier widths. The reason for this is that as the barrier width increases, the donor charge transition state energy also increases, due to the polarization field. This leads to an increase in the energy gain from transferring electrons from the donors to GaN. Consequently, the donor formation energy decreases and the surface defect concentration increases.

For larger barrier alloy compositions, the 2DEG density increases very rapidly at first, followed by a leveling for barrier thicknesses greater than $250 \, \text{Å}$. The initial increase is due to the effect of the polarization field on the donor level in that larger Al fractions in the barrier result in larger polarization fields and a more rapid increase in defect concentration. As the 2DEG density approaches $10^{13} \, \text{cm}^{-2}$, the opposing field due to the transfer of electrons becomes comparable in magnitude to the polarization field. This field acts to reduce the rate at which the defect formation energy is lowered. For a barrier of sufficient thickness, this charge transfer field could completely cancel the polarization field and thereafter, the 2-DEG density would be independent of barrier thickness assuming that the AlGaN barrier remains coherently strained and is not relaxed.

VII.   Low-Field Transport in 2DEG Systems

Calculations of the electron mobility at the AlGaN/GaN interface have been performed (Hsu and Walukiewicz 2001) using methods that have been described previously (Hsu and Walukiewicz 1998). The scattering mechanisms considered were acoustic phonons, Coulomb scattering from both the donor-like defects on the AlGaN barrier surface and unintentional dopants in the GaN, and alloy disorder scattering. In addition, interface roughness scattering was also included.

Figure 17 shows the calculated 2DEG mobilities at low temperature as a function of barrier width for several heterostructures with different Al compositions. Calculations show three distinct regions (Hsu and Walukiewicz 2001). For very small barrier widths, the mobility is quite low and increases slowly with increasing barrier width. Comparing these curves with those in Fig. 16, one can see that the 2DEG density is very small in this

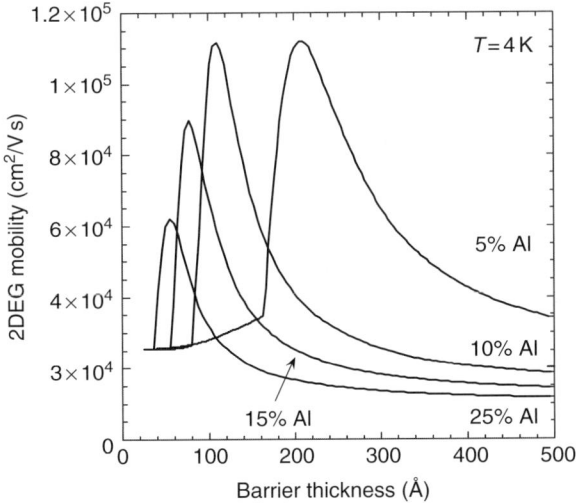

FIG. 17.   Low-temperature 2DEG mobilities as a function of barrier thickness for four different AlGaN/GaN heterostructures with different Al barrier compositions. After Hsu and Walukiewicz (2001).

region, and is due to unintentional dopants in the bulk of the AlGaN barrier. Consequently, Coulomb scattering by charged impurities in the GaN well and the AlGaN barrier keeps the mobility low. In the region with a sudden increase of mobility, the vast majority of the 2DEG electrons originate from donors at the surface of the AlGaN barrier. In this region, the high electron density reduces the efficacy of Coulomb scattering, which results in much higher mobilities. Moreover, most of the Coulomb scattering centers are at the surface of the AlGaN barrier, away from the carriers at the AlGaN/GaN interface. Finally, when the 2DEG density becomes high enough the alloy disorder scattering, which varies with the square of the electron concentration, is the dominant scattering mechanism and the mobility decreases with increasing 2DEG density. Maximum mobilities are generally achieved for barrier thicknesses between 50 Å ($x = 0.25$) and 200 Å ($x = 0.05$). Figure 17 indicates that maximum mobilities are obtained for 2DEG densities in the range of about $3.5 \times 10^{12}\,\mathrm{cm}^{-2}$ ($x = 0.25$) to $5 \times 10^{11}\,\mathrm{cm}^{-2}$ ($x = 0.05$).

Figure 18 shows the overall and individual mobilities as a function of barrier thickness for a 2DEG at the interface of a $\mathrm{Al_{0.07}Ga_{0.93}N/GaN}$ heterostructure. This is the Al alloy fraction that would produce the highest possible low-temperature mobilities for the indicated unintentional doping levels (Hsu and Walukiewicz 2001). The component mobility curves illustrate the dominant scattering mechanisms in each of the three different regions mentioned above. At a barrier thickness of 110 Å, the surface defect ionization energy is pushed above the Fermi energy and there is a rapid increase in both the mobility and the 2DEG concentration. For barriers that are

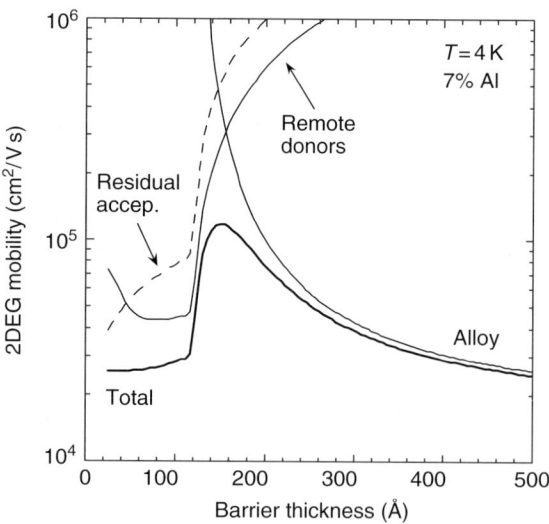

Fig. 18. 2DEG mobility as a function of AlGaN barrier thickness for an $\mathrm{Al_{0.07}Ga_{0.93}N/}$ GaN heterostructure. After Hsu and Walukiewicz (2001).

thicker than 150 Å, alloy disorder becomes the dominant scattering mechanism. A maximum electron mobility somewhat above $10^5 \, cm^2/V \, s$ is predicted for a barrier width of about 150 Å and a corresponding 2DEG density of $1.8 \times 10^{12} \, cm^{-2}$. These calculations indicate that for a standard, undoped AlGaN/GaN heterostructure, the maximum low-temperature mobility is only slightly larger than $10^5 \, cm^2/V \, s$. The principal reasons for the lower values here as compared to those obtained in GaAs/AlGaAs heterostructures are stronger alloy disorder scattering (resulting from the much higher 2DEG density and larger conduction band offsets found in AlGaN/GaN heterostructures) and, to a smaller extent, relatively high levels of residual impurities in the nitride layers.

The calculated mobilities agree well with the experimental values for barrier Al compositions less than about 15%, as shown in Fig. 19 (Hsu and Walukiewicz 2001). However, for higher Al fractions, the measured mobilities are a nearly constant factor of 2–2.5 lower. There are two possible reasons. First, as can be seen in Fig. 15(a), the electron density exceeds $6 \times 10^{12} \, cm^{-2}$ for $x > 0.15$. This corresponds roughly to the electron density at which the higher subbands in the GaN quantum well begins to be occupied, which causes increased scattering. Second, the lattice mismatch at the interface increases with increasing x, leading possibly to increased roughness and thereby reduced electron mobility at the interface. In the end, technological issues may come to be the culprit as was the case in the early development of the GaAs/AlGaAs heterointerface.

The effect of interface roughness scattering on the total mobility has been estimated using 15 Å as the parameter for the characteristic lateral extent of the islands of roughness and adjusting the height of the islands in order to obtain the best fit with experimental data (Hsu and Walukiewicz 2001). A constant island height of 0.9 Å produced the fit shown in Fig. 19(a) (for a constant barrier thickness and variable barrier composition) and Fig. 19(b) (for a constant barrier composition and variable barrier thickness) (Hsu and Walukiewicz 2001). The points represent experimental data and correspond to the samples in Fig. 15(a). The solid line is the calculated mobility neglecting interface roughness scattering. The dashed line is the calculated mobility including interface roughness scattering. One possible interpretation of the data is a sudden degradation of a near perfect interface for heterostructures with a barrier of greater than 15% Al.

The above-mentioned arguments indicate that the highest electron mobilities are expected from heterostructures with 2DEG densities in the range of a few times $10^{12} \, cm^{-2}$. Since the polarization-induced charge transfer increases with increasing width and Al composition of the barrier, the highest mobilities are limited to heterostructures with relatively thin barriers with low Al content for undoped structures. A maximum electron mobility of about $10^5 \, cm^2/V \, s$ at 4 K can be expected from heterostructures with a 15-nm thick $Al_{0.07}Ga_{0.93}N$ barrier (Hsu and Walukiewicz 2001).

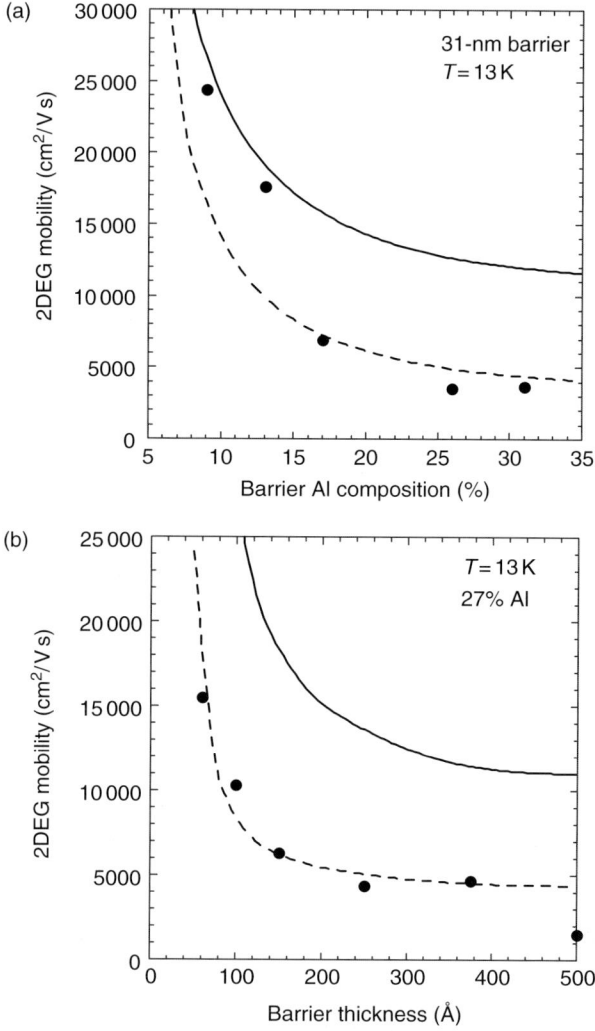

Fig. 19. (a) 2DEG mobilities as a function of Al composition of the barrier. The points represent experimental data and correspond to the samples in Fig. 15(a). (b) 2DEG mobilities as a function of thickness of an $Al_{0.27}Ga_{0.73}N$ barrier along with the experimental data points and correspond to the samples in Fig. 15(b). The solid line is the calculated mobility neglecting interface roughness scattering. The dashed line is the calculated mobility including interface roughness scattering. After Hsu and Walukiewicz (2001).

Figure 20 compares the temperature dependent mobility, both calculated and measured, for an $Al_{0.09}Ga_{0.91}N/GaN$ heterostructure (Hsu and Walukiewicz 2001). The points represent the experimental data. The calculations were performed for a constant 2DEG density of $2.3 \times 10^{12}\, cm^{-2}$. The low-temperature mobility is dominated by alloy disorder scattering with

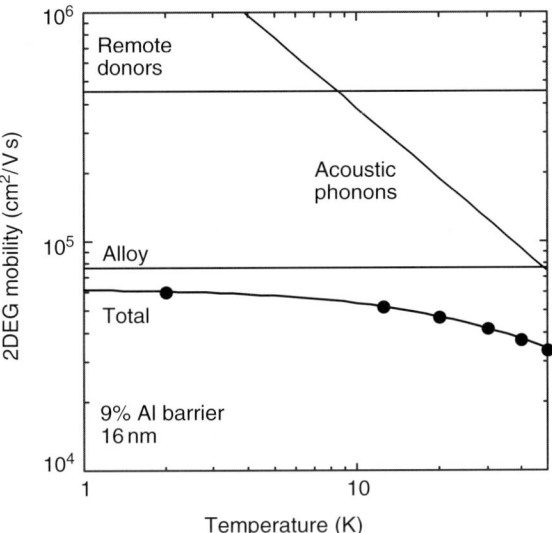

FIG. 20. 2DEG mobilities as a function of temperature in an $Al_{0.09}Ga_{0.91}N/GaN$ heterostructure with a 16-nm thick barrier. After Hsu and Walukiewicz (2001).

only small contributions from ionized centers and acoustic phonon scattering. Above 10 K, however, scattering by acoustic phonons becomes important and it is responsible for a weak but still clearly discernible temperature dependence of the mobility. The acoustic phonon component of the mobility varies roughly as $4.02 \times 10^6 \, (1/T) \, cm^2/V \, s$ and is about an order of magnitude larger in AlGaN/GaN heterostructures than in AlGaAs/GaAs systems.

The calculations reviewed above indicate that scattering by ionized impurities is important only for heterostructures in which the 2DEG density is $10^{12} \, cm^{-2}$ or smaller. Consequently, if one wishes to increase the electron mobility one should reduce the effect of alloy disorder and interface roughness scattering along with improved AlGaN technology. The penetration depth of the 2DEG into an $Al_{0.09}Ga_{0.91}N$ barrier of height $\Delta E_c = 0.2 \, eV$ is $z_p = 0.9 \, nm$, leading to the assertion that the alloy scattering is the dominant mechanism (Hsu and Walukiewicz 2001). One could envision a composite barrier design consisting of a very thin layer of AlN, grown at the interface, followed by a thicker $Al_{0.07}Ga_{0.93}N$ layer. Because the penetration depth of electrons into AlN is only 3 Å and the electron wavefunction decays exponentially, a layer of AlN of only a few angstroms thickness would sufficiently confine the electrons in the binary compounds and nearly eliminate alloy scattering. The thin AlN film at the interface should not significantly affect the charge transfer, which is determined mostly by the composition and thickness of the much thicker remaining part of the barrier. Thus, for a barrier composed of 5 Å of AlN followed by 120 Å of

$Al_{0.07}Ga_{0.93}N$, one could expect to obtain a 2DEG with a density between 2 and $3 \times 10^{12} cm^{-2}$, but with negligible alloy disorder scattering.

VIII.   High-Field Transport

FETs by their nature rely on transport under high electric fields. As such, high-field effects are an integral part of any short-channel FET, realizing that low-field effects too are important in that they determine many of the parasitic resistances with detrimental effects on device performance. Consequently, high-field properties of any semiconductor conptemplated for use in FETs must be scrutinized in terms of its high-field transport and semiconductor nitrides are no exception. High-field transport in nitride semiconductors has recently been reviewed (Farahmand et al. 2001). However, a brief discussion of results is deemed necessary for completeness. While the bulk of the experimental studies and, to some extent, calculations have focused on GaN, the transport properties from the point of electron mobility and velocity of InN are much more conducive for FETs. Holding InN back is, among other factors, its poor quality and high unintentional electron concentration. As has been the case for the GaAs system, it is very likely that some composition, or a range of compositions, of InGaN will be used for FET channels while taking full advantage of GaN and AlGaN in the rest of the structure. Consequently, transport properties, inclusive of low and high field, of nitride semiconductors InN, GaN, AlN and their alloys will be briefly reported following Farahmand et al. (2001).

The steady-state electron drift velocity vs electric field has been calculated for the nitride binaries and ternaries at different temperatures and various doping concentrations (Farahmand et al. 2001). The results for 300 K and an electron concentration of $10^{17} cm^{-3}$ will be shown here. Figure 21 shows the calculated electron steady-state drift velocity vs applied electric field, for GaN, $Al_{0.2}Ga_{0.8}N$, $Al_{0.5}Ga_{0.5}N$, $Al_{0.8}Ga_{0.2}N$, and AlN. This set of calculations is made assuming the maximum alloy scattering rate. The alloy scattering was observed to be the dominant scattering mechanism when the random alloy potential was set equal to the conduction band offset between GaN and AlN. A significantly lower drift velocity is observed when alloy scattering is present, which is simply a result of the higher total scattering rate. Moreover, in the presence of alloy scattering, the peak velocity occurs at a higher applied field compared to the case when alloy scattering is neglected. This is also due to the fact that in the presence of alloy scattering the total scattering rate is higher; thus, a higher field is required to heat the carriers prior to the onset of intervalley transfer.

Figure 22 shows the calculated electron drift velocity vs applied electric field for GaN, $In_{0.2}Ga_{0.8}N$, $In_{0.5}Ga_{0.5}N$, $In_{0.8}Ga_{0.2}N$, and InN. The random

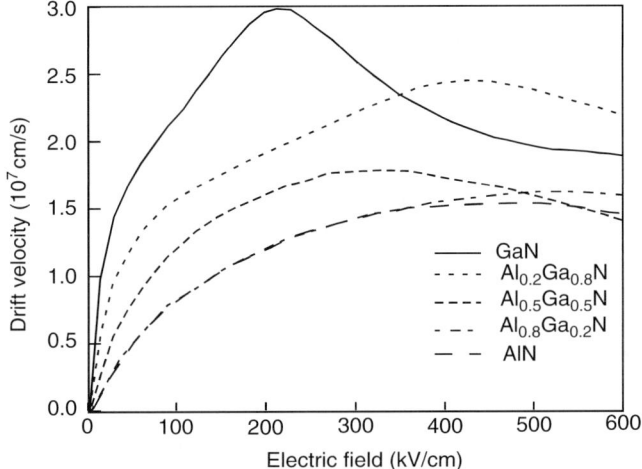

FIG. 21.   Calculated electron drift velocity vs applied electric field for GaN, $Al_{0.2}Ga_{0.8}N$, $Al_{0.5}Ga_{0.5}N$, $Al_{0.8}Ga_{0.2}N$, and AlN. For this calculation, the random alloy potential was set equal to conduction band offsets. Lattice temperature is at 300 K, and electron concentration is equal to $10^{17}\,cm^{-3}$. After Farahmand *et al.* (2001).

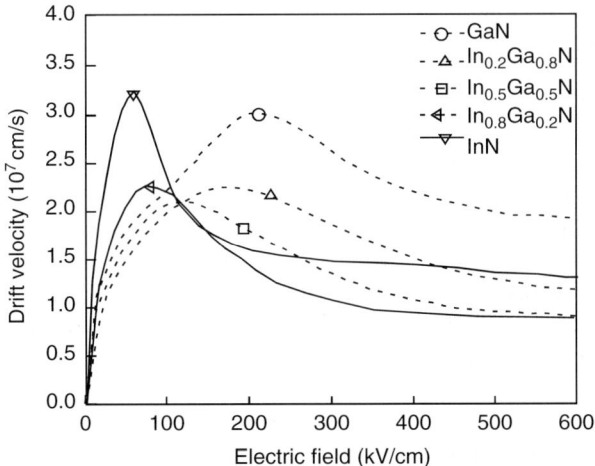

FIG. 22.   Calculated electron drift velocity vs applied electric field for GaN, $In_{0.2}Ga_{0.8}N$, $In_{0.5}Ga_{0.5}N$, $In_{0.8}Ga_{0.2}N$, and InN. For this calculation, the random alloy potential was set equal to conduction band offsets. Lattice temperature is at 300 K, and electron concentration is equal to $10^{17}\,cm^{-3}$. After Farahmand *et al.* (2001).

alloy potential is set equal to the conduction band offsets. It should be mentioned that for $In_xGa_{1-x}N$ compounds the random alloy scattering potential calculated from J. C. Phillips' theory of electronegativity differs in values on the same order as the conduction band offsets.

IX.   Modulation Doped Field-Effect Transistors

With its reduced impurity scattering and unique gate capacitance–voltage characteristics, the MODFET has become the dominant high-frequency device. Among the MODFET's most attractive attributes are close proximity of the mobile charge to the gate electrode and high drain efficiency. As in the case of emitters, GaN-based MODFETs have quickly demonstrated record power levels at high frequencies with very respectable noise performance and large drain breakdown voltages.

In MODFETs, the carriers that form the channel in the smaller bandgap material are donated by the larger bandgap material, and ohmic contacts or both. Since the mobile carriers and their parent donors are spatially separated, short-range ion scattering is nearly eliminated, which leads to mobilities that are characteristic of nearly pure semiconductors. A Schottky barrier is then used to modulate the mobile charge that in turn causes a change in the drain current. Because of this heterolayer construction, the gate can be placed very close to the conducting channel, resulting in large transconductances (Morkoç *et al.* 1991). Figure 23 presents a schematic representation of a GaN/AlGaN MODFET heterostructure in which the carriers are provided by the donors in the wider bandgap AlGaN. In a MODFET device under bias, the carriers can also be provided by the source contact.

1.   MODFET MODEL

The Schrödinger equation and Poisson equation can be used self-consistently in order to study the channel formation and current flow mechanisms in GaN-based MODFETs (Morkoç *et al.* 1991; Sacconi *et al.* 2001). Several approaches have been used to define the system hamiltonian used in the Schrödinger equation, namely effective masses (Bastard 1987), k.p expansion (Oberhuber *et al.* 1998), and tight-binding expansion (Di Carlo *et al.* 1996, 2000; Della Sala *et al.* 1999; Di Carlo 2000). The use of sophisticated models such as k.p or tight-binding is justified, perhaps even made necessary, by the complex wurtzite band structure, particularly for determining the valence band states. Thus, calculations of optical processes involving band-to-band transitions must consider the details of the band structure beyond the simple effective mass approximation (EMA) (Bonfiglio *et al.* 2000; Cingolani *et al.* 2000). However, even when only the conduction band processes are of interest, EMA is still very accurate means of determining the properties of interest. In fact, nitride-based semiconductors in the wurtzite structure posseses a conduction band with a Γ minimum, which can be described reasonably well within such an approximation. Within the effective mass theory, the Schrödinger equation takes the form (Bastard

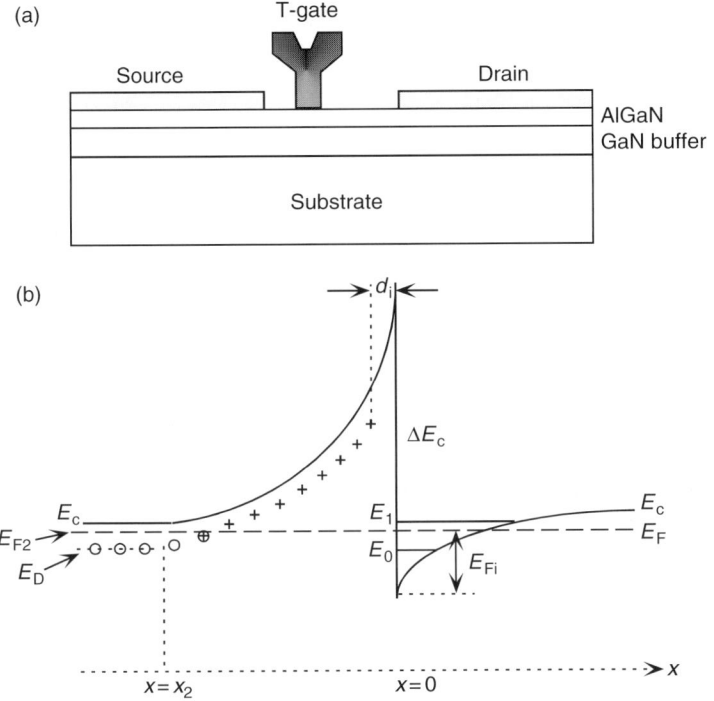

FIG. 23. (a) Schematic representation of an AlGaN/GaN modulation doped field-effect transistor (MODFET). (b) Schematic band structure of an AlGaN/GaN modulation doped heterostructure in which the free carriers are provided to the GaN layer by the dopant impurities placed in the larger bandgap AlGaN barrier layer.

1987; Morkoç *et al.* 1991; Lugli *et al.* 1996):

$$-\frac{h^2}{2}\frac{d}{dz}\left(\frac{1}{m(z)}\frac{d}{dz}\right)\varphi + eV(z)\varphi = E\varphi \qquad (2)$$

where $m(z)$ is the (position dependent) effective mass, V the electrostatic potential, φ the electron wavefunction and E the energy level. Non-parabolicity may induce deviations from the simple parabolic band model; however, this will not substantially change the results that we will present. In the nitride semiconductors with wurtzite structure, spontaneous and piezo-electric polarization effects are present (Fiorentini *et al.* 1999), which necessitates that Poisson equation be solved for the displacement field, $D(z)$:

$$\frac{d}{dz}D(z) = \frac{d}{dz}\left(-\varepsilon(z)\frac{d}{dz}V(z) + P(z)\right) = e\left(p(z) - n(z) + N_D^+ - N_A^-\right) \qquad (3)$$

where ε is the dielectric constant, P the total polarization, n (p) the electron (holes) charge concentration and N_D^+ (N_A^-) the ionized donor (acceptor) density.

In the self-consistent procedure, potential V is obtained Eq. (3) from using an initial guess of the mobile charge concentration, and then inserted into Schrödinger's equation, Eq. (2), which is solved to get the energy levels and wavefunctions of the systems. The new electron charge density is obtained by applying Fermi statistics as follows:

$$n_{2D}(z) = \frac{m(z)k_B T}{\pi\hbar^2} \sum_i |\varphi_i(z)|^2 \ln\left[1 + e^{(E_F - E_i)/k_B T}\right] \tag{4}$$

where E_F is the Fermi level, E_i the energy of the ith quantized level, T the temperature, and k_B the Boltzmann constant. The calculated density is then plugged into the Poisson equation [Eq. (3)] and the iteration repeated until convergence is achieved. Convergence of the self-consistent algorithm can be improved by adopting a special relaxation mechanism. Here, we have used a first-order expansion of the model reported in Trellakis et al. (1997).

In the following, we consider two MODFET structures, namely a single heterojunction AlGaN/GaN normal modulation doped FET (NMODFET) where the AlGaN donor layer is grown on top of the GaN channel layer, and an "inverted" GaN/AlGaN/GaN MODFET (IMODFET) where the channel layer is grown on top of the AlGaN donor layer. The NMODFET structure consists (from the gate to the substrate) of a 150-Å n-doped ($n = 10^{18}$ cm^{-3}) AlGaN, a 50-Å unintentionally doped AlGaN layer, and a thick GaN buffer. The IMODFET consists (from the gate to the substrate) of a 300-Å unintentionally doped GaN layer, a 50-Å unintentionally doped AlGaN, a 150-Å n-doped ($n = 10^{18}$ cm^{-3}) AlGaN, a 300-Å unintentionally doped AlGaN layer, and a thick GaN layer.

We consider a residual doping of 10^{17} cm^{-3} in both GaN and AlGaN layers. We use a Schottky barrier (ϕ_B) of 1.1 eV for the metal–GaN interface and a $\phi_B = 1.2$ eV for the metal–AlGaN interface. Calculations have been performed for Al$_x$Ga$_{1-x}$N with Al content of $x = 0.1$, 0.2, 0.3, 0.4. Both [0001] and [000$\bar{1}$] growth directions are considered. In the simulations, we have used an effective mass of 0.19 for electrons and 1.8 for holes in both GaN and AlGaN layers. The band gaps and band discontinuities of the AlGaN layers used are tabulated in Table III.

As discussed previously, the presence of polarization is quite important in the nitride-based N-MODFET. The conduction bandedge profile for the NMODFET grown in [0001] direction is depicted in Fig. 24 for the cases. (i) with both spontaneous and piezoelectric polarization fields; (ii) without considering the polarization fields; and (iii) with only the piezoelectric polarization fields. The difference in piezoelectric and spontaneous polarization between the AlGaN and GaN layers manifest itself as a fixed

TABLE III

Bandgap and Conduction Band Discontinuities with
Respect to GaN of the $Al_xGa_{1-x}N$ Layer

x (Al)	E_G (eV)	ΔE_C (eV)
0.1	3.62	0.17
0.2	3.85	0.33
0.3	4.09	0.51
0.4	4.35	0.69

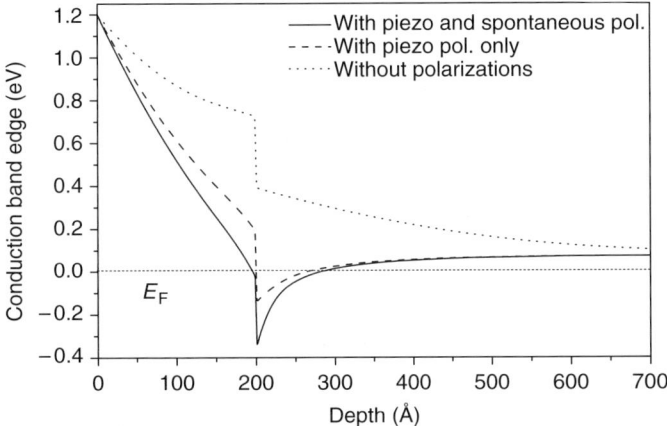

Fig. 24. Calculated conduction band edge for the N-MODFET structure grown in the [0001] direction for $V_G = 0$ with and without polarization fields.

2D-charge density at the interface between the two materials. For the [0001] growth direction considered in Fig. 24, the difference in polarization between the two materials induces a positive charge ($\sigma = +1.12 \times 10^{13}\,cm^{-2}$) at the $Al_{0.2}Ga_{0.8}N/GaN$ interface. Electrons are attracted by this positive charge, tending to accumulate at the interface and thus forming a conductive channel. Moreover, the high electric field due to the interface-charge favors the build up of a large channel density and of a strong channel confinement. Within the AlGaN layer, the strong electric field compensates the space charge contribution coming from the ionized donors. Consequently, it prevents the appearance of the parasitic channel that would otherwise form in the doped AlGaN layer (Lee *et al.* 1984; Oberhuber *et al.* 1998).

The comparison, reported in Fig. 24, between the three cases with different contribution of the polarization fields shows the importance of considering both spontaneous and piezoelectric polarizations in GaN-based device modeling. In fact, neglecting the spontaneous polarization, as was done recently (Gaska *et al.* 1997, 1998; Yu *et al.* 1997; Ramvall *et al.* 1999), the

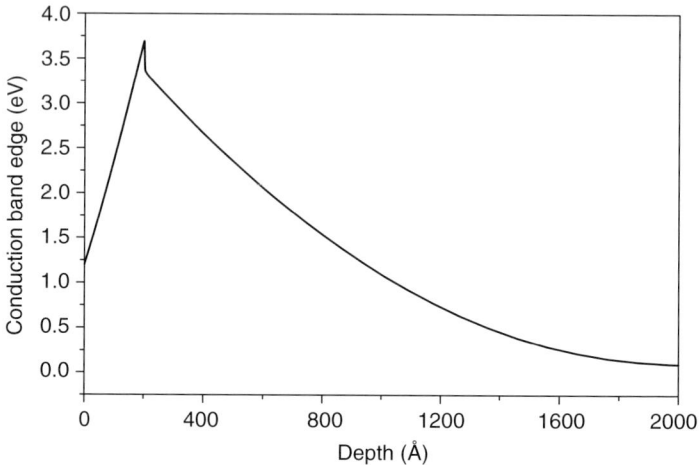

FIG. 25. Calculated conduction band edge for the N-MODFET structure grown in the $[000\bar{1}]$ direction for $V_G = 0$.

channel electron density is underestimated (Oberhuber *et al.* 1998). Clearly, the sign of the polarization charge is crucial. For the same N-MODFET structure grown in the $[000\bar{1}]$ direction, the resulting polarization charge would be negative (with the same magnitude) and electrons would be repelled from the channel as shown in Fig. 25.

The distribution of the free electron charge in the channel is shown in Fig. 26 for several values of the Al concentration of the AlGaN layer. Increasing the Al content induces a larger polarization charge at the GaN/AlGaN interface and consequently a higher channel electron concentration.

The calculations we have shown so far are obtained by considering only the polarization charge at the AlGaN/GaN interface. In reality, however, polarization charges that form at the metal–AlGaN and at the end of the GaN buffer region should be accounted for. The metal–AlGaN charge is completely screened by the charges induced on the metal surface and can, therefore, be neglected. On the other hand, the charges at the end of the buffer region may induce large deviation with respect to the situation depicted above. Oberhuber *et al.* (1998) have considered a $-\sigma/2$ charge at the interface between the GaN buffer layer and the nucleation layer. The exact amount of such charge depends, however, on the morphology of the heterojunction and may differ from the theoretical value $\sigma = \Delta P/e$. The situation is less critical if the bottom interface is far away from the main AlGaN/GaN heterojunction. In this case, the polarization charge that arises can be completely screened by the residual doping of the GaN substrate. On the contrary, if such an interface is close to the AlGaN/GaN heterojunction, the polarization charge can completely deplete the channel. In our simulations,

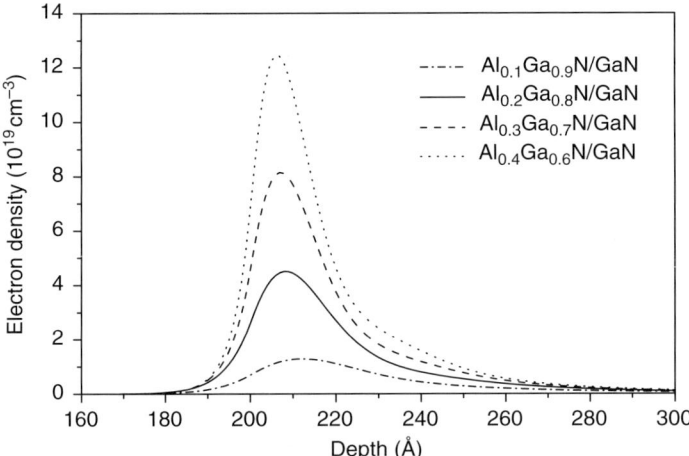

FIG. 26. Electron density distribution in the channel of the [0001] grown N-MODFET for $V_G = 0$ for several Al contents of the $Al_xGa_{1-x}N$ layer.

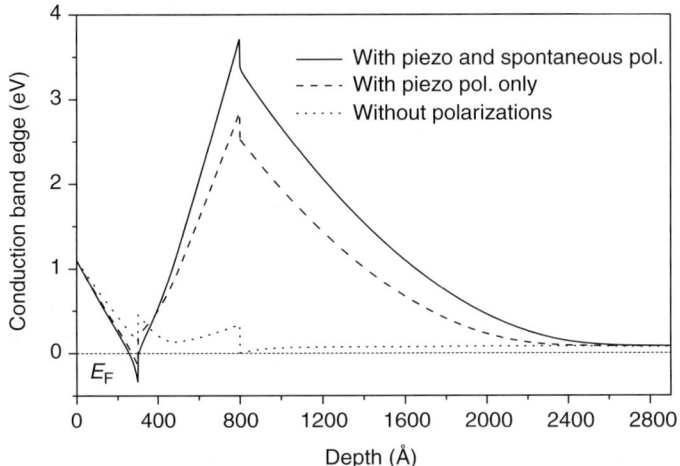

FIG. 27. Conduction bandedge for the I-MODFET structure grown in the $[000\bar{1}]$ direction for $V_G = 0$ with and without polarization fields.

we have considered a thick GaN substrate. Thus, the effect of the polarization charge at the end of the GaN substrate is completely screened.

The bandedge profile and electron densities for the I-MODFET grown in the $[000\bar{1}]$ direction are shown in Figs. 27 and 28, respectively. A comparison of the conduction bandedges with and without polarization charges is also plotted. As for the N-MODFET, the presence of the fixed and positive

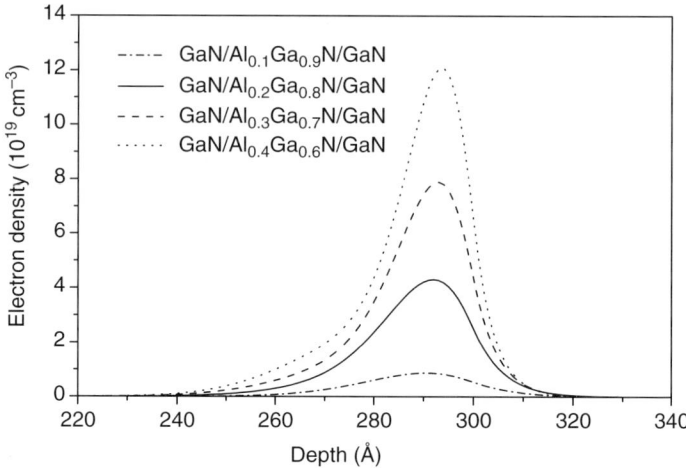

FIG. 28.   Electron density distribution in the channel of the $[000\bar{1}]$ grown I-MODFET for $V_G = 0$ for several Al contents of the $Al_xGa_{1-x}N$ layer.

polarization charge at the GaN/AlGaN interface induces the formation of a channel not present in the absence of the polarization charge. For the I-MODFET a $-\sigma$ polarization charge is also present at the end of AlGaN region (i.e., at the AlGaN/GaN interface). Similar to the [0001] grown N-MODFET, a larger Al content of the AlGaN layer induces a larger polarization charge at the GaN/AlGaN interface and consequently an increase of electron concentration in the channel. Naturally, for [0001] orientation the interface-charge forms below the AlGaN layer, which is not what is desired for an I-MODFET. What is desired is the formation of the electron sheet layer on top of the AlGaN layer, which is possible when the $[000\bar{1}]$ orientation is employed. The structure in its present shape, that is, the [0001] polarity would show FET performance provided that the AlGaN layer is completely depleted but with small transconductance. If AlGaN is not depleted, then the device would function as MESFET dominated by transport in the AlGaN layer unless the gate potential is large enough to deplete the AlGaN layer. To eliminate the formation of an interface electron charge at the bottom of the AlGaN layer, the bottom heterointerface should be graded substantially. In that case, the [0001] polarity would cause the band diagram to accumulate holes at the top interface if they are present. That top interface would accumulate electrons in the $[000\bar{1}]$ polarity.

The channel charge density is, therefore, controlled by two factors: (i) the gate bias as in traditional N-MODFET device and (ii) the Al content of the AlGaN layer, which tailors the polarization field. Charge control in nitride-based devices can be achieved by adjusting two independent parameters and thus offers a wide degree of flexibility with respect to traditional devices. This

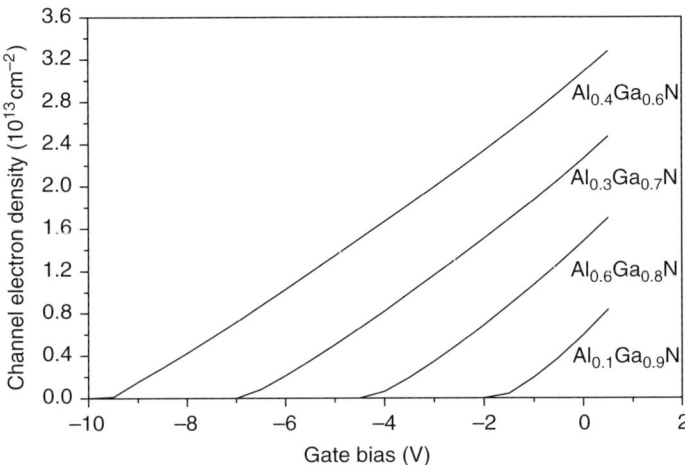

FIG. 29.   Channel electron density as a function V_G (for several AlN mole fractions) for the N-MODFET grown along the [0001] direction.

can be seen from the sheet charge concentration in the channel as obtained by integrating the electron density distribution along the z-direction. Considering the explicit dependence of the sheet charge density on the gate voltage V_G, we have:

$$n_s(V_G) = \int n(V_G, z)\, dz \qquad (5)$$

Figures 29 and 30 show the sheet electron density in the channel as a function of gate bias for several Al contents in the AlGaN layer for both [0001] grown N-MODFET and $[000\bar{1}]$ grown I-MODFET, respectively. Naturally, the channel electron density increases for larger Al contents of the AlGaN layer. We note also that the density is higher for the N-MODFET with respect to the I-MODFET because of the particulars relating to the band bending on the top interface of the I-MODFET. As mentioned earlier, the I-MODFET structure is intended to be used with $[000\bar{1}]$ orientation for investigative purposes only as the body of work in the AlGaAs/GaAs system showed the N-MODFET to be the desired device structure.

2.   Drain Current Model in MODFETs

We have implemented a quasi-2D (Carnez *et al.* 1980; Sandborn *et al.* 1987; Snowden *et al.* 1989) model for the calculation of the current–voltage characteristics of the nitride MODFETs. This model makes use of the exact value of the sheet charge density in a MODFET device channel, obtained from the self-consistent Schrödinger–Poisson solution presented above.

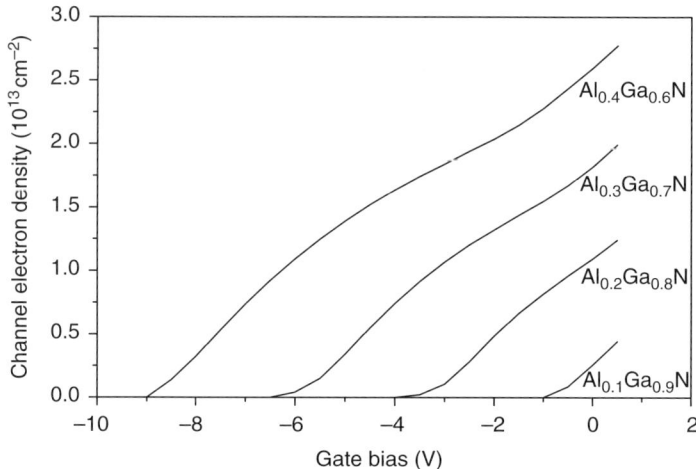

FIG. 30.   Channel electron density as a function V_G (for several AlN mole fractions) for the I-MODFET grown along the $[000\bar{1}]$ direction.

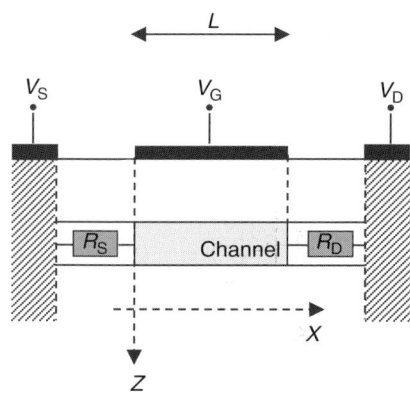

FIG. 31.   Schematic representation of the quasi-2D FET model used.

We assume a FET model shown in Fig. 31 where the x-axis is along the channel and the z-axis is along the growth direction. The model also considers the presence of a drain (R_D) and source (R_S) resistance. When a drain bias (V_D) is applied, the potential along the channel may be considered as varying gradually from the source bias (V_S) to V_D. In this situation, it is still possible to calculate the sheet charge density n_s at every section grid, provided that one considers the $V(x)$ potential (on the top surface) for each point of the channel. Since for n-channel devices V_D is positive and V_S is zero, $V(x)$ contributes to the channel depletion and the sheet charge density n_s for the generic x section of the FET will, therefore, be

$$n_s(x) = n_s(V_G - V(x)) \tag{6}$$

By neglecting diffusion contributions, the source-to-drain current I_{DS} is given by

$$I_{DS} = -qWv(x)n_s(x) \tag{7}$$

where W is the gate width and $v(x)$ the electron mean velocity, supposed independent of the transverse coordinate.

The dependence of the drift velocity on the longitudinal electric field is empirically given by

$$v(x) = \frac{\mu_0 F(x)}{1 + (F(x)/F_C)} \tag{8}$$

where $F(x)$ is the electric field ($= -dV(x)/dx$), μ_0 the low-field mobility and $F_C = v_{sat}/\mu_0$ the electric field at saturation. Parasitic components are included explicitly through the drain and source resistances (R_D, R_S)

$$\begin{cases} V_S^e = V_S + I_{DS}R_S \\ V_D^e = V_D - I_{DS}R_D \end{cases} \tag{9}$$

where V_D^e and V_S^e represent the effective bias boundaries of the gate region on the drain and source sides, respectively. For a certain value of I_{DS}, we can calculate V_D by solving Eq. (7). The explicit equation for the current is

$$I_{DS} = qW \frac{\mu_0 F(x)}{1 + (F(x)/F_C)} n(V_G - V(x)) \tag{10}$$

The numerical solution is based on the discretization of this expression into N sections, each with amplitude h, so that $Nh = L$, where L is the gate length. Given the $(i-1)$th section potential, the ith potential $V_i = V_{i-1} + F_i h$ where F_i is the ith section electric field. We have, then, the N relations:

$$I_{DS} = qW \frac{\mu_0 F_i}{1 + (F_i/F_C)} n(V_G - V_{i-1} - F_i h) \tag{11}$$

Since the $(i-1)$th section potential is known from the previous step, this is a non-linear equation in the unknown F_i. Solving iteratively for all the N sections, one obtains the value of the drain voltage V_D consistent with the assumed current.

Repeating this procedure for a suitable range of values of I_{DS}, one obtains the set of corresponding values of V_{DS} and thus the MODFET I–V characteristics, which are elaborated on below.

3. *I–V* CHARACTERISTICS

In this section, we discuss the simulated *I–V* characteristic of the normal and inverted MODFETs, obtained for a gate length of $L = 0.3\,\mu m$. We have chosen a drain and source contact resistivity of about $1\,\Omega\,mm$, which is consistent with experimentally measured values on these types of devices (Morkoç 1999b). We use a saturation velocity of $2.5 \times 10^7\,cm/s$ (Bhapkar and Shur 1997), while for the low-field mobility we choose a value of $\mu_0 = 1100\,cm^2/V\,s$, slightly higher than the GaN bulk value, according to the experimental and theoretical results for similar devices (Yu *et al.* 1996; Bhapkar and Shur 1997; Oberhuber *et al.* 1998; Murphy *et al.* 1999).

In Fig. 32, we show the I_{DS} vs V_{DS} for the [0001] polarity MODFET for several gate (V_{GS}) voltages. The results are presented for both $x = 0.2$ [Fig. 32(a)] and $x = 0.4$ [Fig. 32(b)] Al concentration of the top layer. For $x = 0.2$, the MODFET reaches pinch-off for a bias voltage of $V_{GS} = -4.4\,V$ while for $x = 0.4$ the pinch-off is reached at $V_{GS} = -9.5\,V$. On the other hand, the saturation drain current for $x = 0.2$ is $I_{DS} = 2.4\,A/mm$ at $V_{GS} = 0$ and it increase up to $5.76\,A/mm$ for an Al content of $x = 0.4$. Thus, the current flowing in the devices depends strongly on the Al content of the top layer. This is essentially due to the increasing of the channel electron density induced by the increase of the polarization charge going from an Al content of 0.2 up to 0.4. This peculiarity of the MODFET should be considered in the design of these devices since fluctuation of the alloy composition of the top layer may induce large variation with respect to nominal electrical values of the device. It should also be pointed out that the gate leakage would determine the extent of gate voltage that can be applied to the gate. For a gate bias of 9.5 V and AlGaN layer thickness of 20 nm, the vertical field under the gate near the source can reach $4.75\,MV/cm$. This means that MODFETs utilizing large mole fractions of Al may require thin AlGaN layers or recessed gates to keep the gate voltage smaller.

A similar situation is obtained for the I-MODFET with the GaN/AlGaN/GaN structure grown in the [000$\bar{1}$] direction, meaning with N polarity. The calculated I_{DS} vs V_{DS} characteristics are reported in Fig. 33(a) and (b) for $x = 0.2$ and $x = 0.4$ Al composition of the AlGaN layer, respectively. Also in this case the pinch-off bias depends critically on the Al composition and varies from $-3.9\,V$ for $x = 0.2$ up to $-9\,V$ for $x = 0.4$. Saturation currents are lower for the I-MODFET at $x = 0.2$ with respect to the equivalent N-MODFET structure. Such difference, however, is negligible for the case with $x = 0.4$.

4. EXPERIMENTAL CONSIDERATIONS

GaN-based FETs are intended primarily for power application at high frequencies. Consequently, traditional small-signal considerations have to be

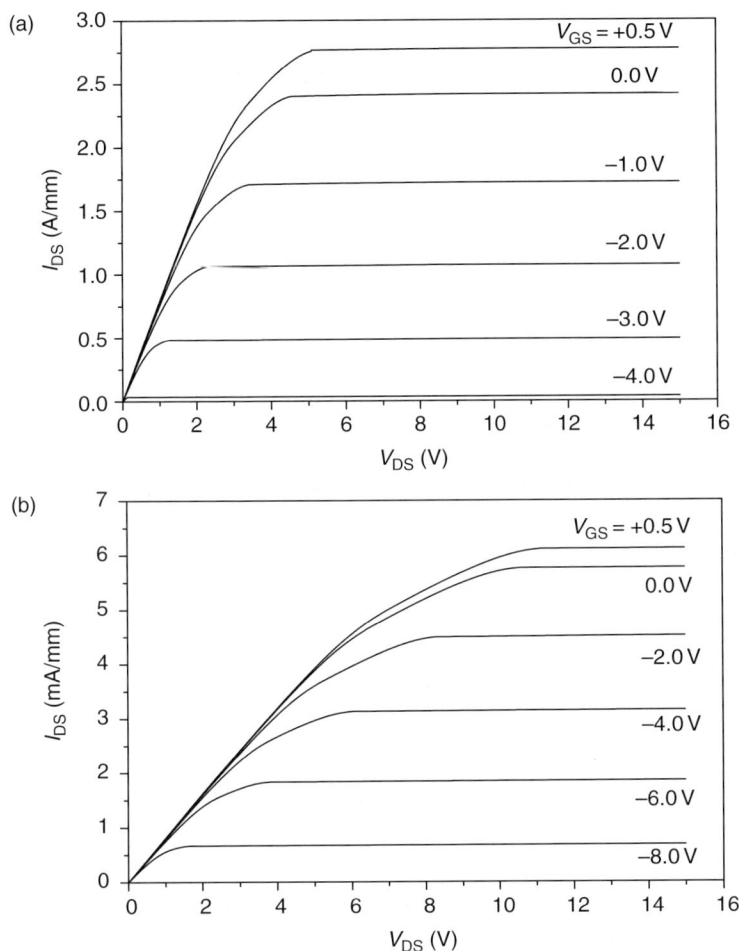

FIG. 32. I_{DS} vs V_{DS} for the [0001] polarity N-MODFET for several gate (V_{GS}) voltages and for (a) $x = 0.2$ and (b) $x = 0.4$.

augmented by large-signal-specific issues. The main parameter facing a power device is the maximum power level that can be obtained and the associated gain. In many applications, the noise figure of the device must also be considered. In simple terms, if the device has large drain breakdown voltage, high gain at high frequencies, and high drain efficiency, the stage is set for a desirable device. Even in a well-designed device in a semiconductor with all the accolades, the thermal wall is a very formidable one. Thus, it is imperative that the effect of temperature on device performance is accounted for accurately. As in small-signal modeling, the first step in power modeling is to establish the basic device geometrical factors that are needed to calculate

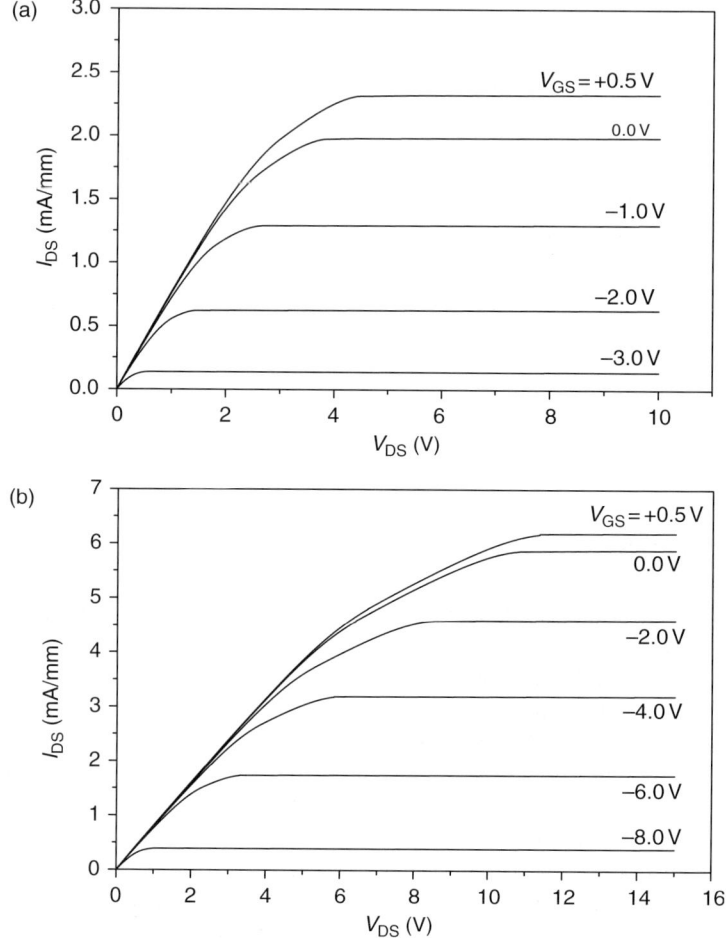

Fig. 33. I_{DS} vs V_{DS} for the $[000\bar{1}]$ polarity I-MODFET for several gate (V_{GS}) voltages and for (a) $x = 0.2$ and (b) $x = 0.4$.

the current–voltage characteristics. Once these are known, the output characteristics superimposed with the load line can be used to estimate the power level that can be obtained from the device provided that it is not limited by the input drive as shown in Fig. 34. In Class-A operation, the maximum power that can be expected from the drain circuit of a device is given by

$$P_{max} = \frac{I_{dson}(V_b - V_{knee})}{8} \tag{12}$$

where I_{dson} is the maximum drain current (this is the drain current with a small positive voltage on the gate electrode), V_b the drain breakdown

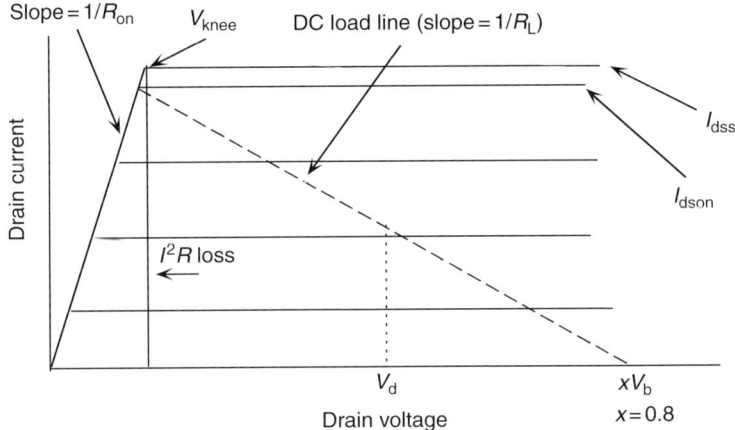

FIG. 34. Schematic representation of I–V characteristics with a load-line for a Class-A operation.

voltage, and V_{knee} the knee voltage as shown in Fig. 34. The allowable positive gate voltage (≈ 1 V) will depend on the channel doping and the work function of the gate metal. The positive gate voltage is limited by the onset of forward Schottky-diode current. The DC load line shown in Fig. 34 would be used in a Class-A RF amplifier with the maximum drain voltage $V_{\text{d}} = V_{\text{b}}/2$. The slope of the load line is $1/R_{\text{L}}$ where R_{L} is the value of the load resistance at the output of the FET. What can be gleaned from Eq. (12) is that V_{b} and I_{dson} must be made as large as possible. The utility of wide bandgap semiconductors such as GaN at this juncture is that the drain breakdown voltage is larger than that in conventional Group III–V semiconductors. In general, the drain can be swung to voltages up to within 80% of the drain breakdown for a 20% margin of safety. It should be pointed out that the maximum drain current in nitride semiconductor-based MODFETs is in the same ballpark as that of more conventional semiconductors. This implies that increased power handling capability is a direct result of large breakdown voltages and thermal conductivity and the fact that higher junction temperatures can be tolerated. Ability to increase drain bias increases the load resistance and makes it easier to impedance match, particularly in devices with large gate widths.

In power devices, power dissipation within the device increases the junction temperature and alters the output characteristics. On the one hand, higher junction temperatures with respect to the case temperature would enhance the heat dissipation to the power of four of the temperature differential, but along with it comes reduced current and increased series resistances, which, in turn, increase the heat dissipation. Moreover, the thermal conductivity of the semiconductor decreases with increased temperature, exacerbating the situation. Consequently, the effect of junction temperature on the output characteristics must be taken into consideration.

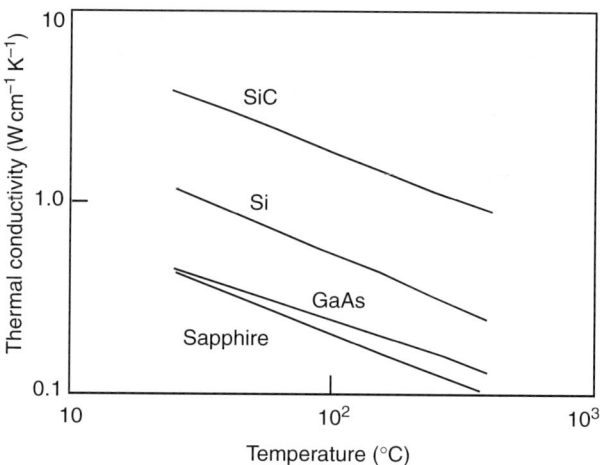

FIG. 35.   Thermal conductivity vs temperature for SiC and sapphire. After Weitzel *et al.*
(1998).

Temperature-dependent material parameters, if known, can be used to calculate the output characteristics with respect to temperature. However, a
more pragmatic approach, particularly when the aforementioned parameters
and or models required are not available, can be taken in which one measures
the output characteristics of the device under consideration as a function of
temperature. The junction temperature is critically dependent on the substrate thermal conductivity that is available for various substrates including
GaN (Morkoç, 1999a). The functional dependence of thermal conductivity
on temperature is

$$\chi(T) = \chi(T_0)(T/T_0)^{-r} \tag{13}$$

where the coefficient r is 0.559, 0.443, 0.524, and 0.544 for Si, GaAs, SiC, and
sapphire, respectively (Weitzel 1998). Thermal conductivity of sapphire, SiC,
GaAs, and Si as a function of temperature are shown in Fig. 35, where $\chi(T_0)$
has also been appropriately reduced to account for the doping of the substrate material.

5.   SCHOTTKY BARRIERS FOR GATES

Any semiconductor device requires metal contacts and MODFETs are no
exception. These devices require ohmic source and drain contacts as well as a
rectifying Schottky barrrier for controlling the charge in the channel.
Schottky barrier-related processes for GaN-based devices are nascent, but
rapid progress is being made. Until recently, it has been difficult to fabricate

good-quality single-crystal films on which a Schottky metal could be deposited, and upon which the properties of Schottky barriers could be studied. However, considerable progress has been made with Pt–GaN Schottky barriers (Mohammad *et al.* 1996; Suzue *et al.* 1996), which have been successfully implemented in GaN-based MODFETs (Aktas *et al.* 1995, 1997; Morkoç 1998b).

Recent successes in growing good-quality single-crystal Group III–V GaN layers prompted the studies of fundamental electrical property of metal–semiconductor barriers on GaN. In order to determine the properties of only the metal–semiconductor junction, one must be able to model the semiconductor. Semiconductors with large defect concentrations are notorious for exhibiting parasitic processes in current–voltage and capacitance–voltage characteristics that cloud the picture. Consequently, good epitaxial layers as well as good metal–semiconductor interfaces are imperative. During the evolutionary period, while the sample quality is acceptable, temperature and frequency dependence of the capacitance–voltage characteristics and temperature-dependent current–voltage characteristics are measured and analyzed for determining effective metal–semiconductor barrier height. To get a large Schottky barrier height for rectifying metal contacts on GaN, which is imperative for low leakage, metals with large work functions such as Au and Pt (Suzue *et al.* 1996) have been explored. Hacke *et al.* (1993) have studied Schottky barriers made of Au on unintentionally doped n-GaN grown by HVPE. The forward current ideality factor was $n_{idl} \sim 1.03$ and the reverse-bias leakage current was $<10^{-10}$ A at a reverse bias of -10 V. While the current–voltage measurement indicated the barrier height to be 0.844 eV, the capacitance measurements led to a value of 0.94 eV.

Suzue *et al.* (1996) and Mohammad *et al.* (1996) have studied the Pt Schottky barriers on unintentionally doped n-GaN. Temperature-dependent current–voltage and capacitance–voltage characteristics in the range of -195 to 42°C were studied to gain insight into the current conduction mechanism. Any excess current observed is traditionally attributed to defects (generation recombination centers) and surface leakage current. The ensuing current is called the Shockley–Read–Hall (SRH) recombination current resulting from the mid-gap states. If one neglects this excess current, a barrier height of about 0.8 eV is deduced as opposed to about 1 eV deduced from the *CV* measurements. Because of the effect of excess current on the slope of the *I–V* curve, *C–V* measurements in this particular case may represent the metal barrier height. An examination of the *C–V* plots, however, indicated that under reverse-bias condition, the capacitance depended insignificantly on the sample temperature and the signal frequency as shown in Fig. 36 (a) and (b). This leads one to conclude that the density of traps has been lowered. The curves corresponding to all temperatures were largely linear, which yielded barrier heights ranging between 0.95 and 1.05 eV. Reduced capacitance with decreasing temperature is consistent with relatively deep donors.

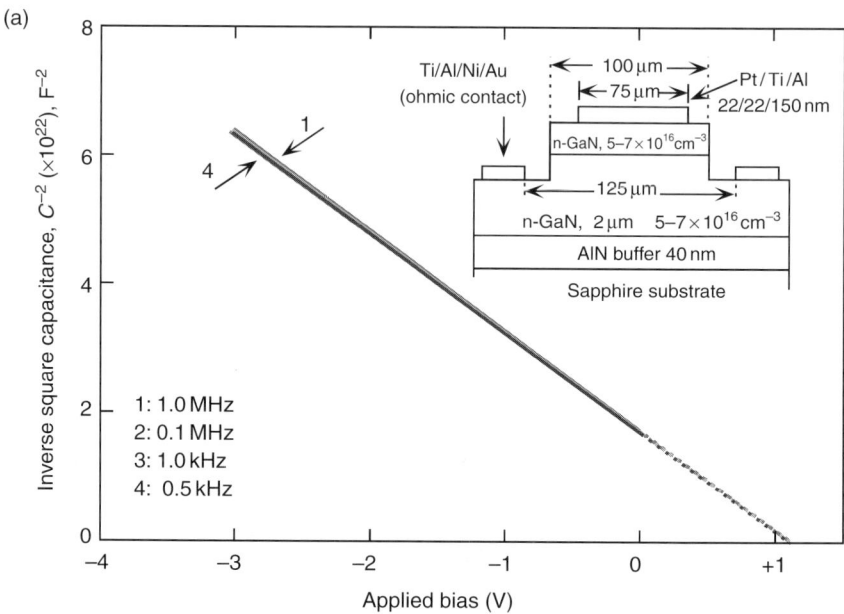

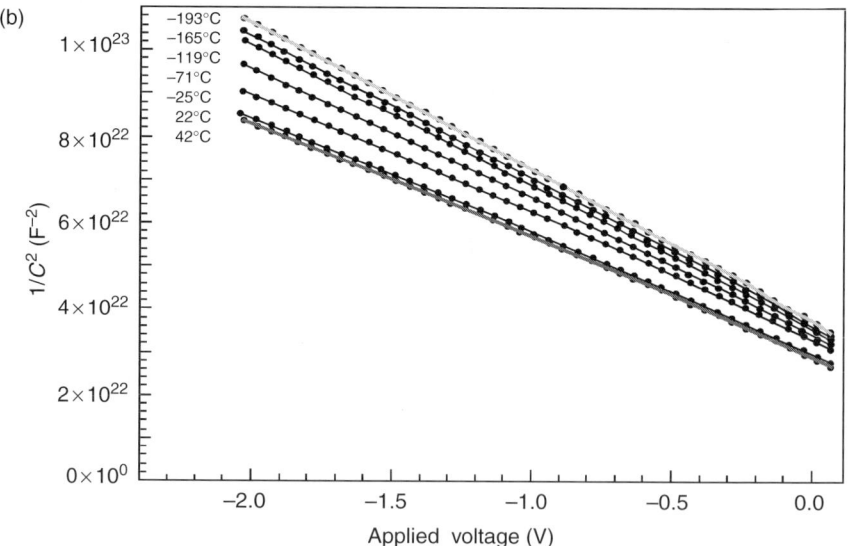

FIG. 36.   (a) Variation of the inverse square capacitance, C^{-2}, vs the applied bias, V, for various signal frequencies, ω, in Pt-GaN Schottky barriers. Curves 2 and 3 lie between curves 1 and 4. (b) Temperature dependence of C^{-2} vs the applied bias V in Pt-GaN Schottky barriers.

Binari *et al.* (1994) determined Ti Schottky barrier heights to be 0.58 and 0.59 eV from the current–voltage and capacitance measurements, respectively. The ideality factor n_{idl} is approximately 1.28. The diode series resistance (R_S) is 100 Ω.

The ternary $Al_xGa_{1-x}N$ is an essential component of nitride-based MODFETs, which makes the investigation of metal $Al_xGa_{1-x}N$ contacts imperative. M. A. Khan *et al.* (1991) reported the fabrication of a Cr/Au Schottky barrier on n-AlGaN. Moreover, M. R. H. Khan *et al.* (1997) studied the Schottky barrier characteristics of the Au–$Al_xGa_{1-x}N$ system. A typical current–voltage characteristic of an $Al_{0.14}Ga_{0.86}N$ Schottky diode had an ideality factor of 1.56 under forward bias and a threshold voltage of about 0.9 V at 0.1 A. The reverse-bias leakage current was recorded to be marginally low (10^{-10} A) for a reverse bias of −10 V. By using the current–voltage method, the barrier height and the electron affinity were determined to be 0.94 and 4.16 eV, respectively. From the C^{-2} vs V plot, the same barrier height and the electron affinity were deduced to be 1.3 ± 0.05 and 3.8 eV, respectively. As the AlGaN quality increases, more in-depth investigations must be undertaken to get an accurate picture of intrinsic parameters. In short, the current conduction mechanism in metal–semiconductor structures is strongly affected by surface and bulk states. Deviations from an ideal ideality factor, such as is the case here, indicates such states. The situation gets more complicated with AlGaN and gets worse as the AlN mole fraction is increased. Likewise, capacitance–voltage measurements also are affected by states that are charged, either at the interface or in the bulk. As is the case in many facets of research and development, insights into the metal–nitride contacts will be gained in an evolutionary manner hinging upon the developments in nitride layers.

6. Contacts to GaN

Ohmic contacts in power devices are extremely important because they affect their efficiency as well as heat dissipation. Initial inferior results helped fuel concerns that GaN-based electronic devices may not perform well. Early specific contact resistivities on n-type GaN using Al and Au metallizations (Foresi and Moustakas 1993) were in the range of 10^{-4} and $10^{-3} \, \Omega \, cm^2$. Major improvements were realized by using Ti/Au (M. A. Khan *et al.* 1993) and Ti/Al (Lin *et al.* 1994), in that specific contact resistivities in the high $10^{-6} \, \Omega \, cm^2$ were obtained with the latter. Carrying the Ti/Al contact work one step further, Wu *et al.* (1997) confirmed that, except at very high annealing temperatures, the ohmic contact suggested by Lin *et al.* (1994) functions very effectively. At very high temperatures, Al in the metal contact melts and tends to ball up, resulting in rough surfaces and increased ohmic contact resistances as pointed out already by Lin *et al.* (1994). In an attempt to circumvent this difficulty, Wu *et al.* (1997) designed a separate layer-metallization method where a realignment and deposition of a second thin Ti layer and a 2000-Å Au overlayer were carried out. Specific contact resistivities were in the range of 3.0×10^{-6} and $5.5 \times 10^{-6}\,\Omega\,cm^2$, depending on the doping concentration in the semiconductor.

In an attempt to obtain improved ohmic contacts, Fan *et al.* (1996) have designed a multilayer ohmic contact method. By using a composite metal layer of Ti/Al/Ni/Au (150 Å/2200 Å/400 Å/500 Å), they obtained very low contact resistivities. Specifically, for n-GaN with doping levels between 2×10^{17} and $4 \times 10^{17}\,cm^{-3}$, they obtained specific contact resistivities in the range of $\rho_S = 1.19 \times 10^{-7}$ and $8.9 \times 10^{-8}\,\Omega\,cm^2$, respectively. The resistance R_T between the two contacts was measured at 300 K using a four-point probe arrangement. The contact resistivity ρ_S was derived from a plot of R_T vs gap length. The method of least squares was used to fit a straight line to the experimental data. These straight lines and the actual experimental results for both alloyed and non-alloyed contacts are shown in Fig. 37. Calculation of the contact resistivity was based on the assumption that the semiconductor sheet resistance underneath the contacts remains unchanged, which is not true for non-alloyed contacts. As for the current conduction mechanism in these ohmic contacts, the large metal–semiconductor barriers diminish the possibility of thermionic-emission-governed ohmic contacts to GaN. The alternative mechanism is some form of tunneling that may take place if GaN

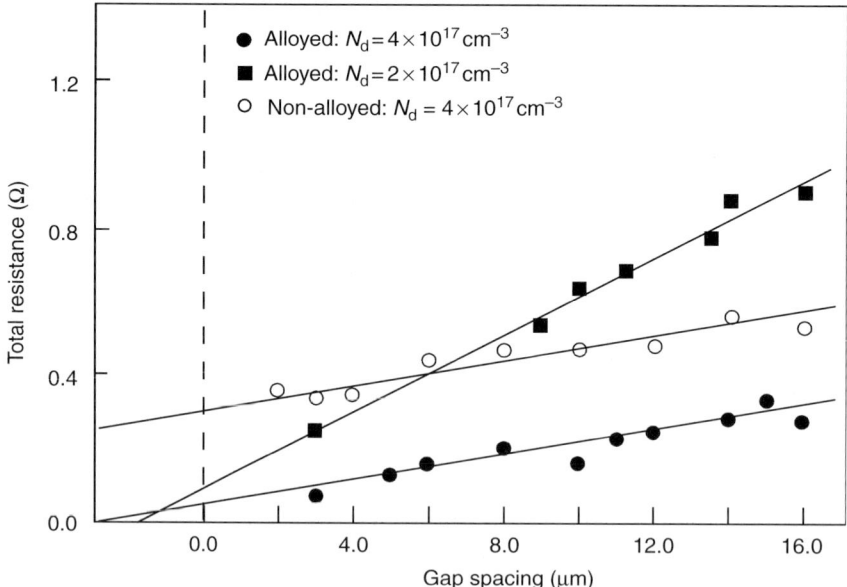

FIG. 37.   Least-squares linear regression of the total resistance between the two adjacent ohmic contact pads in multiple-layer Ti/Al/Ni/Au ohmic contacts on GaN.

is so heavily doped so as to cause a very thin depletion region. Tunneling is possible if, due to annealing (e.g., at 900°C for 30 s), Al and Ti along with Ni undergo substantial interaction with each other and GaN. Investigations showed that Ti receives N from GaN, forming a metallic layer, while the lack of N on GaN provides the desired benefit of increased electron concentration through N vacancy formation (Ruvimov *et al.* 1996). Aluminum passivates the surface and also possibly reacts with Ti to form TiAl.

X.   AlGaN/GaN MODFETs

To reiterate, MODFET's performance is due to the conduction channel that allows large sheet carrier concentrations to be maintained and its unique capacitance–voltage relationship (Moloney *et al.* 1985). Moreover, spatial separation of scattering centers (such as ionized donors) from the electrons leads to low-field transport devoid of ionized impurity scattering. As discussed in detail above, what is somewhat unique to GaN and its alloys is the spontaneous and the strain-induced piezoelectric polarization that causes redistribution of mobile and weakly bound charge and charge collected from metal contacts.

1.   Experimental Performance of GaN MODFETs

Initial GaN MODFETs utilized the background donors in the AlGaN layer, the density of which is not controllable, to say the least, and any other free and weakly bound electrons drawn to the interface. Congruent with the early stages of development and the defect-laden nature of the early GaN and $Al_xGa_{1-x}N$ layers, the MODFETs exhibited very low transconductances (on the order of 20 mS/mm), large on-resistances. In addition, they also exhibited a low-resistance state, which was relatively high to begin with, and a high-resistance state before and after the application of a high-drain voltage (20 V). As in the case of GaAs/AlGaAs MODFETs, hot electron trapping in the larger bandgap material at the drain side of the gate is primarily responsible for the current collapse. The negative electron charge accumulated because of this trapping causes a significant depletion of the channel layer, more probably a pinch-off, leading to a drastic reduction of the channel conductance and the decrease of the drain current. This continues to be effective until the drain-source bias is substantially increased, leading to a space-charge injection and giving rise to an increased drain-source current.

With improvements in the materials quality available, the transconductance, current capacity, and drain breakdown voltage are all increased to the point that GaN-based MODFETs are now strong contenders in the arena of high-power devices/amplifiers, particularly at X band and higher frequencies. As is the case for FET device structure, improved and high-resistivity buffer layers have once again played a pivotal role. For chronological purposes, a brief review of the latest class of MODFETs with high transconductances and current levels is given later in this article.

The first breakthrough in the N-MODFETs based on GaN came during 1994–1995 in the author's laboratory (Aktas *et al.* 1995). These devices with a gate length of 3 μm and gate width of 40 μm exhibited transconductances of about 120 mS/mm with low on-resistances as they sported doped AlGaN donor layers and low resistance ohmic contacts. The *I–V* characteristics of an early N-MODFET device are shown in Fig. 38. Shortly thereafter, devices with a gate length of 2 μm, gate width of 40 μm, and the drain-source separation of 4 μm exhibited drain currents of approximately 500 mA/mm and extrinsic transconductances of approximately $g_{em} = 185$ mS/mm, which are characterized in Fig. 38 (b). The drain breakdown voltage for 1-μm gate-to-drain spacing was approximately 100 V, the exact value depending on the layer design and quality of the layered structure. Soon thereafter, other laboratories achieved similar results in similar structures. What is unique with AlGaN/GaN MODFETs as compared to their GaAs counterparts is the polarization charge discussed earlier. As indicated before, the terminology of polarization, particularly the piezoelectric component, has been used rather liberally. Even terms such as "piezoelectric doping" have been coined, and very high sheet carrier concentrations observed have been ascribed to piezoelectric polarization only. We have to recognize that ultimately, regardless of the source of the carriers, the strength of the electric field that can be accommodated by the semiconductor under the gate without excessive leakage sets an upper limit on the number of carriers at the interface. Use of multi-2DEG structures is one obvious method to increase the current capability of MODFETs, and they have been employed. In those cases, the GaN layer is straddled by two doped AlGaN layers that donate electrons to the channel, thus increasing the number of electrons available for current conduction. By Hall effect measurement, the mobility and sheet carrier densities in the 2DEG were about 304 cm^2/V s and 3.7×10^{13} cm^{-2}, respectively, at room temperature. The sheet carrier concentration may have been affected by piezoelectric effect. A number of double hetero-channel MODFETs (DHCMODFETs) with gate lengths of 1.5–1.75 μm and a gate width of 40 μm have been reported (Fan *et al.* 1997).

The maximum drain saturation current I_{DS} corresponding to a drain–source voltage $V_{DS} = 7$ V and gate–source voltage $V_{GS} = 3.5$ V in a DHCMODFET is about 1100 mA/mm, which is important because in high-power devices, the input is momentarily forward biased. The DHCMODFET has a room-temperature extrinsic transconductance of $g_m = 270$ mS/mm. The value of the total resistance R_T extracted from the linear region of

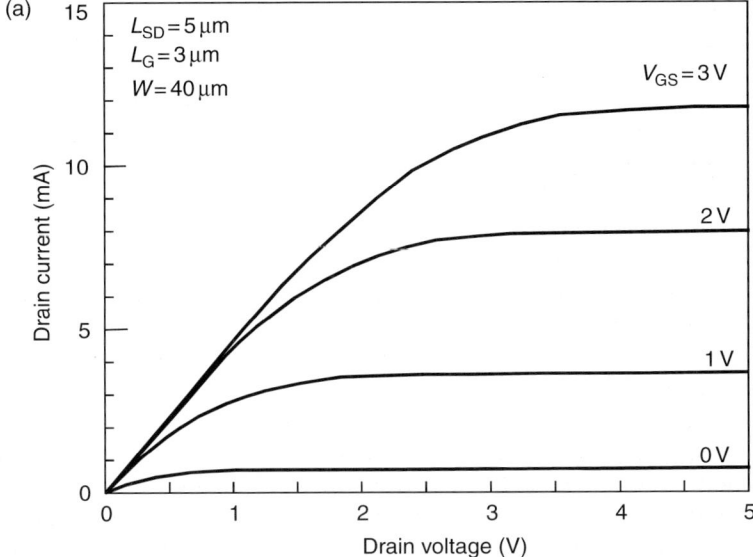

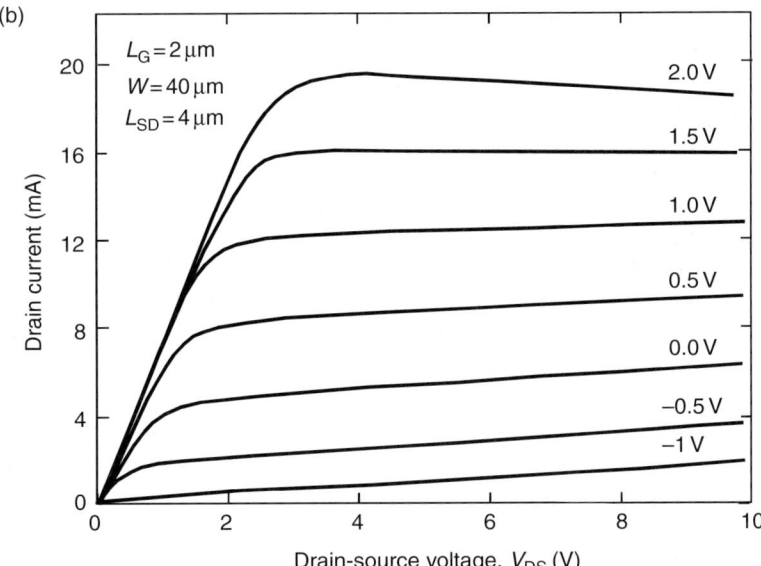

FIG. 38. (a) Output *I–V* characteristics of the first GaN MODFET, which exhibited respectable performance. The 3-μm gate device had a gate length of 3 μm and sported low-resistance ohmic contacts and low-leakage Schottky barriers. (b) Output *I–V* characteristics of a GaN MODFET on sapphire with a 2-μm gate, which exhibits negative output conductance due to thermal effects associated with the relatively low thermal conductivity of sapphire.

the $I-V$ curves is $4\,\Omega/mm$. Near pinch-off, the drain breakdown voltage is about 80 V, indicating excellent power potential of the device. These measurements were made in a nitrogen-pressurized container to avoid possible oxidation of the contacts and probes. The maximum drain–source current and extrinsic transconductance of the DHCMODFET are 500 mA/mm and 120 mS/mm, respectively. These devices maintain reasonable output characteristics at temperatures as high as 500°C with maximum drain current and extrinsic transconductance values of 380 mA/mm and 70 mS/mm, respectively. Cooling to room temperature restored the characteristics, which demonstrates the robustness of this material system and of the metallization employed. It should be noted, however, that high-power operation requires large drain breakdown voltages with the added benefit of having large output resistances, which ameliorates impedance matching.

The heat dissipation is a major problem, however, in GaN MODFETs on sapphire substrates as the thermal conductivity of this substrate is about 0.3 W/cm K (may even be somewhat lower). To make matters worse, the thermal conductivity decreases rapidly as the temperature increases. Consequently, devices show a decreasing drain current (negative differential output conductance) as the drain bias is increased, and, needless to say, the power performance is degraded. To overcome this, one must either remove the sapphire substrate followed by mounting the structure on a substrate with better thermal conductivity, employ flip chip mounting, or grow the structure on a substrate with better thermal conductivity. Among the substrates with better thermal conductivity are Si and in particular SiC. Layers on Si, however, are not of as high quality as one would like, which leaves SiC substrates, which are expensive and suffer from inferior surface characteristics (or smoothness) due to the hardness of SiC. Early attempts in the author's laboratory to grow GaN layers on SiC met with difficulty due to the surface damage roughness, though occasionally very high mobility could be obtained (Lin et al. 1993b).

Two approaches can be employed to remove the surface damage. One is the mechanical/chemical polish, which is very slow in coming, and the other is etching in H and Cl environment at very high temperatures such as 1300°C. GaN layers grown on H treated SiC at high temperature exhibited much lower defect concentration as compared to SiC treated with wet chemical, and wet and dry chemical treatment (Ruterana et al. 1997). The H-cleaning process has been adopted as a standard procedure for MBE growth of GaN on SiC (Lee et al. 2001) with cleaning temperature of about 1700°C. The author in collaboration with the group of E. Janzen at Linköping University was able to H-etch Leyl SiC followed by MODFET growth. These devices did not exhibit the negative differential resistance characteristic of the sapphire substrates. However, SiC substrates prepared by the sublimation method did not appear to survive this high-temperature H-etching process. Researchers have exploited the *in situ* H-etching process

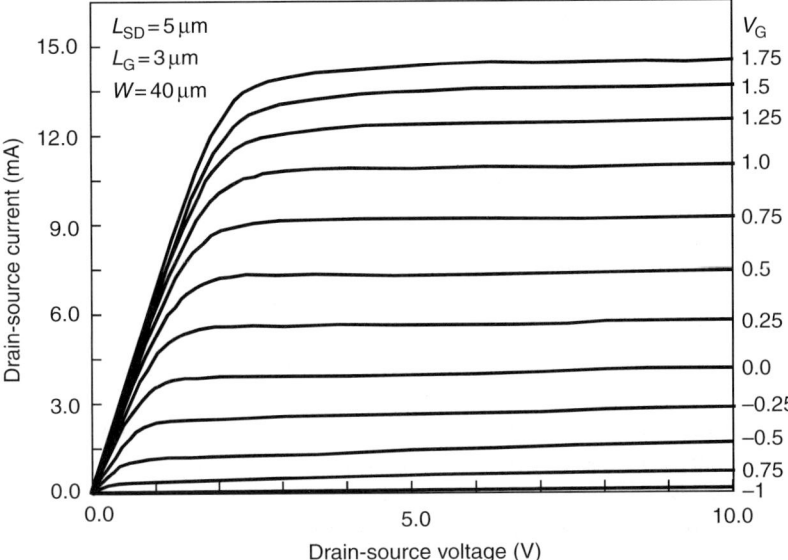

FIG. 39. Output characteristics of a 3-µm gate AlGaN/GaN MODFET grown on Leyl SiC substrate that is void of the output negative conductance. However, the Leyl substrates are highly conductive and not well suited for FETs due to RF shorting/loading. Nevertheless, experiments of this kind serve to prove the point that the negative output conductance observed in devices on sapphire are most likely of thermal origin.

(Powell *et al.* 1998) and HCl-etching process (Powell *et al.* 1994). Reports detailing these processes and their effects have appeared in the open literature already. The *I–V* characteristics of the particular device prepared in 1996 on SiC substrates are shown in Fig. 39. These results were reported in meetings dealing with the development of high-power devices and the case was made for SiC substrates as intrinsically being better for GaN power MODFET applications. Several groups participating in those meetings expended a good deal of effort on SiC substrates with initially comparable high-power performance to that on sapphire. Other groups later propelled GaN MODFETs on SiC substrate to their pinnacle with outstanding performance as will be discussed below. It should be pointed out that the pitch of gates for a power FET on a substrate with very good thermal conductivity can be made smaller than on a substrate with inferior thermal conductivity. Consequently, the chip size can be made much smaller, in addition to other advantages.

MODFETs have progressed to a point where microwave performance has been established for a variety of devices with gate lengths as wide as 2 µm and as narrow as about 0.2 µm. To appreciate the rapid development of the device, its evolution will be succinctly discussed. A typical MODFET structure with 2-µm gate length has been tested for small-signal S-parameters performed at bias conditions used for the power measurements (i.e., 15 V, −2.5 V, and 20 mA for the drain voltage, gate voltage, and drain current, respectively). The unity current gain cut-off frequency (f_T) and maximum frequency of oscillation (f_{max}) were 6 and 11 GHz, respectively, at both 15 and 30 V bias. Values of f_T and f_{max} in excess of 50 and 100 GHz have been reported for short-channel (about 0.2 μm) devices, respectively. As touched upon earlier, devices on sapphire substrates suffer from the low thermal conductivity of sapphire substrates and exhibit negative differential resistance in the output characteristics. Remedies include better heat sinking by flip-chip mounting and the use of high resistivity 4H-SiC substrates, which provide good thermal conductivity but are hard to obtain.

GaN MODFET devices that have been prepared in the author's laboratory on conducting 6H-SiC substrates exhibited output characteristics that lacked the negative resistance (i.e., they exhibited good heat sinking). There have subsequently been a few reports of MODFET power devices on high-resistivity SiC (Binari et al. 1997b; Sullivan et al. 1998; Sheppard et al. 1999) and p-type SiC (Ping et al. 1998) substrates with phenomenal improvement in power-handling capability notwithstanding the rapid progress on sapphire substrates. On sapphire, recent 0.7-μm gate-length $Al_{0.5}Ga_{0.5}N/GaN$ MODFETs exhibited a current density of 1 A/mm, three-terminal breakdown voltages up to 200 V, and CW power densities of 2.84 and 2.57 W/mm at 8 and 10 GHz, respectively, representing a marked performance improvement for GaN-based FETs.

Increasingly outstanding power levels are being achieved with near-half-micrometer or smaller gate lengths. To follow the evolution of the developments, a few examples are cited here. With 0.7-μm gate length devices on SiC substrates, where the gate-source spacing and gate-drain spacing were 0.5 and 0.8 μm, respectively, a total output power of 2.3 W in a device with a 1.28-mm gate periphery has been obtained (Sullivan et al. 1998). The power gain at the 2.3-W output power point was 3.6 dB with a power added efficiency (PAE) of 13.3% for a drain bias of 33 V. The current and power gain cut-off frequencies were 15 and 42 GHz, respectively. The contact resistance, though not the best, was between 2.6 and 3.5 Ω mm. The maximum normalized transconductance was 270 mS/mm and the drain current was 293 mA/mm.

Steady improvement in power performance has led to very recent results at HRL laboratories with record-breaking performance (Micovic et al. 2001). The output I–V characteristics of a 250-nm gate length AlGaN/GaN MODFET device of HRL on SiC is shown in Fig. 40. Typical DC characteristics include 600 mA/mm current performance and > 60 V drain breakdown voltage. The small signal current and power gains as a function of frequency of another device are shown in Fig. 41. The current gain cut-off and maximum power gain cut-off frequencies measured were about 48 and 100 GHz, respectively, for −5.5 and 12.5 V gate and drain bias voltages, respectively. The minimum noise figure and the associated gain of the device

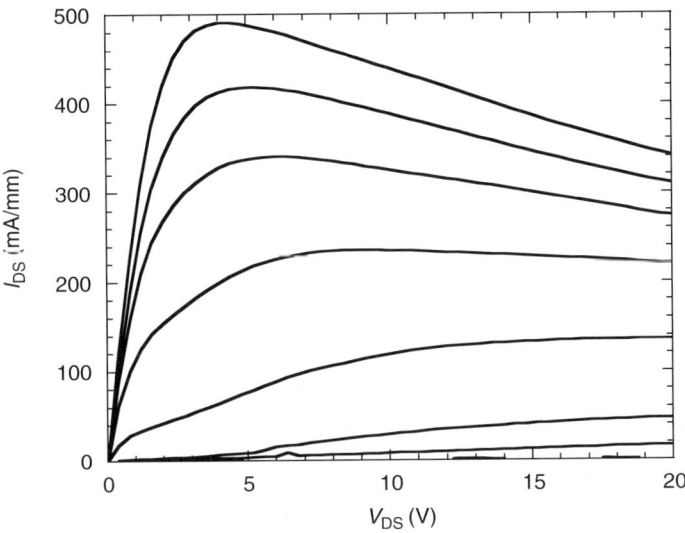

FIG. 40.   The output *I–V* characteristics of a 0.25-μm gate AlGaN MODFET on sapphire fabricated at HRL laboratories. Note the negative differential output conductance due to the poor thermal conductivity of sapphire substrate. Courtesy of Drs N. X. Nguyen and C. Nguyen.

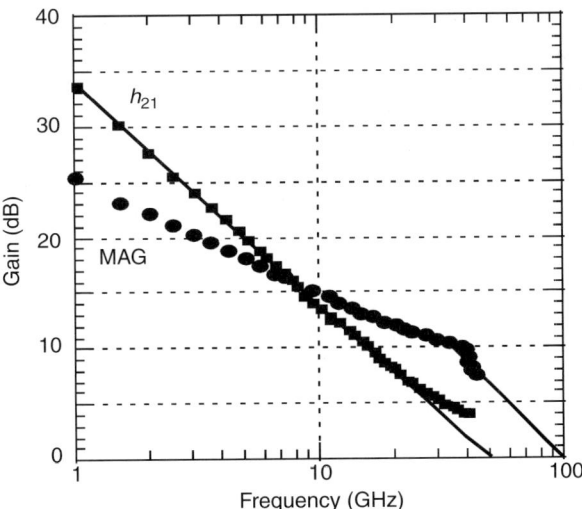

FIG. 41.   Small signal current and power gains as a function of frequency of a 0.25-μm gate AlGaN MODFET on sapphire fabricated at HRL laboratories. Courtesy of Drs N. X. Nguyen and C. Nguyen.

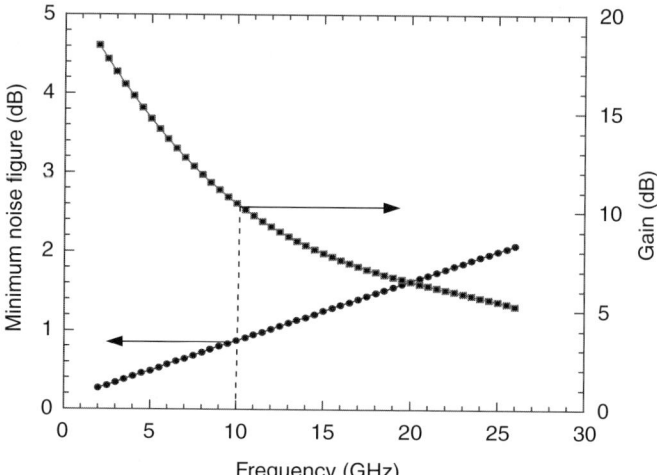

FIG. 42. The minimum noise figure and the associated gain of a 0.25-μm gate AlGaN MODFET on sapphire fabricated at HRL laboratories. Courtesy of Drs N. X. Nguyen and C. Nguyen.

whose *IV* characteristic were mentioned above are shown in Fig. 42. A minimum noise figure of 0.85 dB at 10 GHz with an associated gain of 11 dB is simply remarkable.

At HRL laboratories, 6.3 W of CW output power was obtained at 10 GHz from a 1-mm wide transistor device. More importantly, the power density remained nearly constant as the device size was scaled upward from 0.1-mm width, where the device exhibits 6.5 W/mm, to 1.0 mm. These record-setting transistors were epitaxially grown AlGaN/GaN heterostructures on semi-insulating SiC (silicon carbide) substrates by MBE. HRL has developed a growth process using molecular beam epitaxy (MBE) that has virtually eliminated material defects common to other reported GaN devices, thereby enabling the scaling. MBE growth also produces device characteristics with less than 5% standard deviation over the 2-in. diameter SiC substrate, a six-fold improvement over previously reported results.

As will be discussed in the amplifiers section, the researchers at HRL laboratories have expanded their work to amplifiers with several cells and showed very good power scalability up to 2 mm of total gate periphery (Micovic *et al.* 2001). Using 250-nm gate devices, a CW output power of 22.9 W with an associated power-added efficiency of 37% was measured for an amplifier at 9 GHz with four 1-mm gate periphery devices. Furthermore, Micovic *et al.* (2001) also showed a CW power density of 4 W/mm at 20 GHz, which is the state-of-the-art figure for any three-terminal solid-state device at this frequency.

TABLE IV

THE CURRENT GAIN CUT-OFF AND MAXIMUM OSCILLATION FREQUENCIES VS
GATE LENGTH (AFTER MICOVIC *ET AL.* 2001)

Gate length (nm)	Current gain cut-off (f_T, GHz)	Maximum oscillation (f_{max}, GHz)
250	155	100
150	80	120
50	110	>140

In contrast to the earlier devices whose dc characteristics were mentioned before, the newer devices had a maximum drain current density exceeding 1.4 A/mm, and the peak transconductance of the devices biased at V_{DS} of 15 V was 250 mS/mm. The reverse-bias gate to source breakdown voltage of the devices measured at 1 mA/mm of gate leakage current typically exceeded 80 V. Small-signal RF performance of MODFETs as characterized in the 0.5–40.5-GHz range and the cut-off frequencies estimated will be discussed below. The best results are shown in Table IV for various gate lengths of 200-μm wide devices.

Continuous wave power measurement of 0.1-, 1-, and 2-mm devices was performed at 10 GHz using a load-pull system (Micovic *et al.* 2001). The gate length of the particular device subjected to this particular test was 25 nm. Maximum output power levels of 0.65, 6.3, and 10.5 W were measured for devices with 0.1, 1, and 2 mm total gate periphery, respectively, and scale nearly linearly. Figure 43 shows the output power vs input power of a 2-mm device. Measurements at higher frequencies were also made to determine the device response in terms of its power performance. Using a series of 150 nm × 200 μm gate devices, a CW output power density of 4 W/mm at 20 GHz was obtained. The results of load-pull measurements at 20 GHz are shown in Fig. 44. These are highest reported data to figure for a three-terminal solid-state device at this frequency.

2. POWER AMPLIFIERS

Power-combining single-stage X band power amplifiers have been reported using four 1-mm gate periphery devices mentioned above (Micovic *et al.* 2001). The model used for circuit design was extracted from dc and RF device characteristics (Micovic *et al.* 2001). Parameters needed for modeling were deduced from the small-signal s-parameters in the form of a standard equivalent circuit. Commercial parameter extraction tools that are available can be used for this purpose. The optimum load impedance for a maximum power-combining efficiency was determined by load-pull measurements with

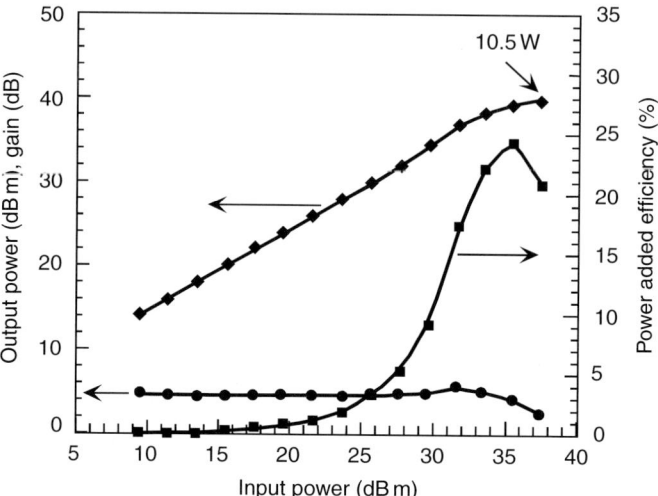

FIG. 43.   Large signal characteristics of 2-mm wide device at 10 GHz. The maximum CW output power of this device was 10.5 W. After Micovic *et al.* (2001).

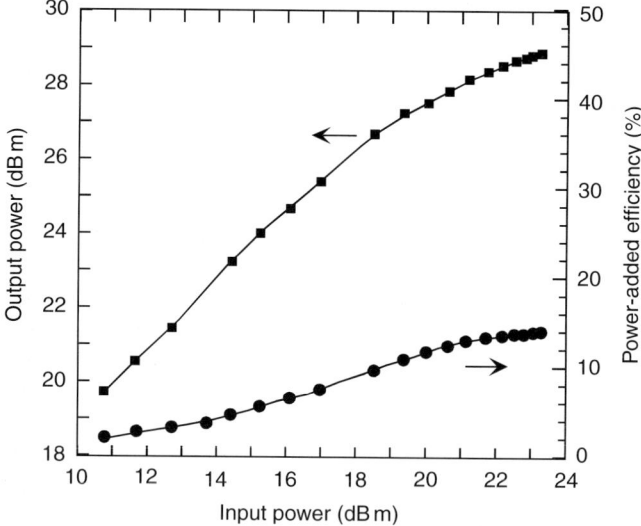

FIG. 44.   Large signal characteristics of 0.15 μm × 200 μm GaN MODFET at 20 GHz. The maximum CW power density of this device was 4 W/mm. After Micovic *et al.* (2001).

the design goals of matching into 50-Ω ports, having a peak power gain of 8 dB at 9 GHz, and 3 GHz bandwidth. Input and output matching networks were built using discrete capacitors and sapphire microstrip elements. Transistors were wire bonded to matching networks. The fixture was

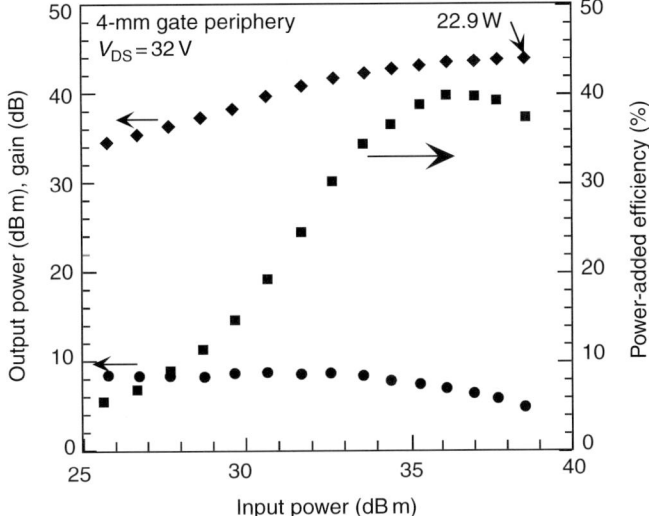

FIG. 45.   Power performance of a single-stage GaN MODFET power amplifier utilizing four discrete 1-mm device cells. Source pads were interconnected by an abridged technique to reduce the source resistance as well as the source inductance. The peak output power of the amplifier is 22.9 W. After Micovic *et al.* (2001).

mounted onto a water-cooled heat sink for heat dissipation (Micovic *et al.* 2001). The power performance of the amplifier at 9 GHz is shown in Fig. 45. A CW output power of 22.9 W with an associated power-added efficiency of 37% was measured for an amplifier at 9 GHz with four 1-mm gate periphery devices.

In power devices, the thermal limitation can never be eliminated completely as is the case in nitride devices; particularly when fabricated on sapphire substrates with a thermal conductivity of only approximately 0.3 W. Inclusion of thermal limitations leads to results shown in Fig. 46 for devices that compete in the high-power device arena (Morkoç 1998b; Weitzel *et al.* 1998). Since new device developments do, in general, compete with existing and alternative technologies, a brief account of competing technologies for power arena is given below. The Si metal semiconductor FET (MESFET) analytical curve, modeled for its simplicity, is slightly above the SiC analytical curve and indicates a maximum power density of 0.35 W/mm at $V_{DS} = 7$ V, which is slightly lower than 0.39 W/mm. Since Si RF MESFETs are unavailable, commercial Si RF metal-oxide semiconductor FET (MOSFET) results were instead used for comparison. At low voltages, the Si MOSFET data parallel the analytical curve, suggesting the validity of the functional dependence of power density on drain voltage. Also shown are two higher-power density data points 0.4 W/mm, $V_{DS} = 28$ V and 0.87 W/mm, $V_{DS} = 48$ V. These higher-power densities were obtained with specially

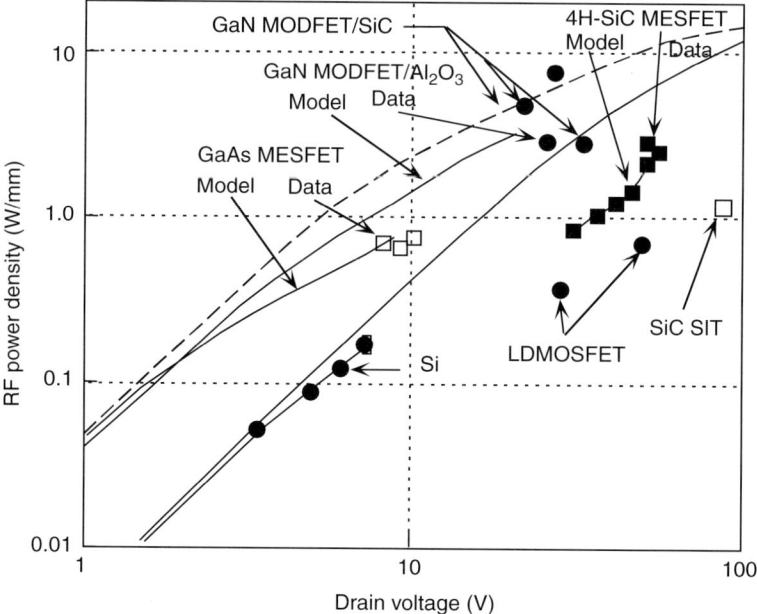

FIG. 46. Simulated and experimental RF power density data for Si, GaAs, SiC, and GaN FETs. After Weitzel *et al.* (1998).

designed RF power MOSFETs that incorporate a lightly doped drain and field plates that significantly increase the breakdown voltage.

The GaAs analytical curve shows the highest power density of all of the devices at the lowest voltages primarily because of the higher electron mobility of GaAs. However, the low-breakdown field limits the GaAs MESFET's drain voltage to about 8 V and power density to 0.63 W/mm including thermal effects. Typical commercially available GaAs MESFET power densities are below 1 W/mm. However, high-performance GaAs FETs with more complex device cross-sections have achieved power densities as high as 1.4 W/mm at 18 V. At 100 V, the SiC MESFET has calculated maximum power densities of 7.96 W/mm with thermal effects and 9.7 W/mm without thermal effects. The highest demonstrated CW power density 3.3 W/mm ($V_{DS} = 50$ V) for a SiC MESFET (Moore *et al.* 1997) is also shown for comparison. Additional SiC data again illustrate the functional dependence of power density on drain voltage. The GaN analytical results are highly dependent on the thermal conductivity of the substrate. With a sapphire substrate, the device is severely thermally limited to 2.24 W/mm at 30 V with a resulting channel temperature of over 400°C. With a SiC substrate, however, the analysis predicts that a GaN MODFET could achieve 15.5 W/mm at 100 V with a channel temperature of about 300°C. We should caution that while the power density figure can be used during the evolution process, eventually the total power figure must prevail. Normalized power density measurements, though frequently reported (a trap the present author also fell into), are often misleading because smaller gate widths naturally lead to larger power densities. This experimental datum point is actually slightly higher than the simulated result, possibly because of the very small size of the experimental device (100 μm width).

The GaN results of analytical models are highly dependent on the thermal conductivity of the substrate. With a sapphire substrate, the device is severely thermally limited to 2.24 W/mm at 30 V with a resulting channel temperature of over 400°C. However, with a SiC substrate (Weitzel *et al.* 1998), the analysis predicts that a GaN MODFET could achieve 15.5 W/mm at 100 V while keeping the channel temperature at about 300°C. The key to further improvements lies with our ability to control the polarity of the films, to prepare inversion domain free material, and to reduce defects. If the past few years are any indication, substantial progress is in the wings.

3.   Anomalies in GaN/AlGaN MODFETs

Field effect transistors, in general, and modulation doped field effect transistors, in particular, exhibit anomalies in their output *I–V* characteristics. Among the causes of these anomalies are channel carriers being trapped in the wide bandgap material and bulk, meaning the buffer layers. In addition, surface states, if present and not passivated, could act to reduce the sheet conducting charge, particularly between the gate and the drain region of the FETs (Binari *et al.* 2001). This is depicted schematically in Fig. 47. Some AlGaAs/GaAs MODFETs, which are the predecessors of the current AlGaN/GaN MODFET, exhibit behavior similar to what was then termed as "current collapse" (Fischer *et al.* 1984). This behavior was attributed to carrier injection from the channel to the AlGaAs at reasonably high fields where they are trapped at low temperatures. With below the gap light

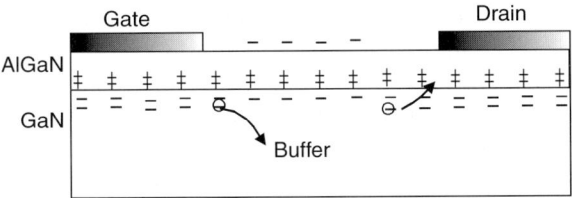

FIG. 47.   Schematic representation of an AlGaN/GaN MODFET indicating how the surface charge and charge injection into traps in the buffer layer and defects in AlGaN could serve in depleting conducting channel carriers. For simplicity, the 2DEG is assumed to be due to donors placed in AlGaN. In addition, the regions from which carrier injection takes place are limited to the spacing between the gate and drain electrodes, the exact location is arbitrarily chosen.

excitation, increasing temperature, and exchange of the source and drain terminals, the effect could be eliminated. The GaAs buffer layer for the AlGaAs/GaAs-based device is of high quality so that its trapping effect was not dominant. The surface states in the AlGaAs/GaAs device were not deemed to have a profound effect on the current–voltage characteristics. However, it is always a prudent approach to passivate the surface states, as was done in the AlGaAs/GaAs device, as they affect the device operation with time greatly. In the AlGaAs/GaAs variety, the trapping effect in the AlGaAs barrier was attributed to DX levels, which are caused by lattice distorting defects, which causes massive change in the bandgap of the semiconductor at the local level; and their behavior could be described by a lattice coordination diagram. Since this effect reduced with lowering the AlAs mole fraction, AlGaAs/InGaAs pseudomorphic modulation doped FETs were developed (Masselink *et al.* 1985), which are the dominant compound semiconductor FETs in industry at the moment, to mitigate the effect of DX centers.

In the AlGaN/GaN system, the surface states and or defects play a much more important role due to polarization fields as the layers are on polar surfaces. Anomalous characteristics, such as the so-called current collapse, kinks in the *I–V* characteristics, and long-term instability, have haunted the device from the time of early development. Preliminary investigations of these phenomena were undertaken some years earlier (Kruppa *et al.* 1995). Now that these devices are strong contenders in the market place for systems applications, these phenomena are getting a good deal of attention. One of the anomalous behaviors is the drain current lag that prevents attainment of RF power congruent with the dc output characteristics of the device. For a maximum drain voltage of 50 V (drain bias of 25 V) and maximum drain current of 1 A, one should normally get 6.25 and 12.5 W in class A and class B operations, respectively, assuming an ideal case with zero saturation voltage and no thermal limitation. However, the observed values in the laboratory are, in general, substantially smaller. This is due to current lag that is basically a failure on the part of drain current to keep up with the gate bias voltage in response to a high-frequency large-signal gate modulation (Nguyen *et al.* 1999), attributed to surface states.

The drain current lag is schematically shown in Fig. 48, where the load line and quiescent operating conditions for class A operation are shown. Also shown are the extremes of dc current as governed by the load line. Current lag is meant to indicate that the RF current (shaded) fails to follow the gate bias and thus the drain current at high frequencies is lower than that measured under dc conditions. The RF current can be determined by the use of the so-called load-pull tuning. It can also be measured under active loading conditions with the use of a high-speed sampling scope for measuring the output RF voltage, in response to an RF input drive, wherein the voltage measured can be converted to current, knowing the load value.

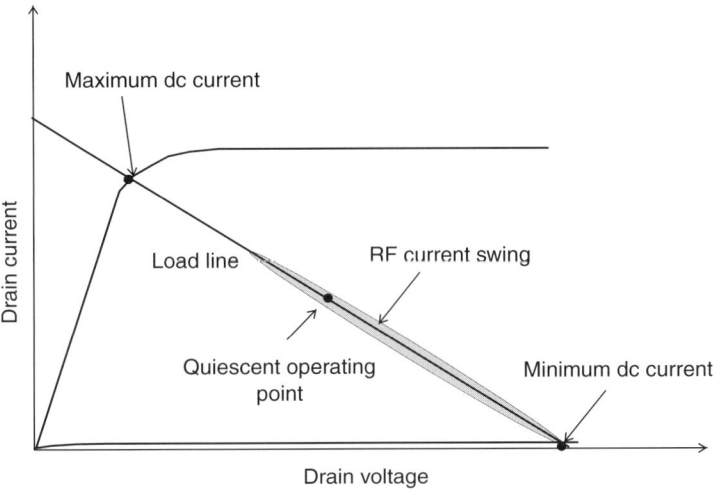

Fig. 48. Schematic representation of RF current lag superimposed on top of dc drain *I–V* characteristics with a load line.

Recently, lattice distortions linked to the minimization of the polarization field have been proposed as a likely cause of current lag, in this particular case current collapse (G. Simin *et al.*, preprint). Through an investigation of time decay of current in response to a pulsed voltage applied between the terminals of adjacent ohmic contacts, in the form of gated transmission line measurement (GTLM) pattern, the authors argue that the time dependence of the current is caused by transient variations of the gate–source and gate–drain resistances, while the channel resistance under the gate remains unaffected. According to the results of these GTLM measurements, the source and drain series resistances are responsible for the current collapse. An increase in the source series resistance should lead to a decrease of current. The same should also cause an increase in the knee voltage, the drain voltage at which the drain current reaches quasi-saturation. One plausible explanation for the increase in series resistance during current transient is the change in strain under and around the gate metal. Increased gate bias from its initial value towards a more negative value causes the electric field in the AlGaN barrier layer to increase, which in magnitude is comparable to the built-in piezoelectric field, several MV cm^{-1}. If the GaN layer underneath the AlGaN layer is not strained, one then surmises that the AlGaN barrier layer undergoes an in-plane tensile strain to match its lattice constant to the underlying GaN layer. If so, the change in the electric field with gate bias would not affect strain in the GaN channel. On the other hand, the surface region of the GaN layer may be somewhat strained. Thus, an increase in the electric field due to increasing gate bias, which is comparable to the piezoelectric field, could increase the tensile strain in AlGaN layer under the gate.

This would expand the AlGaN barrier layer laterally. Processes responding to the piezoelectric induced field, transients of which are slow, could be associated with traps and cause the observed current collapse and lag. In the AlGaN layer under the gate, the gate metal provides a source of electrons in response to the induced piezoelectric charge. Therefore, this region is not expected to contribute to the current collapse (G. Simin *et al.*, preprint).

The current lag can be measured as a function of frequency in the RF regime with an appropriate load line. Since the drain current does not follow the input stimulus due to surface traps, the term "lag" has been coined to describe the phenomenon. The surface must be appropriately passivated to avoid this degradation. The effective methods so far have been the use of low-temperature AlN (J. William *et al.*, in preparation) or Si_3N_4 (Vetury *et al.* 2001) post-growth and fabrication passivation layers. If passivation alone is sufficient to eliminate the current lag, the issue of lattice distortion becomes an interesting one in that it raises the question whether the surface states are involved and if so whether passivation layers alter the strain picture also.

The drain characteristics for a GaN MODFET with a 250-Å Si-doped AlGaN layer that exhibit the aforementioned anomalies are shown in Fig. 49 where two sets of characteristics are included for the same device. The characteristics indicated by the dashed lines are the result when the maximum

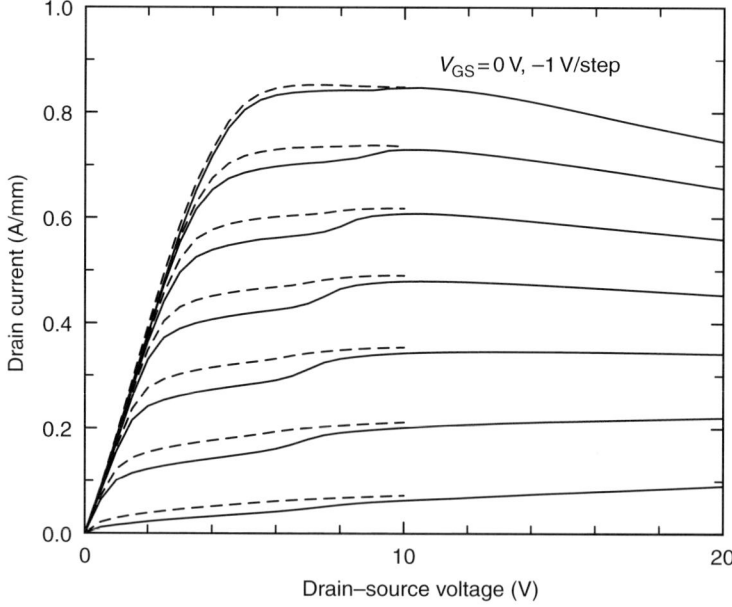

FIG. 49. Output drain current–voltage characteristics for an AlGaN/GaN MODFET with a 250-Å AlGaN layer and a width of 50 μm. The dotted lines are for $V_{DS} < 10$ V and the solid lines are for V_{DS} up to 20 V. The gate length is 0.6 μm. After Binari *et al.* (2001).

V_{DS} is limited to 10 V. On the other hand, the solid lines are for those measured when the maximum V_{DS} is 20 V. By comparing these characteristics, a reduction in drain current for $V_{DS} < 8$ V is noted. This reduction in current after the application of a high-drain voltage is referred to as current collapse. This effect is similar to that reported for GaN MESFETs and is attributed to hot electron injection and trapping in the GaN buffer layer (Binari *et al.* 1997a; Klein *et al.* 1999). As mentioned above, at high-drain voltages, electrons are injected into the GaN buffer layer, where they are trapped. This trapped charge depletes the 2DEG from beneath the active channel and results in a reduction in drain current for subsequent V_{DS} traces. The trapped charge can be released through illumination or thermal emission. The gradual reduction in current for $V_{DS} > 10$ V, seen in Fig. 49, is attributed to self-heating due in part to the sapphire substrate with low thermal conductivity.

The effect of SiN passivation on the drain characteristics is shown in Fig. 50 for the same device before and after passivation. The drain current went up as a result of the increase in n_{sh}. It can be seen that the reduction in current associated with the current collapse phenomenon is unaffected. This is consistent with the proposed mechanism for current collapse, that is, hot electron injection and trapping in the buffer layer without surface involvement.

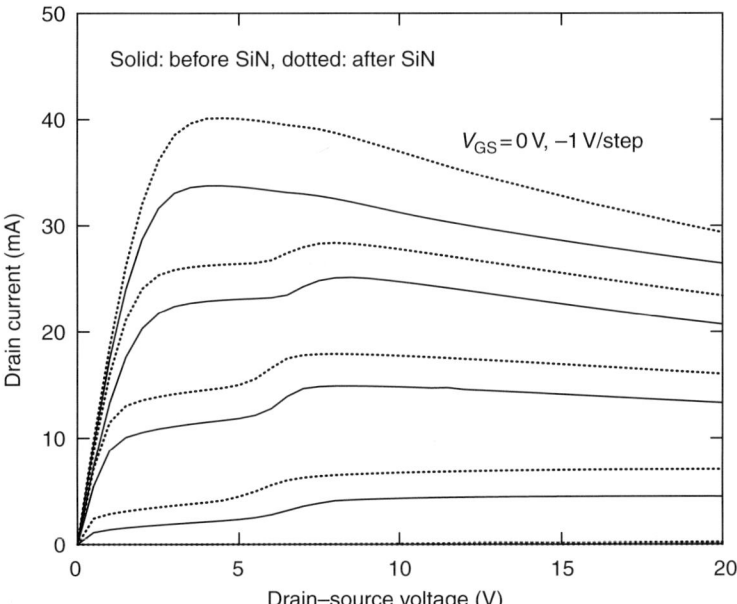

FIG. 50.   Measured drain characteristics of the device of Fig. 51 before and after SiN passivation. After Binari *et al.* (2001).

XI. Heterojunction Bipolar Transistors

Heterojunction bipolar transistors, based on the traditional compound semiconductors and the SiGe/Si sytem, have progressed to the point where their speed and power performance are very attractive for many applications requiring high quality devices. Compared to FETs, higher linearity and larger power densities per unit wafer area can be obtained in bipolar transistors. In addition, being a vertical device, larger breakdown voltages can be obtained, allowing larger load resistances to be used that are easier to impedance match. Unlike FETs, their closest competitors, bipolar devices rely on minority carrier transport and, as such, their critical dimension is the vertical one that is determined by deposition as opposed to lithography. In the silicon world, because of their large current-handling capability, bipolar transistors (BJTs) are even used in unison with field effect devices, CMOS, to drive large capacitive loads for even faster performance (BiCMOS). As alluded to earlier, the basic operation of a bipolar transistor involves minority carrier injection in the forward-biased emitter-base junction, minority carrier transfer through the base, and the collection of minority carriers in the reverse-biased collector junction. Increasing the minority carrier injection efficiency while maintaining a high base doping is a basic requirement for transistors designed for high-frequency and high-speed applications. These design criteria are difficult to achieve using a homojunction emitter, but they may be realized through the use of a heterostructure. Two semiconductors having different bandgaps and a close lattice match form such a heterostructure. A heterojunction in this structure results from an abrupt change in chemical composition during the epitaxial growth process. The heterojunction bipolar transistor (HBT) was first proposed by Shockley (1951) followed by reports (Kroemer 1957, 1982) pointing out potential advantages over conventional homojunction devices (BJTs). Because the semiconductor material with the wider band gap is used as the emitter, this transistor was initially called the Wide Bandgap Emitter Transistor. The structure consists of a lightly doped n^--GaN collector, a p^+-GaN (or a graded AlGaN for field-aided transport across the base), and an n^--AlGaN emitter layer capped with a n^+-GaN contact layer. A schematic diagram of what a fabricated device would look like is shown in Fig. 51. Figure 52 shows a schematic band diagram of an npn HBT under normal bias conditions (forward-biased emitter–base junction and reverse-biased collector-base junction). The arrows indicate the carrier motion. While the forward-injected electrons do not really experience any barrier, the reverse-injected holes from the base experience a large barrier. Consequently, the emitter injection efficiency is high. A good quality base coupled with a very small thickness increases the base transport factor. In cases where the diffusion length is small and/or the surface recombination is severe, grading Al mole fraction in an AlGaN base down toward the collector causes an electric field in the base, helping the

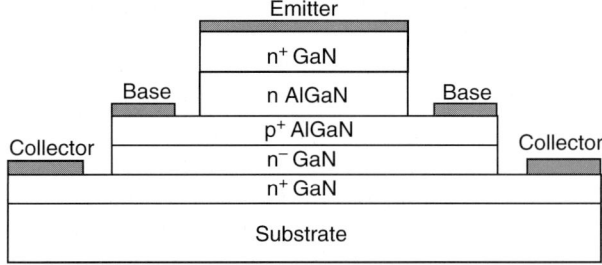

FIG. 51.   Schematic diagram of an AlGaN/GaN heterojunction bipolar transistor (HBT) with or without a compositionally graded base.

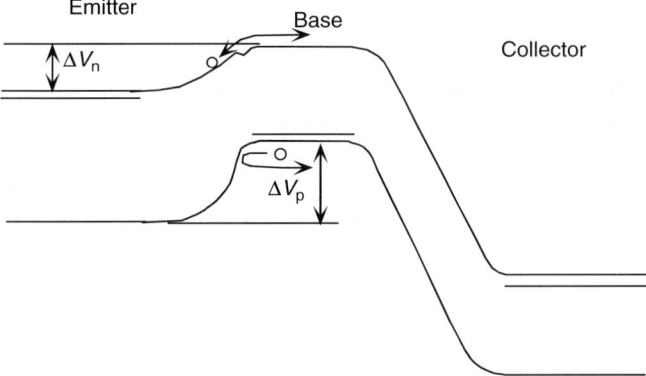

FIG. 52.   Schematic band diagram of an npn HBT under normal bias conditions (forward-biased emitter–base junction and reverse–biased collector-base junction). The arrows indicate the carrier motion.

electron motion in an effort to increase the base transport factor and thus the overall current gain.

Although the HBT concept was proposed some 50 years ago, only in the last 15 years have HBTs recorded dramatic advances. These advances were, to a large extent, fueled by improved crystal growth methods, such as molecular beam epitaxy (MBE), metal-organic chemical vapor deposition (MOCVD) and ultra-high-vacuum chemical vapor deposition (UHVCVD). These technologies provided atomic-level precision in layer thickness and doping concentrations with unprecedented control, ensuring improvements in material quality. Physicists and engineers were thus able to explore new device structures, and to verify non-equilibrium transport mechanisms such as ballistic transport in heterostructures. The current gain h_{FE} of an HBT is sensitive to the material quality, and as the quality improves the HBT current gain increases. The highest current gain cut-off frequencies reported for various conventional compound semiconductors are in the range of about

50–200 GHz (Gao *et al.* 1996). GaAs/AlGaAs layers grown by MBE and MOCVD can be commercially obtained for HBT fabrication. Today's AlGaAs/GaAs digital, analog, and microwave IC chips with more than 10^4 devices are produced in 3-in. production lines.

The high-speed performance of bipolar transistors is represented by the current-gain cutoff frequency f_T and the maximum oscillation frequency f_{max}. High-end commercial Si BJTs with polysilicon emitters exhibit f_T of 20–30 GHz and f_{max} of 15–25 GHz. For Si/SiGe/Si bipolar transistors, the reported f_T is 116 GHz (Crabbé *et al.* 1993; Schüpper *et al.* 1994) or somewhat larger, and the reported f_{max} is 120 GHz (Schüppen *et al.* 1994) or somewhat larger.

GaN-based electronic devices such as high-power and heat-tolerant heterojunction bipolar transistors (HBTs) can be important components of integrated systems designed for high-frequency and high-speed applications, for example, in satellites and all electric aircraft. As mentioned earlier, GaN HBTs could lend themselves to high-power operation with larger operating voltages and better linearity than those that can be attained by FETs. The basic operation of HBTs, involving minority carrier injection in the forward-biased emitter–base junction, minority carrier transfer through the base, and the minority carrier collection in the reverse-biased collector junction, also lead to high-speed performance that is imperative. This high-speed performance is represented by the current-gain cut-off frequency f_T and the maximum oscillation frequency f_{max} (Kroemer 1982). The latter parameter is critically dependent on base resistance, which is a problem for GaN which is known to suffer from low hole concentration in p-type layers, about 10^{18}cm^{-3}. In addition, the deep nature of the most commonly and successfully used dopant, Mg, causes temperature-dependent ionization. This leads to temperature-dependent base resistance, which may actually drop with increasing temperature even though the hole mobility would decrease somewhat.

The first heterojunction bipolar transistor utilizing nitride semiconductors was a hybrid in that both GaN and SiC technologies were used. A GaN/SiC HBT with high current gain has been reported by Pankove *et al.* (1994). The energy bandgap of GaN and SiC are 3.4 and 2.9 eV, respectively. Both GaN and SiC have high thermal conductivities, being 1.3 W/cm K for GaN (higher in GaN with higher quality, as about 2.3 W/cm K) and 4.0 W/cm K for SiC (lower for high-resistivity SiC). Both materials are reasonably lattice matched, in the greater scheme of things. In the HBT reported by Pankove *et al.*, the n-GaN emitter, 0.57-μm thick, had an unintentional doping level $1 \times 10^{18} \text{cm}^{-3}$ (grown by MBE), and the 6-H p-SiC base, 0.2 μm thick, had a doping level $9 \times 10^{18} \text{cm}^{-3}$. The SiC substrate with n-type doping of $1.8 \times 10^{18} \text{cm}^{-3}$ formed the collector. The GaN layer was etched in CCl_2F_2 plasma in unwanted regions during device fabrication. High doping concentrations in the base as well as in the collector led to negligible Early voltage and very small breakdown voltage. The common collector configuration used to get at the current gain in the light of a leaky collector junction relied on differential current gain, which was very high. Using appropriate parameters, one can calculate an emitter injection efficiency of 0.999999. Moreover, using a mobility of $110 \, cm^2/V \, s$ and a lifetime of 5 μs, the diffusion length and the base transport factor were calculated to be 37.7 μm and 0.999987 (both for SiC), respectively (Pankove et al. 1994). The calculated parameters lead to a current gain of 80 409 which is very close to experimental observations.

Base transit time, τ_B, together with emitter transit time, and the transit time at the junction, provides the total transit time in an HBT. The transit time has been widely investigated for well-known and well-established bipolar transistors, including Si homojunction bipolar transistors (Suzuki 1991; Kuo and Lu 1993; Lu and Kuo 1993; Rosenfeld and Alterovitz 1994a,b), heterojunction SiGe HBTs (Gao and Morkoç 1991; Winterton et al. 1993), and AlGaAs/GaAs HBTs (Mazier and Lundstrom 1987; Ritter et al. 1994). These studies suggested that, for the sake of lower base resistance, the base doping concentration of an HBT must be increased. However, such an increase accompanies very minimal increase in the built-in potential and a decrease in the carrier mobility in the base. The increased built-in potential lowers the base transit time τ_B. On the other hand, the decreased mobility increases the base transit time τ_B. Non-uniform doping and the compositional grading of the base also influence the base transit time τ_B. Numerical simulations, in order to determine the room-temperature dependence of the base transit time on various parameters, such as base doping and base compositional grading of Npn $GaN/In_xGa_{1-x}N$ HBTs, have been undertaken some years ago (Mohammad and Morkoç 1995). Additional simulations using a compact simulator of the DC and cut-off frequency performance of GaN/AlGaN have also been performed (Monier et al. 2001; Pulfrey and Fathpour 2001).

As in any calculation, the results depend on the parameters used. In HBTs based on GaN, the major problem is one of a good estimate for the base diffusion length, which critically depends on the layer quality. Other issues include collector breakdown voltage, heterojunction band discontinuities and surface states on the extended regions of the base (beyond the emitter region). Generally speaking, the minority carrier hole diffusion length in GaN would be smaller than conventional compound semiconductors. In very high quality GaN templates, this diffusion length of holes in n-type material was measured to be about 1 μm on the nitrogen face, which represents the early stages of growth on sapphire and is more defective, and 4 μm on the Ga-face, which is some 200 μm away from the initial substrate epitaxial layer interface. Likewise, the minority carrier lifetime ranged from 50 ns near the N-face to 800 ns near the Ga-face (L. Chernyak, Y. J. Park, and H. Morkoç, unpublished results). These numbers are really outstanding and call for extreme caution, as additional measurements must be made to gain confidence in the results.

Most of the conclusions in Monier *et al.* (2001) and Pulfrey and Fathpour (2001) are complementary and predictable. Basically, the current gain is limited by carrier transport across the base, the cut-off frequency is limited by the base transit time that can be enhanced by using a graded junction, and graded base would reduce the transit time. Use of Pnp HBTs, to circumvent the high base resistance plaguing the Npn device, runs into the hole transport limitations across the n-type base. As expected, Pulfrey and Fathpour (2001) also argue that the polarization effects are unlikely to adversely affect the performance of HBTs, current gains in the range of 200–2000 may be possible at room temperature, and the cut-off frequency appears to be around 30 GHz. Despite the odds, experimentalists have been charging ahead. First reports indicated current gains of about 3 (Ren *et al.* 1998; McCharty *et al.* 1999), and 6 with an Early voltage of about 400 V (McCharty *et al.* 2001).

XII.   Conclusions

Modulation doped field-effect transistors (MODFETs) based on the GaN material system have been discussed. Unlike the GaAs-based MODFETs on (100) surfaces, polarization-induced charge in the GaN-based devices on the polar (0001) surfaces is quite large. Consequently, even undoped structures contain sheet electron concentrations in 10^{13} cm^{-2}. An in-depth discussion of polarization effects for normal and inverted MODFETs has been presented. Donor like surface defects most likely provide the charge. Experimental data and theoretical results have been provided on the particulars of the interface charge in relation to parameters such as the AlGaN mole fraction and thickness. In addition, calculation results for electron distribution and current–voltage characteristics of MODFETs have been presented. On the experimental side of MODFET performance, CW power levels of about 22.9 W at 9 GHz have been reported in devices with four 1-mm gate periphery devices in a single-stage power-combining scheme with an associated power-added efficiency of 37%. A minimum noise figure of 0.85 dB with an associated gain of 11 dB at 10 GHz has been obtained. A discussion of the current collapse and lag occurring in GaN-based MODFETs has been presented. In closing, GaN-based MODFETs have made great strides and are continuing to do so despite the less than ideal materials' properties. Anomalies in the current–voltage characteristics at low and high frequencies observed in these devices are attributed to traps in the structure, surface states, and slow trapping processes associated with the field-induced lateral extension of the strain near the gate. It may be only a matter of time for inclusion of these devices in systems. Finally, a brief discussion of heterojunction bipolar transistors, the state of which is really in its infancy, has been provided.

Acknowledgments

This work was supported by the Air Force Office of Scientific Research, the Office of Naval Research, and the National Science Foundation and monitored by G. L. Witt and D. Johnstone (AFOSR), C. E. Wood, Y. S. Park and M. Yoder (ONR), and Verne Hess and U. Varshney (NSF). The author appreciates fruitful discussions with and the assistance of Profs. Aldo Di Carlo, Paolo Lugli, W. Lambrecht, M. S. Shur, F. Bernardini, R. Cingolani, D. Rode, C. Kurdak, and D. Huang, and Drs Fabio Sacconi, C. Nguyen, X. Nguyen, W. Walukiewicz, John Albrecht, and O. Ambacher. Many of these individuals also provided many of their reprints, and preprints. The MODFET model inclusive of polarization charge was developed by Prof. A. Di Carlo and his colleagues at the University of Rome. Finally, the author would like to thank his colleagues and associates for their contributions throughout the evolution of much of the work reported here. Prof. M. Z. Iqbal and a VCU student, J. Spradlin, who read the manuscript very carefully.

References

Aktas, Ö., W. Kim, Z. Fan., S. N. Mohammad, A. Botchkarev, A. Salvador, B. Sverdlov, and H. Morkoç, *Electron. Letts.* **31** (16), 1389 (1995).

Aktas, Ö., Z. Fan, A. Botchkarev, M. Roth, T. Jenkins, L. T. Kehias, and H. Morkoç, Microwave performance of AlGaN/GaN inverted MODFETs, *IEEE Electron Device Lett.* **18**, 293 (1997).

Ambacher, O., Growth and applications of Group III-nitrides, *J. Phys. D: Appl. Phys.* **31**, 2653 (1998).

Ambacher, O., J. Smart, J. R. Shealy, N. G. Weimann, K. Chu, M. Murphy, W. J. Schaff, L. F. Eastman, R. Dimitrov, L. Wittmer, M. Stutzmann, W. Rieger, and J. Hilsenbeck, *J. Appl. Phys.* **85**, 3222 (1999).

Ambacher, O., B. Foutz, J. Smart, J. R. Shealy, N. G. Weimann, K. Chu, M. Murphy, A. J. Sierakowski, W. J. Schaff, L. F. Eastman, R. Dimitrov, A. Mitchell, and M. Stutzmann, Two dimensional electron gases induced by spontaneous and piezoelectric polarization in undoped and doped AlGaN/GaN heterostructures, *J. Appl. Phys.* **87** (1), 334 (2000).

Asbeck, P. M., E. T. Yu, S. S. Lau, W. Sun, X. Dang, and C. Shi, *Solid-State Electron.* **44**, 211 (2000).

Asnin, V. M., F. H. Pollak, J. Ramer, M. Schurman, and I. T. Ferguson, *Appl. Phys. Lett.* **75**, 1240 (1999).

Azuhata, T., T. Sota, K. Suzuki, and S. Nakamura, *J. Phys.: Condens. Matter* **7**, L129 (1995).

Bastard, G., *Wave Mechanics Applied to Semiconductor Heterostructures*, Edition de Physique, Paris, France (1987).

Bellaiche, L., Band gaps of lattice-matched (Ga, In)(As, N) alloys, *Appl. Phys. Lett.* **75** (17), 2578 (1999).

Bellaiche, L., S.-H. Wei, and A. Zunger, Band gaps of GaPN and GaAsN alloys, *Appl. Phys. Lett.* **70**, 3558 (1997).

Bernardini, F., and V. Fiorentini, Macroscopic polarization and band offsets at the nitride heterojunctions, *Phys. Rev. B* **57** (16), 1 (1998).

Bernardini, F., V. Fiorentini, and D. Vanderbilt, Spontaneous polarization and piezoelectric constants in III–V nitrides, *Phys. Rev. B* **56**, R10024 (1997).

Bhapkar, U. V., and M. S. Shur, Monte Carlo calculation of velocity-field characteristics of wurtize GaN, *J. Appl. Phys.* **82**, 1649 (1997).

Binari, S. C., H. B. Dietrich, G. Kelner, L. B. Roland, K. Doverspike, and D. K. Gaskill, Electrical characterisation of Ti Schottky barriers on n-type GaN, *Electron. Lett.* **30**, 909 (1994).

Binari, S. C., K. Ikossi-Anastasiou, J. A. Roussos, W. Kruppa, D. Park, H. B. Dietrich, D. D. Koleske, A. E. Wickenden, and R. L. Henry, *IEEE Electron. Devices* **48**, 465 (2001).

Binari, S. C., W. Kruppa, H. B. Dietrich, G. Kelner, A. E. Wickenden, and J. A. Freitas Jr., Fabrication and characterization of GaN FETs, *Solid-State Electron.* **41**, 1549 (1997a).

Binari, S., J. M. Redwing, G. Kelner, and W. Kruppa, AlGaN/GaN HEMTs grown on SiC substrates, *Electron Lett.* **33** (3), 242 (1997b).

Bonfiglio, A., M. Lomascolo, G. Traetta, R. Cingolani, A. Di Carlo, F. Della Sala, P. Lugli, A. Botchkarev, and H. Morkoç, *J. Appl. Phys.* **87**, 2289 (2000).

Brown, J. D., Z. Yu, J. Matthews, S. Harney, J. Boney, J. F. Schetzina, J. D. Benson, K. W. Dang, C. Terrill, T. Nohava, W. Yang, S. Krishnankutty, Visible-blind UV digital camera based on a 32×32 array of GaN/AlGaN p–i–n photodiodes, *MRS Internet J.* (http://nsr.mij.mrs.org/4/9/).

Bykhovski, A. D., V. V. Kaminski, M. S. Shur, Q. C. Chen, and M. A. Khan, Piezoresistive effect in wurtzite n-type GaN, *Appl. Phys. Lett.* **68**, 818 (1996).

Carnez, B., *et al.*, Modeling of a submicrometer gate field effect transistor including effects of nonstationary electron dynamics, *J. Appl. Phys.* **51**, 784 (1980).

Cheong, M. G., K. S. Kim, C. S. Oh, N. W. Namgung, G. M. Yang, C. H. Hong, K. Y. Lim, E. K. Suh, K. S. Nahm, H. J. Lee, D. H. Lim, and A. Yoshikawa, *Appl. Phys. Lett.* **77**, 2557 (2000).

Chin, V. W. L., T. L. Tansley, and T. Osotchan, *J. Appl. Phys.* **75**, 7365 (1994).

Cingolani, R., A. Botchkarev, H. Tang, H. Morkoç, G. Coli, M. Lomascolo, A. Di Carlo, F. Della Sala, and P. Lugli, *Phys. Rev. B* **61**, 2711 (2000).

Crabbé, E. F., B. S. Meyerson, J. M. C. Stork, D. Harame, Vertical profile optimization very high frequency epitaxial Si- and SiGe-base bipolar transistors, *IEEE IEDM 93 Tech. Dig.*, 83 (1993).

Cui, J., and A. Sun, M. Reshichkov, F. Yun, A. Baski, and H. Morkoç, Preparation of sapphire for high quality III-nitride growth, *MRS Internet J.* (http://nsr.mij.mrs.org/5/7/).

Daudin, B., J. L. Rouviere, and M. Arley, Polarity determination of GaN films by ion channeling and convergent beam electron diffraction, *Appl. Phys. Lett.* **69**, 2480 (1996).

Della Sala, F., A. Di Carlo, P. Lugli, F. Bernardini, V. Fiorentini, R. Scholz, and J. M. Jancu, *Appl. Phys. Lett.* **74**, 2002 (1999).

Dhar, S., and S. Ghosh, *J. Appl. Phys.* **86**, 2668 (1999).

Di Carlo, A., *Phys. Stat. Sol.* **217**, 703 (2000).

Di Carlo, A., S. Pescetelli, M. Paciotti, P. Lugli, and M. Graf, *Solid State Commun.* **98**, 803 (1996).

Di Carlo, A., F. Della Sala, P. Lugli, V. Fiorentini, and F. Bernardini, *Appl. Phys. Lett.* **76**, 3950 (2000).

Duffy, M. T., C. C. Wang, G. D. O'Clock, S. H. McFarlane III, and P. J. Zanzucchi, Epitaxial growth and piezoelectric properties of AlN, GaN, and GaAs on sapphire or spinel, *J. Elect. Mater.* **2**, 359 (1973).

Fan, Z., S. N. Mohammad, W. Kim, Ö. Aktas, A. E. Botchkarev, and H. Morkoç, Very low resistance multi-layer ohmic contact to n-GaN, *Appl. Phys. Lett.* **68**, 1672 (1996).

Fan, Z., C. Lu, A. Botchkarev, H. Tang, A. Salvador, Ö. Aktas, W. Kim, and H. Morkoç, AlGaN/GaN double heterostructure channel modulation doped field effect transistors (MODFETs), *Electron. Lett.* **33**, 814 (1997).

Fang, Z. Q., D. C. Look, W. Kim, Z. Fan, A. Botchkarev, and H. Morkoç, *Appl. Phys. Lett.* **72**, 2277 (1998).

Farahmand, M., C. Garetto, E. Bellotti, K. F. Brennan, M. Goano, E. Ghillino, G. Ghione, J. D. Albrecht, and P. Paul Ruden, Monte Carlo simulation of electron transport in the III-nitride wurtzite phase materials system: binaries and ternaries, *IEEE Trans. Electron Devices* **48**, 535 (2001).

Fiorentini, V., F. Bernardini, F. Della Sala, A. Di Carlo, and P. Lugli, *Phys. Rev. B* **60**, 8849 (1999).

Fischer, M., M. Reinhardt, and A. Forchel, High temperature operation of GaInAsN laserdiodes in the 1.3 μm regime, *58th DRC Device Research Conference*, IEEE, p. 119 (2000a).

Fischer, M., M. Reinhardt, and A. Forchel, Room-temperature operation of GaInAsN/GaAs laser diodes in the 1.5 μm range, *Conference Digest, 2000 IEEE 17th International Semiconductor Laser Conference* (Cat. No.00CH37092), IEEE, Piscataway, NJ, USA, p. 115 (2000b).

Fischer, M., M. Reinhardt, and A. Forchel, A monolithic GaInAsN vertical-cavity surface-emitting laser for the 1.3 μm regime, *IEEE Photonics Technol. Lett.* **12** (10), 1313 (2000c).

Fischer, R., T. J. Drummond, J. Klem, W. Kopp, T. Henderson, D. Perrachione, and H. Morkoç, On the collapse of drain *I–V* characteristics in modulation doped FETs at cryogenic temperatures, *IEEE Trans. Electron. Devices* **ED-31**, 1028 (1984).

Foresi, J. S., and T. D. Moustakas, Metal contacts to gallium nitride, *Appl. Phys. Lett.* **62**, 2859 (1993).

Fritz, I. J., and T. J. Drummond, AlN–GaN quarter-wave reflector stack grown by gas-source MBE on (100) GaAs, *Electron. Lett.* **31**, 68 (1995).

Götz, W., L. T. Romano, J. Walker, N. M. Johnson, and R. J. Molnar, *Appl. Phys. Lett.* **72**, 1214 (1998).

Gao, G. B., and H. Morkoç, *Electron. Lett.* **27**, 1408 (1991).

Gao, G. B., S. N. Mohammad, G. M. Martin, and H. Morkoç, III–V compound semiconductor heterojunction bipolar transistors, in *Compound Semiconductor Electronics: The Age of Maturity*, edited by M. S. Shur. World Scientific Publishing Company, Singapore, pp. 85–174 (1996).

Gaska, R., J. W. Yang, A. Osinsky, A. D. Bykhovski, and M. S. Shur, Piezoeffect and gate current in AlGaN/GaN high electron mobility transistors, *Appl. Phys. Lett.* **71**, 3673 (1997).

Gaska, R., J. W. Yang, A. Osinsky, A. D. Bykhovski, M. S. Shur, V. V. Kaminski, and S. M. Soloviov, The influence of the deformation on the two-dimensional electron gas density in GaN–AlGaN heterostructure, *Appl. Phys. Lett.* **72**, 64 (1998).

Hacke, P., T. Detchprohm, K. Hiramatsu, and N. Sawaki, Schottky barrier on n-type GaN grown by hydride vapor phase epitaxy, *Appl. Phys. Lett.* **63**, 2676 (1993).

Henderson, T., M. I. Aksun, C. K. Peng, H. Morkoç, P. C. Chao, P. M. Smith, K. H. G. Duh, and L. F. Lester, Microwave performance of a quarter micron gate low noise pseudomorphic InGaAs/AlGaAs modulation doped field effect transistor, *IEEE Electron Device Lett.* **EDL-7**, 649 (1986).

Heying, B., R. Averbeck, L. F. Chen, E. Haus, H. Riechert, and J. S. Speck, *J. Appl. Phys.* **88**, 1855 (2000a).

Heying, B., I. Smorchkova, C. Poblenz, C. Elsass, P. Fini, S. Den Baars, U. Mishra, and J. S. Speck, Optimization of the surface morphologies and electron mobilities in GaN grown by plasma-assisted molecular beam epitaxy, *Appl. Phys. Lett.*, **77** (18), 2885 (2000b).

Hsu, L., and W. Walukiewicz, *Phys. Rev.* B **56**, 1520 (1997).

Hsu, L., and W. Walukiewicz, *Appl. Phys. Lett.* **73**, 339 (1998).

Hsu, L., W. Walukiewicz, Effect of polarization fields on transport properties in AlGaN/GaN heterostructures, *J. Appl. Phys.* **89**, 1783 (2001).

Huang, D., F. Yun, P. Visconti, M. A. Reshchikov, D. Wang, H. Morkoç, D. L. Rode, L. A. Farina, Ç. Kurdak, K. T. Tsen, S. S. Park, and K. Y. Lee, Hall mobility and carrier concentration in GaN free-standing templates grown by hydride vapor phase epitaxy with high quality, *Solid State Electron.* **45** (5), 711 (2001).

Jain, S. C., M. Willander, J. Narayan, R. Van Overstraeten, *J. Appl. Phys.* **87**, 963 (2000).

Joshi, R. P., *Appl. Phys. Lett.* **64**, 223 (1994).

Kent, P. R. C., and A. Zunger, Evolution of III–V nitride alloy electronic structure: the localized to delocalized transition, *Phys. Rev. Lett.* **86** (12), 2613 (2001).

Khan, M. A., J. M. Van Hove, J. N. Kuznia, and D. T. Olson, High electron mobility GaN/$Al_xGa_{1-x}N$ heterostructures grown by low-pressure metalorganic chemical vapor deposition, *Appl. Phys. Lett.* **58**, 2408 (1991).

Khan, M. A., J. N. Kuznia, A. R. Bhattarai, and D. T. Olson, Metal semiconductor field-effect transistor based on single-crystal GaN, *Appl. Phys. Lett.* **62**, 1786 (1993).

Khan, M. R. H., H. Nakayama, T. Detchprohm, K. Hiramatsu, and N. Sawaki, A study on barrier height of Au–$Al_xGa_{1-x}N$ Schottky diodes in the range of $0 < x < 0.20$, in *Topical Workshop on III–V Nitrides Proc.*, Nagoya, Japan, 1995. *Solid State Electron.* **41** (2), 259 (1997).

Kitatani, T., M. Kondow, K. Nakahara, M. C. Larson, and K. Uomi, Temperature dependence of the threshold current and the lasing wavelength in 1.3-μm GaInNAs/GaAs single quantum well laser diode, *Opt. Rev.* **5** (2), 69 (1998).

Kitatani, T., K. Nakahara, M. Kondow, K. Uomi, and T. Tanaka, A 1.3-μm GaInNAs/GaAs single-quantum-well laser diode with a high characteristic temperature over 200 K, *Jpn. J. Appl. Phys. Part 2* **39** (2A), L86 (2000).

Klein, P. B., J. A. Freitas Jr., S. C. Binari, and A. E. Wickenden, Observation of deep traps responsible for current collapse in GaN metal–semiconductor field-effect transistors, *Appl. Phys. Lett.* **75**, 4016 (1999).

Kolnik, J., I. H. Oguzman, K. F. Brennan, R. Wang, P. P. Ruden, and Y. Wang, *J. Appl. Phys.* **78**, 1033 (1995).

Kondow, M., T. Kitatani, K. Nakahara, and T. Tanaka, A 1.3-μm GaInNAs laser diode with a lifetime of over 1000 hours, *Jpn. J. Appl. Phys. Part 2* **38** (12A), L1355 (1999).

Kroemer, H., Theory of a wide-gap emitter for transistors, *Proc. IRE* **45**, 1535 (1957).

Kroemer, H., Heterostructure bipolar transistors and integrated circuits, *Proc. IEEE* **70**, 13 (1982).

Kruppa, W., S. C. Binari, and K. Doverspike, Low-frequency dispersion characteristics of GaN HFETs, *Electron. Lett.* **31**, 1951 (1995).

Kuo, J. B., and T. C. Lu, *Solid-State Electron.* **36**, 917 (1993).

Lee, C. D., V. Ramachandran, A. Sagar, R. M. Feenstra, D. W. Greve, W. L. Sarney, L. Salamanca-Riba, D. C. Look, B. S. Bai, W. J. Choyke, and R. P. Devaty, Properties of GaN epitaxial layers grown on 6H-SiC(0001) by plasma-assisted molecular beam epitaxy. *IEEE J. Electron. Mater.* **30** (3), 162 (2001).

Lee, K., M. S. Shur, T. J. Drummond, and H. Morkoç, Parasitic MESFET in (Al,Ga)As/GaAs modulation doped FETs and MODFET characterization, *IEEE Trans. Electron. Devices* **ED-31**, 29 (1984).

Li, N. Y., C. P. Hains, K. Yang, J. Lu, J. Cheng, and P. W. Li, Organometallic vapor phase epitaxy growth and optical characteristics of almost 1.2 μm GaInNAs three-quantum-well laser diodes, *Appl. Phys. Lett.* **75** (8), 1051 (1999).

Lin, M. E., B. Sverdlov, G. L. Zhou, and H. Morkoç, *Appl. Phys. Lett.* **62**, 3479 (1993a).

Lin, M. E., S. Strite, A. Agarwal, A. Salvador, G. L. Zhou, N. Teraguchi, A. Rockett, and H. Morkoç, GaN Grown on hydrogen plasma cleaned 6H-SiC substrates, *Appl. Phys. Letts.* **62** (7), 702 (1993b).

Lin, M. E., Z. Ma, F. Y. Huang, Z. F. Fan, L. H. Allen, and H. Morkoç, *Appl. Phys. Lett.* **64**, 1003 (1994).

Look, D. C., and J. R. Sizelove, *Phys. Rev. Lett.* **82**, 1237 (1999).

Look, D. C., D. C. Reynolds, J. W. Hemsky, J. R. Sizelove, R. L. Jones, and R. J. Molnar, *Phys. Rev. Lett.* **79**, 2273 (1997).

Lu, T. C., and J. B. Kuo, *IEEE Trans. Electron Devices* **ED-40**, 766 (1993).

Lugli, P., M. Paciotti, E. Calleja, E. Munoz, J. J. Sanchez-Rojas, F. Dessenne, R. Fauquembergue, J. L. Thobel, and G. Zandler, HEMT models and simulations, in: *Pseudomorphic HEMTs: Technology and Applications*, edited by R. Lee Ross, S. Swensson, and P. Lugli. Kluwer Press, Dordrecht, p. 141 (1996).

Maeda, N., T. Nishida, N. Kobayashi, and M. Tomizawa, *Appl. Phys. Lett.* **73**, 1856 (1998).

Manfra, M. J., L. N. Pfeiffer, K. W. West, H. L. Stormer, K. W. Baldwin, J. W. P. Hsu, D. V. Lang, and R. J. Molnar, High-mobility AlGaN/GaN heterostructures grown by molecular-beam epitaxy on GaN templates prepared by hydride vapor phase epitaxy, *Appl. Phys. Lett.* **77** (18), 2888 (2000).

Martin, G. A., Semiconductor electronic band alignment at heterojunctions of wurtzite AlN, GaN and InN, PhD thesis, Department of Physics, University of Illinois (1996).

Martin, G. A., S. Strite, A. Botchkarev, A. Agarwal, A. Rockett, H. Morkoç, W. R. L. Lambrecht, and B. Segall, *Appl. Phys. Lett.* **65**, 610 (1994).

Martin, G. A., A. Botchkarev, A. Agarwal, A. Rockett, and H. Morkoç, Valence-band discontinuities of wurtzite GaN, AlN, and InN heterojunctions measured by X-ray photoemission spectroscopy, *Appl. Phys. Lett.* **68**, 2541 (1996).

Masselink, W. T., A. Ketterson, J. Klem, W. Kopp, and H. Morkoç, Cryogenic operation of pseudomorphic AlGaAs/InGaAs single quantum well MODFETs, *Electron. Lett.* **21**, 937 (1985).

Mazier, C. M., and M. S. Lundstrom, *IEEE Electron Devices Lett.* **EDL-8**, 90 (1987).

McCharty, L. S. *et al.*, *IEEE Electron. Devices Letts.* **20**, 277 (1999).

McCharty, L. S., I. S. Smorchkova, H. Xing, P. Kozodoy, P. Fini, J. Limb, D. L. Pulfrey, J. S. Speck, M. J. W. Rodwell, S. P. DenBaars, and U. K. Mishra, *IEEE Trans. Electron Devices* **48** (3), 543 (2001).

Micovic, M., A. Kurdoghlian, P. Janke, P. Hashimoto, D. W. S. Wong, J. S. Moon, L. McCray, and C. Nguyen, *IEEE Trans. Electron Devices* **48** (3), 591 (2001).

Mohammad, S. N., and H. Morkoç, Base transit time in GaN/InGaN heterojunction bipolar transistors, *J. Appl. Phys.* **78**, 4200 (1995).

Mohammad, S. N., and H. Morkoç, Progress and prospects of Group III–V nitride semiconductors, *Progr. Quantum Electron.* **20** (5 and 6), 361 (1996).

Mohammad, S. N., A. Salvador, and H. Morkoç, Emerging GaN based devices, *Proc. IEEE* **83**, 1306 (1995).

Mohammad, S. N., Z. Fan, A. E. Botchkarev, W. Kim, Ö. Aktas, A. Salvador, and H. Morkoç, Near ideal platinum–GaN Schottky diodes, *Electron. Lett.* **32**, 598 (1996).

Molnar, R. J., W. Goetz, L. T. Romano, N. M. Johnson, Growth of gallium nitride by hydride vapor-phase epitaxy, *J. Crystal. Growth* **178** (1–2), 147 (1997).

Moloney, M., F. Ponse, and H. Morkoç, Gate capacitance voltage characteristics of MODFETs: its effect on transconductance, *IEEE Trans. Electron. Devices* **ED-32** (9), 1675 (1985).

Monier, C., F. Ren, J. Han, P-C, Chang, R. J. Shul, K-P, Lee, A. G. Bacca, and S. Pearton, *IEEE Trans. Electron. Devices* **48** (3), 597 (2001).

Moore, K. E., C. E. Weitzel, K. J. Nordquist, L. L. Pond III, J. W. Palmour, S. Allen, and C. H. Carter Jr., *IEEE Electron. Devices Lett.* **18** (2), 69 (1997).

Morkoç, H., Wurtzite GaN based heterostructures by molecular beam epitaxy, *IEEE J. Select. Top. Quantum Electron.* **4** (3), 537 (1998a).

Morkoç, H., Beyond SiC! III–V nitride based heterostructures and devices, in *SiC Materials and Devices*, edited by Y. S. Park, Willardson and Beer Series, Vol. 52, Chapter 8, Academic Press, New York, p. 307 (1998b).

Morkoç, H., *Nitride Semiconductors and Devices*, Springer Verlag, Heidelburg (1999a).

Morkoç, H., GaN-based modulation doped FETs and UV detectors, *Naval Res. Rev.* **51** (1), 28 (1999b).

Morkoç, H., III–Nitride semiconductor growth by MBE: recent issues, *J. Mater. Sci.: MEL*, in press (2002).

Morkoç, H., and S. N. Mohammad, High luminosity gallium nitride blue and blue-green light emitting diodes, *Sci. Mag.* **267**, 51 (1995).

Morkoç, H., and S. N. Mohammad, Light emitting diodes, in *Wiley Encyclopedia of Electrical Engineering and Electronics Engineering*, edited by J. Webster, John Wiley, New York (1999).

Morkoç, H., H. Ünlü, and G. Ji, *Fundamentals and Technology of MODFETs*, Vols. I and II. Wiley, Chichester, UK (1991).

Morkoç, H., S. Strite, G. B. Gao, M. E. Lin, B. Sverdlov, and M. Burns, A review of large bandgap SiC, III–V nitrides, and ZnSe based II–VI semiconductor structures and devices, *J. Appl. Phys. Rev.* **76** (3), 1363 (1994).

Morkoç, H., and R. Cingolani, and B. Gil, *Solid State Electron.* **43** (10), 1909 (1999).

Morkoç, H., A. Di Carlo, and R. Cingolani, *Solid State Electron.* **46** (2), 157 (2002a).

Morkoç, H., A. Di Carlo, and R. Cingolani, GaN-based modulation doped FETs, in *Low Dimensional Nitride Semiconductors*, edited by B. Gil, Oxford University Press, Oxford, in press (2002b).

Murphy, M. J., *et al. MRS Internet J. Nitride Semicond. Res.* **4S1**, G8.4 (1999).

Nakamura, S., T. Mukai, and M. Senoh, *J. Appl. Phys.* **71**, 5543 (1992).

Nakamura, S., M. Senoh, N. Nagahama, N. Iwara, T. Yamada, T. Matsushita, H. Kiyoku, Y. Sugimoto, T. Kozaki, H. Umemoto, M. Sano, and K. Chocho, InGaN/GaN/AlGaN-based laser diodes with modulation-doped strained-layer superlattices, *Jpn. J. Appl. Phys.* **38**, L1578 (1997).

Nakatsuka, S., M. Kondow, T. Kitatani, Y. Yazawa, and M. Okai, Index-guide GaInNAs laser diode for optical communications, *Jpn. J. Appl. Phys. Part 1* **37** (3B), 138 (1998).

Nakamura, S., T. Mukai, and M. Senoh, Candela-class high-brightness InGaN/AlGaN double-heterostructure blue light-emitting diodes, *Appl. Phys. Lett.* **64**, 1687 (1994).

Ng, H. M., D. Doppalapudi, T. D. Moustakas, N. G. Weimann, and L. F. Eastman, *Appl. Phys. Lett.* **73**, 821 (1998).

Nguyen, C., N. X. Nguyen, and D. E. Grider, Drain current compression in GaN MODFETs under large-signal modulation at microwave frequencies, *Electron. Lett.* **35**, 1380 (1999).

Oberhuber, R., G. Zandler, and P. Vogl, *Appl. Phys. Lett.* **73**, 818 (1998).

Ougazzaden, A., S. Bouchoule, A. Mereuta, E. V. K. Rao, and J. Decobert, Room temperature laser operation of bulk InGaAsN/GaAs structures grown by AP-MOVPE using N_2 as carrier gas, *Electron. Lett.*, **35** (6), 474 (1999).

Pankove, J. I., S. S. Chang, H. C. Lee, R. J. Molnar, T. D. Moustakas, and B. Van Zeghbroeck, *IEDM Tech. Dig.*, p. 389 (1994).

Pearton, S. J., J. C. Zolper, R. J. Shul, and F. Ren, GaN: processing, defects, and devices, *J. Appl. Phys.* **86**, 1 (1999).

Ping, A. T., Q. Chen, J. W. Yang, M. A. Khan, and I. Adesida, DC and microwave performance of high-current AlGaN/GaN heterostructure field effect transistors grown on p-type SiC substrates, *IEEE Electron Device Lett.* **19** (2), 54 (1998).

Potin, V., P. Ruterana, G. Nouet, A. Salvador, and H. Morkoç, The atomic structure of the {10–10} inversion domains in GaN/sapphire layers, in *Proc. of Gallium Nitride and Related*

Materials II Symposium, edited by C. R. Abernathy, H. Amano, and J. C. Zolper, Vol. 468. Mater. Res. Soc, Pittsburgh, PA, USA, p. 323 (1997).

Powell, J. A., D. J. Larkin, P. G. Neudeck, J. W. Yang, and P. Pirouz, Investigation of defects in Epitaxial 3C-SiC, 4H-SiC and 6H-SiC Films Grown on SiC Substrates, in *Silicon Carbide and Related Materials*, edited by M. G. Spencer, R. P. Devaty, J. A. Edmond *et al.* Bristol, IOP Publishing, p. 161 (1994).

Powell, J. A., D. J. Larkin, and A. J. Trunek, Use of gaseous etching for the characterization of structural defects in silicon carbide single crystals, in *Silicon Carbide, III-Nitrides, and Related Materials*, edited by G. Pensl, H. Morkoc, B. Monemar, and E. Janzen, Vols. 264–268. Trans Tech Publications, Sweden, p. 421 (1998).

Pulfrey, D. L., and S. Fathpour, *IEEE Trans. Electron Devices* **48** (3), p. 597 (2001).

Ramvall, P., Y. Aoyagi, A. Kuramata, P. Hacke, and K. Horino, Influence of a piezoelectric field on the electron distribution in a double GaN/Al$_{0.14}$Ga$_{0.86}$N heterojuction, *Appl. Phys. Lett.* **74**, 3866 (1999).

Razeghi, M., and A. Rogalski, Semiconductor ultraviolet detectors, *J. Appl. Phys.* **79**, 7433 (1996).

Reinhardt, M., M. Fischer, M. Kamp, and A. Forchel, 7.8 GHz small-signal modulation bandwidth of 1.3 µm DQW GaInAsN/GaAs laser diodes, *Electron. Lett.* **36** (12), 1025 (2000).

Ren, F. *et al.*, *MRS Internet J. Nitride Semicond. Res.* **3**, 41 (1998).

Resta, R., Macroscopic polarization in crystalline dielectrics: the geometric phase approach, *Rev. Mod. Phys.* **66**, 899 (1994).

Ridley, B. K., B. E. Foutz, and L. F. Eastman, *Phys. Rev. B* **61** (24), 1682 (2000).

Ridley, B. K., Exact electron momentum-relaxation times in GaN associated with scattering by polar-optical phonons, *J. Appl. Phys.* **84** (7), 4020 (1998).

Ritter, D., R. A. Hamm, A. Feygenson, and P. R. Smith, *Appl. Phys. Lett.* **64**, 2988 (1994).

Rode, D. L., and D. K. Gaskill, *Appl. Phys. Lett.* **66**, 1972 (1995).

Rode, D. L., in *Semiconductors and Semimetals*, edited by R. K. Willardson and A. C. Beer, Vol. 10. Academic Press, New York, p. 1 (1975).

Rosenfeld, D., and S. A. Alterovitz, *IEEE Trans. Electron Devices* **ED-41**, 848 (1994a).

Rosenfeld, D., and S. A. Alterovitz, *Solid-State Electron.* **37**, 119 (1994b).

Ruterana, P., P. Vermaut, G. Nouet, A. Salvador, and H. Morkoc, Surface treatment and layer structure in 2H-GaN grown on the (0001)$_{Si}$ surface of 6H-SiC by MBE, *MRS Internet J. Nitride Semicond. Res.* **2**, 42 (1997).

Ruvimov, S., Z. Liliental-Weber, J. Washburn, K. J. Duxstad, E. E. Haller, S. N. Mohammad, Z. Fan, and H. Morkoç, Microstructure of Ti/Al and Ti/Al/Ni/Au ohmic contacts for n-GaN, *Appl. Phys. Lett.* **69** (11), 1556 (1996).

Sacconi, F., A. Di Carlo, P. Lugli, and H. Morkoç, Spontaneous and piezoelectric polarization effects on the output characteristics of AlGaN/GaN heterojunction modulation doped FETs, *IEEE Trans. Electron Devices* **TED-48** (3), 450 (2001).

Sandborn, P. A., *et al.* Quasi-two-dimensional modelling of GaAs MESFET's, *IEEE Trans. Electron Devices* **ED-34**, 985 (1987).

Sato, S., and S. Satoh, Room-temperature pulsed operation of strained GaInNAs/GaAs double quantum well laser diode grown by metal organic chemical vapour deposition *Electron. Lett.* **34** (15), 1495 (1998).

Schüppen, A., A. Gruhle, U. Erben, H. Kibbel, and U. König, SiGe-HBTs with high f$_T$ at moderate current densities, *Electron. Lett.* **30**, 1187 (1994a).

Schüppen, A., A. Gruhle, U. Erben, H. Kibbel, and U. König, Multi emitter finger SiGe-HBTs with f$_{max}$ up to 120 GHz, *IEEE IEDM 94 Tech. Dig.*, p. 377 (1994b).

Schetzina, J., *et al.*, in *Proceedings of the International Conference on Silicon Carbide and Related Materials*, October 10–15, Research Triangle Park, NC, USA (1999b).

Schneider, H. C., A. J. Fischer, W. W. Chow, and J. F. Klem, Temperature dependence of laser threshold in an InGaAsN vertical-cavity surface-emitting laser, *Appl. Phys. Lett.*, **78** (22), 3391 (2001).

Seeger, K., *Semiconductor Physics*, 2nd edn, Springer, Berlin (1982).

Sheppard, S. T., K. Doverspike, W. L. Pribble, S. T. Allen, J. W. Palmour, L. T. Kehias, and T. J. Jenkins, High power microwave GaN/AlGaN HEMTs on semi-insulating silicon carbide substrates, *IEEE Electron Device Lett.* **20** (4), 161 (1999).

Shockley, W., US Patent No. 2,569,347 (1951).

Shur, M. S., A. D. Bykhovsky, and R. Gaska, *Solid-State Electron.* **44**, 205 (2000).

Smorchkova, I. P., C. R. Elsass, J. P. Ibbetson, R. Vetury, B. Heying, P. Fini, E. Haus, S. P. DenBaars, J. S. Speck, and U. K. Mishra, *J. Appl. Phys.* **86**, 4520 (1999).

Snowden, C. M., *et al.* Quasi-two-dimensional modelling MESFET simulation for CAD, *IEEE Trans. Electron Devices*, **ED-36**, 1564 (1989).

Steinle, G., F. Mederer, M. Kicherer, R. Michalzik, G. Kristen, A. Y. Egorov, H. Riechert, H. D. Wolf, and K. J. Ebeling, Data transmission up to 10 Gbit/s with 1.3 μm wavelength InGaAsN VCSELs, *Electron. Lett.* **37** (10), 632 (2001).

Strite, S. T., and H. Morkoç, GaN, AlN, and InN: a review, *J. Vac. Sci. Technol. B* **10**, 1237 (1992).

Sullivan, G. J., M. Y. Chen, J. A. Higgins, J. W. Yang, Q. Chen, R. L. Pierson, and B. T. McDermott, High-power 10-GHz operation of AlGaN HFET's on insulating SiC, *IEEE Electron Device Lett.* **19**, 198 (1998).

Suzue, K., S. N. Mohammad, Z. F. Fan, W. Kim, Ö. Aktas, A. E. Botchkarev, and H. Morkoç, Electrical conduction in platinum–gallium nitride Schottky diodes, *J. Appl. Phys.* **80** (6), 4467 (1996).

Suzuki, K., *IEEE Trans. Electron Devices* **ED-38**, 2128 (1991).

Tarsa, E. J., B. Heying, X. H. Wu, P. Fini, S. P. DenBaars, and J. S. Speck, *J. Appl. Phys.* **82**, 5472 (1997).

Trellakis, A., A. T. Halick, A. Pacelli, and U. Ravaioli, *J. Appl. Phys.* **81**, 7880 (1997).

Vermaut, P., P. Ruterana, G. Nouet, A. Salvador, and H. Morkoç, Structural defects due to interface steps and polytypism in III–V semiconducting materials: a case study using high resolution electron microscopy of the 2H-AlN/6H-SiC Interface, *Philos. Mag. A* **75**, 239 (1997).

Vetury, R., N. Q. Zhang, S. Keller, and U. K. Mishra, *IEEE Electron. Devices* **48**, 560 (2001).

Visconti, P., K. M. Jones, M. A. Reshchikov, F. Yun, R. Cingolani, H. Morkoç, S. S. Park, and K. Y. Lee, *Appl. Phys. Lett.* **77**, 3743 (2000a).

Visconti, P., K. M. Jones, M. A. Reshchikov, R. Cingolani, H. Morkoç, and R. J. Molnar, *Appl. Phys. Lett.* **77**, 3532 (2000).

Wagner, J.-M. *et al., Phys. Rev. B* **62**, 4526 (2000).

Walukiewicz, W., *Appl. Phys. Lett.* **54**, 2094 (1989).

Wang, L. W., Large-scale local-density-approximation band gap-corrected GaAsN calculations, *Appl. Phys. Lett.* **78** (11), 1565 (2001).

Wang, Y. J., R. Kaplan, H. K. Ng, K. Doverspike, D. K. Gaskill, T. Ikedo, I. Akasaki, and H. Amano, *J. Appl. Phys.* **79**, 8007 (1996).

Weimann, N. G., L. F. Eastman, D. Doppalapudi, H. M. Ng, and T. D. Moustakas, *J. Appl. Phys.* **83**, 3656 (1998).

Weitzel, C., L. Pond, K. Moore, and M. Bhatnagar, Effect of device temperature on RF FET power density, in *Proc. of Silicon Carbide, III-Nitrides and Related Materials*, ICSI, August 1997, Stockholm, Sweden, Trans Publications, Ltd., Materials Science Forum, Vols. 264–268, p. 907 (1998).

Winterton, S. S., C. J. Peters, and N. G. Tarr, *Solid-State Electron.* **36**, 1161 (1993).

Witek, A., Some aspects of thermal conductivity of isotopically pure diamond-a comparison with nitrides, *Diamond Related Mater.* **7** (7), 962 (1998).

Wook, K., A. E. Botohkarev, H. Morkoç, Z. Q. Fang, D. C. Look, and D. J. Smith, *J. Appl. Phys.* **84**, 6680 (1998).

Wu, Y.-F., B. P. Keller, P. Fini, S. Keller, T. J. Jenkins, L. T. Kehias, S. P. DenBaars, and U. K. Mishra, High Al-content AlGaN/GaN MODFET's for ultrahigh performance, *IEEE Electron Device Lett.* **19** (2), 50 (1998).

Wu, Y., W. Jiang, B. Keller, S. Keller, D. Kapolnek, S. Denbaars, and U. Mishra, Low resistance ohmic contacts to n-GaN with a separate layer method, *Proc. Solid State Electron.* **41** (2), 75 (1997).

Xu, G. Y., A. Salvador, W. Kim, Z. Fan, C. Lu, H. Tang, H. Morkoç, G. Smith, M. Estes, B. Goldenberg, W. Yang, and S. Krishnankutty, Ultraviolet photodetectors based on GaN p–i–n and AlGaN(p)–GaN(i)–GaN(n) structures, *Appl. Phys. Lett.* **71** (15), 2154 (1997).

Yamaguchi, S., M. Kariya, S. Nitta, T. Takeuchi, C. Wetzel, H. Amano, I. Akasaki, Structural properties of InN on GaN grown by metalorganic vapor-phase epitaxy. *J. Appl. Phys.* **85** (11), 7682 (1999).

Yang, K., C. P. Hains, and J. L. Cheng, Efficient continuous-wave lasing operation of a narrow-stripe oxide-confined GaInNAs–GaAs multiquantum-well laser grown by MOCVD, *IEEE Photonics Technol. Lett.* **12** (1), 7 (2000).

Yu, E. T., G. J. Sullivan, P. M. Asbeck, C. D. Wang, D. Qiao, and S. S. Lau, Measurement of the piezoelectrically induced charge in GaN/AlGaN heterostructure field-effect transistors, *Appl. Phys. Lett.* **71**, 2794 (1997).

Yu, Y. F., B. P. Keller, S. Keller, D. Kapolnek, P. Kozodoy, S. P. Denbaars, and U. K. Mishra, *Appl. Phys. Lett.* **69**, 1438 (1996).

Zhu, Q. S., and N. Sawaki, *Appl. Phys. Lett.* **76**, 1594 (2000).

Advanced Semiconductor and Organic Nano-Techniques (Part II)
H. Morkoç (Ed.)

CHAPTER 3

Ultraviolet Photodetectors based on GaN and AlGaN

H. Temkin

TEXAS TECH UNIVERSITY, LUBBOCK, TEXAS

I. Introduction

Semiconductor detectors operating in the ultraviolet part of the spectrum would have a multitude of interesting applications, from astronomy and spectroscopy (Joseph 1995) to terrestrial detection of flame and rocket exhaust (Smith et al. 1999). At altitudes below 20 000 ft, absorption by the ozone layer virtually eliminates solar radiation in the spectral region of 240–280 nm. Photodetectors operating in this range, and not responding to solar illumination outside it, would not be limited by background radiation, that is, would be solar-blind. In order to be useful, and to compete with ultraviolet-sensitive photomultipliers, solar-blind photodetectors would have to satisfy a few difficult requirements:

1. high photoresponse in the 240–280 nm range;
2. low leakage currents resulting in noise equivalent power of $\sim 10^{-15}$ W;
3. complete insensitivity to visible light, that is, out of band rejection on the order of 10^8–10^9, implying very rapid sensitivity roll off.

If a semiconductor-based photodetector with these characteristics could be fabricated, other advantages intrinsic of semiconductor devices would follow. These features, mechanical and environmental robustness, low operating voltages, low power consumption, the ease of forming imaging arrays, etc., are well known from our experience with Si-based visible detectors and infrared detectors based on compound semiconductors (Pearsall and Pollack 1985).

Figure 1 illustrates detectivity of photodetectors available in different parts of the spectrum, from the ultraviolet (250–400 nm), to the visible (400–750 nm), to the near-infrared (750 nm and beyond). For comparison, we include a line denoting background-limited detectivity, that is, detectivity of an ideal photodetector. Si-based ultraviolet detectors suffer from low responsivity, resulting, in part, from visible-rejection filters that have to be included. Only photomultipliers offer high ultraviolet detectivities. In contrast, the visible and infrared regions offer a number of excellent detector choices. In the near-infrared, we have a direct bandgap InGaAs grown on InP, a material developed for telecommunication applications (Panish and Temkin 1993). Further into the infrared, another direct bandgap semiconductor, PbS, offers excellent room-temperature detectivities (Eisenman et al. 1977).

The missing semiconductor system on which high-performance ultraviolet detectors could be based turned out to be GaN–AlN (Morkoc et al. 1994; Mohammad and Morkoc 1996; Ambacher 1998; Morkoc 1998; Nakamura 1998; Nakamura and Fosol 1998; Pearton et al. 1999; Manasreh 2000).

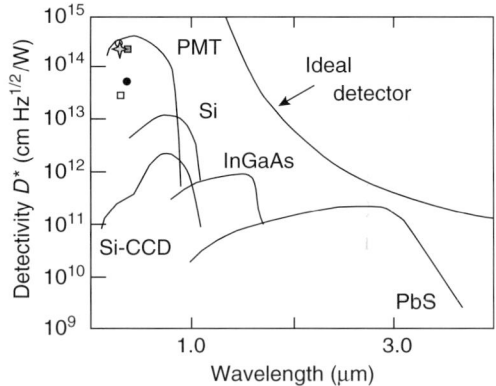

FIG. 1. Specific detectivity D^* plotted as a function of wavelength for a number of photodetectors, from the ultraviolet to the near-infrared. Detectivity of an ideal detector, limited by room-temperature background radiation, is plotted for comparison. Best results for AlGaN detectors are also presented: star (Dupuis 2002), open circle (V. V. Kuryatkov et al., unpublished results), open square (Kuryatkov et al. 2001), full square (Kuryatkov et al. 2001), and open circle (Brown et al. 1999b).

Khan *et al.* (1992) published the first detailed description of detectors based on GaN in 1992. His work was followed by a number of papers describing photoconducting detectors based on GaN and AlGaN (Stevens *et al.* 1995; Walker *et al.* 1996; Wickenden *et al.* 1997). A comprehensive review paper on ultraviolet detectors by Razeghi and Rogalski (1996) included GaN and AlGaN as possible detector materials. The authors recognized the potential of this material system but were concerned about the large density of defects, between 10^7 and 10^{10} cm^{-2}, which they believed would limit performance of III-nitride devices. Rapid progress in the epitaxial growth of III-nitrides and improved device design and fabrication of the last few years, have resulted in high-performance ultraviolet semiconductor photodetectors. Evolution of this field, from Schottky barrier detectors to p–n junction photodiodes and avalanche photodiodes, with a focus on improved electrical characteristics and larger detectivities, is described in this review. A vast majority of detectors described here were fabricated from epitaxial layers grown on sapphire substrates by metalorganic chemical vapor deposition (MOCVD). Only a few relied on techniques such as hydride vapor phase (Molnar *et al.* 1997) or molecular beam (Van Hove *et al.* 1997; Xu *et al.* 1997; Morkoc 1999) epitaxy.

II. Some Definitions

An ideal photodetector produces a current proportional to the incident optical power. The signal current I_p generated in a photodetector illuminated with monochromatic light of wavelength λ can be written as

$$I_\mathrm{p} = q\eta_\mathrm{ext} \frac{P_\lambda}{h\nu} \tag{1}$$

where P_λ is the incident power, $h\nu$ the photon energy, q the electron charge and η_ext the external quantum efficiency. The parameter η_ext includes the effects of the internal quantum efficiency, absorption depth, and surface reflectivity of the device. Current responsivity R_λ defines the photocurrent generated by the detector in response to the incident power P_λ. It can be then written as

$$R_\lambda = \frac{I_\mathrm{p}}{P_\lambda} = \eta_\mathrm{ext} \frac{q}{h\nu} = \eta_\mathrm{ext} \frac{q}{hc} \lambda \tag{2}$$

where c is the speed of light. Responsivity is proportional to the incident wavelength and to η_ext. In the ultraviolet spectral region of interest to us, it varies, for ideal $\eta_\mathrm{ext} = 1$, from 0.161 A/W at 200 nm to 0.294 A/W at 365 nm.

These considerations neglect any currents that might be present in the detector in the absence of any incident optical signal, known as dark currents. Such currents are sources of noise. In a detector under illumination, we can define the signal-to-noise ratio, a ratio of the photocurrent to the randomly varying noise current I_n, as

$$\frac{S}{N} = \frac{I_p}{\sqrt{\langle I_n^2 \rangle}} = \frac{R_\lambda P_\lambda}{\sqrt{\langle I_n^2 \rangle}} \tag{3}$$

Sources of noise common to ultraviolet detectors based on GaN and AlGaN are discussed below. The signal-to-noise ratio is proportional to responsivity and we want to maximize R_λ through better device design and improved materials. At same time, the dark current should be minimized.

In the case of an ideal detector responding to the incident power P_λ by generating photocurrent I_p, the mean-square noise current is $\langle I_n^2 \rangle = 2qI_pB$, where B is the response bandwidth of the post-detector filter or amplifier. In this shot-noise limited case, the signal-to-noise ratio is given by the ratio of squared currents

$$\frac{S}{N} = \eta \frac{P_\lambda}{2h\nu B} \tag{4}$$

and we can write the noise equivalent power (NEP), defined as the signal power required to yield $S/N = 1$, as

$$\text{NEP} = \frac{2h\nu B}{\eta} \tag{5}$$

In a photodiode detector operating at zero bias, and in the absence of any incident light, the thermal noise current arising from random thermal motion of charged carriers dominates the dark current. It can be written as (Chuang 1995)

$$\langle I_n^2 \rangle = \frac{4k_B TB}{R_0} \tag{6}$$

where k_B is the Boltzmann constant and T the temperature. R_0 is the zero bias resistance, or the differential resistance, and it clearly should be maximized in order to minimize dark currents. It can be shown that R_0 is proportional to minority carrier diffusion lengths on both sides of the p–n junction and thus increases in materials free of defects.

III. Noise in Photodetectors

Noise in semiconductors and semiconductor devices is generated by two fundamental processes, random and uncorrelated fluctuations in carrier densities or carrier velocities. These fluctuations are due to the random nature of the thermal generation–recombination (G–R) processes and the thermal motion of carriers, respectively. The noise processes are internal to the detector and noise is generated with and without the external signal, illumination in the case of photodetectors. The physics of noise in semiconductor devices is treated comprehensively in a number of textbooks (Yariv 1985; Motchenbacher and Connelly 1993; Chuang 1995; Donati 2000), monographs (Van Der Ziel 1976, 1986; Kingston 1978), and review papers (Hooge 1969; Vandamme 1994) and the details of it are beyond the scope of this review. Only brief comments and some definitions needed to follow experiments on ultraviolet detectors are discussed.

It is assumed that the noise current (or voltage, or resistance) exhibits frequency dependence. The noise current i_n averages to zero over time, and a power spectrum of noise $\langle i_n^2(f) \rangle$ can be defined instead, in any frequency interval df. It is then convenient to discuss noise in terms of spectral density of the power spectrum $S_I(f)$ defined as a derivative of the mean-square value of the noise current with respect to frequency,

$$S_I = \frac{d(\langle i_n^2 \rangle)}{df} \tag{7}$$

Spectral density can be also defined in terms of voltage noise, $\langle V_n^2(f) \rangle$, or resistance fluctuations, $\langle \delta R_n^2(f) \rangle$, where

$$\frac{S_I}{I^2} = \frac{S_V}{V^2} = \frac{S_R}{R^2} \tag{8}$$

The usual units of $S_I(f)$ and $S_V(f)$ are A^2/Hz and V^2/Hz, respectively. Since $S(f)$ describes the noise within a bandwidth of 1 Hz, the total device noise, needed to obtain the NEP defined by Eq. (5), is obtained by integrating the spectral density function over the spectral band of interest.

Experimentally, noise current of a device is measured over a certain frequency range, with a specified frequency resolution, amplified and analyzed with a fast Fourier transform spectrum analyzer. Specialized equipment now available, such as Stanford Research Model 7870 network signal analyzer, makes noise measurement quite reliable and the noise floor of the entire experimental apparatus can be lower than $10^{-25} A^2/Hz$.

The NEP of a photodetector was defined in Eq. (5) as the incident light power needed to produce a signal current equal to the noise current. In terms of the spectral power density function, the NEP can be defined more generally as

$$NEP = \frac{\sqrt{S_n(f)}}{R_\lambda} \tag{9}$$

where R_λ is the responsivity measured in A/W.

We can now define two figures of merit that are commonly used to describe photodetector performance, D and D^* (Kingston 1978). The detectivity D is simply the inverse of NEP, specified for a bandwidth of 1 Hz and D increases with improved detector performance. Specific detectivity D^* is the value of D normalized to a detector area of 1 cm^2 and the response bandwidth B of 1 Hz:

$$D^* = \frac{\sqrt{AB}}{\text{NEP}} \tag{10}$$

where A is the area of the detector and the unit of D^* is cm Hz$^{1/2}$/W. For a detector with an area of 1 cm^2 and a response bandwidth of 1 Hz, D is equal to D^*.

It is usual to make a distinction between the noise and background radiation limited performance (Kingston 1978). When the thermal noise of the photodiode detector is larger than the noise induced by fluctuations in the background radiation, we have $D_T^* = R_\lambda \sqrt{(R_0 A)/4kT}$, where R_λ is the responsivity and R_0 the differential resistance of the photodetector introduced in Eq. (6). The quantity $R_0 A$ is, thus, a useful figure of merit for photodiodes. When the photocurrent generated by the fluctuating background radiation Φ_B is dominant, specific detectivity becomes $D_{BL}^* = R_\lambda / q(2\eta\Phi_B)^{1/2}$.

Some comments on the origin of noise in photodetectors might be in order at this point. We divide semiconductor device noise into four types: (i) thermal, (ii) shot, (iii) generation–recombination, and (iv) $1/f$ noise. Each of these has a distinct frequency dependence and, therefore, unique spectral power density function, as plotted in Fig. 2. The noise spectrum of a device is

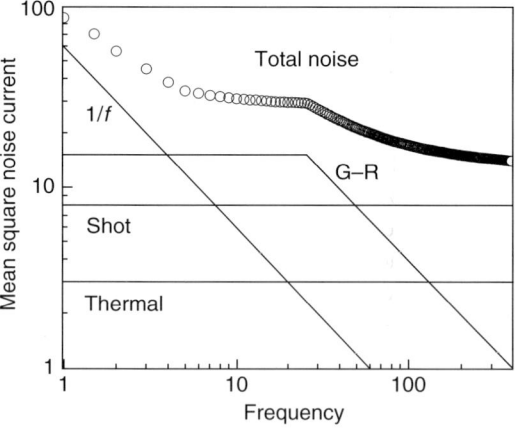

Fig. 2.   A plot of the photodetector noise current as a function of frequency. Specific types of noise are indicated as solid lines, circles indicate a sum of all contributions, the noise spectrum.

usually a weighted sum of all four types of noise. Thermal noise is usually very low and not important close to room temperature or at elevated temperatures; it establishes a high-frequency noise floor of a detector. At room temperature, high-quality detectors are shot-noise limited. The $1/f$ noise is dominant at low frequencies and is often seen in new detector materials and technologies. The G–R noise produces characteristic peaks in the noise spectrum, observable at intermediate frequencies. Measurements of noise as a function of temperature, frequency and current are needed to identify specifying noise contributions.

Thermal noise, also known as Johnson or Nyquist noise, is caused by fluctuations in the velocity of carriers due to random thermal motion. These result in instantaneous current flow seen as noise at terminals. It is a noise common to all conducting elements, including resistors and photodetectors, and its spectral density function can be written as, see Eq. (4), $S_I = 4k_BT/R$, where R is the resistance and T the temperature of the device. Resistance of photodiodes is usually high enough to limit the thermal noise to a low value. Ultimately, thermal noise imposes a floor on photodetectors that can be lowered only by reducing the ambient temperature.

Shot noise is generated by a series of discrete random events, shots, independent of each other. Its presence in semiconductors is usually associated with some kind of a potential barrier. Thus, phenomena as different as thermionic emission or diffusive motion of carriers, electrons and holes across a p–n junction give rise to shot noise. Absorption of a photon in a photodetector, which involves transition of a hole across the bandgap, is another example of a random process. When the detector is followed by a filter with the bandwidth B, the resulting mean-square photocurrent is $\langle i_n^2 \rangle = 2qIB$, where I is the average current (Kingston 1978). The spectral density function is then $S_I(f) = 2eI$, a characteristic signature of shot noise. Neither thermal nor shot noise are frequency dependent, both of these types of noise are known as "white".

The process of generation and recombination of free-carriers in a semiconductor gives rise to G–R noise. These processes are random in nature and could be considered as shot-noise type. However, the distinguishing characteristic is the presence of a well-defined time constant τ; hence, the carrier density changes with time as $dN/dt = N/\tau$. For a constant bias applied to the device, changes in the carrier density result in resistance fluctuations dR, which can be detected as current fluctuations $dI = V/dR$. It is, thus, convenient to describe G–R noise by

$$\frac{S_R}{R^2} = \frac{S_N}{N^2} = \frac{\langle(\Delta N)^2\rangle}{N^2}\frac{4\tau}{1+\omega^2\tau^2} \qquad (11)$$

The spectral density function $S_N(f)$ is of the low pass filter type, and is frequency dependent above the frequency ω corresponding to the characteristic time constant(s) τ, after which it decreases rapidly, with a slope of 20 dB/decade. Experimentally, carrier trapping and recombination at deep levels can be observed through measurements of $S_N(f)$.

Power spectrum of the $1/f$ noise, as the name would suggest, depends on frequency as $1/f^\gamma$, where α is a constant close to unity. At low frequencies, the $1/f$ noise might be the dominant noise component. This type of noise is known under many different names, for example, excess noise (Motchenbacher and Connelly 1993), flicker noise (Van Der Ziel 1976), and there appear to be many distinct physical phenomena producing (almost) the same $1/f$ characteristics. For instance, G–R noise associated with a large number of different recombination time constants would produce the $1/f$ signature (Van Der Ziel 1986). Devices as different as bipolar and field effect transistors are known to exhibit this type of noise. The $1/f$ noise is described by a characteristic spectral density function $S_n(f)$:

$$S_n = S_0 \frac{I_d^2}{f^\gamma} \tag{12}$$

where and S_0 and γ (which is usually close to unity) are fitting parameters.

IV. Spectral Response of GaN and AlGaN Detectors

Spectral response of semiconductor photodetectors is largely determined by their bandgap. The dependence of the bandgap of AlGaN on composition can be described as (Ambacher 1998)

$$E_g(x) = E_g(\text{GaN})(1-x) + E_g(\text{AlN})x - bx(1-x) \tag{13}$$

where $E_g(\text{GaN})$ is the bandgap of GaN, equal to 3.41 eV at room temperature, and $E_g(\text{AlN})$ the bandgap of AlN, equal to ~ 6.2 eV. The bowing parameter b is in the range of 1.0–1.5 eV. The bandgap of AlGaN can be, thus, continuously tuned from 3.4 to 6.2 eV, or from 365 to 230 nm. The response of photodetectors based on AlGaN can be tuned over this energy (wavelength) range by adjustments in the alloy composition. This is illustrated in Fig. 3, which plots normalized responsivity of $Al_xGa_{1-x}N$ photodetectors with the Al content x increasing from $x=0$ (GaN) to $x=0.70$ (Sandvik et al. 2001). Each of the photoresponse spectra, representing a different AlGaN composition, peaks at the bandedge. Responsivity drops abruptly below the edge, at longer wavelengths, by as much as four orders of magnitude. Responsivity also shows a slow drop at shorter wavelengths. These features of the response spectrum are typical to AlGaN detectors.

Performance of semiconductor-based photodetectors is inextricably connected to the perfection of materials in which they are implemented. Two

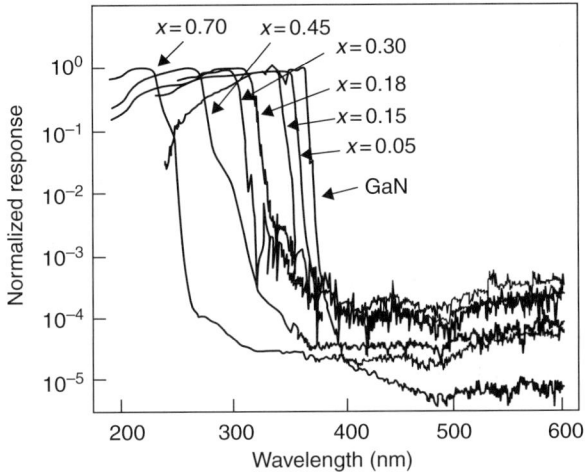

FIG. 3. Normalized spectral response of AlGaN photodetectors for a number of alloy compositions, from GaN to $Al_{0.7}Ga_{0.3}N$. After Sandvik *et al.* (2001).

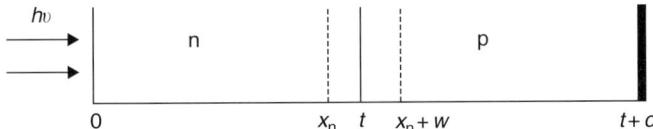

FIG. 4. A schematic cross-section of a photodiode structure used to model spectral response as a function of materials parameters.

parameters describing the semiconductor are of particular use in the description of semiconductors. These are minority carrier diffusion lengths, $L_{p(n)}$ for holes(electrons), and surface recombination velocity $s_{p(n)}$, in p- and n-type semiconductors. Hole diffusion length in n-type GaN increases from 1.2 to 3.4 µm, for carrier concentrations decreasing from 5×10^{15} to 2×10^{18} cm^{-3} (Chernyak *et al.* 1996). Diffusion length of minority electrons in p-type GaN appears considerably shorter, down to ~ 0.2 µm in the material with hole concentration of $1-4 \times 10^{17}$ cm^{-3}, reflecting the difficulty of obtaining high-quality p-type layers (Bandic *et al.* 1998). In semiconductors such as GaAs, diffusion length is greatly reduced by the presence of dislocations, which act as radiative recombination centers, but the role of dislocations in GaN is not well understood at this time.

The spectral response of semiconductor photodetectors can be described as a function of materials parameters based on a one-dimensional model shown in a schematic cross-section in Fig. 4. In this type of a diffusion-limited model, the diode structure is divided into n- and p-type, and depletion regions. A similar model can be used to describe Schottky barrier detectors.

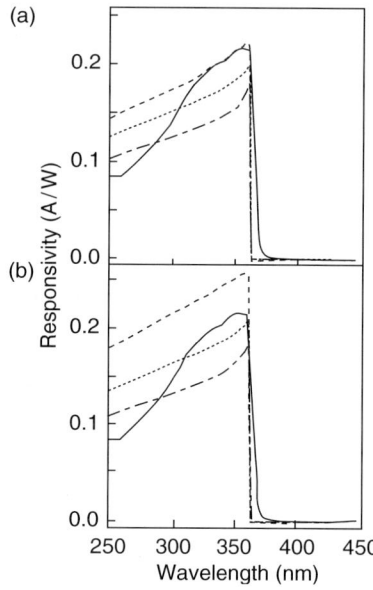

FIG. 5. Comparison of measured spectral response of a GaN-based photodetector and simulated spectra plotted for (a) constant surface recombination velocity s_p and different values of the minority hole diffusion length L_p, and (b) constant L_p and different values of s_p.

The spatial charge of width w surrounds the junction at $x = t$, and the two quasi-neutral regions $(0, x_n)$ and $(x = x_n + w, t + d)$ are uniformly doped. The depletion width w depends on the doping levels. The dark current consist of electrons injected from the n-side and holes injected from the p-side. The saturation current density can be then calculated from standard expressions (Razeghi and Rogalski 1996). A finite surface recombination velocity is assigned to each surface and interface of the device. The saturation current density is a function of minority carrier diffusion lengths, minority carrier diffusion coefficients, minority carrier concentration, detailed junction design, and surface recombination velocity. The total current in the photo-detector is the difference between the dark current and photocurrent, the latter given by Eq. (1); the two current components are assumed independent of each other. The wavelength dependence of responsivity is entered through the wavelength dependence of the absorption coefficient (Muth et al. 1999). The results are shown in Fig. 5, which also shows the experimental respon-sivity curve. Figure 5(a) shows simulated spectra plotted for a constant surface recombination velocity $s_p = 100$ cm/s and different values of the minority hole diffusion length. Figure 5(b) plots responsivity for a fixed value of the minority hole diffusion length $L_p = 2.0$ μm and different values of s_p. Longer diffusion lengths result in higher responsivity and less peaking at the bandedge. Lower surface recombination velocities also result in improved

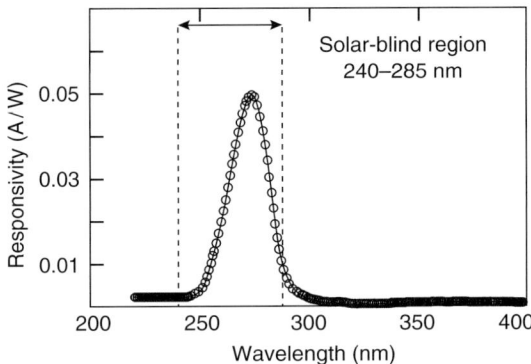

FIG. 6. Spectral response of a solar-blind photodiode based on AlGaN. After Brown *et al.* (2000).

responsivity and less of a decrease above the bandedge. Diffusion lengths increase in material with lower defect density while the effective surface recombination velocity can be controlled through passivation layers.

A solar-blind photodetector based on AlGaN, developed in the last few years, has a very different photoresponse spectrum, as illustrated in Fig. 6. A schematic cross-section of this device is shown in Fig. 12(c). The device is designed for illumination through the substrate. It consists of three layers. The middle, absorbing, layer has an Al content of ~ 0.5 in order to produce peak responsivity near 280 nm. The composition of this layer also determines the long wavelength response of the detector. The lower, n-type, layer of AlGaN determines the short wavelength response. This layer should be transparent at 280 nm and this is assured by increasing its Al content to about 0.6, corresponding to the bandedge of 4.66 eV or 266 nm. The bandgap of the upper p-type layer is designed for efficient collection of photogenerated carriers, slightly larger than that of the absorbing layer. This design results in photodetector response limited to the 260–280-nm range and high quantum efficiency (Brown *et al.* 2000).

V.   GaN-based Schottky Barrier Detectors

Schottky barrier photodetectors are simple to grow, since only a single dopant is required, with the *n*-type material used most often, and to fabricate. This makes them an excellent vehicle for evaluation of detector materials. However, despite the relative simplicity, these detectors have the potential of reaching high quantum efficiency and high speed. Considerable progress has been already achieved in the development of GaN-based Schottky barrier detectors with high unity gain responsivities and response bandwidth in the megahertz range (Khan *et al.* 1993; Binet *et al.* 1997; Carrano *et al.* 1997a,b; Chen *et al.* 1997; Osinsky *et al.* 1998b; Adivarahan *et al.* 2000; Zhao *et al.* 2000). Improved performance correlates with reduction in leakage currents, obtained through improved materials and more sophisticated processing methods, leading to lower noise.

GaN Schottky barrier photodetectors with high quantum efficiencies of $\eta \sim 50\%$ and responsivities of 0.18–0.19 A/W, in the absence of internal gain, could be fabricated by devoting careful attention to crystal growth and device fabrication (Carrano *et al.* 1997a, 1998; Adivarahan *et al.* 2000). The photoresponse of these detectors is fairly flat above the bandedge of ~ 350 nm, it drops by as much as three orders of magnitude below the bandedge. Schottky barrier detectors based on AlGaN are at this point somewhat less advanced than their GaN counterparts. Nevertheless, detectors prepared on alloys of AlGaN, with the Al content up to 0.35, exhibited ultraviolet spectral response peaking at 280–290 nm with the responsivity of 0.1 A/W (Omnes *et al.* 1999; Pau *et al.* 2000). A very large drop in the sub-band response (four orders of magnitude) was demonstrated, even in the absence of sophisticated post-growth fabrication. High insensitivity in the visible region was maintained even for AlGaN devices prepared on Si(111) substrates (Pau *et al.* 2000).

A number of implementations have been used to fabricate Schottky diodes on GaN and AlGaN. A general cross-section is shown in Fig. 7. The fabrication process starts by defining a Schottky mesa, by dry etching down to a heavier doped, n+, epitaxial contact layer prepared under the low-doped device layer. Once the mesa is formed, an ohmic contact is fabricated on the exposed surface, usually by a lift-off process, and annealed. The mesa-defined device is often called a "vertical" geometry diode. If properly fabricated, this type of a diode should exhibit low series resistance. It is also possible to form the ohmic contact on the top surface of the device, without mesa etching. Since both ohmic and Schottky contacts are placed side-by-side, such a device is often called a "lateral" diode. Once the ohmic contact is formed and annealed, the Schottky contact is prepared next, typically using Pd or Pt. This contact is not annealed. The Schottky metal need not be very thick; semi-transparent metal layers, less than 10 nm in thickness, are commonly applied for this purpose. As shown in Fig. 7,

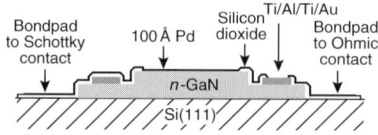

FIG. 7.   Schematic cross-section of a Schottky diode structure. The AlN buffer layer, at the substrate interface, is not shown. Schottky diodes grown on sapphire and silicon substrates are discussed here.

the diode can be illuminated either through the Schottky contact or, if formed on sapphire, through the transparent substrate. It is also possible to form the diode as two Schottky contacts back-to-back, often done in the form of interdigitated metal fingers, separated from each other by a few micrometers. The advantage of this process is its simplicity, since only one deposition step is needed, but the device geometry is not well defined. In the device illustrated in Fig. 7, the fabrication process also includes more robust connections to the Schottky and ohmic contacts, to be used for probing or wire bonding. These need to be isolated from the device with a layer of SiO_2.

A detailed study of Schottky diodes on GaN and mechanisms leading to excess leakage currents was carried out by Carrano et al. (1997b). Back-to-back Schottky diodes used in their study were fabricated as interdigitized metal–semiconductor–metal structures consisting of 2 μm wide fingers separated by 2, 5, and 10 μm in different samples. The metal contact started with 5 nm of Ti, deposited to assure good adhesion to GaN, followed by 80 nm of Pt, the Schottky metal with a relatively high work function (5.6 eV). Contact fingers were deposited into openings in a SiO_2 passivation layer, used also as an antireflection coating, deposited first on the surface of GaN.

Figure 8 shows current–voltage characteristics of Schottky diodes obtained on two samples of GaN, 1.5 and 4.0 μm thick, grown by MOCVD on sapphire substrates. The layers of GaN were not intentionally doped and had n-type carrier concentrations of $\sim 6 \times 10^{16}\,cm^{-3}$, as determined by Hall measurements. At low voltages Schottky diodes exhibit very low dark currents and current densities, in the mid-$10^{-6}\,A/cm^2$ range (Carrano et al. 1997a). Furthermore, diodes fabricated on thicker layers showed markedly improved I–Vs, both in terms of significantly lower dark current and lower breakdown voltage. Figure 8 also shows, as dashed lines, results of model calculations in which current transport at the Schottky contact was assumed to be either due to conventional thermionic emission, including barrier

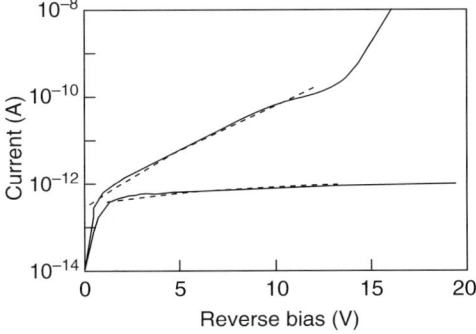

Fig. 8.   Current–voltage characteristics of Schottky diodes fabricated on GaN epitaxial layers of different thicknesses. Dashed lines show calculated I–Vs. After Carrano et al. (1997a).

lowering due to image force effects (Bhattacharya 1994), or to thermionic field emission (Tiwari 1992). The latter includes tunneling currents. Excellent agreement with the thermionic emission model is evident for devices fabricated on thicker GaN. The I–V of a device fabricated on a thin GaN layer shows a considerably larger slope of the leakage current, suggestive of larger tunneling component. The thermionic emission model provides a good fit to the data, confirming the importance of tunneling currents, and yielding a donor concentration of $N_d \sim 1 \times 10^{17}\,\mathrm{cm}^{-3}$, in good agreement with the measured value.

The thickness dependence of the leakage current was interpreted in terms of deep levels associated with threading dislocations. Compared to other compound semiconductors, dislocation densities in GaN are remarkably high, between 10^7 and $10^{10}\,\mathrm{cm}^{-2}$. This is true even in high-quality device structures. While in the early years of GaN dislocations were essentially ignored, there is now a growing body of evidence, some of which is reviewed here, that they are a serious problem affecting GaN-based devices.

The lattice mismatch between GaN and substrates such as sapphire is very large, $\sim 14.6\%$, and epitaxial layers nucleate by forming isolated three-dimensional islands. Threading dislocations are formed when these island coalesce into a continuous film. The density of threading dislocations is, thus, highest at, or close to, the substrate interface. The mechanisms that reduce the density of threading dislocations in a growing layer, either by annihilation or fusion of adjacent dislocations, are a subject of considerable theoretical interest (Mathis et al. 2001). Experimentally, the dislocation density drops off quite rapidly, by at least an order of magnitude in the first micron of growth (Wu et al. 1996). It is, thus, reasonable to assume, following Carrano et al. (1998), a substantially lower dislocation density in thicker layers of GaN, with the ensuing difference in the magnitude of the leakage current as well as its detailed mechanism.

While the work of Carrano et al. (1997b, 1998) stressed the importance of improved materials properties and lower defect densities, Adivarahan et al. (2000) and Deelman et al. (2001) concentrated on the importance of high-quality device fabrication in achieving low dark-current densities. Figure 9 shows current–voltage characteristics of Schottky diodes fabricated with and without a SiO_2 surface passivation layer (Adivarahan et al. 2000). The n-GaN layers, grown by MOCVD on sapphire substrates, were 1.2-μm thick and their carrier density was estimated at $n \sim 3 \times 10^{16}\,\mathrm{cm}^{-3}$. Epitaxial growth also included a buffer layer of AlN, approximately 80-nm thick. The basic device consisted of a semi-transparent Schottky contact surrounded by an annular ohmic contact, both deposited on to the top surface of GaN. The ohmic contact consisting of Ti/Al/Ni/Au was deposited first and annealed at 650 °C for 1 min. Schottky barrier contacts, with diameters from 50 to 400 μm, consisted of $\sim$5.0–7.5-nm-thick Pt layers, defined by a lift-off process. The separation between the ohmic and Schottky contacts was kept

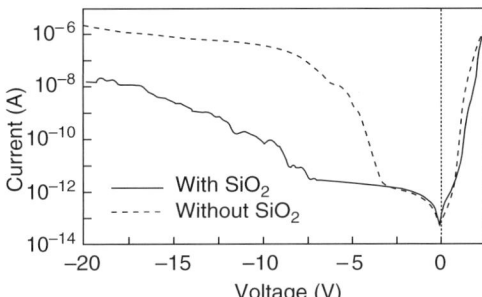

FIG. 9.   Current–voltage characteristics of Schottky diodes fabricated with and without SiO_2 surface passivation layer. After Adivarahan *et al.* (2000).

at 10 μm. A control process was carried out on a part of the same epitaxial wafer. It included deposition of a thin, 0.1 μm, layer of SiO_2 as a first step in the fabrication sequence. Schottky barriers were, thus, formed both directly on GaN and over the SiO_2 layer. Ohmic contacts were deposited into the opening in the SiO_2 layer. Plasma-enhanced CVD (PECVD) process used for SiO_2 deposition is characterized by low deposition temperature and results in low density of interfacial steps on GaN (Casey *et al.* 1996). The same type of SiO_2 was used by Carrano *et al.* (1997a) to form a passivation layer over their interdigitated detectors.

The SiO_2 passivation has a dramatic effect on leakage currents, particularly for bias above 5 V, wherein the dark current is reduced by factors as large as 10^2–10^4. The effect on dark currents at lower bias is less pronounced. At 5 V, a dark current as small as 1 pA was measured in a 200-μm diameter device. This corresponds to a very low leakage current density of $\sim 3 \times 10^{-7}\,A/cm^2$. The geometry of interdigitated detectors reported by Carrano *et al.* (1997a) makes direct comparisons difficult but their leakage currents, at 5 V, appear to be similarly low. It is clear that leakage current densities in diodes prepared on high-quality GaN can be reduced to the $10^{-7}\,A/cm^2$ range.

Noise performance of GaN detectors was first investigated by Chen *et al.* (1997). Their epitaxial structure consisted of a $\sim$40-nm-thick buffer layer of AlN, followed by a $\sim$1-μm-thick n^+-GaN layer doped n-type to $\sim 3 \times 10^{18}\,cm^{-3}$ and a $\sim$0.4-μm thick n^--GaN layer doped to $\sim 3 \times 10^{16}\,cm^{-3}$. Vertical-geometry diodes with semi-transparent Pd Schottky barrier contacts were fabricated by defining 4 mm $\times$ 4 mm mesas using reactive ion etching. The bottom ohmic contact was made to the n^+ layer and annealed at 800 °C for 30 s. The dark current of the completed devices was linearly dependent on the bias, suggesting a leakage path parallel to the Schottky barrier and ohmic contacts. The leakage was attributed to high density of threading dislocations, estimated at $\sim 10^9\,cm^{-2}$. At a reverse bias of $\sim$5 V, these diodes showed a dark current density of $\sim 3.8 \times 10^{-7}\,A/cm^2$.

The noise of GaN Schottky diodes is predominantly of the $1/f$ character, with the noise spectral density given by Eq. (12). The coefficient γ was close to unity. The characteristic value S_0 was proportional to the square of the average dark current, once again underscoring the importance of minimizing such currents. At all frequencies, up to the photodetector bandwidth $B \sim 3$ MHz, the $1/f$ noise power density S_n was much higher than the shot-noise density.

The mean square noise current was obtained by integrating the spectral density over the frequency range from 0 to B Hz:

$$\langle i_n^2 \rangle = \int_0^B S_n \, df = \int_0^1 S_0 \, df + \int_1^B \frac{S_0}{f} \, df = S_0 \left[\ln \left(\frac{B}{1 \text{ Hz}} \right) + 1 \right] \qquad (14)$$

where the limit of 1 Hz is imposed by the experimental setup. Assuming a load resistor of $R_L = 600 \, \Omega$, the noise power $P_n = \langle i_n^2 \rangle R_L \sim 2.9 \times 10^{-15}$ W was obtained. The NEP is found by dividing the mean-square noise current by the responsivity of the device, as defined by Eq. (9). Given the experimental $R \sim 0.18$ A/W at 360 nm, the NEP was calculated as 1.2×10^{-8} W. NEP normalized by the square root of the bandwidth was calculated as $\sim 7 \times 10^{-12}$ W/Hz$^{1/2}$. Very similar values of NEP were also obtained by Monroy et al. (1998) in lateral Schottky barrier detectors, from which specific detectivity of 6.1×10^9 cm Hz$^{1/2}$/W was calculated. These values for NEP and D^* are approaching the range of parameters typical of commercial Si-based ultraviolet detectors.

Noise characteristics of AlGaN Schottky barrier detectors with the Al content up to $x \sim 0.22$–0.26 were investigated by Osinsky et al. (1998b) and Monroy et al. (1998). Their devices, prepared on sapphire substrates with structures similar to those used for GaN, exhibited cutoff wavelengths as short as 290 nm. Vertical geometry diodes showed leakage current densities of $\sim 2.5 \times 10^{-3}$ A/cm^2, considerably larger than those of GaN devices. In order to understand the source of the excess current, diodes were fabricated with dimensions ranging from 50×50 to $500 \times 500 \, \mu$m^2. However, the dark current was observed to scale as 1.3 power of the detector size, about halfway between bulk leakage (power ~ 2) and surface leakage (power ~ 1), suggesting that both current components were contributing. Bulk leakage through dislocations and surface leakage due to damage during the mesa-etching process were suggested as possible sources (Osinsky et al. 1998b). The lateral devices of Monroy et al. (1998) also appeared quite leaky, as judged by the low diode resistance quoted, suggesting a surface-related leakage. With the 290-nm responsivity of 0.07 A/W and relatively high dark currents, the noise equivalent power of 6.6×10^{-9} W was obtained.

Schottky barrier diodes fabricated on GaN grown on Si(111) substrates by gas-source molecular beam epitaxy (Nikishin et al. 1999a, 2000) were investigated by Deelman et al. (2001). The detector structure consisted of a 40-nm-thick AlN buffer and a 1.5-μm-thick n-GaN layer. The Si surface was prepared prior to growth using a two-step etching procedure that produces an atomically clean hydrogenated surface (Nikishin *et al.* 1999b). The AlN buffer was grown in a two-dimensional growth mode. The deposition of GaN initially started in a three-dimensional mode, and transitioned within the first $\sim$0.1 μm to a two-dimensional growth mode. The background carrier concentration in the undoped GaN layers was $n \sim 5 \times 10^{17}$ cm^{-3}, as determined from capacitance–voltage measurements.

Vertical-geometry Schottky diodes, with a structure as illustrated in Fig. 7, were defined with Cl-based reactive ion etching. Mesa dimensions ranging from 50 to 500 μm^2 were used. The lower, ohmic, contact consisted of 30 nm Ti/70 nm Al/10 nm Ti/200 nm Au layers annealed at 600°C for 20 s in nitrogen ambient. The semi-transparent Schottky barrier was formed by a 10-nm-thick layer of Pd. The optical transmission spectrum of this layer is nearly flat in the wavelength range 200–365 nm, and the Schottky barrier height is approximately 0.9 eV (Chen *et al.* 1997). A 0.1-μm-thick SiO$_2$ layer was subsequently deposited by PECVD to provide both surface passivation and electrical isolation between ohmic and Schottky contacts. There was no SiO$_2$ under the Schottky metal. Windows over the Schottky electrodes were defined in the SiO$_2$ layer using standard dry etching. Finally, Ti/Au bond-pads to the electrodes were deposited.

The forward-bias current ($\sim$10^{-11} A at +1 V) of Schottky devices formed on GaN/Si(111) layers were comparable to that of other III-nitride Schottky diodes (Deelman *et al.* 2001). Assuming conventional thermionic emission and an effective Richardson constant $A^{**} = 24$ A/cm^2/K^2, the Schottky barrier height ϕ_{Bn} was determined from the forward-bias *I–V*s to be 0.9–1.1 eV, in agreement with previous results (Wang *et al.* 1996; Chen *et al.* 1997). An ideality factor $n \sim 15$ was obtained by fitting the low-level injection regime. This high value of n indicates the ohmic contact may retain some Schottky character (Krishnankutty *et al.* 1998). In addition to nonlinearity of the ohmic contact, interface contamination, insulating layers, and deep impurity levels are possible mechanisms for the high value of n that are not included in the idealized model of the Schottky diode (Sze 1981). Measurements of 86 × 86 μm^2 devices show dark currents of 4 pA at a reverse bias of 5 V, resulting in the very low dark-current density of 5.4×10^{-8} A/cm^2.

Measurements of the noise power density as a function of diode current are shown in Fig. 10. As is common for Schottky diodes prepared on GaN, at low frequencies the noise is 1/f limited (Chen *et al.* 1997). At zero bias, the noise spectral density measured at 1 Hz is 9×10^{-29} A^2/Hz. As expected, the noise power density scales with the dark current. For the lowest dark currents, noise power density of these Schottky diodes begins to approach the shot noise [$I_{shot} = (2eI)^{1/2}$] limit. This is a tremendous improvement over the initial performance of Schottky diodes on GaN (Chen *et al.* 1997).

Further reduction of low-bias dark currents and noise will require better understanding of the tunneling frequently encountered in Schottky diodes.

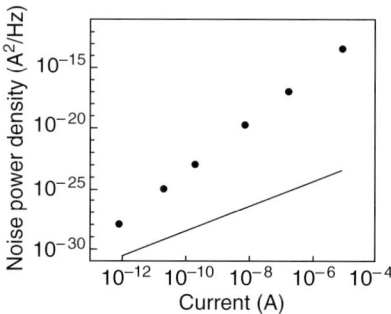

FIG. 10.   Noise power density measured as a function of reverse dark current for Schottky diodes prepared on GaN/Si(111). The solid line shows the expected shot-noise dependence. After Deelman *et al.* (2001).

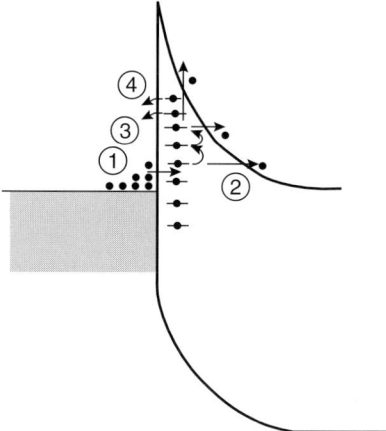

FIG. 11.   Band diagram of a Schottky diode interface illustrating specific trap-assisted tunneling processes invoked to explain observed *IV* anomalies. After Carrano *et al.* (1997b).

The presence of tunneling current also appears to correlate with rapid degradation of the current–voltage characteristics under voltage stress. The mechanism of tunneling, the signature of which is strong voltage dependence of the dark current at low bias, was investigated by Carrano *et al.* (1997b, 1998), who postulated a sequence of deep-level assisted processes in order to explain their experimental results. A schematic band diagram of the metal–semiconductor interface is shown in Fig. 11. With the initial application of external bias, electrons tunnel through the Schottky barrier to an interfacial state labeled 1 in Fig. 11. At least two possibilities are then available, tunneling through the remaining barrier, process 2, or thermal excitation through a set of deep states, process 3. Most of the defect states available would be filled upon completion of the first voltage sweep. This leaves only a few empty states available for deep-level assisted tunneling and the second voltage sweep would show a "degraded" current–voltage curve, lower but unstable dark current. Once the population of filled states reaches equilibrium, the current–voltage characteristic becomes stable. It is also possible to release the trapped electrons, process 4, by tunneling back to the Schottky metal. This gives rise to excess dark current at low bias. Some of the detrapping processes appear to be strongly influenced by illumination with white light. This frees defect states and results in increased leakage when dark-current measurements are made following illumination.

Another sign on interfacial trapping affecting Schottky barrier devices is the appearance of gain at forward bias. This can be a significant effect and current gains as large as 50, at a forward bias of 0.7 V, have been observed in GaN Schottky barrier detectors (Adivarahan *et al.* 2000). The effect is usually attributed to trapping of photogenerated holes at the metal–semiconductor barrier interface. Under reverse bias, applied field separates electrons and holes and gain disappears. Large gains observed in GaN photoconductive detectors are also attributed to carrier trapping at deep levels. This results in highly nonlinear response to the incident light, significant photoresponse to sub-bandgap light, and very long response times. The effect has been modeled in some detail by Garrido *et al.* (1998).

VI. GaN Homojunction Detectors

p–n junction based photodetectors are considerably more difficult to prepare than the Schottky diodes discussed above. Introduction of p-type dopant, Mg, into GaN is now well understood but the preparation of p-type AlGaN, particularly Al-rich alloys, still presents difficult problems. However, the possibility of significant reduction of dark currents in p–n junction devices is quite compelling and makes research on such devices worthwhile. Furthermore, recent advances in simultaneous introduction of both n- and p-type dopants (Adivarahan *et al.* 2001), known as co-doping, and the advent of doping superlattices (Waldron *et al.* 2001) promise a solution to the p-type doping problem in the near future. This will assure a dominant role of p–n junction detectors in most applications.

We discuss here three types of p–n junction photodiodes, as shown in Fig. 12: (a) homojunction of GaN (Chen *et al.* 1995; Osinsky *et al.* 1997, 1998a; Van Hove *et al.* 1997; Kuksenkov *et al.* 1998; Li *et al.* 1999; Tarsa *et al.* 2000); (b) heterojunction of GaN and AlGaN (Krishnankutty *et al.* 1998; Brown *et al.* 1999a,b; Collins *et al.* 1999; Kuryatkov *et al.* 2001); and (c) heterojunctions of AlGaN (Parish *et al.* 1999; Brown *et al.* 2000; Sandvik *et al.* 2001).

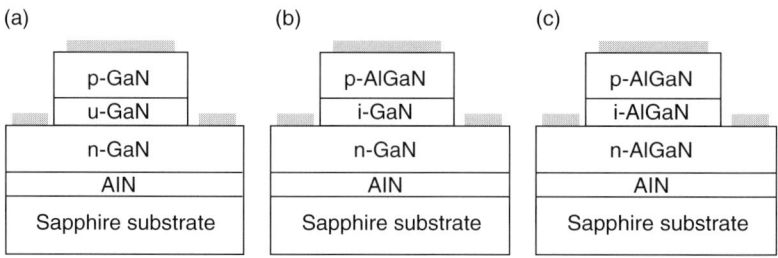

FIG. 12. Schematic cross-sections of the photodiode structures discussed: (a) GaN homojunction, (b) AlGaN/GaN single heterojunction, and (c) double heterojunction of AlGaN.

1. CURRENT–VOLTAGE CHARACTERISTICS

GaN homojunction detectors of the type illustrated in Fig. 12(a) were first investigated in detail by Kuksenkov et al. (1998). Photodiode structures, grown by MOCVD on sapphire, consisted of a thin AlN buffer layer, followed by a 0.8-μm-thick n-type layer of GaN (doped $n \sim 10^{18}\,\mathrm{cm}^{-3}$) and a 2.2-μm-thick p-type layer of GaN doped with Mg. The p-doping level varied along the growth direction from $\sim 10^{16}\,\mathrm{cm}^{-3}$ at the p–n junction interface to $\sim 7 \times 10^{17}\,\mathrm{cm}^{-3}$ at the contact surface. This resulted in a p–π–n structure, where π stands for a lightly doped compensated p-type layer sandwiched between more heavily doped p and n layers. Reactive ion etching was used to define rectangular mesas of individual diodes prior to ohmic contacts deposition.

The photodiodes responsivity reached $R \sim 0.14\,\mathrm{A/W}$ at 363 nm, similar to that of GaN Schottky barrier detectors described above. The responsivity dropped by more than three orders of magnitude above the cut-off wavelength. The photodiode bandwidth of $\sim 32\,\mathrm{MHz}$ was determined from detector response to ~ 1-ns-long laser pulse.

Reverse-bias current–voltage characteristics of a $200 \times 200\,\mu\mathrm{m}^2$ GaN photodiode, measured in the temperature range of 290–570 K, are shown in Fig. 13. All of the I–V curves show an exponential dependence of the dark current on the bias voltage. An increase in the dark current with the device temperature for a given value of reverse bias is also exponential.

The exponential dependence of current on both the voltage and temperature is difficult to explain by conventional models of reverse-bias conductivity (Razeghi and Rogalski 1996). Measurements on devices with different sizes showed that the current scaled with the device area, rather than the perimeter of the active region, thus ruling out any significant contribution of leakage currents at the device mesa sidewalls. Due to the large bandgap of GaN and extremely small thermal excitation rate, diffusion currents in the neutral region and G–R currents in the depletion region should be negligibly small. Even though the room-temperature dark current for low reverse bias

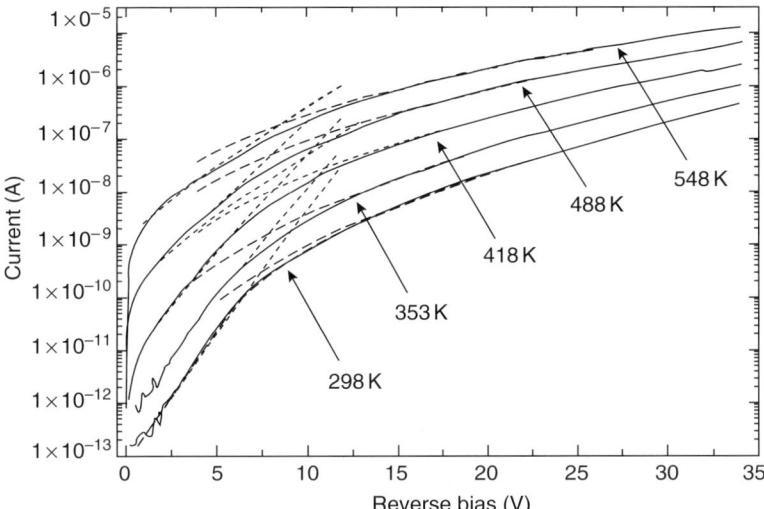

Fig. 13. Dark current as a function of reverse bias for a $200 \times 200\,\mu m$ GaN homojunction photodiode measured at different temperatures. Fitting results are shown by dashed lines for the low-bias data and by dotted lines for high-bias data (Kuksenkov *et al.* 1998).

was less than 1 pA, it is still 10^6 times higher than the saturation current of $\sim 2 \times 10^{-18}$ A expected from thermal excitation of carriers in GaN. The onset of impact ionization in the photodiodes under test was observed only at a bias greater than -40 V and, therefore, it was not expected to contribute significantly to the measured *I–V* characteristics. Finally, direct band-to-band or trap-assisted tunneling current has exponential voltage dependence, but should be only weakly dependent on temperature.

   The conductivity of GaN diodes was modeled by assuming that the current is due to hopping of charged carriers via localized defect-related states (traps) in the depletion region. Hopping was studied extensively in amorphous semiconductors, where it was found to be the dominant mechanism of current flow, especially at low temperatures (Morgan and Walley 1971). Hill (1971) and Pollak and Riess (1976) provided a description of the current-field dependence, for moderate electric fields, as

$$j = j(0)\, \exp\left(C\frac{eFa}{2kT}\left(\frac{T_0}{T}\right)^{1/4}\right) \qquad (15)$$

where j is the current density, $j(0)$ the low-field current density, T_0 a characteristic temperature parameter, T the temperature, k the Boltzmann constant, F the electric field, e the electron charge, C a constant of the order of unity, and a the localization radius of the electron wave function. At very low fields, hopping conductivity $\sigma(F) = j(F)/F$ is expected to follow Mott's law for variable-range hopping (Mott 1969):

$$\sigma(0) \propto \exp\left(-\left(\frac{T_0}{T}\right)^{1/4}\right) \tag{16}$$

where T_0 is the same parameter as in Eq. (15). The validity of Eq. (15) was experimentally demonstrated for many amorphous solids (Aspley et al. 1978; Mackintosh et al. 1983).

Equation (15) was used to fit the I–V data for reverse bias of −1 to −6 V. For this bias range, the total current through the p–n junction was equal to the saturation current originating in the depletion region and its density given by Eq. (15). The depletion region width w was extracted from the measured capacitance–voltage (C–V) characteristics. The average electric field at a given reverse bias V was then found from $F = (V + V_i)/w$, where V_i is the built-in junction voltage (estimated at ~ 1 V), and substituted into Eq. (15). Assuming a localization radius of $a = 10$ Å (Knotek et al. 1973), the best fit to the experimental data at all temperatures, shown in Fig. 13 by dashed lines, was obtained for the constant $C = 0.4$, a median value between $C = 0.8$ obtained by Hill (1971) and $C = 0.17$ obtained by Pollak and Riess (1976).

The characteristic temperature T_0, extracted from the temperature dependence of the extrapolated zero-field current $j(0)$, is $T_0 = 1.16 \times 10^{10}$ K. The density of states, per unit energy, taking part in the hopping conduction was then estimated from $N \sim 18/(kT_0 a^3)$ (Seager and Pike 1974) and then the value of $N \sim 1.8 \times 10^{16}/\text{cm}^3/\text{eV}$ was calculated. The T_0 obtained through this analysis is approximately three orders of magnitude higher than the $T_0 \sim 10^7$ observed in amorphous semiconductors such as germanium, silicon and heavily implanted GaAs (Kuriyama 1997). Correspondingly, the value of N is three orders of magnitude lower, which explains the extremely strong temperature dependence of the reverse-bias current observed in GaN photodiodes.

Hopping conductivity, previously observed only in amorphous solids or at very low temperatures, was attributed by Kuksenkov et al. (1998) to a large density of threading dislocations in epitaxial GaN. In a 200×200 μm^2 device, given a typical dislocation density of 10^8 cm^{-2}, one could expect on the average $\sim 40\,000$ dislocations. The dangling bonds at dislocation boundaries are likely candidates for the equivalent of localized states taking part in hopping conduction (Bagraev et al. 1989).

Equation (15) is expected to apply for fields $F \ll (2kT/ea) \sim 5 \times 10^5$ V/cm at room temperature. It is evident from Fig. 13 that the slope of the measured I–V characteristics changed noticeably for reverse-bias voltages higher than ~ 7 V, corresponding to the field of $\sim 2.2 \times 10^5$ V/cm. At higher bias, the diode conductivity was attributed to a combination of hopping and Poole–Frenkel effect, or field-assisted thermal ionization of carriers from traps in the depletion region. Similar behavior was previously observed in amorphous germanium (Morgan and Walley 1971). For sufficiently high electric fields, a single hop can take a charge carrier into the conduction band, instead of taking it to the other localized state. Poole–Frenkel conduction was observed in thin films of various wide-bandgap materials, including AlN and diamond (Khan *et al.* 1994; Gonon *et al.* 1996). It was shown that the field dependence of the current density could be written as (Simmons 1971)

$$j = j_0 \exp\left(\frac{\beta_{PF} F^{1/2}}{kT}\right), \quad \beta_{PF} = \left(\frac{e^3}{\pi \varepsilon \varepsilon_0}\right)^{1/2} \tag{17}$$

where j_0 is the low-field current density, β_{PF} the Poole–Frenkel constant, ε_0 the vacuum permittivity and ε the high-frequency dielectric constant of the material. For GaN, taking $\varepsilon(\infty) = 5.35$, $\beta_{PF} = 3.3 \times 10^{-4}\,eV/V^{1/2}\,cm^{1/2}$ was calculated (Kuksenkov *et al.* 1998).

The *I–V* curves calculated according to Eq. (17) (shown by dotted lines in Fig. 13), demonstrate an excellent agreement with the experimental data for high bias. At all temperatures, the best fit was obtained for the Poole–Frenkel constant $\beta_{PF} = 4.5 \times 10^{-4}\,eV/V^{1/2}\,cm^{1/2}$. It is interesting to note that for high bias Eq. (17) predicts currents much higher than those observed experimentally. This can be explained assuming that the width of the energy band for localized states near the Fermi level is limited to a certain value ΔE. The hopping conduction with electric field will saturate when the energy acquired by a charge carrier moving against the field eFR, where R is the length of the jump, becomes comparable to ΔE. At room temperature, for $F = 2.2 \times 10^5\,V/cm$, the estimated ΔE was $\sim 0.1\,eV$. This width is typical of dislocation-related energy bands (Bagraev *et al.* 1989). Consequently, $N' = N \Delta E \sim 2 \times 10^{15}\,cm^{-3}$ was obtained for the spatial density of hopping conduction centers. Another conclusion was that even at high bias, hopping is responsible for a significant part of the diode current, which can explain the experimentally estimated β_{PF} being ~ 1.4 times its theoretical value.

The detailed analysis of *I–V* characteristics of GaN diodes provided by Kuksenkov *et al.* (1998) pointed out the importance of dislocations as a source of excess dark conductivity. This was confirmed by elegant experiments of Kodozoy *et al.* (1998) on GaN diodes prepared by lateral epitaxial overgrowth (LEO).

LEO is a method of reducing the high density of threading dislocation in GaN (Zheleva *et al.* 1997), illustrated schematically in Fig. 14. Epitaxial layers of GaN, grown under standard growth conditions and, therefore, with high dislocation density are masked with a layer of SiO_2. A second epitaxial growth is then carried out in windows opened in the SiO_2 mask. In MOCVD growth, there is no decomposition of metalorganics on the mask layer and, therefore, no growth. GaN grown in the windows is highly dislocated, it replicates the first layer of GaN. However, once the growth clears the mask,

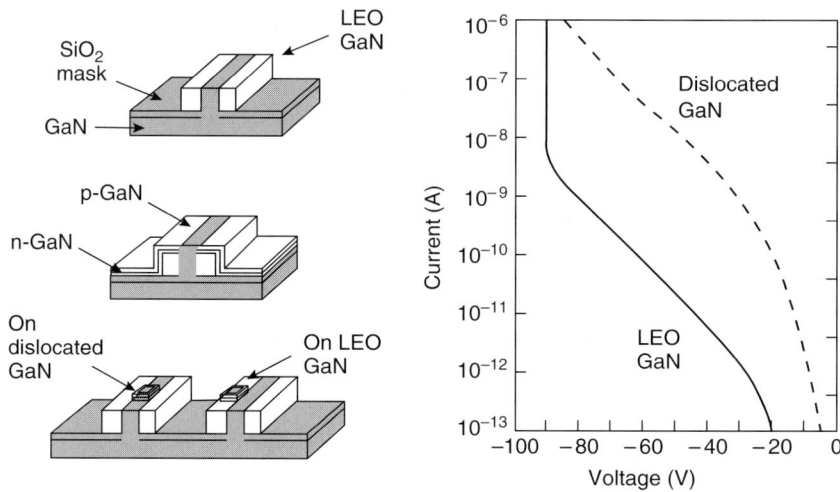

FIG. 14. Lateral epitaxial overgrowth (LEO) experiments demonstrating the importance of dislocations. Diodes fabricated on LEO material show a significant reduction in leakage currents. After Kozodoy *et al.* (1998).

GaN starts growing in the lateral direction and the laterally grown GaN shows dramatically reduced density of threading dislocations. For instance, while the dislocation density in the window region can be as high as $4 \times 10^8 \, \text{cm}^{-2}$, in the laterally grown GaN wings it drops to less than $10^6 \, \text{cm}^{-2}$. Given a suitable window pattern and high growth rates, laterally grown layers from adjacent windows can be made to coalesce forming a planar structure.

In the experiment of Kozodoy *et al.* (1998) discussed here, the mask windows were 5-μm wide and separated by 45-μm. Once the growth through windows was carried out and the laterally grown GaN produced, resulting in a mesa of about 8 μm high and 8 μm wide, the wafer was removed from the epitaxial apparatus and the mask removed. The subsequent growth of GaN n- and p-type layers was then carried out over the entire surface of the wafer. The p–n junction consisted of a 1-μm-thick layer of n-type GaN, with the electron concentration of $5 \times 10^{16} \, \text{cm}^{-3}$, followed by a 0.5-μm-thick layer of Mg doped GaN, with a hole concentration of $1 \times 10^{18} \, \text{cm}^{-3}$. Two types of small diodes, $2 \times 20 \, \mu\text{m}^2$ in size, were fabricated on top of laterally grown mesas, as illustrated in Fig. 14. One, placed directly over the window in the first SiO$_2$ mask was expected to have as many as 200 dislocations each. The other, placed to the side of the window, on laterally grown GaN, was expected to have no threading dislocations.

A comparison of *I–V* curves obtained on dislocated and dislocation-free diodes shown in Fig. 14 is quite revealing. Elimination of threading dislocations has resulted in greatly reduced dark current, by some three orders of magnitude. In dislocation-free devices, dark currents at bias below 20 V were too low to be measured. Their dark current density was estimated to be lower than 2.5×10^{-7} A/cm^2. In the absence of dislocations, the low-bias current showed activated temperature dependence, suggesting a deep trap.

2.   NOISE MEASUREMENTS UNDER REVERSE BIAS

Measurements of noise power density carried out on GaN homojunction diodes as a function of frequency and temperature are shown in Fig. 15 (Kuksenkov *et al.* 1998). The reverse bias was varied from -5 to -30 V, and the sample temperature from 298 to 523 K. The data shown in Fig. 15 were obtained at a reverse bias of 10 V. The equipment was calibrated by measuring the room-temperature noise current of a conventional 600 Ω resistor. The noise floor of the setup, determined by the noise parameters of the preamplifier, is $\sim 10^{-25}$ A^2/Hz. At elevated temperatures and/or high values of the reverse bias, where the dark current of the photodiode increased above ~ 1 nA, $1/f$ noise was found dominant. All of the measured noise spectra satisfy Eq. (12), the standard expression for the $1/f$ noise, with $1.0 \leq \gamma \leq 1.1$. The values of s_0, the parameter relating the noise power to the dark current are plotted in Fig. 16 as a function of bias and temperature. The s_0 decreased slightly with increasing bias and rapidly with increasing temperature.

Two models of the $1/f$ noise in hopping conductivity have been proposed, both of which are related to the presence of localized states separated from

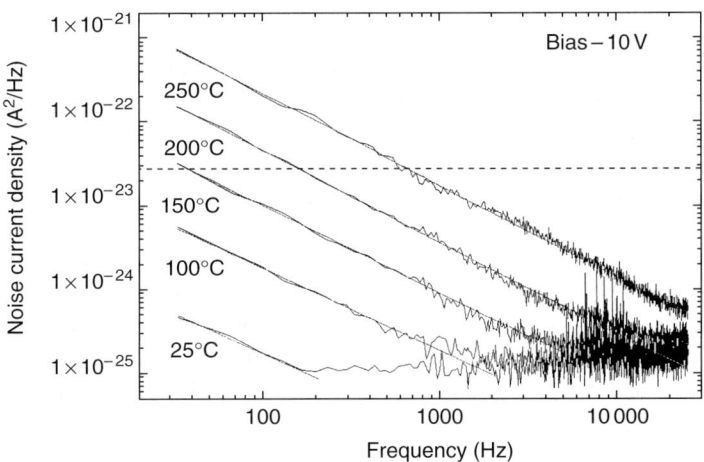

FIG. 15.   Dark-current noise spectra and corresponding fits at five different temperatures. Noise current density of a 600 Ω resistor is shown for comparison as a dashed line (Kuksenkov *et al.* 1998).

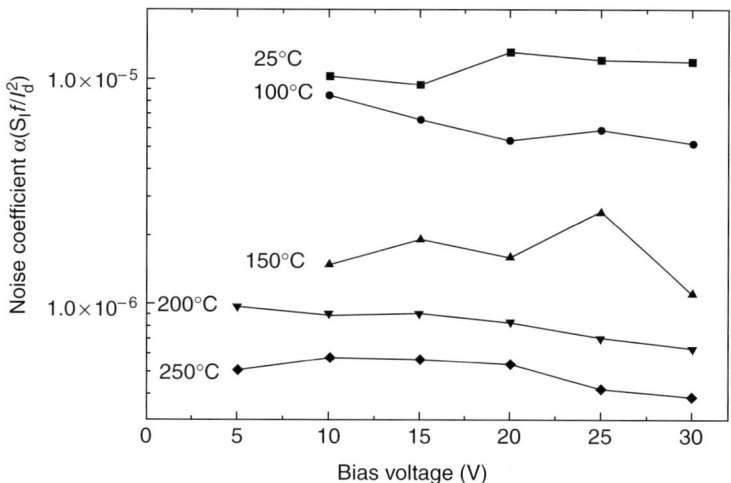

FIG. 16. Bias and temperature dependence of the proportionality coefficient between dark current and noise power.

each other by distances longer than the characteristic electron jump length. Such sites are not incorporated into the so-called critical current-carrying network, but function as "dead ends", slowly exchanging electrons with sites within the network. In the number fluctuation theory (Shklovskii 1980), this leads to slow fluctuations in the total number of electrons taking part in hopping conduction. The "mobility" fluctuation theory (Kozub 1996) assumes that trapping of an electron at a "dead end" changes the distribution of charge in the material and shifts the energy levels of neighboring traps. This leads to slow fluctuations in the effective "mobility" of hopping electrons.

The "mobility" fluctuation theory predicts a very weak temperature dependence of the low-frequency hopping noise and it does not appear applicable to GaN homojunctions. The number fluctuation theory results in the frequency dependence of the noise spectral density given by a specialized form of Eq. (12) proposed by Hooge (1969):

$$S_n = \alpha \frac{I_d^2}{f\overline{N}} \tag{18}$$

where $\overline{N}$ is the average number of electrons taking part in the conduction process and the value of the Hooge parameter α is expected to be of the order of unity. The parameter α can be estimated from the following consideration. From Eqs. (12) and (18), one has $\alpha = s_0 \overline{N}$. At room temperature and reverse bias of 10 V, s_0 is $\sim 1.3 \times 10^{-5}$. At the same temperature and bias, the dark current of the photodiode is ~ 1 nA, corresponding to the diode resistance

$R \sim 1 \times 10^{10}\,\Omega$. The number of electrons $\overline{N}$ can be then found from $R = L^2/(e\mu\overline{N})$, where the length L is taken to be equal to the depletion layer width w, obtained from the C–V data. The effective mobility μ is estimated from the theory of hopping conduction (Hill 1971) using parameters obtained from the fit to the I–V data at room temperature. This analysis yields $\mu \sim 5 \times 10^{-6}\,\text{cm}^2/\text{V}^1\text{s}^1$, $\overline{N} \sim 2.3 \times 10^5$, and, finally, $\alpha \sim 3$, consistent with expectations of the number fluctuation theory. Both the bias and temperature dependence of the parameter s_0 can be explained by the change in the number of hopping electrons $\overline{N}$. With increasing bias, $\overline{N}$ goes up due to increasing width of the depletion layer, and with temperature, due to decreasing length of a typical jump.

At room temperature and for reverse bias below 10 V, the dark-current noise spectra of GaN diodes disappear completely under the noise floor of the measurement apparatus. The device performance can be estimated by extrapolating noise data taken at higher bias. The dark current at −3 V was 2.7 pA, resulting in S_n (at 1 Hz) $= 7.3 \times 10^{-29}\,\text{A}^2/\text{Hz}$. At this bias, the magnitude of the $1/f$ noise for frequencies above $\sim$85 Hz is lower than that of the shot noise $S_{\text{shot}} \approx 8.6 \times 10^{-31}\,\text{A}^2/\text{Hz}$. The corresponding NEP ($f \geq 100$ Hz) was $6.6 \times 10^{-15}\,\text{W}/\text{Hz}^{1/2}$, comparable to the best results reported for ultraviolet-enhanced silicon photodiodes. Low-noise GaN photodetectors were also prepared by gas source molecular beam epitaxy (Xu *et al.* 1997; Morkoc 1999).

3.   NOISE MEASUREMENTS UNDER FORWARD BIAS

Under a forward bias, the built-in electric field decreases, hopping is no longer a dominant conductivity mechanism and the dark current is dominated by diffusion currents in the neutral regions (Kuksenkov *et al.* 1998). The $1/f$ current noise disappears and instead most devices show a very clear G–R noise component with a Lorentzian spectrum of the form:

$$S_n(\omega) = A\frac{\tau_0}{1 + (\omega\tau_0)^2} \tag{19}$$

where A is a constant and τ_0 the characteristic time of the G–R process. For the accurate determination of τ_0 from experimental data, it is convenient to multiply the measured noise density S_n/I_d^2 by $\omega = 2\pi f$ and plot it with a linear scale as a function of frequency f (Jones 1994). The Lorentzian then appears as a symmetrical peak at f_0, and τ_0 is found from $\tau_0 = 1/2\pi f_0$.

Figure 17 plots the frequency dependence of the normalized noise density under forward bias, at three different temperatures. It is clear that f_0 is increasing and, therefore, τ_0 is decreasing with temperature. For a thermally

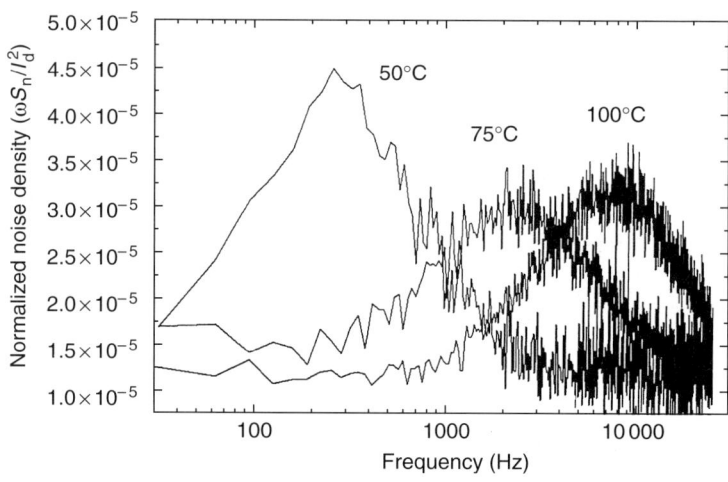

FIG. 17.   Example of noise spectroscopy: normalized noise spectra of a $200 \times 200\,\mu m$ GaN homojunction detector, at a *forward bias* of 2 V, at different temperatures.

activated *G–R* process, the characteristic time τ_0 is expected to follow an Arrhenius equation:

$$\tau_0 = \tau_{00}\,\exp\left(\frac{\varepsilon}{kT}\right) \tag{20}$$

where τ_{00} is a constant and ε the activation energy. The value of ε can be estimated from the slope of $\log(\tau_0 T^2)$ plotted against the inverse temperature (Copeland 1971). A value of $\varepsilon \sim 0.49\,eV$ was obtained (Kuksenkov *et al.* 1998), in agreement with the expected activation energy for centers associated with Ga antisite defects in GaN (Tansley and Egan 1993).

4.   Noise Measurements Under Illumination

Noise measurements on GaN photodiodes under illumination were performed by Kuksenkov *et al.* (1998). The photodiode bias was kept at $-5\,V$, and the intensity of light produced by a Xe arc lamp was adjusted by changing the lamp current. At low frequencies, the noise of the lamp itself was dominant. Above $\sim 10\,kHz$, the measured noise spectrum was completely flat. The corresponding noise current density is plotted in Fig. 18 as a function of the photocurrent I_p. The noise changes with illumination as $\sqrt{2eI_p}$, expected from the shot noise theory (Van Der Ziel 1986). No other noise sources appear to be present.

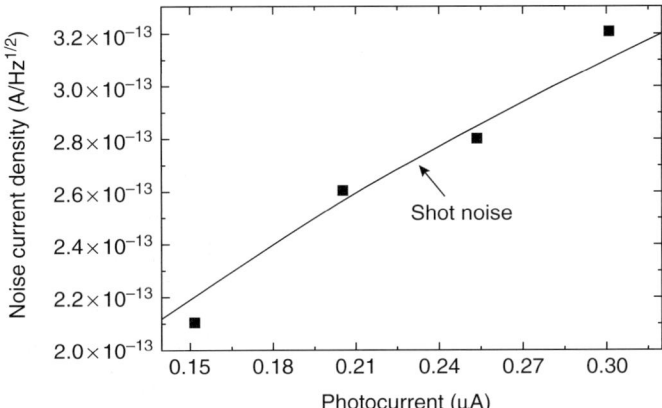

FIG. 18.   Measured spectral noise density at 20 kHz, under illumination, as a function of photocurrent. Solid line shows the calculated shot-noise level.

VII.   AlGaN–GaN Heterojunctions

GaN homojunction photodiodes frequently suffer from relatively low external quantum efficiencies and nonuniform spectral response that shows a peak at the band edge $\sim$360 nm and a fall-off at shorter wavelengths (Chen et al. 1995). At shorter wavelengths, the absorption coefficient of GaN increases and the light is absorbed very close to the surface. Most of the electrons and holes that are absorbed within $\sim$0.14 μm of the surface will recombine before they can diffuse to the depletion region. As a result, the efficiency at these shorter wavelengths decreases. Addition of a p-AlGaN window layer allows more of the incident light to propagate to the intrinsic absorption layer, which results in improved quantum efficiency (Li et al. 1999). However, higher resistivity of AlGaN results in an electric field crowding effect, leading to spatial variations in the responsivity and the time response. To reduce the field crowding, a GaN/AlGaN heterojunction p–i–n photodiode with a semi-transparent p contact was designed (Collins et al. 1999). A schematic cross-section of such a single heterojunction device is illustrated in Fig. 12(b).

Noise measurements of recessed-window $Al_{0.13}Ga_{0.87}N/GaN$ p–i–n photodiodes with semi-transparent p-contacts were carried out by Kuryatkov et al. (2001). The diode structure was grown on sapphire substrate by metalorganic vapor phase epitaxy. A thin p metal of 50 Å Ni/100 Å Au covered the entire mesa, resulting in a uniform distribution of the electric field throughout the device-active volume. Quantum efficiency of these photodiodes reached 38% at 15 V. The response remained relatively flat in the spectral region from 360 to down to 330 nm. The 300-nm efficiency was lower, about 22% at −15 V.

Room-temperature I–V characteristics typical of AlGaN/GaN single heterojunctions are shown in Fig. 19. Current density is plotted as a function of bias for two devices, with mesa diameters of 50 and 250 μm. All the diodes measured showed very low leakage currents at low-bias voltages, less than 1×10^{-10} A/cm^2 at -5 V. In a diode with 50-μm diameter, the dark current at -5 V was below 10^{-14} A. Room-temperature currents at lower bias could not be thus measured very accurately and the zero-bias performance was estimated by extrapolation. The current density curves for the two device sizes overlap, indicating scaling with the area of the device. Also shown is the I–V curve obtained with the photodiode illuminated with ~360 nm light from a GaN light emitting diode. The purpose is to show that detectors discussed here exhibit flat photoresponse down to zero bias.

I–V measurements carried out at higher temperatures, up to 250°C, showed the dark-current activation energy of 0.4 eV. Single activation energy was obtained for the reverse-bias range from -1 to -20 V. The same activation energy was obtained previously from the gate leakage measurements on AlGaN/GaN field effect transistors (Rumyantsev *et al.* 2000).

The zero-bias detector resistance, R_0, is estimated from room-temperature I–Vs and their temperature dependence. The quantity R_0A, a product of zero-bias resistance and the area of the diode is a commonly used figure of merit, see Eq. (6). For the larger diodes, 250 μm in diameter, $R_0 = 2.5 \times 10^{14}$ Ω was measured, with a corresponding $R_0A \sim 1.13 \times 10^{11}$ Ω cm^2. The dark current of the smaller device was too low to make a direct measurement; the estimated R_0 was greater than 4×10^{15} Ω. Taking the measured

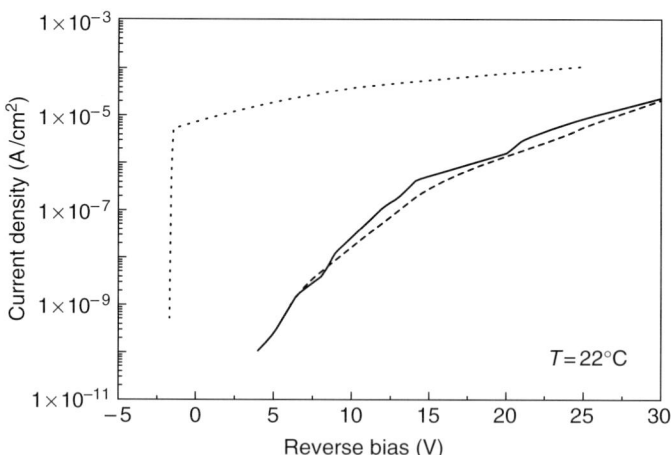

Fig. 19. Reverse current density of a GaN/AlGaN single heterojunction diode plotted as a function of the bias voltage. Dark current density is shown for two devices, solid line represents a 50-μm diode and the dashed line a diode with 250 μm diameter. The dotted line shows current–voltage characteristics under illumination. After Kuryatkov *et al.* (2001).

$R_\lambda \sim 0.1$ A/W and the R_0As discussed above, room-temperature thermal noise-limited detectivity $D^* \sim 3.2 \times 10^{14}$ cm Hz$^{1/2}$/W was obtained at 360 nm. This detectivity, illustrated in Fig. 1 as a solid square, is close to that of a UV photomultiplier. High values of $R_0A \sim 1.5 \times 10^9\,\Omega\,\mathrm{cm}^2$ were reported for similar single heterostructure AlGaN/GaN diodes by Brown *et al.* (1999b), resulting in room-temperature detectivities of $D^* \sim 6.3 \times 10^{13}$ cm Hz$^{1/2}$/W.

Figure 20 shows the noise power density measured for AlGaN/GaN devices at 1 Hz, as a function of reverse bias. A $1/f$ type noise was observed for reverse bias below 10 V. For a higher bias level, the dependence changed to the $1/f^\gamma$ type, where $\gamma = 3$. This suggests that the noise is due to a number of G–R centers (Kim *et al.* 1999). At frequencies higher than 1 kHz, only the thermal and shot noise were detected. The overall noise levels were low and, at room temperature and for frequencies greater than $\sim$200 Hz, the signal was below the experimental floor ($\sim 2 \times 10^{-29}$ A^2/Hz). Measurements of the temperature dependence of the noise power density showed activated behavior, with the activation energy of 0.37–0.42 eV, consistent with I–V measurements. The data of Fig. 20 show a well-defined exponential dependence of S_n with bias, changing by eight and five orders of magnitude, as the bias was changed from -5 to -30 V, in larger and smaller devices, respectively. This well-behaved bias dependence made it possible to extrapolate S_n to zero bias, an operating point of photodetectors in most imaging

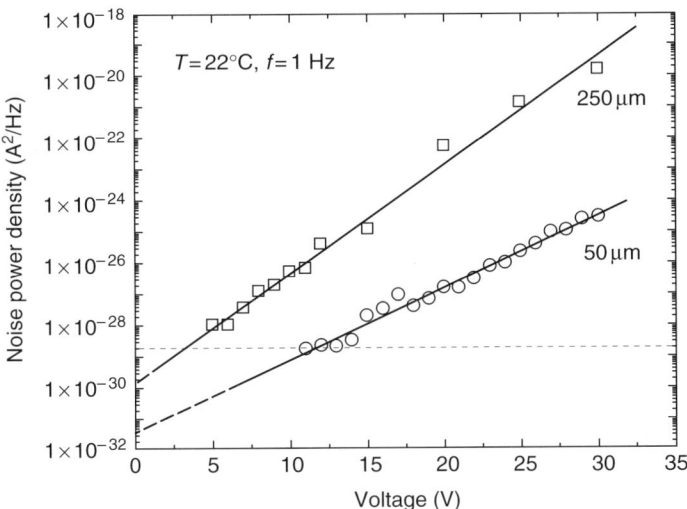

FIG. 20.   Noise spectral density of a GaN/AlGaN single heterojunction diode as a function of the bias voltage. Measurements were carried out at room temperature, at 1 Hz. The zero-bias noise spectral density is extrapolated from higher bias measurements. Also indicated is the experimental floor of the measuring apparatus (Kuryatkov *et al.* 2001).

applications. The extrapolated zero-bias S_n was 1.5×10^{-30} and $3.6 \times 10^{-32} \, A^2/Hz$ for the 250- and 50-μm diameter diodes, respectively.

The 1-Hz data of Fig. 20, replotted as a function of dark current, could be fitted with a phenomenological expression $S_i^* = (I_{dark}^2/f^\gamma)\sqrt{A/A_0}$, where A_0 is the characteristic area parameter and A the area of the diode, both measured in cm^2. The area dependence of the noise power density indicates a bulk, and not surface, scaling of the noise. This is in good agreement with experiments on GaN photodiodes in which noise was assumed to be due to conduction through threading dislocations (Kuksenkov et al. 1998).

The noise measurements of AlGaN/GaN diodes result in the NEP of 1.2×10^{-14} and $1.9 \times 10^{-15} \, W$ for the two detector sizes, respectively. As defined in Eq. (9), the NEP is the noise current density divided by the responsivity and Δf is the electrical bandwidth of the receiver, included since only the noise seen by the receiver is important. Using Eq. (10), and assuming a receiver bandwidth of 1 kHz, we calculate detectivities D^* of 3.5–$5.2 \times 10^{13} \, cm \, Hz^{1/2}/W$ for device diameters between 50 and 250 μm.

VIII. Heterojunctions of AlGaN

True solar-blind performance can be obtained only from heterojunctions of AlGaN, the type of device structure illustrated in Fig. 12(c). Preparation of solar-blind photodetectors, with the photoresponse illustrated in Fig. 6, and optimization of their performance is, thus, of considerable current interest.

The device structure is usually grown on sapphire substrates by MOCVD. The thin nucleation layer of AlN is followed by a ~1-μm-thick n-type layer of AlGaN, an undoped absorbing layer of AlGaN with a somewhat lower Al content, and a ~0.5-μm-thick p-type layer of AlGaN. In order to produce peak responsivity near 280 nm, the absorbing layer has Al content of about ~0.5. The Al content of the n- and p-type layers should be adjusted depending on whether the top or through-the-substrate illumination is desired. In the case of through-the-substrate illumination, the lower AlGaN layer should be transparent to the wavelength of interest and its Al content is increased accordingly to about 0.6. With the Al content of the p-type layer fixed at 0.45–0.5, the photodetector response in the range of 250–280 nm will be obtained, as discussed in Section IV. Larger bandgaps of p-type layers are limited by the difficulty of doping in Al-rich alloys and top illumination is rarely used. The responsivity curve typical of double heterostructure devices designed for through-the-substrate illumination is shown in Fig. 6. With careful adjustment of composition of the n- and p-type layers, and high-quality epitaxial growth, this design yields high peak responsivities and quantum efficiencies. For instance, $R_\lambda \sim 0.051 \, A/W$ could be obtained at

273 nm, corresponding to the internal quantum efficiency of 27% (Brown *et al.* 2000). Responsivity of 0.041 A/W was obtained at 243 nm, at zero bias, using double heterostructure photodetectors optimized for either top or through-the-substrate illumination (Sandvik *et al.* 2001). This corresponds to a quantum efficiency of 23%. Solar-blind rejection of four orders of magnitude was measured for these devices. Surprisingly high internal quantum efficiency of 90% was reported by Walker *et al.* (2000), at 232 nm, in top-illuminated AlGaN diodes modified by the addition of a thin p-GaN contact layer. This layer was introduced to improve carrier collection efficiency. Unexpectedly, this device had a fairly low zero-bias resistance $R_0 \sim 2 \times 10^{10}\,\Omega$ and low detectivity. The highest detectivity in AlGaN photodetectors of the type illustrated in Fig. 12(c), $D^* = 3.2 \times 10^{14}\,cm\,Hz^{1/2}/W$ at 275 nm was described recently by Dupuis *et al.* (2002). This device achieved a remarkable combination of excellent optical and electrical characteristics, a zero-bias external quantum efficiency of 53% ($R_\lambda \sim 0.12\,A/W$), and low leakage current density of $8.5 \times 10^{-11}\,A/cm^2$ at a reverse bias of 5 V. This result is entered in Fig. 1 as a data point (a star).

I–V characteristics and the zero-bias resistance of a double heterostructure photodetector, a square mesa $200 \times 200\,\mu m^2$ in size, are shown in Fig. 21 (Brown *et al.* 2000). Very low leakage currents observed in double heterostructures of AlGaN indicate high quality of ternary layers and excellent fabrication process. The dynamic resistance, dV/dI, peaks near the zero bias point at $R_0 = 2 \times 10^{11}\,\Omega$, corresponding to the product $R_0A \sim 8 \times 10^7\,\Omega\,cm^2$. This yields room-temperature thermal noise-limited detectivity $D^* \sim 3.3 \times 10^{12}\,cm\,Hz^{1/2}/W$ at a wavelength of 273 nm. This detectivity is entered in Fig. 1 as a data point (full circle).

Large arrays of AlGaN/GaN photodetectors for imaging applications have been grown on sapphire wafers (Brown *et al.* 1999a,b). Following a

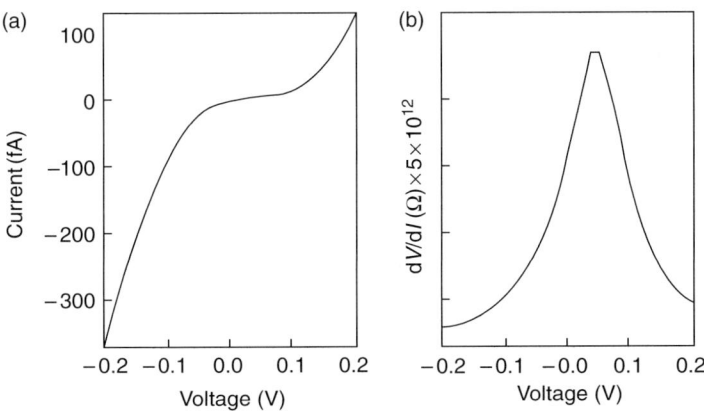

FIG. 21.   (a) Current–voltage characteristics and (b) differential resistance near zero-bias of an AlGaN heterostructure photodiode (Brown *et al.* 2000).

conventional device process of mesa etching and ohmic contact formation, In bump-bonds were fabricated over p-contacts. These were used to connect the diode array to a silicon readout circuit. Photodiodes were illuminated through the sapphire substrate polished on both sides. Arrays as large as 128×128, consisting of 16 384 individual diodes, could be prepared and used in a focal plane array camera.

Figure 22 plots the zero-bias resistance of a double heterostructure photodiode as a function of temperature (V. V. Kuryatkov, H. Temkin, J. C. Campbell, and R. D. Dupuis, unpublished results). Low leakage currents of these devices often require measurements at elevated temperatures. The inset shows spectral response peaking at 280 nm, where $R_\lambda \sim 0.03$ A/W, with the shape typical of solar-blind, double heterostructure devices. The diode illustrated here, 50 μm in diameter, has $R_0 \sim 1 \times 10^{11}$ Ω at 520 K. This can be accurately measured with standard equipment. R_0 increases rapidly with decreasing temperature, reaching $\sim 3 \times 10^{14}$ Ω at 350 K, where the measurement becomes quite difficult. Extrapolation to room temperature yields $R_0 \sim 2.5 \times 10^{16}$ Ω. This is a very high resistance but considerably smaller that that obtained by a direct measurement, which is not reliable due to exceedingly small dark current. The product $R_0 A \sim 6.1 \times 10^{12}$ Ω cm^2 yields thermal noise-limited detectivity $D^* \sim 3.8 \times 10^{14}$ cm Hz$^{1/2}$/W at a wavelength of 280 nm. Detectivity of this device is entered in Fig. 1 as a data point (open circle). This type of electrical characteristics, when combined with improved quantum efficiency, should yield specific detectivities in the $\sim 10^{15}$ cm Hz$^{1/2}$/W range.

Hopping conductivity through dislocations in GaN p–n junctions, discussed above, gives rise to exponential dependence on temperature at low

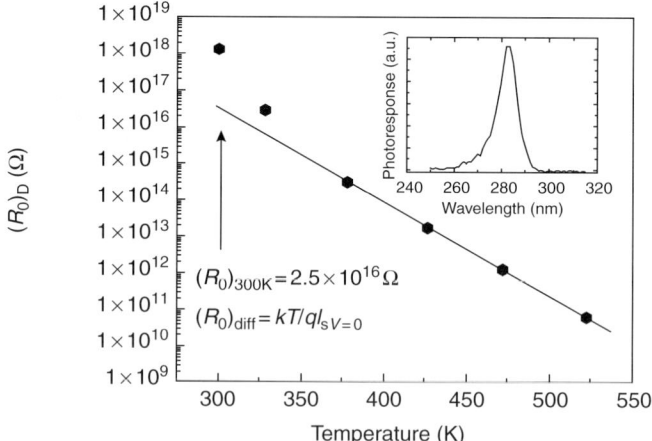

FIG. 22. Zero-bias resistance R_0 of an AlGaN heterostructure photodiode measured as a function of temperature (V. V. Kuryatkov *et al.*, unpublished results). Inset shows spectral response of the photodetector.

electric fields. The data of Fig. 22, obtained on p–n junctions of AlGaN, strongly suggests a similar mechanism. This is supported by the field dependence of the dark current of AlGaN heterojunctions plotted in Fig. 23 for a range of temperatures (V. V. Kuryatkov, H. Temkin, J. C. Campbell, and R. D. Dupuis, unpublished results). The data, obtained for fields ranging from $\sim 1.6 \times 10^5$ to $\sim 2 \times 10^6$ V/cm, shows clear Poole–Frenkel dependence, as described by Eq. (16). Excellent fit to the data, for the entire range of fields and temperatures, can be well fitted with the Poole–Frenkel constant of $\beta_{PF} = 3.3 \times 10^{-4}$ eV/V$^{1/2}$ cm$^{1/2}$, very similar to that obtained for GaN. The basic conductivity mechanism appears to be the same as in GaN, despite formation of a ternary alloy by addition of Al.

The importance of dislocations in dark conductivity of AlGaN was addressed by lateral epitaxial growth experiments, similar to those illustrated in Fig. 14 and applied to GaN devices (Parish *et al.* 1999). Lateral growth of GaN from the seed layer was continued until coalescence of adjacent wings, resulting in a planar wafer. Homojunction-type photodetectors based on $Al_{0.33}Ga_{0.67}N$ were formed on this template wafer. Diodes were fabricated on dislocated areas, grown directly over the seed layer, on low dislocation density wing areas, and on coalesced areas that contain some defects and dislocations. Diodes fabricated over the wing areas had very low leakage currents, 10 nA/cm^2 at a reverse bias of 5 V. Leakage currents in diodes formed over coalesced areas were about an order of magnitude higher. Leakage currents in dislocated diodes were larger again, by as much as six orders of magnitude.

Another important consequence of reduced dislocation density is improved spectral response, specifically a more rapid responsivity roll-off

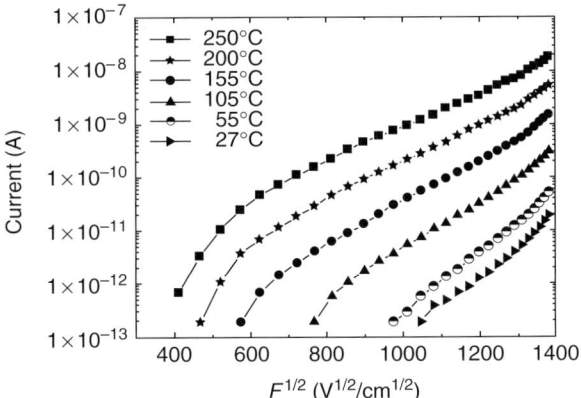

FIG. 23.   Dark current of an AlGaN double heterostructure photodiode plotted as a function of the square root of applied electric field, measured as a function of temperature. Dashed line delineates the region in which data can be fitted assuming Poole–Frenkel effect.

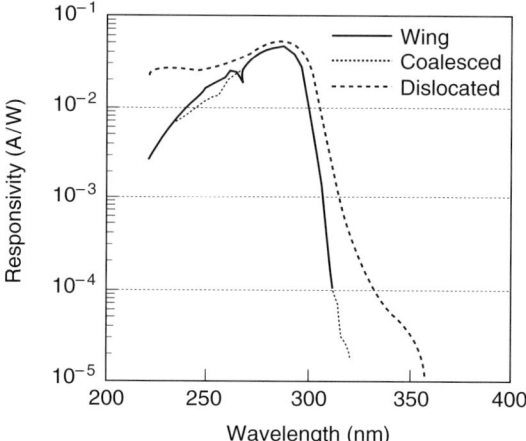

FIG. 24. Changes in spectral response of AlGaN homojunction detectors with reduced dislocation density. Notice greatly reduced sub-bandgap response of diodes fabricated on low dislocation density wing regions. After Kozodoy *et al.* (1998).

below the bandedge (Parish *et al.* 1999). The spectral response curves obtained on AlGaN homojunction photodiodes are illustrated in Fig. 24. These diodes were fabricated on GaN template wafers grown by lateral epitaxial overgrowth. The peak responsivity of $R_\lambda \sim 0.057$ A/W, at 287 nm, did not appear to be affected by the dislocation density. However, significant differences were observed in responsivity curves below the band edge. In low-dislocation density devices responsivity falls off very rapidly, three orders of magnitude within $\sim$54 nm. This is important in assuring solar-blind operation. The longer wavelength light, not absorbed in the AlGaN layer, reaches the underlying GaN template layer and is absorbed in it. Because of the bandgap discontinuity between AlGaN and GaN holes photogenerated in GaN do not contribute to the photocurrent. Dislocations in GaN act as leakage paths for photogenerated holes, resulting in excess longer wavelength photoresponse. Similar photoresponse of the dislocated structure contributes to above-bandgedge responsivity.

IX. Avalanche Photodiodes

In photodiodes based on GaN and AlGaN, reduced density of dislocations results in improved performance but devices are operable even with large numbers of dislocations. In the case of avalanche photodiodes, a minimum dislocation density threshold must be reached before gain can be obtained. High electric fields needed to initiate avalanche multiplication, greater than $\sim 1.6\,MV/cm^2$, simply cannot be reached in materials with high dislocation density. A competing process of microplasma breakdown occurs first, at much lower electric fields. This is illustrated in the micrograph in Fig. 25, which shows a GaN p–i–n diode, $200 \times 200\,\mu m^2$ in size, under a pulsed reverse bias of 80 V, corresponding to a field of $\sim 1.2\,MV/cm^2$ (A. Osinky, unpublished results). The bright points correspond to regions of microplasma breakdown, each associated with a cluster of dislocations. Excellent processing technology is needed, in addition to high-quality material, to prepare devices without sidewall leakage or excessive sidewall fields that may cause premature breakdown or result in spatially localized high gain.

Until very recently, no epitaxial growth method was capable of producing uniform, dislocation free layers of GaN or AlGaN. McIntosh *et al.* (1999, 2000) described the first GaN avalanche photodiodes prepared by hydride vapor phase epitaxy. This method is capable of growing GaN at very high rates and can, therefore, produce thick layers with low dislocation density. The device structure was placed on top of a 10-μm-thick GaN buffer layer. Dry etched mesa diodes with diameters between 30 and 60 μm were prepared. Resulting devices were characterized by low dark current, $\sim 100\,fA$ at an unspecified low bias; but this value was measurement limited. With the reverse bias as large as 150 V, dark currents were still below 200 nA. A maximum gain of 10, corresponding to an external quantum efficiency of 350%, was measured just below the reverse breakdown bias of 220 V. This gain was spatially uniform, as demonstrated by two-dimensional raster images of the photocurrent.

Much larger avalanche gains can be realized by operating the device in a photon-counting or Geiger mode (McIntosh *et al.* 2000; Verghese *et al.* 2001). In this mode of operation, reverse bias greater than the breakdown voltage is applied for a short time, a few microseconds. When the incident

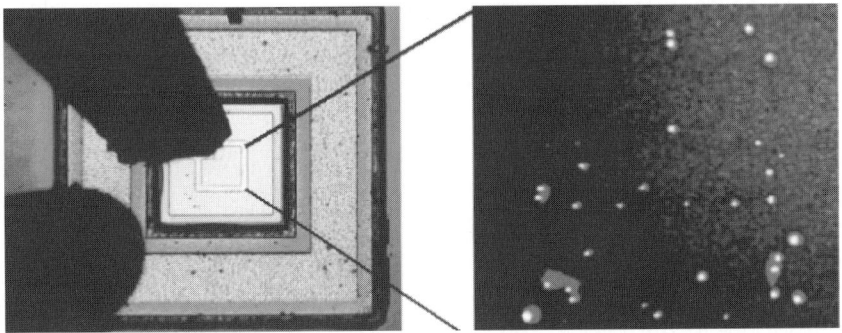

FIG. 25. A GaN avalanche photodetector, $200 \times 200\,\mu m^2$ in size; magnified image shows regions of microplasma breakdown, assumed to be associated with high dislocation regions. Device was operated at a reverse bias of 80 V. Courtesy of Dr A. Osinsky.

light is present during the pulse, the voltage across the device is reduced by 1–2 V. The voltage is monitored by a decision circuit connected to the diode that counts the number of pulses above a preset threshold. Effective gains as large as 10^7 have been reached with GaN avalanche photodiodes.

Avalanche breakdown in hydride-grown GaN diodes was analyzed in some detail through measurements of its temperature dependence (Aggarwal *et al.* 2001; Verghese *et al.* 2001). At the breakdown voltage, each carrier injected into the depletion region generates, on the average, one electron–hole pair. This corresponds to a condition $\alpha W = 1$, where α is the ionization coefficient and W the depletion width (Stillman and Wolfe 1973). In GaN, the electron and hole ionization coefficients are expected to be similar. Oguzman *et al.* (1997) calculated bulk ionization rates of $\alpha = 5 \times 10^4\,\mathrm{cm}^{-1}$ and $\beta = 4 \times 10^4\,\mathrm{cm}^{-1}$ (at a field of $4\,\mathrm{MeV/cm}$), for electrons and holes, respectively. This is in excellent agreement with the experimentally determined electron impact ionization coefficient $\alpha = 2.9 \times 10^8 \exp(-3.4 \times 10^7 / E)\,\mathrm{cm}^{-1}$, where E is the electric field (Kunihiro *et al.* 1999). The depletion width in a GaN avalanche photodiode at breakdown can be then written as $W \sim (\alpha\beta)^{-1/2} \sim 0.22\,\mu\mathrm{m}$, in excellent agreement with the measured value of the depletion width $W \sim 0.26\,\mu\mathrm{m}$. The depletion width condition is not satisfied in devices with defect-induced breakdown.

Between 100 and 250 K, avalanche breakdown voltage was found to increase from about 72 to 92 V. Above 200 K the temperature dependence of the breakdown was linear and could be described by a coefficient $dV_B/dT \sim 0.2\,\mathrm{V/K}$. The temperature dependence of the breakdown could be modeled by assuming avalanche gain limited by carrier-phonon scattering (Aggarwal *et al.* 2001). In this model, the energy gained by an accelerating charge is proportional to the applied field eV_b/W, where V_b is the breakdown voltage, and the mean free path between collisions, λ. Since the energy gain can be at most equal to the bandgap, one gets $\lambda(eV_b/W) = E_g$. For $\lambda = \lambda_0(2N+1)^{-1}$, where λ_0 is the $T = 0$ mean free path and N the photon occupation number, the breakdown voltage can be written as

$$V_b = \frac{E_g W}{e\lambda_0}(2N + 1) \qquad (21)$$

With the measured $dE_g/dT = -0.67\,\mathrm{meV/K}$, $E_g = 3.4\,\mathrm{eV}$, phonon energy of $42.3\,\mathrm{meV}$, and assuming $\lambda_0 = 13\,\mathrm{nm}$, an excellent fit was obtained to the measured temperature dependence of the breakdown voltage. The agreement between the measurement and the model confirms the avalanche nature of the breakdown and imposes a limit on the temperature dependence of V_b. Defect-related breakdown tends to have much weaker temperature dependence.

These first reports of avalanche photodiodes in GaN provide an important feasibility demonstration. However, hydride vapor phase epitaxy is somewhat limited (at least at present) in the growth of AlGaN. In this

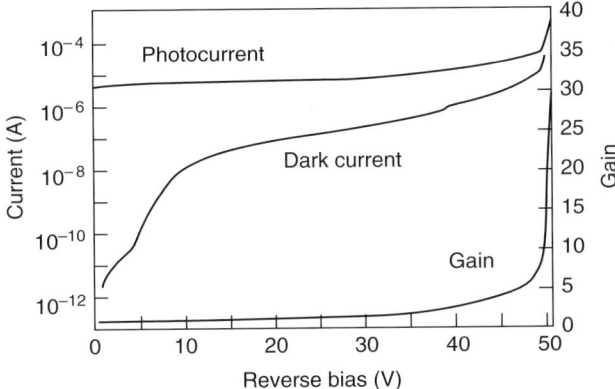

FIG. 26.   Current–voltage and gain–voltage plots for a GaN avalanche photodetector. After Carrano *et al.* (2000).

respect, reports of first GaN avalanche photodiodes grown by the more standard and versatile method of MOCVD are of considerable interest (Carrano *et al.* 2000; Yang *et al.* 2000). Figure 26 shows the room temperature *I–V* characteristics of a GaN avalanche photodiode. The device diameter was 24 µm. Very low leakage currents were measured, $\sim 20 \times 10^{-12}$ A/cm at a reverse bias of 5 V (Carrano *et al.* 2000). At low bias, the responsivity of the device was independent of the voltage, down to zero-bias. Figure 26 also plots the dependence of gain on the applied voltage. A clear onset of gain was observed close to 35 V and the maximum gain of ~ 25 was achieved close to the breakdown of ~ 48–50 V. The field at breakdown was estimated at ~ 3.5 MV/cm, consistent with the results obtained on hydride-grown devices.

X.   Conclusions

Advances in epitaxy of III-nitrides, improved structure designs, and more sophisticated fabrication have turned ultraviolet photodetectors from laboratory curiosities into high-performance devices. Mechanisms of dark conductivity and noise have been investigated in considerable detail and the relationship between device noise and threading dislocations is now well established. Solar-blind photodiode detectors based on AlGaN show high quantum efficiencies, large zero-bias resistances, and high room-temperature detectivities. Their performance is now comparable to that of photomultipliers and these detectors have already been demonstrated in large-format imaging arrays. Further advances are promised by recent demonstrations of avalanche gain.

XI. Acknowledgments

The work at Texas Tech was made possible through generous and consistent support from DARPA, under contracts administered by Drs A. Hussain, R. A. Leheny, E. Martinez, and J. C. Carrano, and the Jack F Maddox Foundation. Our interest in the problem of ultraviolet detectors was stimulated by Dr Asif Khan and we would like to thank him for it. Dr D. Kuksenkov, now with Corning, carried out much of the original work on the dislocation-related noise and we would like to thank him for his contribution. Very rewarding collaboration with Profs. R. D. Dupuis and J. C. Campbell is noted and much appreciated.

REFERENCES

Adivarahan, V., G. Simin, J. W. Yang, A. Lunev, M. A. Khan, N. Pala, M. Shur, and R. Gaska, SiO_2-passivated lateral-geometry GaN transparent Schottky-barrier detectors, *Appl. Phys. Lett.* **77**, 863 (2000).

Adivarahan, V., G. Simin, G. Tamulaitis, R. Srinivasan, J. Yang, M. Asif Khan, M. S. Shur, and R. Gaska, Indium–silicon co-doping of high-aluminum-content AlGaN for solar blind photodetectors, *Appl. Phys. Lett.* **79** (12), 1903 (2001).

Aggarwal, R. L., I. Melngailis, S. Verghese, R. J. Molnar, M. W. Geis, and L. J. Mahoney, Temperature dependence of the breakdown voltage for reverse-biased GaN p–n–n+ diodes, *Solid State Commun.* **117**, 549 (2001).

Ambacher, O., Growth and applications of Group Ill-nitrides, *J. Phys. D: Appl. Phys.* **31**, 2653 (1998).

Aspley, N., E. A. Davis, A. P. Troup, and A. D. Ioffe, Electronic properties of ion-bombarded evaporated germanium and silicon, *J. Phys. C: Solid State Phys.* **11**, 4983 (1978).

Bagraev, N. T., A. I. Gusarov, and V. A. Mashkov, Spin-correlated electron transfer along broken bonds in semiconductors, *Sov. Phys. JETP* **68** (4), 816 (1989).

Bandic, Z. Z., P. M. Bridger, E. C. Piquette, and T. C. McGill, Electron diffusion length and lifetime in p-type GaN, *Appl. Phys. Lett.* **73** (22), 3276 (1998).

Bhattacharya, P., *Semiconductor Optoelectronic Devices*, Prentice Hall, Englewood Cliffs, NJ (1994).

Binet, F., J. T. Duboz, N. Laurent, E. Rosencher, O. Briot, and R. L. Aulombard, *J. Appl. Phys.* **81**, 6449 (1997).

Brown, J. D., J. Boney, J. Matthews, P. Srinivasan, T. Nohava, W. Yang, and S. Krishnankutty, UV-specific (320–365 nm) digital camera based on 128 × 128 focal plane array of GaN/AlGaN p–i–n photodiodes, *MRS Internet J. Nitride Semicond. Res.* **5**, 6 (1999a).

Brown, J. D., Z. Yu, J. Matthews, S. Harney, J. Boney, J. F. Schetzina, J. D. Benson, K. W. Dang, C. Terril, T. Nohava, W. Yang, and S. Krishnankutty, Visible-blind UV digital camera based on a 32 × 32 array of GaN/AlGaN p–i–n photodiodes, *MRS Internet J. Nitride Semicond. Res.* **4**, 9 (1999b).

Brown, J. D., J. Li, P. Srinivasan, J. Matthews, and J. F. Schetzina, Solar-blind AlGaN Heterostructure Photodiodes, *MRS Internet J. Nitride Semicond. Res* **5**, 9 (2000).

Carrano, J. C., P. A. Grudowski, C. J. Eiting, R. D. Dupuis, and J. C. Campbell, Very low dark current metal–semiconductor–metal ultraviolet photoconductors fabricated on single-crystal GaN epitaxial layers, *Appl. Phys. Lett.* **70** (15), 1992 (1997a).

Carrano, J. C., P. A. Grudowski, C. J. Eiting, R. D. Dupuis, and J. C. Campbell, Current mechanisms in GaN-based metal–semiconductor–metal photodetectors, *Appl. Phys. Lett.* **72** (5), 542 (1997b).

Carrano, J. C., T. Li, P. A. Grudowski, C. J. Eiting, R. D. Dupuis, and J. C. Campbell, Comprehensive characterization of metal–semiconductor–metal-ultraviolet photodetectors fabricated on single-crystal GaN, *J. Appl. Phys.* **83** (11), 6148 (1998).

Carrano, J. C., D. J. H. Lambert, C. J. Eiting, C. J. Collins, T. Li, S. Wang, B. Yang, A. L. Beck, R. D. Dupuis, and J. C. Campbell, GaN avalanche photodiodes, *Appl. Phys. Lett.* **76** (7), 924 (2000).

Casey, H. C., G. G. Fountain, R. G. Alley, B. P. Keller, and S. P. DenBaars, *Appl. Phys. Lett.* **68**, 1850 (1996).

Chen, Q., M. A. Khan, C. J. Sun, and J. W. Yang, Visible-blind ultraviolet photo-detectors based on GaN p–n junctions, *Electron. Lett.* **31** (20), 1781 (1995).

Chen, Q., J. W. Yang, A. Osinsky, S. Gangopadhyay, B. Lim, M. Z. Anwar, M. A. Khan, D. Kuksenkov, and H. Temkin, Schottky barrier detectors on GaN for visible-blind ultraviolet detection, *Appl. Phys. Lett.* **70** (17), 2277 (1997).

Chernyak, L., A. Osinsky, H. Temkin, J. W. Yang, Q. Chen, and M. A. Khan, Electron beam induced current measurements of minority carrier diffusion length in gallium nitride, *Appl. Phys. Lett.* **69**, 2531 (1996).

Chuang, S. L., *Physics of Optoelectronic Devices*, Wiley Series in Pure and Applied Optics, Wiley, New York (1995).

Collins, C. J., T. Li, A. L. Beck, R. D. Dupuis, J. C. Campbell, J. C. Carrano, M. J. Schurman, and I. A. Ferguson, Improved device performance using a semi-transparent p-contact AlGaN/GaN heterojunction positive–intrinsic–negative photodiode, *Appl. Phys. Lett.* **75** (14), 2138 (1999).

Copeland, J. A., Semiconductor impurity analysis from low-frequency noise spectra, *IEEE Trans. Electron Devices* **ED-18** (1), 50 (1971).

Deelman, P. W., R. N. Bicknell-Tassius, S. Nikishin, V. Kuryatkov, and H. Temkin, Low-noise GaN Schottky diodes on Si(111) by molecular beam epitaxy, *Appl. Phys. Lett.* **78** (15), 2172 (2001).

Donati, S., *Photodetectors*, Prentice Hall, Englewood Cliffs, NJ (2000).

Dupuis, R. D., Wocsemmad 2002, Austin, Texas (2002).

Eisenman, W. L., J. D. Merriam, and R. F. Potter, Operational characteristics of infrared photodetectors, in *Semiconductors and Semimetals*, Vol. 12, edited by R. K. Willardson and A. C. Beer, Academic Press, New York (1977).

Garrido, J. A., E. Munroy, I. Izpura, and E. Munoz, Photoconductive gain modeling of GaN photodetectors, *Semicond. Sci. Technol.* **13**, 563 (1998).

Gonon, P., Y. Boiko, S. Prawer, and D. Jamieson, Poole–Frenkel conduction in polycrystalline diamond, *J. Appl. Phys.* **79** (7), 3778 (1996).

Hill, R. M., Hopping conduction in amorphous solids, *Philos. Mag.* **24**, 1307 (1971).

Hooge, F. N., 1/f noise is no surface effect, *Phys. Lett. A* **29**, 139 (1969).

Jones, B. K., Low-frequency noise spectroscopy, *IEEE Trans. Electron Devices* **41** (11), 2188 (1994).

Joseph, C. L., *Exp. Astron.* **6**, 97 (1995).

Khan, M. A., J. N. Kuznia, D. T. Olson, J. M. van Hove, M. Blasingame, and L. F. Reitz, *Appl. Phys. Lett.* **60**, 2917 (1992).

Khan, M. A., J. N. Kuznia, D. T. Olson, M. Blasingame, and A. R. Bhattarai, *Appl. Phys. Lett.* **63**, 2455 (1993).

Khan, A. H., J. M. Meese, T. Stacy, E. M. Charlson, E. J. Charlson, G. Zhao, G. Popovici, and M. A. Prelas, Electrical characterization of aluminum nitride films on silicon grown by chemical vapor deposition, *Diamond, SiC and Nitride Wide Bandgap Semiconductors, Symposium Proceedings*, San Francisco, USA, p. 637 (1994).

Kim, J. S., C. H. Park, H. S. Min, and I. Y. Park, *AIP Conf. Proc.* **466**, 123 (1999).

Kingston, R. H., *Detection of Optical and Infrared Radiation*, Springer, Berlin (1978).

Knotek, M. L., M. Pollak, T. M. Donovan, and H. Kurtzmann, *Phys. Rev. Lett.* **30**, 854 (1973).

Kozodoy, P., J. P. Ibbetson, H. Marchand, P. T. Fini, S. Keller, J. S. Speck, S. P. DenBaars, and U. K. Mishra, Electrical characterization of GaN p–n junctions with and without threading dislocations, *Appl. Phys. Lett.* **73** (7), 975 (1998).

Kozub, V. I., Low-frequency noise due to site energy fluctuations in hopping conductivity, *Solid State Commun.* **97** (10), 843 (1996).

Krishnankutty, S., W. Yang, T. Nohava, and P. P. Ruden, Fabrication and characterization of GaN/AlGaN ultraviolet-band heterojunction photodiodes, *MRS Internet J. Nitride Semicond. Res.* **3**, 7 (1998).

Kuksenkov, D., H. Temkin, A. Osinsky, R. Gaska, A. Khan, Origin of conductivity and low frequency noise in reverse-biased GaN p–n junction, *Appl. Phys. Lett.* **72** (11), 1365 (1998).

Kunihiro, K., K. Kasahara, Y. Takahashi, and Y. Ohno, Experimental evaluation of impact ionization coefficients in GaN, *IEEE Electron Devices Lett.* **20** (12), 608 (1999).

Kuriyama, K., K. Kazama, T. Koyama, T. Takamori, T. Kamijoh, Photoquenching of the hopping conduction in arsenic-ion-implanted MBE grown GaAs, *Solid State Commun.* **103** (3), 145 (1997).

Kuryatkov, V. V., H. Temkin, J. C. Campbell, and R. D.Dupuis, Low-noise photodetectors based on heterostructures of AlGaN–GaN, *Appl. Phys. Lett.* **78** (21), 3340 (2001).

Li, T., A. L. Beck, C. Collins, R. D. Dupuis, J. C. Campbell, J. C. Carrano, M. J. Shurman, and I. A. Ferguson, Improved ultraviolet quantum efficiency using a semitransparent recessed window AlGaN/GaN heterojunction p–i–n photodiode, *Appl. Phys. Lett.* **75** (16), 2421 (1999).

Mackintosh, A. J., R. T. Phillips, and A. D. Ioffe, The electrical conductivity of amorphous antimony and its dependence on film thickness, *Physica B* **117–118**, 1001 (1983).

Manasreh, D., editor, *III-Nitride Semiconductors: Electrical, Structural, and Defect Properties*, Elsevier, Amsterdam (2000).

Mathis, S. K., A. E. Romanov, L. F. Chen, G. E. Beltz, W. Pompe, and J. S. Speck, Modeling of threading dislocation reduction in growing GaN layers, *J. Crystal Growth* **231**, 371 (2001).

McIntosh, K. A., R. J. Molnar, L. J. Mahoney, A. Lightfoot, M. W. Geis, K. M. Molvar, I. Melngailis, R. L. Aggarwal, W. D. Goodhue, S. S. Choi, D. L. Spears, and S. Verghese, GaN avalanche photodiodes grown by hydride vapor-phase epitaxy, *Appl. Phys. Lett.* **75** (22), 3485 (1999).

McIntosh, K. A., R. J. Molnar, L. J. Mahoney, K. M. Molvar, I. Melngailis, N. Efremov Jr., and S. Verghese, Ultraviolet photon counting with GaN avalanche photodiodes, *Appl. Phys. Lett.* **76** (26), 3938 (2000).

Mohammad, S. N., and H. Morkoc, *Prog. Quantum Electron.* **20**, 361 (1996).

Molnar, R. J., W. Gotz, L. T. Romano, and N. M. Johnson, *J. Crystal Growth* **178**, 147 (1997).

Monroy, E., F. Calle, E. Muñoz, F. Omnès, P. Gibart, and J. A. Muñoz, $Al_xGa_{1-x}N$:Si Schottky barrier photodiodes with fast response and high detectivity, *Appl. Phys. Lett.* **73** (15), 2146 (1998).

Morgan, M., and P. A. Walley, Localized conduction processes in amorphous germanium, *Philos. Mag.* **23**, 661 (1971).

Morkoc, H., *Wide Band Gap Nitrides and Devices*, Springer, Berlin (1998).

Morkoc, H., GaN-based modulation doped FETs and UV photodetectors, *Naval Res. Rev.* **51** (1), 26 (1999).

Morkoc, H., S. Strite, G. B. Gao, M. E. Lin, B. Sverdlov, and M. Burns, *J. Appl. Phys.* **76**, 1368 (1994).

Motchenbacher, C. D., and J. A. Connelly, *Low-noise Electronic System Design*, Wiley Interscience, New York (1993).

Mott, N. F., Conduction in non-crystalline materials III. Localized states in a pseudogap and near extremities of conduction and valence bands, *Philos. Mag.* **19**, 835 (1969).

Muth, J. F., J. D. Brown, M. A. L. Johnson, Z. Yu, R. M. Kolbas, J. W. Cook Jr., and J. F. Schetzina, Absorption coefficient and refractive index of GaN, AlN, and AlGaN alloys, *MRS Internet J. Nitride Semicond. Res.* **4S1**, G5.2 (1999).

Nakamura, S., *IEEE J. Select. Areas Commun.* **4**, 483 (1998).

Nakamura, S., and G. Fosol, *The Blue Laser Diode*, Springer, Berlin (1998).

Nikishin, S. A., N. N. Faleev, V. G. Antipov, S. Francoeur, L. Grave de Peralta, G. A. Seryogin, H. Temkin, T. I. Prokofyeva, M. Holtz, and S. N. G. Chu, High quality GaN grown on Si(111) by gas source molecular beam epitaxy with ammonia, *Appl. Phys. Lett.* **75**, 2073 (1999a).

Nikishin, S. A., V. G. Antipov, S. Francoeur, N. N. Faleev, G. A. Seryogin, V. A. Elyukhin, H. Temkin, T. I. Prokofyeva, M. Holtz, A. Konkar, and S. Zollner, High-quality AlN grown on Si(111) by gas source molecular beam epitaxy with ammonia, *Appl. Phys. Lett.* **75**, 484 (1999b).

Nikishin, S. A., N. N. Faleev, A. S. Zubrilov, V. G. Antipov, and H. Temkin, Growth of AlGaN on Si(111) by gas source molecular beam epitaxy, *Appl. Phys. Lett.* **76** (21), 3028 (2000).

Oguzman, I. H., E. Bellotti, K. Brennan, J. Kolnik, R. Wang, and P. P. Ruden, Theory of hole initiated impact ionization in bulk zincblende and wurtzite GaN, *J. Appl. Phys.* **81** (12), 7827 (1997).

Omnes, F., N. Marenco, B. Beaumont, Ph. De Mierry, E. Monroy, F. Calle, and E. Munoz, Metalorganic vapor-phase epitaxy-grown AlGaN materials for visible-blind ultraviolet photodetector applications, *J. Appl. Phys.* **86** (9), 5289 (1999).

Osinsky, A., S. Gangopadhyay, R. Gaska, B. Williams, M. A. Khan, D. Kuksenkov, and H. Temkin, Low noise p–π–n GaN ultraviolet photodetectors, *Appl. Phys. Lett.* **71** (16), 2334 (1997).

Osinsky, A., M. S. Shur, R. Gaska, and Q. Chen, *Electron. Lett.* **34**, 691 (1998a).

Osinsky, A., S. Gangopadhyay, J. W. Yang, R. Gaska, D. Kuksenkov, H. Temkin, I. K. Shmagin, Y. C. Chang, J. F. Muth, and R. M. Kolbas, Visible-blind GaN Schottky barrier detectors grown on Si(111), *Appl. Phys. Lett.* **72** (5), 551 (1998b).

Panish, M. B., and H. Temkin, *Gas Source Molecular Beam Epitaxy*, Springer, Berlin (1993).

Parish, G., S. Keller, P. Kozodoy, J. P. Ibbetson, H. Marchand, P. T. Fini, S. B. Fleischer, S. P. Denbaars, U. K. Mishra, and E. J. Tarsa, High performance (Al,Ga)N-based solar-blind ultraviolet p–i–n detectors on laterally epitaxially overgrown GaN, *Appl. Phys. Lett.* **75** (2), 247 (1999).

Pau, J. L., E. Monroy, F. B. Naranjo, E. Munoz, F. Calle, and M. A. Sanchez-Garcia, High visible rejection AlGaN photodetectors on Si(111) substrates, *Appl. Phys. Lett.* **76** (19), 2785 (2000).

Pearsall, T. P., and M. A. Pollack, Compound semiconductor photodiodes, in *Semiconductors and Semimetals*, Vol. 22d, edited by W. T. Tsang, Academic Press, New York, (1985).

Pearton, S. J., J. C. Zolper, R. J. Shul, and F. Ren, GaN: processing, defects, and devices, *J. Appl. Phys.* **86** (1), 1 (1999).

Pollak, M., and I. Riess, A percolation treatment of high-field hopping transport, *J. Phys. C* **9**, 2339 (1976).

Razeghi, M., and A. Rogalski, Semiconductor ultraviolet detectors, *Appl. Phys. Rev., J. Appl. Phys.* **79** (10), 7433 (1996).

Rumyantsev, S., M. Levinshtein, R. Gaska, M. S. Shur, J. W. Yang, and M. A. Khan, *J. Appl. Phys.* **87**, 1849 (2000).

Sandvik, P., K. Mi, F. Shahedipour, R. McClintock, A. Yasan, P. Kung, and M. Razeghi, AlGaN for solar-blind UV detectors, *J. Crystal Growth* **231**, 366 (2001).

Seager, C. H., and G. E. Pike, Percolation and conductivity: a computer study, *Phys. Rev. B* **10** (4), 1435 (1974).

Shklovskii, B. I., Theory of $1/f$ noise for hopping conduction, *Solid State Commun.* **33** (3), 273 (1980).

Simmons, J. G., Conduction in thin dielectric films, *J. Phys. D: Appl. Phys.* **4**, 613 (1971).

Smith, G., J. Van Nostrand, P. J. Schreiber, M. J. Estes, T. Dang, H. Temkin, and J. Hoelscher, UV Schottky barrier detector development for possible air force applications, *Proc. SPIE* **3629**, 184 (1999).

Stevens, K. S., M. Kinniburgh, and R. Beresford, Photoconductive ultraviolet sensor using Mg-doped GaN on Si(111), *Appl. Phys. Lett.* **66** (25), 3518 (1995).

Stillman, G. E., and C. M. Wolfe, in *Semiconductors and Semimetals*, edited by R. K. Willardson and A. C. Beer, Vol. 12, Academic Press, New York (1973).

Sze, S. M., *Physics of Semiconductor Devices*, Wiley, New York (1981).

Tansley, T. L., and R. J. Egan, *Phys. Rev. B* **45**, 10 (1993).

Tarsa, E. J., P. Kozodoy, J. P. Ibbetson, B. P. Keller, G. Parish, and U. K. Mishra, *Appl. Phys. Lett.* **77**, 316 (2000).

Tiwari, S., *Compound Semiconductor Devices*, Academic Press, New York (1992).

Vandamme, L. K. J., Noise as a diagnostic tool for quality and reliability of electronic devices, *IEEE Trans. Electron Devices*, **41** (11), 2176 (1994).

Van Der Ziel, A., *Noise in Measurements*, Wiley Interscience, New York (1976).

Van Der Ziel, A., *Noise in Solid State Devices and Circuits*, Wiley Interscience, New York (1986).

Van Hove, J. M., R. Hickman, J. J. Klassen, P. P. Chow, and P. P. Ruden, Ultraviolet-sensitive, visible-blind GaN photodiodes fabricated by molecular beam epitaxy, *Appl. Phys. Lett.* **70** (17), 2282 (1997).

Verghese, S., K. A. McIntosh, R. J. Molnar, L. J. Mahoney, R. L. Aggarwal, M. W. Geis, K. M. Molvar, E. K. Duerr, and I. Melngailis, GaN avalanche photodiodes operating in linear gain mode and Geiger mode, *IEEE Trans. Electron Devices* **48** (3), 502 (2001).

Waldron, E., J. W. Graff, and E. F. Schubert, Improved mobilities and resistivities in modulation-doped p-type AlGaN/GaN superlattices, *Appl. Phys. Lett.* **79** (17), 2737 (2001).

Walker, D., X. Zhang, P. Kung, H. Saxler, S. Javadpour, J. Xu, and M. Razeghi, AlGaN ultraviolet photoconductors grown on sapphire, *Appl. Phys. Lett.* **68**, 2100 (1996).

Walker, D., V. Kumar, K. Mi, P. Sandvik, P. Kung, X. H. Zhang, and M. Razeghi, Solar-blind AlGaN photodiodes with very low cutoff wavelength, *Appl. Phys. Lett.* **76** (4), 403 (2000).

Wang, L., M. I. Nathan, T. H. Lim, M. A. Khan, and Q. Chen, *Appl. Phys. Lett.* **68**, 1267 (1996).

Wickenden, D. K., Z. Huag, D. Brent Mott, and P. Shu, Development of GaN photoconductive detectors, *Johns Hopkins APL Tech. Dig.* **18** (2), 217 (1997).

Wu, X. H., L. M. Brown, D. Kapolnek, S. Keller, B. Keller, S. P. DenBaars, and J. S. Speck, Defect structure of metal-organic chemical vapor deposition-grown epitaxial (0001) GaN/Al$_2$O$_3$, *J. Appl. Phys.* **80** (6), 3228 (1996).

Xu, G. Y., A. Salvador, W. Kim, Z. Fan, C. Lu, H. Tang, H. Morkoç, G. Smith, M. Estes, B. Goldenberg, W. Yang, and S. Krishnankutty, High speed low noise ultraviolet photodetectors based on GaN p–i–n and AlGaN(p)–GaN(i)–GaN(n) structures, *Appl. Phys. Lett.* **71**, 2154 (1997).

Yang, B., T. Li, K. Heng, C. Collins, S. Wang, J. C. Carrano, R. D. Dupuis, J. C. Campbell, M. J. Schurman, and I. T. Ferguson, Low dark current GaN avalanche photodiodes, *IEEE J. Quantum Electron.* **36** (12), 1389 (2000).

Yariv, A., *Optical Electronics*, Holt, Reinhart and Winston, p. 345 (1985).

Zhao, Z. M., R. L. Jiang, P. Chen, D. J. Xi, Z. Y. Luo, R. Zhang, B. Shen, Z. Z. Chen, and Y. D. Zheng, *Appl. Phys. Lett.* **77**, 444 (2000).

Zheleva, T. S., O. H. Nam, M. D. Bremser, and R. F. Davis, *Appl. Phys. Lett.* **71**, 2638 (1997).

Advanced Semiconductor and Organic Nano-Techniques (Part II)
H. Morkoç (Ed.)

CHAPTER 4

Organic Field-Effect Transistors for Large-Area Electronics

C. D. Dimitrakopoulos

IBM CORPORATION, YORKTOWN HEIGHTS, NEW YORK

I. Introduction

Organic semiconductors have been studied for more than half a century (Pope and Swenberg 1999). However, until recently they had failed to have a significant practical impact in optoelectronic applications, despite of the fact that a very large number of experimental and theoretical studies have been published. Initially, industrial applications of organic semiconductors exploited their photoconductive properties in xerography. However, the potential for applications of organic semiconductors with much broader impact became clear with the initial demonstrations of organic electro-luminescent diodes (Tang and Van Slyke 1987; Burroughes et al. 1990) and organic thin-film field-effect transistors (OTFTs) (Ebisawa et al. 1983; Kudo et al. 1984; Tsumura et al. 1986) based on either small organic molecules (Kudo et al. 1984; Tang and Van Slyke 1987) or conjugated polymers (Ebisawa et al. 1983; Tsumura et al. 1986; Burroughes et al. 1990). The impressive improve-ments in performance and efficiency of organic devices during the last decade (Yu and Heeger 1996; Dimitrakopoulos et al. 1998; Friend et al. 1999;

Hellemans 1999; Forrest 2001) attracted the interest of the optoelectronics industry and opened the way to practical applications for organic semiconductors. In this chapter, we will focus mainly on the field of OTFTs. We will make an effort to provide a fairly complete account of the broad spectrum of materials, fabrication processes, designs, and applications of OTFTs, with an emphasis on recent developments. Older papers that, in the authors' opinion, played a pivotal role in shaping the OTFT field will also be emphasized, but the reader is encouraged to look up a number of previously published review papers that cover that earlier period in more detail (Greenham and Friend 1995; Lovinger and Rothberg 1996; Brown *et al.* 1997; Katz 1997; Garnier 1998; Horowitz 1998; Katz and Bao 2000). Reported results from single crystal organic field effect transistors will be used to define the upper limit of performance of OTFTs and to gain a better understanding of the underlying device physics.

As in traditional inorganic semiconductors, organic semiconductors function either as p-type, in which the majority charge carriers are holes, or n-type, in which the majority charge carriers are electrons. The most widely studied organic semiconductors have been p-type. However, in the last decade, several new reports on OTFTs based on n-type organic semiconductors have appeared, and thus we have devoted a small fraction of this chapter to n-type OTFTs.

OTFTs based on conjugated polymers, oligomers, or fused aromatics have been envisioned as a viable alternative to more traditional, mainstream thin-film transistors (TFTs) based on inorganic materials. Because of the relatively low mobility of organic semiconductors, OTFTs cannot rival the performance of field-effect transistors based on single-crystalline inorganic semiconductors, such as Si, Ge, and GaAs, which have charge carrier mobilities (μ) of three or more orders of magnitude higher (Taur and Ning 1998). Consequently, OTFTs are not suitable for use in applications requiring very high switching speeds. However, the unique processing characteristics and already demonstrated performance of OTFTs suggest that they can be competitive candidates for existing or novel TFT applications requiring large area coverage, low-temperature processing, structural flexibility, and especially low cost. Such applications include switching devices for active matrix flat panel displays (AMFPDs) based on liquid crystal pixels (AMLCDs), organic light emitting diodes (AMOLEDs), and "electronic paper" displays (Wisnieff 1998) based on pixels comprising either electrophoretic ink-containing microcapsules (Comiskey *et al.* 1998) or "twisting balls" (Sheridon 1978). Additionally, organic sensors (Crone *et al.* 2001), low-end smart cards, radio-frequency identification (RFID) tags, and electronic tickets consisting of organic integrated circuits have been proposed and prototype all-polymer integrated circuits have been demonstrated (Drury *et al.* 1998).

Each application will require different performance standards for OTFTs. At present, the entrenched technology in large area electronics applications, especially backplanes of AMLCDs, is based on TFTs comprising hydro-genated amorphous silicon (a-Si:H) active layers. However, OTFTs can enable applications that are not possible using the a-Si:H TFTs, taking advantage of the fact that OTFTs, can be processed at or close to room temperature and thus are compatible with transparent plastic substrates. Because of the high processing temperature used during a-Si:H deposition (ca. 360°C), it is not possible to make an AMLCD based on such TFTs on a transparent plastic substrate. For OTFTs to compete with a-Si:H TFTs for applications pre-sently addressed by the use of a-Si:H TFTs, the former should exhibit device performance similar to that of the latter, that is, field effect mobility $\mu = 1\,\text{cm}^2\,\text{V}^{-1}\,\text{s}^{-1}$, and current modulation (or on/off ratio, $I_{\text{on/off}}$) of 10^6 or higher at a maximum operating voltage of about 15 V or less. Additionally, they should be stable after prolonged exposure to ambient conditions, and should not exhibit large threshold voltage shifts.

II. Origin of Semiconductivity in Organics

The origin of semiconducting properties in organic compounds is most usually related to the existence of extended bonding and anti-bonding molecular orbitals, called π and π^*, respectively. These orbitals are generated as follows: the electronic structure of carbon may be represented as $1s^2\,2s^2\,2p^2$, which becomes $1s^2\,2s^1\,2p_x^1\,2p_y^1\,2p_z^1$ by promotion of one of the 2s electrons to the 2p orbital. When C forms four single bonds, the four outer orbitals become equivalent by hybridization, and each is called an sp^3 orbital. This is the tetrahedral, sp^3-hybridized single bond configuration of the C atom, and organic materials that contain exclusively such carbon atoms are insulating. However, when two C atoms form a double or triple bond between them, as in the cases of ethylene and acetylene, one or two π and π^* orbital pairs are formed, respectively.

In the case of a C atom pair bonded with a double bond, each C atom is bonded to three other atoms. Three equivalent co-planar sp^2 orbitals arise from the hybridization of the $2s^1$, $2p_x^1$, and $2p_y^1$ electrons, creating σ-bonds with the three nearest neighbor atoms. Each C atom of the pair has one remaining electron in its $2p_z$ orbital, which is perpendicular to the plane of the hybridized sp^2 orbitals. These two parallel $2p_z$ orbitals, which are initially degenerate, can overlap forming the bonding and anti-bonding molecular orbitals discussed above. In the ground state, both electrons are in the bonding orbital (highest occupied molecular orbital or HOMO) and the anti-bonding orbital (lowest unoccupied molecular orbital or LUMO) is vacant. This situation is called a π bond. Electrons can be promoted from the ground state (HOMO) to an unoccupied state by optical or thermal excitation. In the case of larger conjugated molecules such as benzene, further energy level splitting occurs. Each one of the six C atoms of benzene uses sp^2 orbitals to form three σ-bonds, as described above, and has one remaining p orbital containing one electron. Each one of these six p orbitals can overlap equally with the two adjacent p orbitals producing six new orbitals that extend over the whole molecule. Three of these are bonding π orbitals. One of these π orbitals has the lowest energy and is the ground state of the molecule. The six π electrons establish a delocalized electron density above and below the plane of the σ-bonds, with no π electron density in this plane. This configuration enhances the stability of the benzene molecule over the configuration comprising a fixed alternation of localized single and double bonds.

In the case of a C atom pair bonded with a triple bond, each C atom is bonded to two other atoms. Hence, sp hybridization is used and the four atoms (the C atom pair plus the two atoms with which it is bonded) are in a straight line. Each C atom has two remaining p orbitals remaining ($2p_y$ and $2p_z$) with one electron in each. These orbitals are perpendicular to each other and to the C–C axis. Each pair of parallel 2p orbitals can overlap forming two pairs of π and π^* orbitals. Thus, the triple bond between C atoms is composed of one σ-bond and two π-bonds.

There are three main types of structure that exhibit electron delocalization: (a) double (or triple) bonds in conjugation; (b) double (or triple) bonds in conjugation with a p orbital on an adjacent atom; and (c) hyperconjugation. The details of this subject are out of the scope of this chapter and can be found in advanced organic chemistry textbooks (March 1985).

Intermolecular interactions in a solid consisting of molecules with delocalized π and π^* molecular orbitals, lead to still further splitting of these molecular energy levels, resulting to the formation of narrow bands. The existence of such bands is the basis of electrical conduction in organic solids and the origin of their semiconducting properties. Figure 1 (Karl 1974; Karl 2001) demonstrates schematically the dependence of the energy gap, E_g, between the filled π extended states (or valence band) and π^* extended states (conduction band) in molecular single crystals for different molecules in the acene series. E_g becomes smaller as the size of the molecule and thus the size of the π and π^* molecular orbitals becomes larger. In the limiting case of graphite, which consists of extended sheets of carbon atoms with honeycomb structure, the very large and approximately two dimensional π and π^* molecular orbitals form energy bands with substantial dispersion, and are responsible for the semimetallic character of this material.

III.   Common Semiconducting Organic Materials

Polymers, their oligomer homologs, fused aromatics, and other small molecules that posses extended π and π^* molecular orbitals, as described in

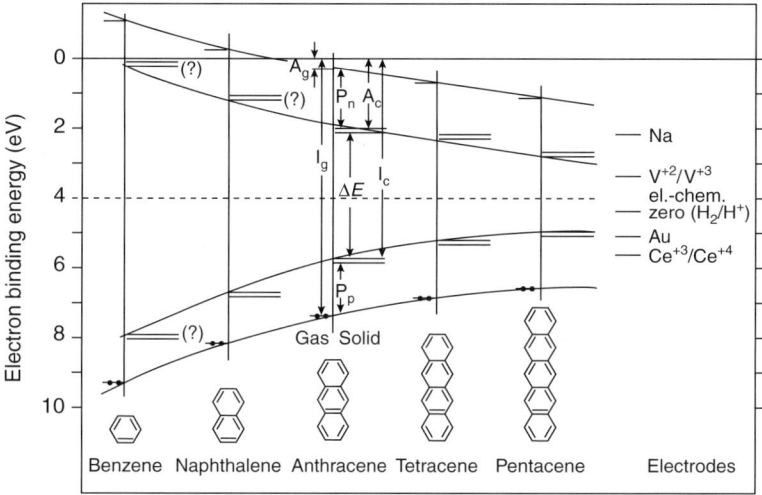

FIG. 1. This diagram depicts the energy levels of single ionic states of the acene series from benzene to pentacene (HOMO and LUMO). The left side of the vertical lines that correspond to each molecule refers to the free molecule in the gas state (g), while the right side refers to the molecule in the crystal (c). $-P_p$ and $+P_n$ are the polarization shifts of the hole and the electron state, respectively, and correspond to the shift of the HOMO level and the LUMO level, respectively, of the free ion when brought from vacuum into a neutral crystal. Additionally, the Fermi level of Na and Au, and the redox potential of electrolytic redox systems are given. Reprinted with permission from Karl (2001). This is an updated version of Fig. 5 of Karl (1974).

the previous section, can exhibit semiconducting properties. Figures 2 and 3 exhibit some of the most representative p- and n-type organic semi-conductors, respectively. The numbers representing each molecular structure correspond to the numbers in column 4 of Tables I and II.

IV. Charge Transport Mechanisms in Organic Semiconductors

The upper limit of microscopic mobilities in organic molecular crystals, determined at 300 K by time-of-flight experiments, falls between 1 and $10 \, cm^2 \, V^{-1} \, s^{-1}$ (Pope and Swenberg 1999). The weak intermolecular inter-action forces in organic semiconductors, typically van der Waals interactions with energies smaller than $10 \, kcal \, mol^{-1}$, may be responsible for this limit, since the vibrational energy of the molecules reaches a magnitude close to that of the intermolecular "bonding" energies at or above room temperature. In contrast, in inorganic semiconductors such as Si, Ge, and GaAs, the atoms are held together with very strong covalent bonds, which in the case of Si have energies as high as $76 \, kcal \, mol^{-1}$. In these semiconductors, charge carriers move as highly delocalized plane waves in wide bands and have a very high mobility at room temperature ($\mu \sim 10^3 \, cm^2 \, V^{-1} \, s^{-1}$). In this case,

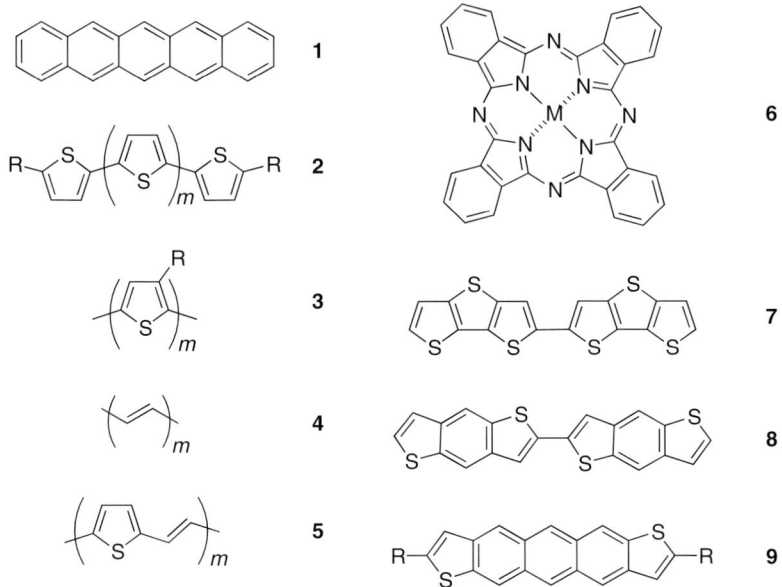

FIG. 2. Molecular structures of common p-type organic semiconductors. (1) pentacene; (2) α,ω-dialkyl-oligothiophene, m = 2–6, R = C_nH_{2n+1}, n = 0–8; (3) HT-poly(3-alkyl)thiophene, R = C_nH_{2n+1}, n = 6; (4) polyacetylene; (5) polythienylenevinylene; (6) metal-phthalocyanine; (7) bis(dithienothiophene); (8) bis(benzodithiophene); (9) dihexyl-anthradithiophene.

the mobility is limited by lattice vibrations (phonons) that scatter the carriers and thus it is reduced as the temperature increases.

Band transport is not applicable to disordered organic semiconductors, where carrier transport takes place by hopping between localized states and carriers are scattered at every step. Hopping is assisted by phonons and the mobility increases with temperature, although typically it remains very low overall ($\mu \ll 1\ \mathrm{cm^2\ V^{-1}\ s^{-1}}$). The boundary between band transport and hopping is defined by materials having room temperature mobilities of the order of $1\ \mathrm{cm^2\ V^{-1}\ s^{-1}}$ (Horowitz 1998; Nelson et al. 1998; Pope and Swenberg 1999). Thin films of highly ordered organic semiconductors, such as several members of the acene series including pentacene, have room temperature mobilities in this intermediate range (Lin et al. 1997b; Nelson et al. 1998; Dimitrakopoulos et al. 1999b). In some cases, temperature, independent mobility has been observed (Pope and Swenberg 1999), even in polycrystalline thin films of pentacene (Nelson et al. 1998). This observation was used to argue that a simple thermally activated hopping mechanism can be excluded as a transport mechanism in high-quality polycrystalline thin films of pentacene despite of the fact that in some samples containing a large concentration of traps related to structural defects and chemical impurities the mobility increases with temperature (Nelson et al. 1998). Trapping at the

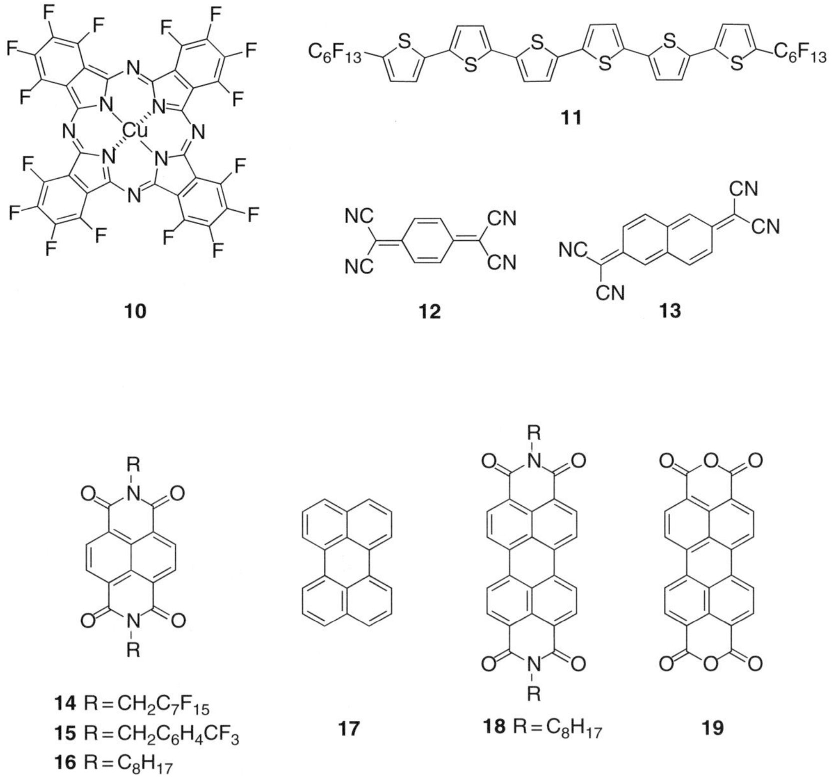

FIG. 3.   Molecular structures of common n-type organic semiconductors. (**10**) F16CuPc; (**11**) DFH-6T; (**12**) TCNQ; (**13**) TCNNQ, (**14**) NTCDI-C8F; (**15**) NTCDI-BnCF3; (**16**) NTCDI-C8H; (**17**) perylene; (**18**) PTCDI-C8H; (**19**) PTCDA.

grain boundaries in polycrystalline films of pentacene and the dependence of trap concentration on the film deposition conditions has been suggested as the main cause of the observed variability of the temperature dependence of mobility (Schön *et al.* 2000c).

Understanding the transport mechanism in single crystals of organic semiconductors will facilitate our understanding of transport in the technologically more relevant polycrystalline thin films of these materials. At low temperatures, coherent band-like transport of delocalized carriers becomes the prevalent transport mechanism in single crystals of pentacene, tetracene, and other acenes. Very high mobility values have been measured using time of flight experiments (up to $400\,cm^2\,V^{-1}\,s^{-1}$ for holes in single crystals of naphthalene at 4.2 K) (Warta *et al.* 1985; Karl *et al.* 1991) and field effect experiments (up to $\sim 10^5\,cm^2\,V^{-1}\,s^{-1}$ for holes in single crystals of tetracene and pentacene at 1.7 K) (Schön *et al.* 2000b,c). In the latter experiments, the mobility increases from its room temperature value of approximately

TABLE I

Highest Reported Field-effect Mobility (μ) Values Measured from p-type OTFTs, Annually from 1984 to the Present Time, for Each One of the Most Promising p-type Organic Semiconductors. For a p-type Organic Semiconductor that Already has a Data Entry in Table I from a Previous Year, a New Mobility Value is Entered Only if it is Higher than the Preceding Entry

Year	Mobility[a] ($cm^2 V^{-1} s^{-1}$)	Material (deposition method); (v) = vacuum deposition; (s) = from solution	Compound no.	I_{on}/I_{off}[b]	W/L	Reference
1964	NR[c]	Cu-phthalocyanine (v) (first demonstration of field effect in small organic molecules)	6	NR	NR	Heilmeier and Zanoni (1964)
1983	NR	Polyacetylene (s) (first demonstration of field effect in polymers)	4	NR	200	Ebisawa et al. (1983)
1984	1.5×10^{-5}	Merocyanine		NR	7000	Kudo et al. (1984)
1986	10^{-5}	Polythiophene (s)	3	10^3	NR	Tsumura et al. (1986)
1988	10^{-4}	Polyacetylene (s)	4	NR	750	Burroughes et al. (1988)
	10^{-3}	Phthalocyanine (v)	6	NR	3	Clarisse et al. (1988)
	10^{-4}	Poly(3-hexylthiophene) (s)	3	NR	NR	Assadi et al. (1988)
1989	10^{-3}	Poly(3-alkylthiophene) (s)	3	NR	NR	Paloheimo et al. (1989)
	10^{-3}	α-Hexathiophene (v)	2, $m=4$	NR	NR	Horowitz et al. (1989)
1992	0.027	α-Hexathiophene (v)	2, $m=4$	NR	100	Horowitz et al. (1992)
	2×10^{-3}	Pentacene (v)	1	NR	NR	Horowitz et al. (1992)
1993	0.05	α,ω-Dihexyl-hexathiophene (v)	2, $m=4$	NR	100–200	Garnier et al. (1993)
	0.22[d]	Polythienylenevinylene (s)	5	NR	1000	Fuchigami et al. (1993)
1994	0.06	α,ω-Dihexyl-hexathiophene (v)	2, $m=4$	NR	50	Garrier et al. (1994)
1995	0.03	α-Hexathiophene (v)	2, $m=4$	$>10^6$	21	Dodabalapur et al. (1995)
	6×10^{-3}	α-Quaterthiophene (V)	2, $m=2$	NR	NR	Katz et al. (1995)

Year		Compound		I_{on}/I_{off}[b]		Reference
1996	0.038	Pentacene (v)	**1**	140	1000	Dimitrakopoulos et al. (1996)
	0.02	Phthalocyanine (v)	**6**	2×10^5	NR	Bao et al. (1996a)
	0.045	Poly(3-hexylthiophene) (s)		340	20.8	Bao et al. (1996b)
	0.62	Pentacene (v)		10^8	11	Lin et al. (1996)
1997	1.5	Pentacene (v)	**1**	10^8	2.5	Lin et al. (1997b)
	0.012	Trans-trans-2,5-bis-[2-{5-(2,2'-bithienyl)}ethenyl] thiophene		10^3	2.3	Dimitrakopoulos et al. (1997)
	0.13	α,ω-dihexyl-hexathiophene (v)	**2**, $m=4$	$>10^4$	7.3	Dimitrakopoulos et al. (1998)
	0.04	Bis(benzodithiophene) (v)	**8**	NR	NR	Laquindanum et al. (1997)
	0.05	Bis(dithienothiophene) (v)	**7**	10^8	500	Sirringhaus et al. (1997)
	1.5×10^{-3}	α-quinquethiophene (v)	**2**, $m=3$	NR	NR	Hajlaoui et al. (1997)
1998	0.1	Poly(3-hexylthiophene) (s)	**3**	$>10^6$	20	Sirringhaus et al. (1998)
	0.23	α,ω-Dihexyl-quaterthiophene (v)	**2**, $m=2$	NR	1.5	Katz et al. (1998a)
	0.15	Dihexyl-anthradithiophene	**9**	NR	1.5	Laquindanum et al. (1998)
2000	0.1	α,ω-Dihexyl-quinquethiophene (s)	**2**, $m=3$	NR	NR	Katz et al. (2000a)
	2.4	pentacene (v)		10^8	10–100	Schön et al. (2000c)

[a]Measured at room temperature.

[b]Values for I_{on}/I_{off} correspond to different gate voltage ranges and thus are not readily comparable to each other. The reader is encouraged to read the details of the experiments in the cited references.

[c]NR = not reported.

[d]This result has not been reproduced yet.

199

TABLE II

Highest Reported field-effect Mobility (μ) Values Measured from n-type OTFTs, Annually from 1990 to the Present Time, for Each One of the Most Promising n-type Organic Semiconductors. For a n-type Organic Semiconductor that Already has a Data Entry in Table II from a Previous Year, a New Mobility Value is Entered Only if it is Higher than the Preceding Entry

Year	Mobility[a] ($cm^2 V^{-1} s^{-1}$)	Material	Compound no.	I_{on}/I_{off}[b]	W/L	Reference
1990	2×10^{-4}	Pc$_2$Lu (v)		NR[c]	20	Guillaud et al. (1990)
	1.4×10^{-3}	Pc$_2$Tm (v)		NR	20	Guillaud et al. (1990)
1993	5×10^{-4}	C$_{60}$/C$_{70}$ (9:1) (v)		NR	16000	Kastner et al. (1993)
1994	3×10^{-5}	TCNQ (v)	12	450	2000	Brown et al. (1994)
1995	0.08	C$_{60}$ (v)		10^6	400	Haddon et al. (1995)
	0.3			22	400	ibid.
1996	1.5×10^{-5}	PTCDI-Ph (v)		NR	100	Horowitz et al. (1996)
	0.003	TCNNQ (v)	13	NR	NR	Laquindanum et al. (1996)
	10^{-4}	NTCDI (v)		NR	NR	Laquindanum et al. (1996)
	0.003	NTCDA (v)		NR	21	Laquindanum et al. (1996)
1997	10^{-4}–10^{-5}	PTCDA (v)	19	NR	21	Ostrick et al. (1997)
1998	0.03	F16CuPc (v)	10	5×10^4	21	Bao et al (1998)
2000	0.06	NTCDI-C8F (v)	14	10^5	17	Katz et al. (2000a)
	0.1	NTCDI-C8F (v)	14	10^5	1.5	Katz et al. (2000a)
	0.01	NTCDI-C8F (s)	14	NR	NR	Katz et al. (2000a)
	0.12	NTCDI-BnCF3 (v)	15	NR	17	Katz et al. (2000b)
	0.16[d]	NTCDI-C8H (v)	16	>100	17	Katz et al. (2000a,b)
	0.02	DFH-6T (v)	11	10^5	20	Facchetti et al. (2000)
	0.5	Pentacene (v)	1	NR	10–100	Schön et al. (2000c)
2001	0.6	PTCDI-C8H (v)	18	10^5	16	Malenfant et al. (2001) Malenfant et al. (2002)

[a]Measured at room temperature.

[b]Values for I_{on}/I_{off} correspond to different gate voltage ranges and thus are not readily comparable to each other. The reader is encouraged to read the details of the experiments in the cited references.

[c]NR = not reported.

[d]After pumping for 4 days in high vacuum. Tested in vacuum.

$3\,cm^2\,V^{-1}\,s^{-1}$ to about $10^5\,cm^2\,V^{-1}\,s^{-1}$ at 1.7 K following a power law ($\mu \propto T^{-n}$, $n = 2.7$) (Schön et al. 2000a,b; Schön and Batlogg 2001), in agreement with earlier work in which $n = 2.9$ was reported (Karl et al. 1991; Karl 2001). At 1.7 K, an effective electronic bandwidth of $\sim 0.5\,eV$ ($\gg k_BT$, where k_B is the Boltzmann constant) is estimated from the low effective mass for holes, measured to be of the order of $1–1.5\,m_e$, where m_e is the free electron mass (Schön et al. 2000b). Furthermore, both the integer and fractional quantum Hall effects were observed in acene single crystals (Schön et al. 2000b). All this is clear evidence of band transport, at low temperatures, in such crystals. In 1974, Burland observed cyclotron resonance of holes in high quality single crystals of anthracene at 2 K, the first time such an observation was made in a wide band gap molecular crystal (Burland 1974; Burland and Konzelmann 1977). From the reported values of the effective mass of holes, m^*, in these crystals, determined to be $11 m_e$, where m_e is the free electron mass, and the hole scattering time, τ, measured to be $\sim 7 \times 10^{-11}$ and $4 \times 10^{-10}\,s$, respectively in two different anthracene crystals, one could calculate hole mobility values of about 11 200 and $64\,000\,cm^2\,V^{-1}\,s^{-1}$, using the formula: $\mu = (e/m^*)\tau$, where e is the elementary charge. This was the first time that evidence of such enormously high charge carrier mobility values in organic semiconductors was presented (Burland 1974). The temperature dependence of the electron mobility in pentacene and tetracene single crystals was shown to follow the same power law that describes the hole mobility temperature dependence in the same materials, from 1.7 to 300 K. From all the work presented above, it is apparent that extremely high quality, ultra pure single crystals of organic semiconductors are required to obtain such high mobilities and observe the described phenomena. This behavior is rather impossible to observe in polycrystalline films, in which traps attributed to structural defects dominate transport.

The temperature dependence of the electron mobility in naphthalene single crystals below 100 K also follows a power law, although with a lower exponent ($\mu \propto T^{-n}$, $n \approx 1.5–1.7$), consistent with the band model (Warta et al. 1985; Silinsh and Čápek 1994; Karl 2001). However, between 100 and 300 K, the electron mobility remains practically constant (Karl 1985; Silinsh and Čápek 1994). The $\mu(T) = $ constant region has been described phenomenologically as the superposition of two independent carrier mechanisms. According to one interpretation the first mechanism is described using the concept of an adiabatic, nearly small molecular polaron (MP) (Silinsh and Čápek 1994). According to this model, the carriers are treated as heavy polaron-type quasiparticles, which are formed as a result of the interaction of the carriers with intramolecular vibrations of the local lattice environment, and move coherently via tunneling. In this model, the mobility follows the power law $\mu_{MP} = aT^{-n}$ (Silinsh and Čápek 1994). The second mechanism involves a small lattice polaron (LP), which moves by thermally activated hopping and thus exhibits a typical exponential dependence of mobility on temperature:

$\mu_{LP} = b \cdot \exp[-E_a/kT]$. The superposition of these two mechanisms can reproduce the experimentally measured temperature dependence of mobility from just a few Kelvins to room temperature (Silinsh and Čápek 1994).

Band-like transport of delocalized carriers is shown to be the prevalent transport mechanism along the crystal directions with high $\pi-\pi^*$ orbital overlap, in single crystals of $\alpha,\text{-}\omega$-hexathiophene (6T) and α,ω-quaterthiophene (4T) (Schön and Batlogg 2001).

In the following sections, we will discuss carrier transport in various polycrystalline thin films of organic semiconductors, assuming that transport in individual crystallites in the film takes place according to the mechanisms described in this section.

At this point, we can propose two possible ways for eliminating the potential fundamental upper limit of about $10\,\mathrm{cm}^2\,\mathrm{V}^{-1}\,\mathrm{s}^{-1}$ for the room temperature mobility of OTFTs, imposed by the weak intermolecular forces existing among nearest-neighbor (nn) molecules. One is to strengthen such interactions. This can be done by creating a stronger interaction between nn molecules. However, this has to take place without breaking the conjugation of the molecules, and without reducing the intermolecular overlap between nn molecules. Stronger intermolecular interactions would result in more rigid crystalline structures, and thus it would take temperatures higher than room temperature to generate substantial scattering of highly delocalized carriers by lattice vibrations. Using such a strategy, one, in effect, could produce at room temperature, mobility values comparable to the high mobility that exists at very low temperatures in crystals of the acene series, or at least considerably higher than their room temperature mobilities. A second way involves a more drastic change in the conduction path and mechanism. It involves carrier transport via an array of single molecules, such as nanotubes or polymer chains that would bridge the gap between the source and drain electrodes of a TFT. Intermolecular transport is replaced by intramolecular transport. This would require a drastic reduction in the size of the TFT channel from micrometer- to nanometer-size so that it is shorter than the length of a single molecule. Recently, TFTs based on carbon nanotubes have been constructed and mobilities are of the order of $100\,\mathrm{cm}^2\,\mathrm{V}^{-1}\,\mathrm{s}^{-1}$ (Martel et al. 1998; Collins et al. 2001). This is an excellent example of the potential effectiveness of this strategy. The successful execution of any one of these strategies would prove that the existing performance limits are partially imposed by the design and size of OTFTs and not by the inherent properties of organic materials.

V. Organic Transistor Operation and Modeling

Figure 4 shows two common device configurations used in OTFTs. The $I\text{--}V$ characteristics of OTFTs can be adequately described by models developed for inorganic semiconductors (Sze 1981) as shown earlier (Torsi

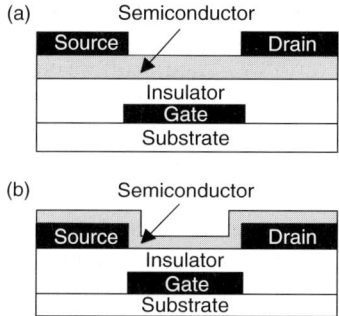

FIG. 4. OTFT device configurations: (a) Top-contact device, with source and drain electrodes deposited onto the organic semiconducting layer. (b) Bottom-contact device, with the organic semiconductor deposited onto the gate insulator and the prefabricated source and drain electrodes.

et al. 1995; Dimitrakopoulos *et al.* 1996, 1998; Brown *et al.* 1997; Horowitz 1998; Horowitz *et al.* 1999b). Polycrystalline pentacene OTFTs are used here to demonstrate typical *I–V* characteristics of OTFTs and the methods used to calculate the field effect mobility and other device parameters such as the current modulation (the ratio of the current in the accumulation mode, I_{on}, over the current in the depletion mode, I_{off}, also referred to as I_{on}/I_{off} ratio), and the threshold voltage, V_T. Polycrystalline pentacene OTFTs exhibit p-type behavior (the majority carriers are holes). Thus, when the gate electrode is biased positively with respect to the grounded source electrode, they operate in the depletion mode, and the channel region is depleted of carriers resulting in high channel resistance (off state). When the gate electrode is biased negatively, they operate in the accumulation mode and a large concentration of carriers is accumulated in the transistor channel, resulting in low channel resistance (on state). For n-type TFT operation, the electrode polarity is reversed and the majority carriers are electrons instead of holes. The accumulation layer is limited to the first few organic monolayers from the organic semiconductor/insulator interface, since it has been shown that organic layer thickness equivalent to only few such monolayers is sufficient for proper transistor operation and additional thickness does not substantially increase I_{on} (Ostoja *et al.* 1993; Dodabalapur *et al.* 1995).

A typical plot of drain current I_D vs drain voltage V_D at various gate voltages V_G is shown in Fig. 5(a), which corresponds to a top-contact OTFT [Fig. 4(a)] using a polycrystalline pentacene film as the semiconductor, 5000 Å thermally grown SiO_2 as the gate insulator, a heavily doped n-type Si wafer as the gate, and gold source and drain electrodes. At low V_D, I_D increases linearly with V_D (linear regime) and is approximately determined from the following equation:

$$I_D = \frac{WC_i\mu}{L}\left(V_G - V_T - \frac{V_D}{2}\right)V_D$$

where L is the channel length, W the channel width, C_i the capacitance per unit area of the insulating layer, V_T the threshold voltage, and μ the field effect mobility, which can be calculated in the linear regime from the transconductance,

$$g_m = \left.\frac{\partial I_D}{\partial V_G}\right|_{V_D=\text{const.}} = \frac{W C_i}{L}\mu V_D$$

by plotting I_D vs V_G at a constant low V_D, with $-V_D \ll -(V_G - V_T)$, and equating the value of the slope of this plot to g_m. The calculated linear mobility value is $0.49\,\text{cm}^2\,\text{V}^{-1}\,\text{s}^{-1}$ at $V_D = -10\,\text{V}$. The value of V_D is chosen so that it lies in the linear part of the I_D vs V_D curve. For this device $L = 15.4\,\mu\text{m}$ and $W = 1\,\text{mm}$.

For $-V_D > -(V_G - V_T)$, I_D tends to saturate (saturation regime) due to the pinch-off of the accumulation layer, and is modeled by the equation $I_D = (W C_i \mu / 2L)(V_G - V_T)^2$. In the saturation regime, μ can be calculated from the slope of the plot of $\sqrt{|I_D|}$ vs V_G. Figure 5(b) shows a semi-logarithmic plot of I_D vs V_D at various V_G and provides more information about the low V_G sweeps (close to the turn-on voltage), that do not appear clearly in Fig. 5(a).

Figure 6(a) shows a graph that contains a plot of I_D vs V_G at a constant $V_D = -100\,\text{V}$ (saturation regime), from the same device as in Fig. 5. Figure 6(b) shows a graph that contains a semilogarithmic plot of I_D vs V_G and a plot of $\sqrt{|I_D|}$ vs V_G. The mobility calculated in the saturation regime was $0.60\,\text{cm}^2\,\text{V}^{-1}\,\text{s}^{-1}$, which is slightly larger than the linear regime mobility. This result is comparable to reported hole mobilities from OTFTs with polycrystalline pentacene film channels grown on SiO_2 using a substrate temperature (T_{sub}) of 120°C during deposition (Lin *et al.* 1997a). The $I_{\text{on}}/I_{\text{off}}$ ratio was approximately 5×10^6 when V_G was scanned from -100 to $+5\,\text{V}$. The V_T, calculated by the intersection of the dashed slope line in Fig. 6(b) with the x-axis, was $-28\,\text{V}$. However, the turn-on voltage for this device was about $+5\,\text{V}$. It is important to note that W/L must be at least 10 in order to minimize the effects of fringe currents flowing outside the channel on the calculated mobility value, otherwise this value is overestimated. Alternatively, accurate mobility measurements could be obtained by patterning the semiconductor such that its width does not exceed the width of the OTFT channel.

The mobility values of pentacene OTFTs most often reported in the literature are calculated in the saturation regime (Lin *et al.* 1996, 1997a,b; Nelson *et al.* 1998). Typically, the mobility calculated in the saturation regime is much higher than the mobility calculated in the linear regime, because the linear regime mobility is more negatively affected by departures from linearity in the I_D vs V_D curves, at low V_D. Gold and other metals with even higher work functions (e.g., Pd, Pt) that are most commonly used for source and drain contacts in p-type OTFTs, form ohmic contacts with pentacene, in the top contact configuration at least, as expected by comparing gold's work function to the valence band (or HOMO) energy level of

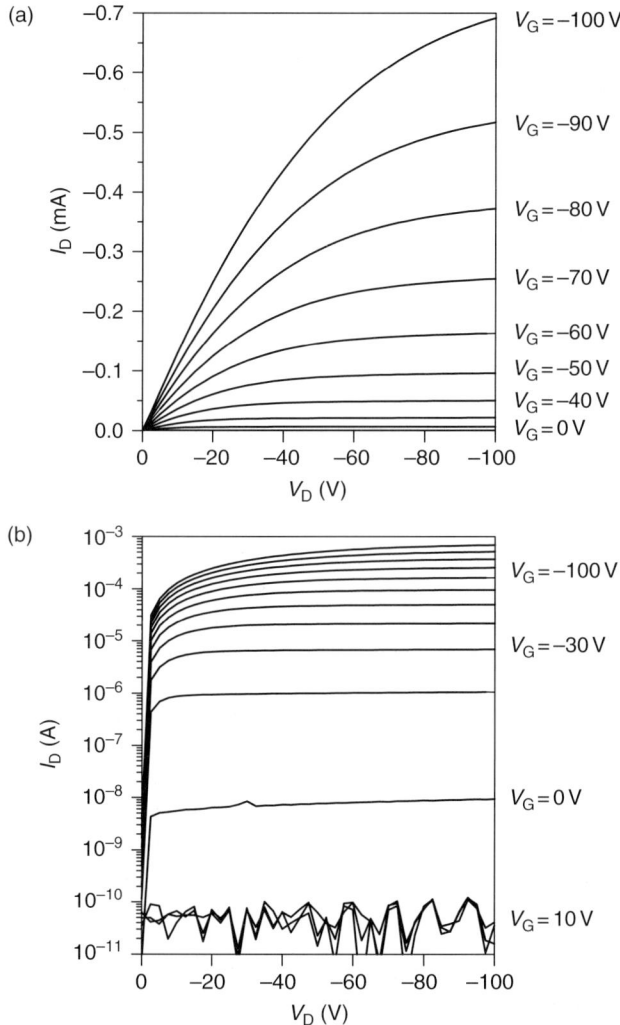

FIG. 5.   (a) Plot of drain current I_D versus drain voltage V_D at various gate voltages V_G from a top-contact OTFT comprising a polycrystalline pentacene thin film channel, a 5000 Å thick SiO_2 gate insulator layer, a heavily doped n-type Si wafer as the gate, and Au source and drain electrodes. $L = 15.4\,\mu m$ and $W = 1\,mm$. The linear regime mobility is $0.49\,cm^2\,V^{-1}\,s^{-1}$ at $V_D = -10\,V$. (b) Semilogarithmic plot of I_D versus V_D at various gate voltages V_G for the same device.

pentacene crystals. A proof of this has been provided by comparisons of two- and four-point resistance data from pentacene with Au contacts (Schön and Batlogg 2001). In addition, hole mobilities from the linear and saturation regimes calculated from single crystal FETs of pentacene, are identical (Schön *et al.* 2000c). The origin of large differences between the two mobilities

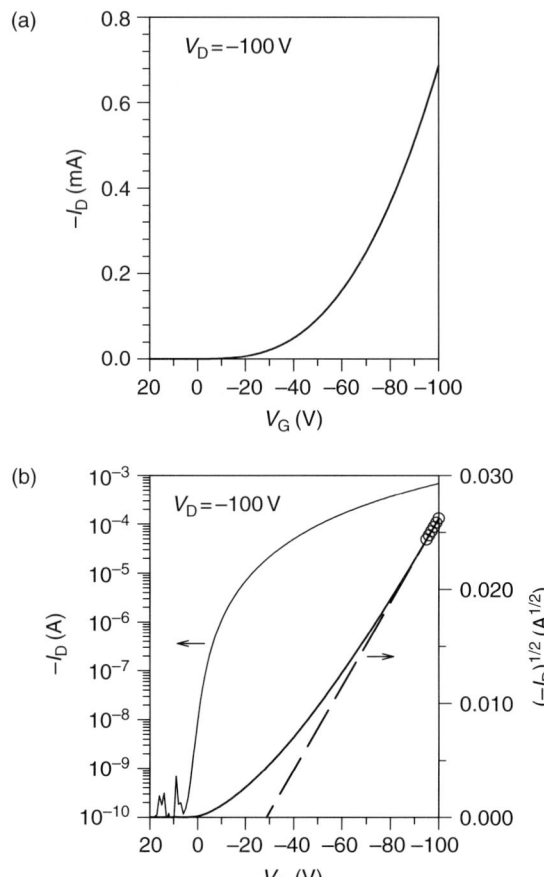

FIG. 6. (a) Plot of I_D versus V_G at $V_D = -100$ V (saturation regime) from the device corresponding to Fig. 5. (b) Semilogarithmic plot of I_D versus V_G (left y-axis) and plot of $\sqrt{|I_D|}$ versus V_G (right axis) from the same device. The field-effect mobility μ, calculated in the saturation regime, is 0.60 cm^2 V^{-1} s^{-1}. The I_{on}/I_{off} ratio was more than 10^6 when V_G was scanned from -100 V to $+5$ V.

could, however, be the existence of large concentrations of trap states in the channel (e.g., related to grain boundaries).

VI. Organic Transistor Fabrication Methods

1. VACUUM DEPOSITED ORGANIC SEMICONDUCTOR FILMS

Organic semiconductor films can be deposited by sublimation in a variety of vacuum deposition systems, utilizing techniques in which the deposition parameters vary widely. OTFTs with good charge carrier transport properties have been fabricated using most of these techniques. The base pressure of the deposition system is an important deposition parameter, since it determines, among other things, the mean free path of the sublimed organic semiconductor molecules, and the presence of unwanted atoms and molecules in the vicinity of the substrate surface during film formation. It ranges from less than 10^{-9} torr, in ultrahigh vacuum (UHV) organic molecular beam deposition (OMBD) (Dimitrakopoulos *et al.* 1998), to about 10^{-7} torr for common high vacuum (HV) bell-jar deposition systems, to more than 10^{-3} torr for simple, glass-wall vacuum sublimation systems (Malenfant *et al.* 2001; 2002).

Substrate temperature and deposition rate are two other deposition parameters that can influence dramatically thin film morphology and thus the transport characteristics of OTFTs (Dimitrakopoulos *et al.* 1996; Jentzsch *et al.* 1998).

Purity of the organic source material is also very important and together with substrate cleanliness they can determine in large part the quality of an OTFT. The latter is very important since the carrier accumulation layer is occurring in the first few monolayers of the organic semiconductor at the interface with the insulator (Ostoja *et al.* 1993; Dodabalapur *et al.* 1995). Impurities can affect the mobility, the on/off ratio and in some cases even the polarity of the OTFT. For example, iodine-doped pentacene is a p-type semiconductor (Minakata *et al.* 1992) while alkaline metal-doped pentacene is an n-type semiconductor (Minakata *et al.* 1993).

As an alternative to HV and UHV deposition systems, organic semiconductor thin films have been grown from the vapor phase in a stream of flowing gas (hydrogen or forming gas), at pressures of the order of 1 torr in an apparatus similar to the one described in Burrows *et al.* (1995) and Baldo *et al.* (1998). The general morphology of the resulting films is similar to that of films deposited with HV and UHV deposition methods.

2. SOLUTION PROCESSED ORGANIC SEMICONDUCTOR FILMS

The technology that is believed to have the potential to produce the highest impact on manufacturing costs is the use of soluble organic semiconductors, both polymers and oligomers, combined with large area stamping or screen printing or even inkjet printing techniques that could eliminate conventional lithography.

Spin coating is one of the most widely used methods for deposition of thin films of soluble organic semiconductors from solution. An appropriate quantity of solution is applied on a substrate that is then rotated at a speed of a few hundred to few thousand rounds per minute (rpm). A film with a uniform thickness is produced when spin coating process parameters are optimized. The film thickness is proportional to the viscosity of the solution and inversely proportional to the rotation speed. Close to the edges of the substrate the film thickness is usually different than the rest of the substrate. After spin coating, the substrate is usually heated at moderate temperatures to remove the solvent from the film.

Drop casting (or solution casting), and dip casting are other techniques used to produce organic semiconductor films.

While spin coating, solution casting and printing are perhaps the most commercially feasible processing techniques for soluble organic semiconductors, the Langmuir–Blodgett (LB) technique and its variations have also been explored for the preparation of thin films of polymeric semiconductors (Paloheimo et al. 1990, 1992; Bjørnholm et al. 1998; Xu et al. 2000).

VII. Organic Transistor Performance

1. PROGRESS IN PERFORMANCE OF p-TYPE OTFTs

Table I lists the highest field-effect mobility (μ) values measured from p-type OTFTs as reported in the literature, annually from 1984 to the present time, for each one of the most promising p-type organic semiconductors. For each p-type organic semiconductor that already has a data entry in Table I and Fig. 7 for a previous year, a new mobility value is entered only if it is higher than the value of the preceding entry. We can observe an impressive increase in mobility, which is the result of either improvements in the processes used for the fabrication of OTFTs or the synthesis of new organic semiconductor materials. A typical path to performance increase could be described as a three-stage process: (i) a new organic semiconductor is synthesized or a known one is used for the first time as the active layer in an OTFT; (ii) the film deposition parameters for the semiconducting organic layer are optimized to obtain the most advantageous structural and morphological characteristics for improved performance until no more improvement seems possible; and (iii) injection from the source and drain contacts is optimized. After this point, another incremental improvement in mobility is usually obtained from the synthesis and/or first OTFT application of a new organic semiconductor. Today, we have reached an important point in the performance vs time plot. The most widely used organic semiconductors, such as pentacene, thiophene oligomers, and regioregular poly(3-alkyl-thiophene), seem to have reached "maturity" as far as their performance is concerned. Their individual performance versus time curves seem to have saturated (when a new, higher value is not reported in the years following the last entry for a material, it means that there was no improvement in mobility during those years). In the past, each time such a performance saturation occurred, a new material was introduced whose performance broke the

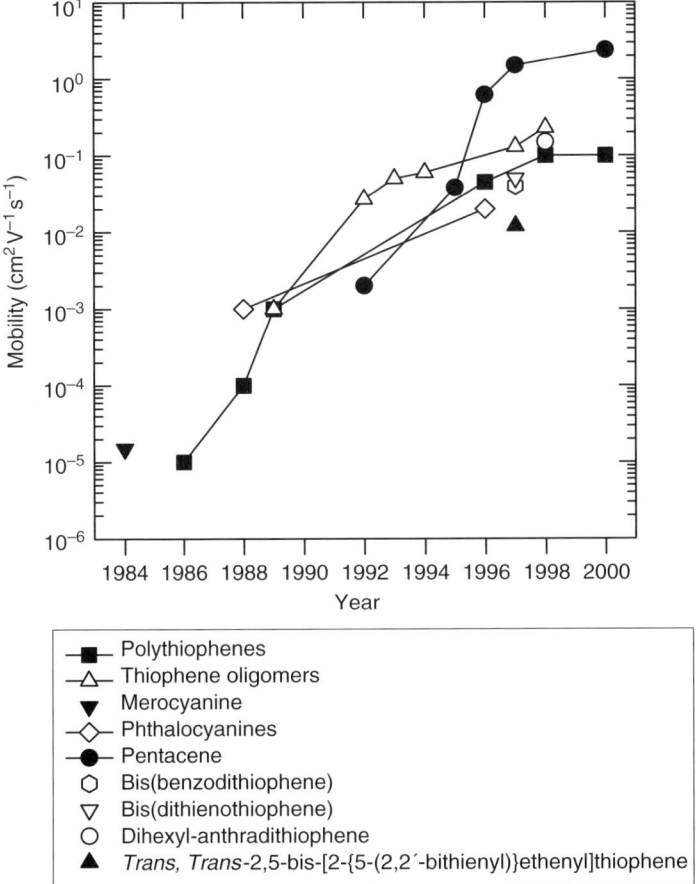

FIG. 7. Evolution of OTFT hole mobility for the most common p-type organic semiconductors. The various p-type materials are grouped together into families of similar molecules taking into account only the core part of each molecule.

temporarily established upper limit in performance (see, e.g., the mobility vs time curve for OTFTs based on oligothiophene oligomers in Fig. 7 and the emergence of pentacene OTFTs at the time that the performance of the former started to saturate).

For instance, one can argue that today's maximum hole field-effect mobility of $2.4\,\mathrm{cm^2\,V^{-1}\,s^{-1}}$ for OTFT (measured at room temperature in pentacene) (Schön *et al.* 2000c) has approached a fundamental limit, at least as far as the known classes of semiconducting organic materials are concerned. This pentacene thin film was grown from the vapor phase in a stream of flowing gas (hydrogen or forming gas), in an apparatus similar to the one described in Burrows *et al.* (1995), Kloc *et al.* (1997), Baldo *et al.*

(1998), and Laudise *et al.* (1998) and its grain size exceeded the channel length (L) of the TFTs used to determine the field-effect mobility. Consequently, this value is very close to room temperature field-effect mobility values of 3.2 and 2.7 cm^2 V^{-1} s^{-1} reported by Schön *et al.* (2000a,c) for holes in pentacene single crystals grown from the vapor phase under hydrogen or forming gas flow, while the substrate was held at a relatively high temperature (Kloc *et al.* 1997; Laudise *et al.* 1998).

The synthesis, fabrication, and characterization of OTFTs based on polycrystalline, vapor deposited α,ω-hexathiophene and α,ω-dihexyl-hexathiophene films by Horowitz *et al.* (1989, 1992) and Garnier *et al.* (1993) respectively played a very important role in the evolution of the field of organic transistors as evidenced by the large number of entries in Table I. That work showed that relatively high mobilities are attainable by polycrystalline organic semiconductors. Later, the work of Dodabalapur *et al.* (1995) showed that OTFTs with current modulation of 10^6 were attainable by carefully purifying the evaporation source material. As shown in Table I, the highest values of field-effect mobility (μ) reported in different years for α,ω-dihexyl-hexathiophene (DH6T) films were 0.06 cm^2 V^{-1} s^{-1} (Garnier *et al.* 1994) and 0.13 cm^2 V^{-1} s^{-1} (Dimitrakopoulos *et al.* 1998) when a polymeric gate insulator is used. With a SiO$_2$ gate insulator the two highest mobility values were 0.02 cm^2 V^{-1}s^{-1} (Horowitz *et al.* 1993) or 0.03 cm^2 V^{-1} s^{-1} (Katz *et al.* 1995). For unsubstituted α-hexathiophene and SiO$_2$ gate insulator, the highest value was 0.03 cm^2 V^{-1} s^{-1} (Dodabalapur *et al.* 1995).

Pentacene TFTs have produced the highest field-effect mobility values reported from OTFTs. From 1992 (first use of pentacene in an OTFT) (Horowitz *et al.* 1992) to 1997 the mobility (in cm^2 V^{-1} s^{-1}) was raised from 2×10^{-3} (Horowitz *et al.* 1992) to 0.038 (Dimitrakopoulos *et al.* 1996) to 0.62 (Lin *et al.* 1996) to 1.5 (Lin *et al.* 1997b).

In the following paragraphs, we summarize recent developments in the performance of p-type OTFTs fabricated by solution processes. One of the first solution-processable organic semiconductors used for field-effect transistors was poly(3-hexylthiophene) or P3HT, in which the addition of alkyl side-chains enhanced the solubility of the polymer chains (Assadi *et al.* 1988). P3HT films spun from a chloroform solution had mobilities in the range of 10^{-5}–10^{-4} cm^2 V^{-1} s^{-1}. These mobilities were comparable to the mobilities obtained from electrochemically prepared polythiophene field-effect transistors (Tsumura *et al.* 1991), indicating that the incorporation of insulating alkyl side-chains was not detrimental to the electronic properties of polythiophene. A comparison study of poly(3-alkylthiophene)s with side chains ranging in length from butyl to decyl showed that field-effect mobility decreases with increasing chain length (Paloheimo *et al.* 1991). For films spun from chloroform, mobilities ranged from 1–2×10^{-4} cm^2 V^{-1} s^{-1} for poly(3-butylthiophene) and poly(3-hexylthiophene) (P3HT) down to 6×10^{-7} cm^2 V^{-1} s^{-1} for poly(3-decylthiophene).

When regioregular P3HT (Chen *et al.* 1992, 1995; McCullough *et al.* 1992; McCullough 1998) (Fig. 2) consisting of 98.5% or more head-to-tail (HT) linkages was used to fabricate field-effect transistors, a dramatic increase in mobility was observed relative to regiorandom poly(3-alkylthiophene)s (Bao *et al.* 1996b). Mobilities as high as $0.045\,cm^2\,V^{-1}\,s^{-1}$ were achieved in films drop-cast from a chloroform solution (Bao *et al.* 1996b). Drop-cast films of highly regioregular P3HT self-orient into a well-ordered lamellar structure with an edge-on orientation of the thiophene rings relative to the substrate (Chen *et al.* 1992, 1995; McCullough *et al.* 1992; McCullough 1998). This is in contrast to solution-cast films of regiorandom poly (3-alkylthiophene)s, which are essentially amorphous (Chen *et al.* 1992, 1995; McCullough *et al.* 1992; McCullough 1998). Spin-coated films of regioregular P3HT are also well ordered, but the lamellae adopt different orientations depending on the degree of regioregularity (Sirringhaus *et al.* 1999a). Highly regioregular P3HT (greater than 91% HT linkages) also forms lamellae with an edge-on orientation (π–π stacking direction in the plane of the substrate) when spun from chloroform. Mobilities of 0.05–$0.1\,cm^2\,V^{-1}\,s^{-1}$ were obtained for 96% regioregular P3HT (Sirringhaus *et al.* 1999a). In contrast, spin-coated films of P3HT with low regioregularity (81% HT linkages) consisted of lamellae having a face-on orientation (π–π stacking direction perpendicular to the substrate) and resulted in mobility of $2 \times 10^{-4}\,cm^2\,V^{-1}\,s^{-1}$. Drop-cast films of 81% regioregular P3HT adopted an edge-on lamellar structure, resulting in an order-of-magnitude increase in mobility compared to spin-coated films. This study indicates that, in addition to the degree of order in the polymer film, the π–π^* stacking direction relative to the substrate, which depends on the film deposition method, greatly affects the field-effect mobility (Sirringhaus *et al.* 1999a). This is reasonable, since the edge on lamellar structure ensures that delocalized intermolecular states are formed in the direction parallel to the substrate, which happens to be the transport direction in OTFT devices.

The mobility of regioregular P3HT has been found to vary by two orders of magnitude depending on the solvent used, with chloroform giving the highest mobility (Bao *et al.* 1996b). Modification of the substrate surface prior to deposition of regioregular poly(3-alkylthiophene) has also been found to influence film morphology. For example, treatment of SiO_2 with hexamethyldisilazane (HMDS) or an alkyltrichlorosilane replaces the hydroxyl groups at the SiO_2 surface with methyl or alkyl groups. The apolar nature of these groups apparently attracts the hexyl side-chains of P3HT, favoring lamellae with an edge-on orientation. Mobilities of 0.05–$0.1\,cm^2\,V^{-1}\,s^{-1}$ from highly regioregular P3HT have been attributed to this surface modification process (Sirringhaus *et al.* 1998, 1999b). In addition, it has been shown that top contact devices yield mobilities that are typically larger by a factor of two compared to bottom contact devices (Bao *et al.* 1997; Sirringhaus *et al.* 1998). Exposure of poly(3-alkylthiophene) films to air causes an increase in conductivity and a subsequent degradation of the transistor on/off ratio.

This is the result of doping with oxygen (doping by water is a less probable but possible cause), since it is possible to achieve high on/off ratios by preparing and testing devices in a dry N_2 atmosphere (Bao et al. 1997; Sirringhaus et al. 1998).

Sirringhaus et al. have recently demonstrated direct inkjet printing of OTFTs based on solution-processed polymer electrodes (water-soluble poly[3,4-ethylenedioxythiophene] doped with polystyrene sulfonic acid (PEDOT/PSS), insulators (polyimide), and active organic semiconducting layer (poly[9,9-dioctylfluorene-co-bithiophene], called F8T2, from xylene solution) (Sirringhaus et al. 2000b). It was shown that F8T2, which is a nematic liquid crystalline conjugated polymer semiconductor, can be preferentially oriented by rubbed polyimide layers, and when used as the active channel in OTFTs it exhibits mobility of $0.02\,cm^2\,V^{-1}\,s^{-1}$ and an on/off ratio of 10^5 (Sirringhaus et al. 2000a,b).

The LB technique and its variations have also been employed for the preparation of poly(3-alkylthiophene) field-effect transistors (Paloheimo et al. 1990, 1992; Bjørnholm et al. 1998; Xu et al. 2000). The maximum mobility measured from multilayer LB films of P3HT was $0.02\,cm^2\,V^{-1}\,s^{-1}$ (Bjørnholm et al. 1998).

Unsubstituted quinquethiophene and end-substituted quater-, quinque-, and hexathiophene display enough solubility in organic solvents to allow fabrication of field-effect devices by solution processing techniques. Initial studies of solution-processed oligothiophene transistors gave mobilities of $\sim 5 \times 10^{-5}\,cm^2\,V^{-1}\,s^{-1}$ for quinquethiophene and α,α'-diethylquaterthiophene (DE4T) (Akimichi et al. 1991). More recently, mobilities in the range of 0.01–$0.1\,cm^2\,V^{-1}\,s^{-1}$ have been achieved using solution-processed substituted oligothiophenes (Katz et al. 1998b, 1999, 2000a; Garnier et al. 1998). The mobility was found to depend strongly on film morphology, which can be controlled by processing conditions, such as solution concentration, substrate temperature during casting, solvent choice, and environmental conditions during film drying.

Other soluble organic oligomers have also been investigated as semiconducting materials for field-effect transistors. Anthradithiophene, a fused heterocycle compound, is soluble in its dihexyl end-substituted form. Dihexylanthradithiophene (DHADT) transistors were fabricated by solution casting from hot chlorobenzene, followed by evaporation of the solvent in a vacuum oven at various temperatures. The electrical characteristics of the films were strongly dependent on the temperature during film drying. The highest mobilities, 0.01–$0.02\,cm^2\,V^{-1}\,s^{-1}$, were obtained for a drying temperature of $100°C$ (Laquindanum et al. 1998; Katz et al. 1999). In comparison, vacuum evaporated films of DHADT gave mobilities as high as $0.15\,cm^2\,V^{-1}\,s^{-1}$ (Laquindanum et al. 1998).

Transistors utilizing another thiophene-containing oligomer, trans-trans-2,5-bis-[2-{5-(2,2'-bithienyl)}ethenyl]thiophene (BTET), were fabricated by spin coating from hot N-methyl pyrrdidone (NMP). The mobility of such a device was $1.4 \times 10^{-3} \, cm^2 \, V^{-1} \, s^{-1}$ compared to $0.012 \, cm^2 \, V^{-1} \, s^{-1}$ for a vacuum evaporated device (Dimitrakopoulos *et al.* 1997).

Although a great deal of success has been achieved with soluble oligomers, the solubility of these oligomers is low, requiring the solvents to be heated. Drop casting and spinning yield films that are non-uniform in thickness, morphology, and electrical properties. To circumvent such problems, another approach to solution-processable oligomeric materials has been developed that involves a soluble precursor molecule that is not semiconducting, and which can be converted to its semiconducting insoluble form upon heating. This approach has been realized for pentacene, with initial reported mobilities of $0.01-0.03 \, cm^2 \, V^{-1} \, s^{-1}$ (Brown *et al.* 1995, 1996, 1997). The pentacene precursor is soluble in dichloromethane and forms continuous, amorphous films when spun onto substrates. The conversion to pentacene is accomplished by heating the films to a temperature of $140-220°C$ in vacuum. Tetrachlorobenzene is eliminated in the conversion process. In a more recent study, a mobility of $0.2 \, cm^2 \, V^{-1} \, s^{-1}$ was achieved by treatment of the SiO_2 substrate with HMDS prior to spin coating the precursor, and by optimizing the conversion conditions. Precursor films were formed by spin coating from a $1.5 \, wt\%$ solution in dichloromethane, and heating the film for $5 \, s$ at $200°C$ followed by rapid quenching (Herwig and Müllen 1999).

Recently the one-step synthesis of a novel, soluble pentacene precursor was reported and pentacene based OTFTs were fabricated in this way and tested (Afzali *et al.* 2001, 2002). This synthetic approach used an efficient Lewis acid-catalyzed Diels-Alder reaction of pentacene with a hetero dienophile under moderate conditions to form an adduct that is highly soluble in chlorinated solvents and ethers. Heating spin-coated thin films of this adduct at moderate temperatures (up to $200°C$) causes a retro Diels-Alder reaction that converts the adduct film to a pentacene thin film. Mobilities up to $0.9 \, cm^2 \, V^{-1} \, s^{-1}$ were reported from devices fabricated using this synthetic approach. (Afzali *et al.* 2001, 2002).

The precursor approach has also been applied to polymers. In fact, one of the first reported transistors based on organic polymer channels used precursor-route polyacetylene as the semiconducting layer (Burroughes *et al.* 1988, 1989). Another polymeric semiconductor that has been processed from a soluble precursor polymer is polythienylenevinylene (PTV) (Tsumura *et al.* 1991; Brown *et al.* 1995, 1997; Drury *et al.* 1998).

2.   PROGRESS IN PERFORMANCE OF n-TYPE OTFTs

Reports on n-type OTFTs started appearing in the literature about 10 years ago. Table II lists the highest field-effect mobility (μ) values measured from n-type OTFTs as reported in the literature, annually from 1990 to

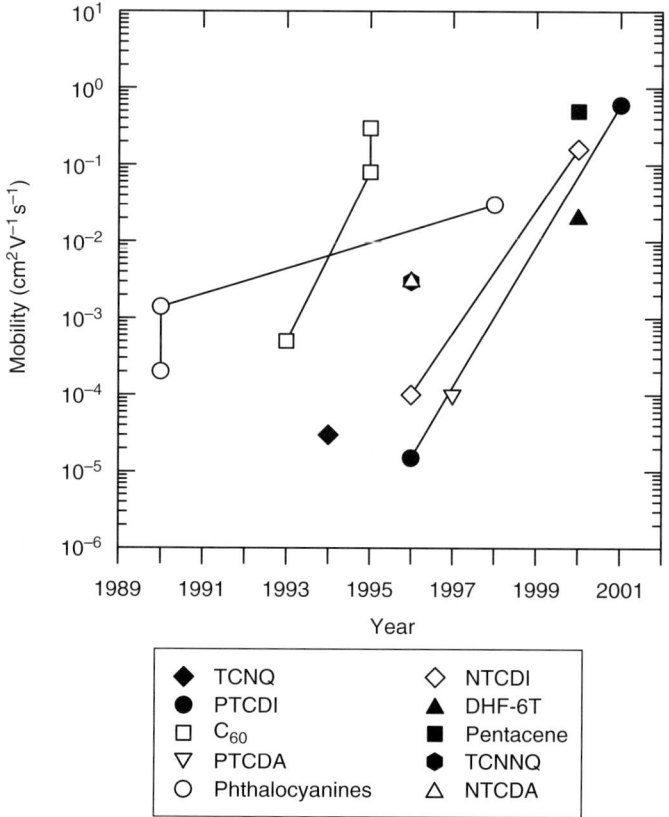

FIG. 8. Evolution of OTFT hole mobility for the most common n-type organic semiconductors. The various n-type materials are grouped together, when possible, into families of similar molecules considering only the core part of each molecule.

the present time, for each one of the most promising n-type organic semiconductors. Figure 8 conveys graphically most of the information contained in Table II. The various n-type materials are grouped together, when possible, considering only the core part of each molecule and not the specific substituents.

Early work in this area reported some n-type OTFTs with respectable mobilities and on/off ratios but they were unstable in air. Some materials provided devices with limited air stability but charge transport properties were poor. More recently, researchers have been able to fabricate n-type OTFTs with high mobility and current on/off ratio as well as fairly stable operation in air and occasionally solution processability.

An early study on n-type OTFTs was reported by Guillaud et al. (1990) and involved Lutetium (Pc$_2$Lu) and thulium (Pc$_2$Tm) bisphthalocyanines.

When electrical measurements were made *in situ* at room temperature and without breaking vacuum at any point during the process of device fabrication and measurement, electron mobilities between 2×10^{-4} and $1.4 \times 10^{-3} \, cm^2 \, V^{-1} \, s^{-1}$ for both Pc_2Tm and Pc_2Lu were obtained. However, upon exposure to air, only p-type activity was observed in those devices.

Tetracyanoquinodimethane (TCNQ) **12** was used as the active channel layer in n-type OTFTs by Brown *et al.* (1994). Bottom contact OTFTs [Fig. 4(b)] were fabricated with gold source and drain electrodes. The reduction potential of TCNQ is $+0.19 \, V$ vs the standard calomel electrode. This lies in the same region as the oxidation potential of many p-type materials, which are known to make ohmic contacts with gold. Hence, the LUMO level of TCNQ crystal is expected to be in the vicinity of the Fermi level of the gold contacts, thus the energy barrier (if any) for electron injection from gold into TCNQ should be very small. A mobility of $3 \times 10^{-5} \, cm^2 \, V^{-1} \, s^{-1}$ and an on/off ratio of 4–450 ($I_{on/off}$ increased upon exposure to air) were reported.

C_{60} and C_{60}/C_{70} fullerenes were used as the active channel layer in n-type OTFTs initially by Kastner *et al.* (1993), who reported mobilities up to $5 \times 10^{-4} \, cm^2 \, V^{-1} \, s^{-1}$. In a later report Haddon *et al.* reported on C_{60} OTFTs with improved transport characteristics (Haddon *et al.* 1995). In that work, the C_{60} films were deposited by sublimation in UHV and consisted of random polycrystalline grains of about $60 \, Å$ in size. C_{60} devices were shown to be sensitive to amine exposure and pre-treatment of the substrate with tetrakis (dimethylamino)ethylene (TDAE) prior to deposition of C_{60} moves the threshold voltage to negative values (the device becomes "normally-on") and increases the FET mobility from 0.08 to $0.3 \, cm^2 \, V^{-1} \, s^{-1}$, while it reduced the I_{on}/I_{off} ratio. The resistivity of C_{60} OTFTs quickly increased by four to five orders of magnitude upon exposure to ambient atmosphere but their transport properties could be restored upon placing the devices in UHV at elevated temperature for about 12 h. Mobilities of $0.5 \, cm^2 \, V^{-1} \, s^{-1}$ were measured from C_{60} single crystals (Frankevich *et al.* 1993) using time-of-flight measurements. Thus, the presence of grain boundaries in evaporated, polycrystalline thin films of C_{60} lowers their mobility.

Compared to materials with anisotropic carrier transport properties (e.g., pentacene), which require highly ordered structures with specific orientations in order to produce high field-effect mobilities along a desired direction, transport characteristics in films of C_{60} are pretty isotropic. This is probably the result of the approximately spherical shape of C_{60} as opposed to the rigid rod, or more generally, elongated shape of most of other organic semiconductors.

Laquindanum *et al.* (1996) and Katz *et al.* (2000a) have explored several materials based on the naphthalene framework (see Fig. 3). Initially they explored 1,4,5,8-naphthalene tetracarboxylic dianhydride (NTCDA) in which mobilities of $1–3 \times 10^{-3} \, cm^2 \, V^{-1} \, s^{-1}$ were measured for materials deposited onto substrates at held at $T_{sub} = 55°C$, but the mobility decreases by 1–2 orders of magnitude upon exposure to air. Lower mobilities are observed when devices are exposed to atmospheric moisture after sublimation. For devices with NTCDA deposited at $T_{sub} = 25°C$ mobilities of 10^{-4} cm^2 V^{-1}s^{-1} were obtained. Although grain sizes (200 nm) were similar for films grown at both substrate temperatures, films deposited at 55°C were more continuous, which explains the higher mobility of the latter. Devices constructed from 1,4,5,8-naphthalene tetracarboxylic diimide (NTCDI) provided mobilities on the order of 10^{-4} cm^2 V^{-1}s^{-1}. 11,11,12,12-Tetracyanonaphtho-2,6-quinodimethane (TCNNQ) **13** OTFTs displayed higher mobility (10^{-3} cm^2 V^{-1}s^{-1}) than (TCNQ) (Brown *et al.* 1994) (see above), while having better air stability than NTCDA but it exhibited a poor on/off ratio.

In a recent study, Katz *et al.* reported the transport characteristics of n-type OTFTs comprising a series of vacuum deposited films based on *N*-substituted naphthalene-1,4,5,8-tetracarboxylic diimide derivatives (see Fig. 3) (Katz *et al.* 2000a,b). Transport properties and environmental stability varied greatly with substitution. Only the diimides with fluorinated end-substituents (NTCDI-C8F, NTCDI-BnCF3, **14** and **15**, respectively in Fig. 3) showed high mobilities in air, whereas linear, alkyl functionalized diimides NTCDI-C8H (**16** in Fig. 3), NTCDI-C12H, NTCDI-C18H, gave mobilities of 0.16, 0.005–0.01, and 0.005 cm^2 V^{-1}s^{-1}, respectively, but only under vacuum. The highest mobility in air (0.12 cm^2 V^{-1}s^{-1}) was obtained with NTCDI-BnCF3, while the highest on/off ratio in air ($>10^5$) was achieved with NTCDI-C8F, which had mobility up to 0.1 cm^2 V^{-1}s^{-1}. Mobilities were much higher when the material was deposited onto a substrate held at an elevated temperature. Alternative source and drain electrodes were tested and carbon electrodes gave similar results to gold electrodes yet the former were not as reproducible as the latter. Carbon source and drain electrodes could be used in a bottom contact configuration [Fig. 4(b)] whereas gold electrodes, even if cleaned with oxygen plasma, generally produced either low-performance devices or devices that did not operate at all. Interestingly, aluminum electrodes did not provide active devices, despite the fact that aluminum has a lower work function than gold and thus should be more suitable for electron injection into these electron transporters. Although this is likely due to the oxidation of aluminum, which creates an insulating layer of aluminum oxide, another mechanism could be at work as well. In the following paragraph we will summarize some important results that shed some light on this issue.

The efficiency of injection of electrons (holes) from a metal contact into the conduction band or LUMO (valence band or HOMO) of an n-type (p-type) organic semiconductor depends on the energy barrier ϕ_{Be} (ϕ_{Bh}) that electrons (holes) have to overcome at the metal/organic semiconductor interface [Fig. 9, taken from Hill *et al.* (1998)]. At the metal/organic semiconductor interface, ϕ_{Be} and ϕ_{Bh} depend, respectively, on the position of the LUMO and HOMO relative to the Fermi level (E_F) of the metal. If the assumption of a common

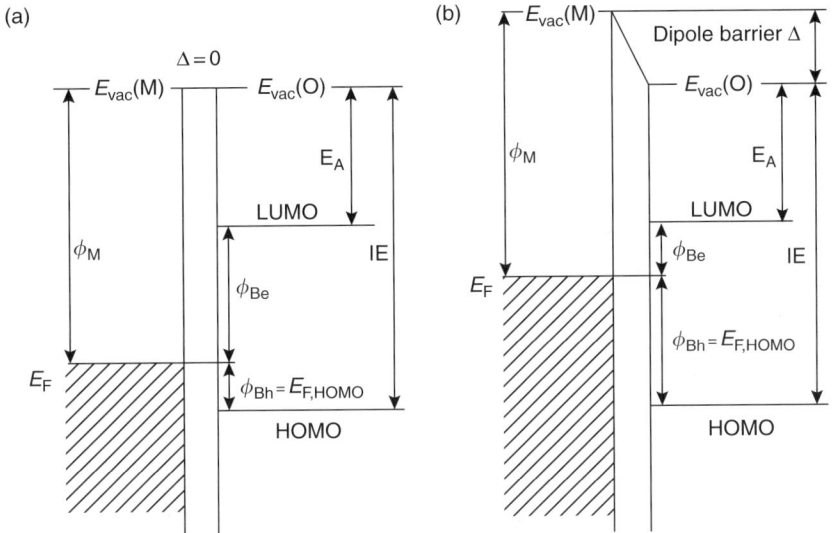

FIG. 9.   Schematic of an organic-metal interface energy diagram (a) without and (b) with an interface dipole Δ. See text for symbol definition. Reprinted with permission from Hill *et al.* (1998).

vacuum level at the metal/organic semiconductor interface (analogous to the Schottky–Mott limit for inorganic semiconductor interfaces) was valid, ϕ_{Be} would be the difference between the metal work function (ϕ_M) and the electron affinity of the organic semiconductor (E_A), and ϕ_{Bh} would be the difference between the ionization potential (IE) of the organic semiconductor and ϕ_M. For a long time this was considered a valid assumption due to the generally weak interaction at metal/organic semiconductor interfaces. However, recent studies have shown that this assumption is not always valid (Nariola *et al.* 1995; Ishii and Seki 1997; Hill *et al.* 1998). These results indicate the existence of a vacuum level shifts at the metal/organic semiconductor interface, whose magnitude and sign depend on the specific metal/organic combination, and can be attributed to an ultrathin, interfacial electric dipole layer that can be as high as 1.5 eV (Campbell *et al.* 1996, 1997; Ishii and Seki 1997; Hill *et al.* 1998, 2000). To complicate matters, chemical interactions at the interface also affect charge injection barriers. Such interactions depend not only on the constituents of the interface but also on the sequence and method of formation of the interface (Hill *et al.* 1998, 2000). In some organic semiconductors, such as 3,4,9,10-perylenetetracarboxylic dianhydride (PTCDA), the position of E_F in the band gap of the organic semiconductor is essentially pinned near the top of the band gap for several metal contacts (Hill *et al.* 1998). As a result, the electron injection barrier does not depend on the metal contact used.

It is apparent from the previous paragraph that simply trying to match the contact metal work function with the LUMO level (in n-type OTFT) or the HOMO level (in p-type OTFT) of the organic semiconductor is not always an adequate device design rule. Most of the work done to date has employed gold electrodes as the source and drain. Gold has a work function of $\sim$5.0 eV against vacuum and since many attractive n-type organic semiconductors have solid-state electron affinity levels $\sim$4.0 eV, this energy barrier of approximately 1 eV would be expected to severely limit charge injection into the semiconductor. Interestingly, this is often not the case. For example, in the case of N,N'-diphenyl-1,4,5,8-naphthyltetracarboxylimide (DP-NTCDI), the vacuum level in the organic semiconductor side of the interface with Au moves downwards by 0.9 eV thus bringing the E_F of the gold very close to the LUMO of the DP-NTCI, thus facilitating electron injection (Ishii and Seki 1997). When Al is used as the metal contact, however, the vacuum level in the organic semiconductor side of the interface moves upwards by -0.2 eV, which could be attributed to charge transfer from Al to DP-NTCDI (Ishii and Seki 1997). As a result, the electron injection barrier is higher for the Al contact than for the Au contact.

It has been shown in the past that the energy barrier for carrier injection between a metal and a conjugated organic material can be manipulated by the insertion of an appropriate oriented dipole layer, such as a self-assembled monolayer (SAM), between the metal electrode and the organic semiconductor (Campbell *et al.* 1996, 1997). The discussion in the previous two paragraphs explains this result. In their n-type OTFT work with N-substituted naphthalene$-$1,4,5,8-tetracarboxylic diimide derivatives (see Fig. 3), Katz *et al.* (2000b) tried this approach in order to obtain working bottom contact OTFTs with Au electrodes, since devices with unmodified Au electrodes rarely produced transistor activity. Good results were obtained using a 3,4-dichlorobenzyl mercaptan SAM. Relatively high currents (up to 100 μA) were obtained from devices with NTCDI-BnCF3 active channels, providing a mobility of approximately 0.1 cm^2 V^{-1} s^{-1} for bottom contact devices ($T_{dep} = 90$–100°C). It is possible that the modification of the energy barrier ϕ_{Be} by deposition of an appropriate thiol SAM on the surface of gold (Campbell *et al.* 1996, 1997) resulted in more efficient electron injection, which in turn improved device performance. Furthermore, the presence of the thiol SAM may also result in an increased grain size of the semiconductor on the electrode and in the channel area close to the electrode edge, such that the concentration of grain boundaries and thus charge carrier traps is reduced, as previously reported for pentacene devices (see discussion in later section) (Kymissis *et al.* 2001). NTCDI-C8F is soluble in hot α,α,α-trifluorotoluene, and solution casting resulted in morphologically non-uniform films, with some regions of the films showing mobilities greater than 0.01 cm^2 V^{-1} s^{-1} (Katz *et al.* 2000b).

In NTCDI-C8F OTFTs, the gate and drain voltage could be cycled repeatedly in air without any serious deterioration of device performance until failure of the gold contacts occured (Katz *et al.* 2000a). As mentioned above, OTFTs based on linear, alkyl functionalized diimides NTCDI-.18 C8H, NTCDI-C12H, NTCDI-C18H, operated only under vacuum. Interestingly, the reduction potential data does not show a significant difference between fluorinated and non-fluorinated derivatives. Also, the potentials measured for NTCDI-C8F and other fluorinated NTCDI derivatives are formally outside the stability window described by de Leeuw *et al.* (1997) and Katz *et al.* (2000a), thus indicating that air stability can be obtained even with materials that are expected to be thermodynamically unstable. Since molecular orientation and layer spacings are similar in thin films of fluorinated and non-fluorinated derivatives, by examining the structure of the thin films it was concluded (Katz *et al.* 2001) that in the case of NTCDI substituted with linear side-chains, the denser packing of the fluorinated side-chains vs the aliphatic ones could provide a "kinetic" barrier to atmospheric contaminants such as oxygen. Consistent with this hypothesis, is the relatively lower stability of the less densely packed NTCDI-BnCF3 derivative vs. NTCDI-C8F. However, such an argument is not so straightforward when comparing NTCDI-BnCF3 to NTCDI-BnCH3, in which the former provides relatively air stable devices while the later has mobilities on the order of 10^{-5} cm^2 V^{-1} s^{-1} even under vacuum. It is also interesting to note the dramatic stability difference observed between single crystals of NTCDI-BnCF3 and NTCDI-BnCH3. The subtle packing differences in the single crystal may account for the significantly lower activation energy required to fill trapping levels in the NTCDI-BnCF3 derivative, thus providing better device performance (Katz *et al.* 2001). All compounds appeared to have similar film morphologies (Katz *et al.* 2001) and thus, the observed stability and performance differences cannot be attributed to differences in film morphology.

n-Type OTFTs with N,N'-diphenyl-3,4,9,10 perylenetetracarboxylic diimide (PTCDI-Ph) as the active layer have been fabricated and tested and their electron mobility was measured to be 1.5×10^{-5} cm^2 V^{-1} s^{-1} (Horowitz *et al.* 1996). Gold and aluminum source and drain contacts were used in a bottom contact device configuration. OTFTs with aluminum electrodes were found to have three times lower performance. Devices degraded rapidly in air and the field effect could not be observed after two to three days in devices with gold contacts, and even faster in devices with aluminum contacts.

n-Type OTFTs based on PTCDA (**19**) have exhibited field-effect mobilities of 10^{-4}–10^{-5} cm^2 V^{-1} s^{-1} (Ostrick *et al.* 1997). Films of PTCDA were found to grow quasiepitaxially under specific conditions (especially low substrate temperature, $T_{sub} = 90$ K) on various substrates, thus resulting in highly ordered films in which the molecular plane of PTCDA is approximately parallel to the substrate and the molecules pack in stacks that are almost perpendicular to the substrate (Zang *et al.* 1991; Fenter *et al.* 1995). It has been reported that the conductivity is very anisotropic in such films, with the in-plane conductivity being at least six orders of magnitude lower that the conductivity perpendicular to the film plane (Zang *et al.* 1991). Thus, the low mobilities can be ascribed to limited electronic orbital overlap in the direction of transport, which in the case of OTFT is parallel to the substrate. PTCDA OTFTs do not operate in wet air, yet in vacuo or under dry oxygen they perform as described above. The effect of moisture on the devices is reversible. Karl and Marktanner have shown using time-of-flight experiments that the electron mobility in polycrystalline thin films of PTCDA is inversely proportional to the width of the X-ray rocking curve from the thin film (Karl and Marktanner 1998; Karl *et al.* 1999). The maximum time-of-flight mobility reported for PTCDA thin films was $3 \times 10^{-2} \, cm^2 \, V^{-1} \, s^{-1}$ for measurements performed under vacuum (Karl and Marktanner 1998; Karl *et al.* 1999).

Time-of-flight mobility measurements from perylene single crystals have been reported (Karl 1985, 2001; Warta *et al.* 1985). Electron mobilities up to $5.5 \, cm^2 \, V^{-1} s^{-1}$ were measured at room temperature in perylene single crystal devices, while at low temperatures ($\sim 40 \, K$) reached approximately $150 \, cm^2 \, V^{-1} s^{-1}$. The charge transport of electrons in perylene can be described as coherent band-like transport below room temperature (Karl 2001). On the contrary, the dominant charge transport mechanism for holes in perylene single crystals is incoherent hopping transport (Karl 2001). In perylene single crystals the electron mobility is much higher than the hole mobility, and it seems that trapping states affect hole transport much more severely that electron transport, in contrast to what is usually observed in most other organic semiconductors (Karl 2001).

Recently, it was reported that *N,N'*-dioctadecyl-3,4,9,10-perylenetetracarboxylic diimide (PTCDI-C18H) forms several crystalline and liquid crystalline phases, as demonstrated by X-ray diffraction experiments that show a high degree of order in all three dimensions (Struijk *et al.* 2000). X-ray experiments on the liquid crystalline phases suggest a smectic phase, in which the linear alkyl chains are interdigitated. Charge carrier mobilities measured by pulse-radiolysis time-resolved microwave conductivity were higher than $0.1 \, cm^2 \, V^{-1} \, s^{-1}$ for liquid crystalline phases and higher than $0.2 \, cm^2 \, V^{-1} \, s^{-1}$ for crystalline phases (Struijk *et al.* 2000).

We have recently shown that OTFTs based on *N,N'*-dioctyl-3,4,9,10-perylenetetracarboxylic diimide (PTCDI-C8H) **18** as the organic semiconductor provides bottom contact devices with mobilities as high as $0.6 \, cm^2 \, V^{-1} \, s^{-1}$ in the saturation regime and current on/off ratios $>10^5$ (Malenfant *et al.* 2001, 2002). High threshold voltages ($>75 \, V$) were observed which can be attributed to the existence of traps related to structural defects, especially in the region of the channel close to the Au contacts. X-ray studies in reflection mode revealed a (001) plane spacing of approximately $20 \, \text{Å}$, which is consistent with a spacing of $21 \, \text{Å}$ that resulted from a modeled structure based on interdigitated alkyl chains (Malenfant *et al.* 2001, 2002).

This structure is also consistent with the smectic ordering observed earlier for PTCDI-C18H (Struijk *et al.* 2000).

The work of Bao *et al.* has explored a variety of fluorine-substituted metallophthalocyanines as channels in n-type OTFTs (Bao *et al.* 1998). Mobilities were affected by the substrate temperature during deposition as well as the metal center of the phthalocyanine. F16CuPc (**10**)-based OTFTs with the channel material deposited at $T_{dep} = 125°C$ exhibited the best performance with a mobility of $0.03\,cm^2\,V^{-1}\,s^{-1}$ and on/off ratio in the range of 10^4–10^5. It is known that electron withdrawing groups tend to lower the LUMO level of conjugated organic molecules. By fluorinating the phthalocyanine ring, the LUMO level drops by ca. 1.6 eV relative to the unsubstituted molecule, as shown by UPS measurements and UV–Vis data. This makes it less susceptible to oxidation (Karmann *et al.* 1996). Additionally, the lower LUMO level is more accessible for electron injection from metal contacts. The fluorinated metallophthalocyanine devices were very stable in air. They could be stored in air for half a year without showing any decrease in mobility or on/off ratio (Bao *et al.* 1998). All fluorinated metallophthalocyanines adopt an edge on stacking configuration, effectively resulting in a fluorinated barrier at the film surface (Bao *et al.* 1998). This could play an important role in the high air stability observed in these devices, effectively providing a barrier for the diffusion of oxygen towards potential oxidation sites, as discussed previously for NTCDI-C8F (see above).

Schön *et al.* have studied the electron transport in single crystals of $F_{16}CuPc$ in air as a function of time and temperature (Schön *et al.* 2000d; Schön and Bao 2001). Space charge limited current (SCLC) spectroscopy was used to estimate the intrinsic mobility in single crystals. Charge transport was measured in the π-stacking direction for both single crystals and thin films, in order to eliminate the effect of transport anisotropy, while focusing on the effect of grain boundaries, disorder, and chemical doping and other extrinsic parameters. Mobilities as high as $1.7\,cm^2\,V^{-1}\,s^{-1}$ at room temperature were measured using SCLC spectroscopy in single crystals with minimal shallow trap concentrations. At temperatures below 100–150 K, the effective mobility is dominated by trapping and thus it is thermally activated, whereas at higher temperatures, the effective mobility follows a power law dependence suggesting that the dominant mechanism is band transport limited by phonon scattering. This is in contrast to what is observed with thin films, in which thermally activated transport over an energy barrier related to grain boundaries is the dominant transport mechanism. Studies on single crystals in air revealed that the mobility reaches stability after approximately 500 h. This degradation is reversed by annealing the crystals in hydrogen atmosphere. Thin films, prepared by thermal evaporation onto 125°C substrates were also examined. The FET mobility is temperature independent at low temperature where tunneling becomes the dominant transport mechanism where as it is thermally activated at higher temperatures.

Recently, Facchetti *et al.* reported on the use of α,ω-diperfluorohexyl-sexithiophene (DFH-6T) (**11**) as a novel n-type organic semiconductor (Facchetti *et al.* 2000). Once again, a known p-type material (i.e., α,ω-dihexylsexithiophene-DH-6T) was converted to n-type by functionalizing it with appropriate electron-withdrawing groups. Although the band gap remains the same for both DFH-6T and DH-6T ($\sim$2.4 eV), redox studies reveal that both the HOMO and the LUMO levels are shifted by 0.27 eV below the corresponding levels of DH-6T. Elevated substrate temperatures produced larger grains and pretreatment of the gate insulator surface with $CF_3(CF_2)_5CH_2CH_2SiCl_3$ led to approximately 10% larger crystallite size and grain to grain and grain to substrate interconnectivity were considerably enhanced. Top contact devices were prepared by pre-treating SiO_2 surfaces with either hexamethyldisilazane (HMDS) or $CF_3(CF_2)_5CH_2CH_2SiCl_3$ prior to the deposition of DFH-6T to alter nature of the oxide surface. Mobilities of $0.02\,cm^2\,V^{-1}\,s^{-1}$ and on/off current ratio of 10^5 were obtained in the saturation regime from devices using gold source and drain electrodes, in nitrogen atmosphere. Similar results were obtained with aluminum electrodes. Air stability was not addressed. These devices were obtained via depositions where the substrate temperature was 80–100°C. However, mobilities of 10^{-4} were obtained when the substrate temperature during deposition was 50°C and a reduced mobility was also obtained for $T_{sub} = 120$°C. The turn-on voltage is fairly high (25–35 V) and increases with time, yet it can be stabilized and reduced by post-growth annealing of the film.

VIII. Relation between Morphology and Electrical Properties

The fabrication of OTFTs based on polycrystalline, vapor deposited α-hexathiophene (Horowitz *et al.* 1985, 1992) and α,ω-dihexyl-hexathiophene films by Garnier *et al.* (1993) not only showed that relatively high mobilities are attainable by polycrystalline organic semiconductors, but also delineated the strategies that should be followed in order to increase the performance of OTFTs. In the case of chain- or rod-like molecules, such as thiophene oligomers, large π-conjugation length along the long axis of the molecule and close molecular packing of the molecules along at least one of the short molecular axes (π-stacking) are two important conditions for high carrier mobility. Carrier transport is easiest parallel to the π-stacking direction, thus this direction should be parallel to the gate insulator surface in OTFTs, which means that each molecule should be positioned with its long molecular axis almost perpendicular to this surface. α,ω-Dihexyl-hexathiophene films grow with such orientation of the molecules and there is a large anisotropy in conductivity between the direction parallel and perpendicular to the substrate surface, with the conductivity being much higher in the former case.

These principles are also in operation in OTFTs based on polycrystalline, vapor deposited pentacene thin films (Horowitz *et al.* 1992; Dimitrakopoulos *et al.* 1996; Lin *et al.* 1996, 1997b; Schön *et al.* 2000c). Figure 10 (Dimitrakopoulos and Mascaro 2001) contains proof of the above claims. By growing amorphous films of pentacene, which is achieved by keeping the substrate temperature at $-196°C$ during deposition, a film that is practically insulating is produced (Dimitrakopoulos *et al.* 1996). This is due to the fact that the overlap of the molecular orbitals of nearest-neighbor molecules is very limited because of the disorder in the solid. When the substrate temperature is kept at room temperature (27°C) during deposition, a very highly ordered film is deposited and the mobility measured at room temperature is $0.6\,cm^2\,V^{-1}\,s^{-1}$ (Dimitrakopoulos and Mascaro 2001). The well-defined X-ray diffraction peaks correspond to (00*l*) reflections, with a *d*-spacing of about 15.4 Å, which correspond to a crystalline structure with the long axis of the pentacene molecule being almost perpendicular to the substrate surface. The structure of this thin film is different than the structure of single crystals of pentacene, thus we must distinguish between the "thin-film phase" (Dimitrakopoulos *et al.* 1996; Jentzsch *et al.* 1998) and the "single-crystal phase" (Campbell *et al.* 1961) of pentacene. One of the main differences of these phases is that the *d*-spacing of the (00*l*) reflections is different. It is

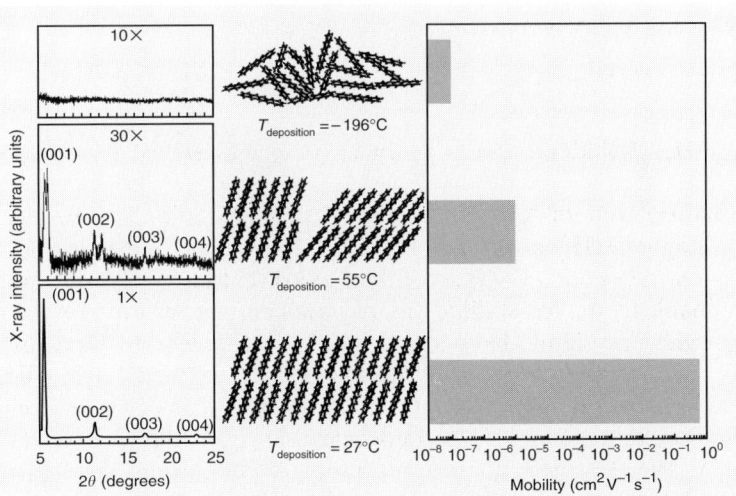

FIG. 10.   X-ray diffractograms, schematic representations of structural order, and field-effect mobilities corresponding to three different thin films of pentacene. An amorphous phase is achieved using a substrate temperature, $T_{sub} = -196°C$, and a deposition rate, DR, of $0.5\,\text{Å}\,sec^{-1}$. A single, "thin-film phase" resulted for $T_{sub} = 27°C$ and $DR = 1\,\text{Å}\,sec^{-1}$. Setting $T_{sub} = 55°C$ and $DR = 0.25\,\text{Å}\,sec^{-1}$ yielded a film consisting of two phases, the "thin film phase" and the "single crystal phase". Reprinted with permission from Dimitrakopoulos and Mascaro (2001). Data is partially taken from Dimitrakopoulos *et al.* (1996).

slightly smaller (14.5 Å) in the "single-crystal phase", thus in this phase the long axis of the pentacene molecule makes a larger angle with the direction that is perpendicular to the substrate surface than in the "thin-film phase". When a mixture of the thin film phase and the single crystal phase is grown (Dimitrakopoulos *et al.* 1996), the mobility is very low, possibly due to the high defect concentration resulting from the coexistence of the two phases.

Figure 11 shows the morphology of a pentacene film grown on SiO_2 at room temperature (Dimitrakopoulos and Graham, unpublished results). We observe that pentacene forms single crystal islands with a size up to more than 1 μm. Subsequent layers growing on top of these islands are smaller in size leading to a terrace-and-step morphology. Similar morphology is observed even when an SAM of a molecule such as the 1-diethoxy-1-sila-cyclopent-3-ene is grown on the SiO_2 surface (Kosbar *et al.* 2001). The angles formed by the sides of some of the uppermost pentacene islands seem to be consistent—at the extent of accuracy that the photograph permits—with the angles of the *ab* plane of the triclinic unit cell of pentacene (Campbell *et al.* 1961). Recently, photoelectron emission microscopy (PEEM) was employed to study pentacene thin film growth in real time and the resulting understanding contributed significantly to the identification of the pentacene growth mechanism on various surfaces (Meyer zu Heringdorf

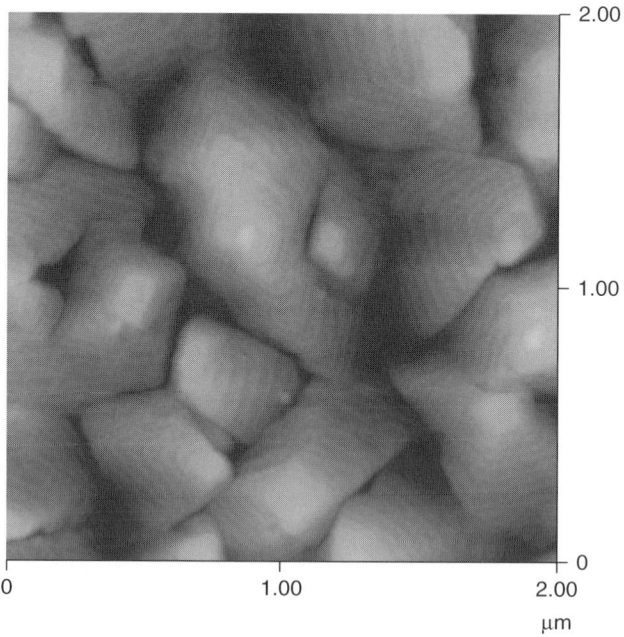

Fig. 11. AFM micrograph showing the morphology of a pentacene film grown on SiO_2 at room temperature.

et al. 2001). Polycrystalline films of pentacene with grain sizes approaching 100 μm were fabricated on clean Si(001) surfaces passivated with a cyclohexene layer (Meyer zu Heringdorf *et al.* 2001). Such large grain growth can be attributed to the relatively low nucleation density of pentacene grains on such surfaces under the conditions used, which is of the order of $10^{-3}\,\mu m^{-2}$, and the absence of heterogeneous nucleation on the carefully cleaned Si substrates (Meyer zu Heringdorf *et al.* 2001). However, the nucleation density of pentacene grains on—the technologically more important for OTFT applications—SiO$_2$ surface was reported to be 100 times higher than on Si(001) and cyclohexene-modified Si(001) surfaces, under the same deposition parameters (Meyer zu Heringdorf *et al.* 2001). Higher nucleation density generally translates to smaller grain sizes. Schön *et al.* have deposited polycrystalline pentacene films on polyimide substrates with grain sizes of 100 μm using a substrate temperature below 200°C (Schön *et al.* 2000c)

In two other interesting reports, the effect of gate insulator surface structure and order on pentacene (Chen *et al.* 2001; Swiggers *et al.* 2001) and other organic semiconductor (Chen *et al.* 2001) film orientation and device performance were investigated.

IX.   Dependence of Mobility on Gate Voltage

Up to the present time, pentacene OTFTs have shown the highest hole mobilities among TFTs with an organic semiconducting channel (see Table I). However, the operating voltage required to produce such performance (approximately 100 V) is often considered too high, especially for portable, battery-powered device applications. The required maximum voltage depends on the thickness of the gate insulator, but because the subject matter of OTFT applications is large-area electronics, we should take on account that there is a minimum in the thickness of the gate insulator, imposed by reliability considerations such as the requirement for pinhole-free films and low leakage currents. Thus, it is safe to assume that it would probably be impractical to reduce the thickness of the dielectric below 100 nm with a more realistic thickness being 300 nm. We have studied the gate voltage dependence of mobility in pentacene devices, and used our understanding to demonstrate high-performance pentacene TFTs exhibiting mobility up to $0.4\,\mathrm{cm^2\,V^{-1}\,s^{-1}}$ at low operating voltages (~5 V) (Dimitrakopoulos *et al.* 1999a,b). An example of the dependence of mobility on V_G is shown in Fig. 12, which shows plots of mobility and $\sqrt{|I_D|}$ vs V_G for the same device as in Figs. 5 and 6. The mobility is continuously calculated in the saturation regime for various maximum values of V_G. The mobility depends strongly on V_G. At low V_G, it increases linearly from very low values (about $0.016\,\mathrm{cm^2\,V^{-1}\,s^{-1}}$ at $V_G = -1\,V$) and at higher V_G it starts to saturate

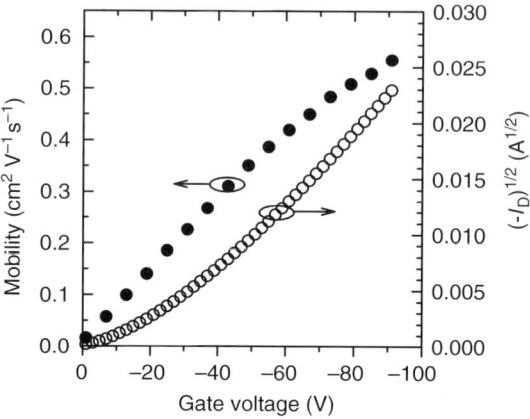

FIG. 12.   Plots of mobility and $\sqrt{|I_D|}$ vs. V_G for the same device as in Figs 5 and 6.

($\mu = 0.59 \, \text{cm}^2 \, \text{V}^{-1} \, \text{s}^{-1}$ at 91 V). At even higher V_G, the mobility should reach a plateau. The voltage limit of 100 V of our HP 4145 B semiconductor parameter analyzer did not allow us to reach this mobility plateau with this OTFT that comprised a relatively thick (5000 Å) SiO_2 gate insulator. When the insulator is thinner, the mobility plateau is reached at $V_G \cong -80 \, \text{V}$ (Dimitrakopoulos *et al.* 1999b). Furthermore, improvements in some of the deposition parameters of the organic semiconductor, which were discussed above, have resulted in higher mobilities in the device of Figs. 5, 6, and 12, relative to earlier pentacene devices with the same gate insulator (see, e.g., the open circles in Fig. 13 that correspond to mobilities of 0.012 and $0.124 \, \text{cm}^2 \, \text{V}^{-1} \, \text{s}^{-1}$ at $V_G = 20$ and 100 V, respectively). For comparison, we note that the mobility measured at maximum gate voltage $V_G = 19$ and 100 V is 0.140 and $0.604 \, \text{cm}^2 \, \text{V}^{-1} \, \text{s}^{-1}$, respectively.

In order to lower the operating voltage of pentacene OTFTs, we have employed gate insulators with a relatively high dielectric constant (ε), such as metal oxide films of barium zirconate titanate (BZT) (Dimitrakopoulos *et al.* 1999b), or barium strontium titanate (BST) (Dimitrakopoulos *et al.* 1999a) Additionally, we have demonstrated the full compatibility of low operating voltage OTFTs with transparent plastic substrates by making devices on polycarbonate substrates using an all-room-temperature process sequence (Dimitrakopoulos *et al.* 1999b). It is important to note here that the operating voltage can be reduced by reducing the thickness of the dielectric layer instead of increasing its dielectric constant. However, the fact that the envisioned applications for OTFTs involve almost exclusively large-area electronics, a minimum dielectric layer thickness of 1000 Å or more is dictated by reliability and manufacturing yield considerations. Thicker dielectric layers are more suitable for large area electronics applications since they suppress the formation of pinholes and the problems associated with step coverage. Thus, a higher dielectric constant gate insulator is considered the appropriate solution for high-mobility pentacene OTFT with low operating voltage. Later, Gundlach *et al.* (1999) demonstrated pentacene OTFT devices on plastic with mobility up to $1.1\,\mathrm{cm^2\,V^{-1}s^{-1}}$ at operating voltage of about 25 V.

Figure 13(a) and (b) shows the dependence of field effect mobility, μ, on the charge per unit area on the semiconductor side of the insulator, Q_S, and the gate field, E, respectively (Dimitrakopoulos *et al.* 1999a). Both E and Q_S are proportional to V_G. The black circles correspond to a pentacene-based device with a 1200 Å thick SiO_2 gate insulator thermally grown on the surface of a heavily doped n-type Si wafer that acted as the gate electrode. The open circles correspond to a similar device with a 5000 Å thick SiO_2 gate insulator. The mobility for the SiO_2-based devices is calculated in the saturation regime using a gate sweep, as explained in Fig. 6, and is then plotted vs the maximum V_G used in each gate sweep. The maximum V_G is varied from -20 to -100 V. During all sweeps, V_D is kept constant at -100 V in order to eliminate any effects that source and drain contact imperfections might have on our results. The mobility increases linearly with increasing Q_S and E and eventually saturates [Fig. 13(a) and (b), respectively]. Q_S is a function of the concentration of accumulated carriers in the channel region (N). Since the accumulation region has been shown in the past to be two-dimensional and confined very close to the interface of the insulator with the organic semiconductor (Ostoja *et al.* 1993; Dodabalapur *et al.* 1991), all of this charge is expected to be localized within the first few semiconductor monolayers from this interface, screening the field out of the rest of the pentacene thickness. By replacing SiO_2 with an insulator having a similar thickness but a much higher dielectric constant an accumulated carrier concentration similar to the SiO_2

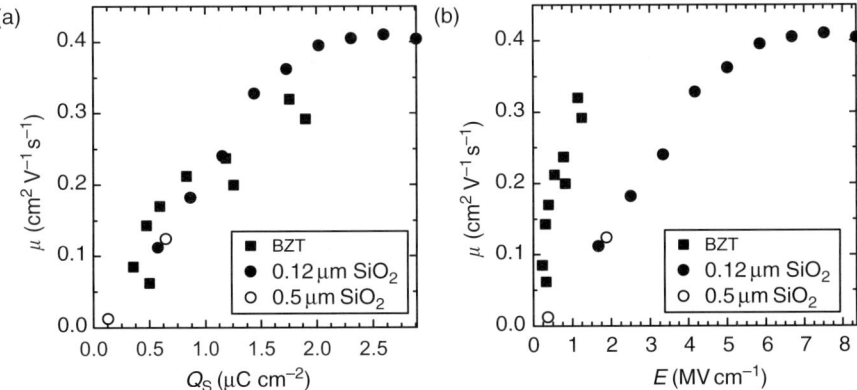

FIG. 13. Dependence of field effect mobility, μ, on (a) charge per unit area on the semiconductor side of the insulator, Q_S, and (b) gate field, E.

case could be attained at much lower V_G, and hence E, with all the other parameters being similar. The black squares in Fig. 13 correspond to devices comprising room temperature sputtered BZT as the gate insulator. From Fig. 13(b), it is obvious that in devices with BZT gate insulator the applied gate field used to obtain mobility values similar to those of the devices with SiO_2 gate insulator was about five times lower compared to the fields used in the latter devices. This clearly proves that high field is not required to obtain high mobility. Thus, the gate voltage dependence of mobility in these devices is due to the higher concentration of holes accumulated in the channel. Figure 13(a), which plots μ vs charge per unit area, Q_S, corroborates this conclusion. The values of Q_S and N required to reach a certain mobility value are practically the same for devices with SiO_2 and BZT gate insulators although much different gate voltage and gate field values were required to obtain such a mobility in each case.

The gate voltage dependence of vacuum deposited pentacene OTFTs was first reported in Dimitrakopoulos et al. (1996). A similar behavior was later reported for pentacene OTFTs deposited from solution via a tetra-chlorobenzene-containing precursor of pentacene (Jarret et al. 1997). The multiple trapping and release (MTR) model (Le Comber and Spear 1970), which is widely used to model the behavior of a-Si:H TFT seems to be a mechanism that explains reasonably well the observed characteristics in vapor deposited polycrystalline films of pentacene (Dimitrakopoulos et al. 1999a,b). This model, which is based on the assumption that the intrinsic charge transport mechanism is one involving extended states, has been suc-cessfully used in the past to model the field dependence of mobility in α-6T and DH6T OTFTs (Horowitz et al. 1995, 1999a; Horowitz 1998; Horowitz and Hajlaoui 2000). According to this model, a distribution of traps exists in the forbidden gap above the valence band edge. At low gate bias, most of the holes injected in the semiconductor are trapped into these localized states. The deepest traps are filled first and carriers can be released thermally. As the negative gate bias increases, the Fermi level approaches the valence band edge as more traps are filled (p-type material). At an appropriately high gate voltage, all trap states are filled and subsequently injected carriers move with the microscopic mobility associated with carriers in the delocalized (valence) band (Dimitrakopoulos et al. 1999a,b). Several trap levels have been reported for thin polycrystalline vapor deposited films of pentacene at depths ranging from 0.06 to 0.68 eV (Muzicante and Silinsh 1995), which could account for the traps described in the MTR model. Traps can be linked to impurities, and various structural defects in the crystalline structure of the pentacene film, including point defects, dislocations, and most importantly, grain boundaries (Karl 1990). The concept of charged grain boundaries has been used to explain the gate voltage dependence of mobility in poly-crystalline pentacene (Schön et al. 2000c) and oligothiophene (Horowitz and Hajlaoui 2000; Horowitz et al. 2000) films. An energy barrier is created at the grain boundaries and is a function of the charged trapping states at the boundaries, the carrier concentration within the grains and the temperature (Schön et al. 2000c). The effective mobility across two grains of tetracene that are separated by a grain boundary is given by:

$$\frac{1}{\mu} = \frac{1}{\mu_G} + \frac{1}{\mu_{GB}},$$

where μ_G is the single crystal mobility (intragrain mobility) and μ_{GB} is the mobility across the grain boundary (Schön and Kloc 2001). The charge transport is dominated by thermionic emission over the potential barrier at the grain boundary from 20 to 150 K, and by the intragrain transport (μ_G) above 150 K. From 4 to about 20 K transport is dominated by tunneling of charge carriers through the potential barrier at the grain boundary. Chwang and Frisbie have shown that carrier transport in α,ω-dihexyl-hexathiophene (6T) OTFTs is limited by the presence of grain boundaries in the channel (Chwang and Frisbie 2001). Their experiments involved transport measurements through single grain boundaries in vapor deposited 6T and showed that transport is dependent on carrier concentration exhibiting decreasing activation energy with increasing carrier concentration. Another important conclusion of that work is that larger threshold voltages in OTFTs correlate with larger trap densities at grain boundaries (Chwang and Frisbie 2001).

In the case of OTFTs comprising amorphous organic semiconductor channels, the experimentally obtained gate voltage and temperature dependence of the field-effect mobility has been studied and successfully modeled theoretically (Vissenberg and Matters 1998). The mobility dependence on temperature in such OTFTs exhibits a simple Arrhenius behavior, with a gate voltage dependent activation energy. Yu et al. (2000) have proposed a model that successfully explains field and carrier density dependences of the mobility in films of conjugated organics, both in low-field/high carrier concentration cases (OTFTs), and in high-field/low carrier concentration cases (organic diodes). Their model was based on the assumptions that thermal fluctuations modify the energy levels of polaronic electronic states, and that the primary restoring force for these fluctuations is steric, which leads to spatial correlation in the energies of the localized electronic states (Yu et al. 2000).

Pentacene transistor drain–source contacts can be made in one of two configurations as discussed before [Figs. 4(a) and (b)]: top and bottom contact, respectively. It is well established that the performance of pentacene devices with the bottom contact configuration is inferior to that of devices with the top contact configuration. Consequently, most high-performance pentacene TFTs reported in the literature have the top contact configuration, and shadow masking is generally used to pattern the source and drain contacts on top of the pentacene. Unfortunately, this is a process that cannot be used in manufacturing, hence a protocol that allows the patterning of the source and drain electrodes on the insulator before the deposition of pentacene, according to the schematic shown in Fig. 4(b), had to be developed. This should be done either with photolithography or with some other patterning technique, such as microcontact printing, stamping or screen-printing. Furthermore, the performance of devices fabricated with such a process should be similar if not better than that of top contact devices [Fig. 4(a)].

Figure 14 shows a pentacene layer as it was grown on SiO_2 and a Au electrode (Dimitrakopoulos et al. 2000). The edge between SiO_2 and Au is marked by the end of the white area in the middle photograph that corresponds to Au (due to variations in image contrast, the pentacene-covered Au appears different in the top two pictures). On SiO_2, far away from the Au edge, pentacene consists of fairly large grains (having sizes between 0.2 and 0.5 µm). On Au, the grain size is dramatically reduced. This small crystal growth persists into the channel region (on SiO_2) (Kymissis 1999; Dimitrakopoulos et al. 2000, 2002; Kymissis et al. 2001). Close to the Au edge but on the SiO_2 side there is a transition region where the grain size increases with increasing distance from the edge. It is the morphology of the pentacene film in the OTFT channel region close to the electrode edge that causes the performance limitation of the bottom contact TFT. Right at the edge of the Au electrode, there is an area with very small crystals and thus a large number of grain boundaries. Grain boundaries are high-volume and low-order regions that contain many morphological defects, which in turn are linked to the creation of charge carrier traps with levels lying in the band gap. The creation of an unusually large concentration of defects in the region of the channel close to the electrode edge can be considered responsible for the reduced performance of bottom contact pentacene TFTs. The reduction of their concentration to levels similar to those in the area at the center of the channel (lower part of Fig. 14) should result in bottom contact devices with performance similar to or better than that of top contact devices. In a typical bottom contact pentacene TFT, the mobility is equal to or less than $0.16\,cm^2\,V^{-1}\,s^{-1}$. We have used a SAM of 1-hexadecane thiol to modify the surface energy of the Au electrode in an effort to improve the crystal size and ordering of the overgrown pentacene layer (Kymissis et al. 2001; Dimitrakopoulos et al. 2002). Mobilities calculated in saturation from such devices were up to $0.48\,cm^2\,V^{-1}\,s^{-1}$, which is three times larger than the mobility of devices with untreated Au electrodes. Mobilities calculated in the linear regime were up to five times higher in devices treated with a SAM vs untreated devices (Kymissis et al. 2001; Dimitrakopoulas et al. 2002). The pentacene layers for both devices were deposited in the same deposition run. Figure 15 provides an explanation for the improvement in device performance (Kymissis et al. 2001). The SAM deposited on Au resulted in a pentacene grain size on Au similar to the large grains grown on the SiO_2 in the center of the channel. There is no transition region at the Au edge, hence the trap concentration must have been drastically reduced.

FIG. 14. SEM image of a pentacene thin film grown on SiO_2 and a Au electrode. The grain size is much smaller on Au than on SiO_2 far from the Au edge. The pentacene grain size on SiO_2 in the region close to the Au edge is similar to that on Au and increases with increasing distance from the edge. Reprinted with permission from Dimitrakopoulos *et al.* (2000).

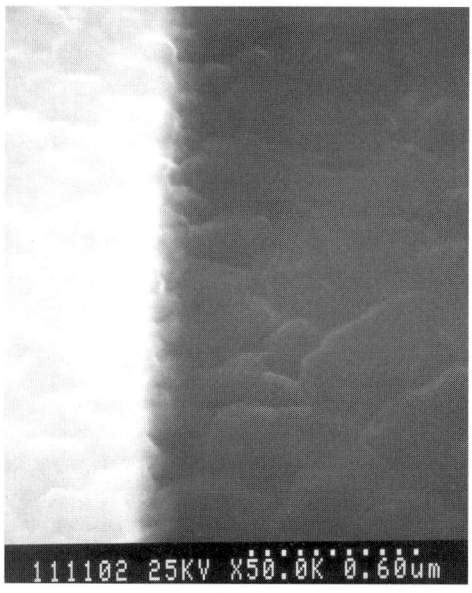

111102 25KV X50.0K 0.60um

FIG. 15. SEM image of a pentacene thin film grown on SiO_2 and a Au electrode covered with a SAM of 1-hexadecane thiol. A similar grain size on both the SiO_2 and the Au/SAM surface is observed. The pentacene grain size transition region at the Au edge is eliminated. Reprinted with permission from Kymissis *et al.* (2001), © 2001 IEEE.

X. Conclusions and Outlook

OTFT performance has improved in an impressive pace during the last 15 years. Our understanding of organic semiconductor synthesis, structure, and transport properties has improved significantly. Advances in OTFT performance characteristics have been the result of the introduction of novel materials obtained either by the chemical modification of existing molecules or by the synthesis of completely new ones, followed by optimization of their morphology and structural order.

Presently, products based on organic semiconductors have started trickling into the market. As an example, car-stereo displays with OLED pixels can be mentioned. Active matrix display backplanes based on OTFTs are currently being developed in several industrial and university labs. Organic semiconductors, such as pentacene, deposited by vacuum sublimation remain the best performers due not only to their molecular electronic properties, but also to their very well-ordered structures, which result from the use of this highly controllable deposition method. However, substantial improvements have taken place in solution processed organic semiconductors too, and their field effect mobilities are approaching those of vapor deposited OTFTs. There is a potentially important cost-advantage associated with the solution processing of organic TFTs, as it eliminates the need for expensive vacuum chambers and lengthy pump-down cycles. However, for this advantage to be realized, all or at least most of the layers comprising the OTFT device should be deposited using methods that do not involve vacuum deposition. Reel-to-reel processing, which has obvious advantages over batch fabrication processes for reducing costs, can be applied to both vacuum and solution deposited organic semiconductors, but if one considers the ability to stamp or inkjet print solution-based organics, thus eliminating traditional lithographic steps, then their potential cost advantage over vacuum deposited organics becomes more apparent. Good device stability and long lifetimes are two other requirements that will allow the full realization of the advantages of organic electronics. Encapsulation techniques can and have worked in the past, but they can substantially reduce the cost advantages that organic semiconductors, which are stable during processing and during operation, would offer.

The flexibility offered by molecular engineering of organic semiconductor molecules, is the major advantage of this technology. Due to this, the future seems very promising for inexpensive organic electronics that will address new and existing application needs, as new and improved materials find their way into organic electronic devices.

REFERENCES

Afzali, A., C. D. Dimitrakopoulos, and T. L. Breen. Synthesis and application of a novel soluble pentacene precursor in thin film transistors, *2001 MRS Fall Meeting Abstracts*, Symposium BB8.11, p. 553.

Afzali, A., C. D. Dimitrakopoulos, and T. L. Breen. High-performance, solution-processed organic thin film transistors from a novel pentacene precursor, *J. Am. Chem. Soc. (Communication)* 124, 8812 (2002).

Akimichi, H., K. Waragai, S. Hotta, H. Kano, and H. Sakaki, Field-effect transistors using alkyl substituted oligothiophenes, *Appl. Phys. Lett.* **58**, 1500 (1991).

Assadi, A., C. Svensson, M. Willander, and O. Inganäs, field-effect mobility of poly(3-hexylthiophene), *Appl. Phys. Lett.* **53**, 195 (1988).

Baldo, M., M. Deutsch, P. Burrows, H. Gossenberger, M. Gerstenberg, V. Ban, and S. Forrest, Organic vapor phase deposition, *Adv. Mater.* **10**, 1505 (1998).

Bao, Z., A. J. Lovinger, and A. Dodabalapur, Organic field-effect transistors with high mobility based on copper phthalocyanine, *Appl. Phys. Lett.* **69**, 3066 (1996a).

Bao, Z., A. Dodabalapur, and A. J. Lovinger, Soluble and processable regioregular poly(3-hexylthiophene) for thin film field-effect transistor applications with high mobility, *Appl. Phys. Lett.* **69**, 4108 (1996b).

Bao, Z., Y. Feng, A. Dodabalapur, V. R. Raju, and A. J. Lovinger, High-performance plastic transistors fabricated by printing techniques, *Chem. Mater.* **9**, 1299 (1997).

Bao, Z., A. J. Lovinger, and J. Brown, New air-stable n-channel organic thin film transistors, *J. Am. Chem. Soc.* 207 (1998).

Bjørnholm, T., D. R. Greve, N. Reitzel, T. Hassenkam, K. Kjaer, P. B. Howes, N. B. Larsen, J. Bøgelund, M. Jayaraman, P. C. Ewbank, and R. D. McCullough, Self-assembly of regioregular, amphiphilic polythiophenes into highly ordered-stacked conjugated polymer thin films and nanocircuits, *J. Am. Chem. Soc.* **120**, 7643 (1998).

Brown, A. R., D. M. de Leeuw, E. J. Lous, and E. E. Havinga, Organic n-type field-effect transistors, *Synth. Met.* **66**, 257 (1994).

Brown, A. R., A. Pomp, C. M. Hart, and D. M. de Leeuw, Logic gates made from polymer transistors and their use in ring oscillators, *Science* **270**, 972 (1995).

Brown, A. R., A. Pomp, D. M. de Leeuw, D. B. M. Klaassen, E. E. Havinga, P. T. Herwig, and K. Müllen, Precursor route pentacene metal-insulator-semiconductor field-effect transistors, *J. Appl. Phys.* **79**, 2136 (1996).

Brown, A. R., C. P. Jarrett, D. M. de Leeuw, and M. Matters, Field effect transistors made from solution processed organic semiconductors, *Synth. Met.* **88**, 37 (1997).

Burland, D. M., Cyclotron resonance in a molecular crystal—anthracene, *Phys. Rev. Lett.* **33**, 833 (1974).

Burland, D. M. and U. Konzelmann, Cyclotron resonance and carrier scattering processes in anthracene crystals, *J. Chem. Phys.* **67**, 319 (1977).

Burroughes, J. H., C. A. Jones, and R. H. Friend, New semiconductor device physics in polymer diodes and transistors, *Nature* **335**, 137 (1988).

Burroughes, J. H., R. H. Friend, and P. C. Allen, Field-enhanced conductivity in poly-acetylene—construction of a field-effect transistor, *J. Phys. D: Appl. Phys.* **22**, 956 (1989).

Burroughes, J. H., D. D. Bradley, A. R. Brown, R. N. Marks, K. Mackay, R. H. Friend, P. L. Burn, and A. B. Holmes, Light-emitting diodes based on conjugated polymers, *Nature* **347**, 539 (1990).

Burrows, P. E., S. R. Forrest, L. S. Sapochak, J. Schwartz, P. Fenter, T. Buma, V. S. Ban, and J. L. Forrest, Organic vapor phase deposition: a new method for the growth of organic thin films with large optical non-linearities, *J. Crystal Growth* **156**, 91 (1995).

Campbell, R. B., J. Monteath Robertson, and J. Trotter, The crystal and molecular structure of pentacene, *Acta Crystallogr.* **14**, 705 (1961).

Campbell, I. H., S. Rubin, T. A. Zawodzinski, J. D. Kress, R. L. Martin, D. L. Smith, N. N. Barashkov, and J. P. Ferraris, Controlling Schottky energy barriers in organic electronic devices using self-assembled monolayers, *Phys. Rev. B* **54**, 14321 (1996).

Campbell, I. H., J. D. Kress, R. L. Martin, D. L. Smith, N. N. Barashkov, and J. P. Ferraris, *Appl. Phys. Lett.* **71**, 3528 (1997).

Chen, T.-A., and R. D. Rieke, The first regioregular head-to-tail poly(3-hexylthiophene-2,5-diyl) and a regiorandom isopolymer: Ni vs. Pd catalysis of 2(5)-bromo-5(2)-(bromozincio)-3-hexylthiophene polymerization, *J. Am. Chem. Soc.* **114**, 10087 (1992).

Chen, T.-A., X. Wu, and R. D. Rieke, Regiocontrolled synthesis of poly(3-alkylthiophenes) mediated by Rieke zink: their characterization and solid-state properties, *J. Am. Chem. Soc.* **117**, 233 (1995).

Chen, X. L., A. J. Lovinger, Z. Bao, and J. Sapjeta, Morphological and transistor studies of organic molecular semiconductors with anisotropic electrical characteristics, *Chem. Mater.* **13**, 1341 (2001).

Chwang, A. B. and C. D. Frisbie, Temperature and gate voltage dependent transport across a single organic semiconductor grain boundary, *J. Appl. Phys.* **90**, 1342 (2001).

Clarisse, C., M. T. Riou, M. Gauneau, and M. Le Contellec, Field-effect transistors with diphthalocyanine thin film, *Electron. Lett.* **24**, 674 (1988).

Collins, P. G., M. R. Arnold, and Ph. Avouris, Engineering carbon nanotubes and nanotube circuits using electrical breakdown, *Science* **292**, 706 (2001).

Comiskey, B., J. D. Albert, H. Yoshizawa, and J. Jacobson, An electrophoretic ink all-printed reflective electronic displays, *Nature* **394**, 253 (1998).

Crone, B., A. Dodabalapur, A. Gelperin, L. Torsi, H. E. Katz, A. J. Lovinger, and Z. Bao, Electronic sensing of vapors with organic transistors, *Appl. Phys. Lett.* **78**, 2229 (2001).

de Leeuw, D. M., M. M. J. Simenon, A. R. Brown, and R. E. F. Einerhand, Stability of n-type doped conducting polymers and consequences for polymeric microelectronic devices, *Synth. Met.* **87**, 53 (1997).

Dimitrakopoulos C. D., and D. J. Mascaro, Organic thin film transistors: A review of recent advances, *IBM J. Res. Dev.* **45**, 11 (2001).

Dimitrakopoulos, C. D., A. R. Brown, and A. Pomp, Molecular beam deposited thin films of pentacene for organic field effect transistor applications, *J. Appl. Phys.* **80**, 2501 (1996).

Dimitrakopoulos, C. D., A. Afzali-Ardakani, B. Furman, J. Kymissis, and S. Purushothaman, *Trans-trans* 2,5-bis (2-{5-2,2′ bithienyl}ethenyl)thiophene: synthesis, characterization, thin film deposition and fabrication of organic field effect transistors, *Synth. Met.* **89**, 193 (1997).

Dimitrakopoulos, C. D., B. K. Furman, T. Graham, S. Hegde, and Purushothaman, Field-effect transistors comprising molecular beam deposited α,ω-dihexyl-hexathienylene and polymeric insulators, *Synth. Met.* **92**, 47 (1998).

Dimitrakopoulos, C. D., J. Kymissis, S. Purushothaman, D. A. Neumayer, P. R. Duncombe, and R. B. Laibowitz, Low-voltage, high-mobility pentacene transistors with solution-processed high dielectric constant insulators, *Adv. Mater.* **11**, 1372 (1999a).

Dimitrakopoulos, C. D., S. Purushothaman, J. Kymissis, A. Callegari, and J. M. Shaw, Low-voltage organic transistors on plastic comprising high dielectric constant gate insulators, *Science* **283**, 822 (1999b).

Dimitrakopoulos, C. D., J. Kymissis, and S. Purushothaman, Organic semiconductor thin-film field-effect transistors, in *Final Program and Proceedings of The International Conference on Digital Printing Technologies, NIP16*, Society of Imaging Science and Technology, October 15–20, Vancouver, Canada, p. 493 (2000).

Dimitrakopoulos, C. D., J. Kymissis, and S. Purushothaman, Method for improving performance of organic semiconductors in bottom electrode structure, US Patent No. 6335539 issued 1/1/2002.

Dodabalapur, A., L. Torsi, and H. E. Katz, Organic transistors: Two-dimensional transport and improved electrical characteristics, *Science* **268**, 270 (1995).

Drury, C. J., C. M. J. Mutsaers, C. M. Hart, M. Matters, and D. M. de Leeuw, Low-cost all-polymer integrated circuits, *Appl. Phys. Lett.* **73**, 108 (1998).

Ebisawa, F., T. Kurokawa, and S. Nara, Electrical properties of polyacetylene/polysiloxane interface, *J. Appl. Phys.* **54**, 3255 (1983).

Facchetti, A., Y. Deng, A. Wang, Y. Koide, H. Sirringhaus, T. J. Marks, and R. H. Friend, Tuning of semiconductive properties of sexithiophene by α,ω-substitution—α,ω-diper-fluorohexylsexithiophene: the first n-type sexithiophene for thin film transistors, *Angew. Chem. Int. Ed.* **39**, 4547 (2000).

Fenter, P. E., P. E. Burrows, P. Eisenberger, and S. R. Forrest, Layer-by-layer quasi-epitaxial growth of a crystalline organic thin film, *J. Crystal Growth* **152**, 65 (1995).

Forrest, S., Science and technology at the nanometer scale using vacuum-deposited organic thin films, *MRS Bull.* **26**, 108 (2001).

Frankevich, E., Y. Maruyama, and H. Ogata, Mobility of charge carriers in vapor-phase grown C_{60} single crystal, *Chem. Phys. Lett.* **214**, 39 (1993).

Friend, R. H., J. Burroughes, and T. Shimoda, Polymer diodes, *Phys. World (UK)* **12**, 35 (1999).

Fuchigami, H., A. Tsumura, and H. Koezuka, Polythienylenevinylene thin-film transistor with high carrier mobilitym, *Appl. Phys. Lett.* **63**, 1372 (1993).

Garnier, F., Thin film transistors based on organic conjugated semiconductors, *Chem. Phys.* **227**, 253 (1998).

Garnier, F., A. Yassar, R. Hajlaoui, G. Horowitz, F. Dellofre, B. Servet, S. Ries, and P. Alnot, Molecular engineering of organic semiconductors: Design of self-assembly properties in conjugated thiophene oligomers, *J. Am. Chem. Soc.* **115**, 8716 (1993).

Garnier, F., R. Hajlaoui, A. Yassar, and P. Srivastava, All-polymer field-effect transistors realized by printing techniques, *Science* **265**, 1684 (1994).

Garnier, F., R. Hajlaoui, A. E. Kassmi, G. Horowitz, L. Laigre, W. Porzio, M. Armanini, and F. Provasoli, *Chem. Mater.* **10**, 3334 (1998).

Greenham, N., and R. H. Friend, Semiconductor device physics of conjugated polymers, in *Solid State Physics: Advances in Research and Applications*, edited by H. Ehrenreich and F. Spaepen, Vol. 49, Academic Press, San Diego, CA, p. 1 (1995).

Guillaud, G., M. Al Sadound, and M. Maitrot, Field-effect transistors based on intrinsic molecular semiconductors, *Chem. Phys. Lett.* **167**, 503 (1990).

Gundlach, D. J., H. Klauk, C. D. Sheraw, C.-C. Kuo, J.-R. Huang, and T. N. Jackson, High-mobility, low voltage organic thin film transistors, *1999 IEEE International Electron Devices Meeting (IEDM)—Technical Digest*, p. 111, IEEE, Piscataway, NJ (1999).

Haddon, R. C., A. S. Perel, R. C. Morris, T. T. M. Palstra, A. F. Hebard, and R. M. Fleming, C_{60} thin film transistors, *Appl. Phys. Lett.* **67**, 121 (1995).

Hejlaoui, R., G. Horowitz, F. Garnier, A. Arce-Brouchet, L. Laigre, A. El Kassmi, F. Demanze, and F. Kouki, Improved field-effect mobility in short oligothiophenes: quaterthiophene and quinquethiophene, *Adv. Mater.* **9**, 389 (1997).

Heilmeier G. H., and L. A. Zanoni, Surface studies of a-copper phthalocyanine films, *J. Phys. Chem. Solids* **25**, 603 (1964).

Hellemans, A., Insulator gives plastic transistor a boost, *Science* **283**, 771 (1999).

Herwig, P. T., and K. Müllen, A soluble pentacene precursor: Synthesis, solid-state conversion into pentacene and application in field-effect transistor, *Adv. Mater.* **11**, 480 (1999).

Hill, I. G., A. Rajagopal, and A. Kahn, Molecular level alignment at organic semiconductor-metal interfaces, *Appl. Phys. Lett.* **73**, 662 (1998).

Hill, I. G., J. Schwartz, and A. Kahn, Metal-dependent charge transfer and chemical interaction at interfaces between 3,4,9,10-perylenetetracarboxylic bisimidazole and gold, silver and magnesium, *Org. Electron.* **1**, 5 (2000).

Horowitz, G., Organic field effect transistors, *Adv. Mater.* **10**, 365 (1998).

Horowitz, G., and M. E. Hajlaoui, Mobility in polycrystalline oligothiophene field-effect transistors dependent on grain size, *Adv. Mater.* **12**, 1046 (2000).

Horowitz, G., D. Fichou, X. Peng, Z. Xu, and F. Garnier, Horowitz, G., Fichou, D., Peng, X., Xu, Z., and Garnier, F. (1989). A field-effect transistor based on conjugated alpha-sexithienyl, *Solid State Commun.* **72**, 381 (1989).

Horowitz, G., X. Peng, D. Fichou, and F. Garnier, Role of the semiconductor/insulator interface in the characteristics of π-conjugated-oligomer-based thin-film transistors, *Synth. Met.* **51**, 419 (1992).

Horowitz, G., F. Dellofre, F. Garnier, R. Hajlaoui, M. Hmyene, and A. Yassar, All-organic field-effect transistors made of π-conjugated oligomers and polymeric insulators, *Synth. Met.* **54**, 435 (1993).

Horowitz, G., R. Hajlaoui, and P. Delannoy, Temperature dependence of the field-effect mobility of sexithiophene. Determination of the density of traps, *J. Phys. III France* **5**, 355 (1995).

Horowitz, G., F. Kouki, P. Spearman, D. Fichou, C. Nogues, X. Pan, and F. Garnier, Evidence for n-type conduction in a perylene tetracarboxylic diimide derivative, *Adv. Mater.* **8**, 242 (1996).

Horowitz, G., R. Hajlaoui, D. Fichou, and A. El Kassmi, Gate voltage dependent mobility of oligothiophene field effect transistors, *J. Appl. Phys.* **85**, 3202 (1999a).

Horowitz, G., R. Hajlaoui, R. Bourgouiga, and M. Hajlaoui, Theory of organic field-effect transistors, *Synth. Met.* **101**, 401 (1999b).

Horowitz, G., M. E. Hajlaoui, and R. Hajlaoui, Temperature and gate voltage dependence of hole mobility in polycrystalline oligothiophene thin film transistors, *J. Appl. Phys.* **87**, 4456 (2000).

Ishii, H., and K. Seki, Energy level alignment at Organic/metal interfaces studied by UV photoemission: Breakdown of traditional assumption of common vacuum level at the interface, *IEEE Trans. Electron Devices* **44**, 1295 (1997).

Jarret, C. P., A. R. Brown, R. H. Friend, M. G. Harrison, D. M. de Leeuw, P. Herwig, and K. Mullen, Field-effect transistor studies of precursor-pentacene thin films, *Synth. Met.* **85**, 1403 (1997).

Jentzsch, T., H. J. Juepner, K. W. Brzezinka, and A. Lau, Efficiency of optical second harmonic generation from pentacene films of different morphology and structure, *Thin Solid Films* **315**, 273 (1998).

Jérome, D., and K. Bechgaard, Superconducting plastic, *Nature* **410**, 162 (2001).

Karl, N., Organic semiconductors, in *Festkörperprobleme XIV/Advances in Solid State Physics*, edited by H. J. Queisser, Vol. XIV, Pergamon/Vieweg, Braunschweig, p. 261 (1974).

Karl, N., Organic semiconductors, in *Landolt-Börnstein, Group III Semiconductors*, edited by O. Madelung, Vol. 17, Springer, Berlin, p. 106 (1985).

Karl, N., Getting beyond impurity-limited transport in organic photoconductors, in *Defect Control in Semiconductors*, edited by K. Sumino, Vol. II, North Holland, Amsterdam, p. 1725 (1990).

Karl, N., Charge-carrier mobility in organic crystals, in *Organic Electronic Materials*, edited by R. Farchioni and G. Grosso, Springer Series in Materials Science, Vol. 41, Springer, Berlin, p. 283 (2001).

Karl, N., and J. Marktanner, Structural order and photoelectric properties of organic thin films, *Mol. Cryst. Liq. Cryst.* **315**, 163 (1998).

Karl, N., J. Marktanner, R. Stehle, and W. Warta, High-field saturation of charge carrier drift velocities in ultrapurified organic photoconductors, *Synth. Met.* **41–43**, 2473 (1991).

Karl, N., K.-H. Kraft, J. Marktanner, M. Münch, F. Schatz, R. Stehle, and H.-M. Uhde, Fast electronic transport in organic molecular solids? *J. Vac. Sci. Technol. A* **17**, 2318 (1999).

Karmann, E., J. P. Meyer, D. Schlettwein, and N. I. Jaeger, Photoelectrochemical effects and (photo)conductivity of "n-type" phthalocyanines, *Mol. Cryst. Liq. Cryst.* **283**, 283 (1996).

Kastner, J., J. Paloheimo, and H. Kuzmany, Fullerene field-effect transistors, in *Electronic Properties of High-T_c Superconductors*, edited by H. Kuzmany, M. Mehring, and J. Fink, Springer Series in Solid State Sciences, Vol. 113, Springer, Berlin, p. 512 (1993).

Katz, H. E., Organic molecular solids as thin film transistor semiconductors, *J. Mater. Chem.* **7**, 369 (1997).

Katz, H. E., and Z. Bao, The physical chemistry of organic field effect transistors, *J. Phys. Chem. B* **104**, 671 (2000).

Katz, H. E., A. Dodabalapur, L. Torsi, and D. Elder, Precursor synthesis, coupling, and TFT evaluation of end-substituted thiophene hexamers, *Chem. Mater.* **7**, 2238 (1995).

Katz, H. E., A. J. Lovinger, and J. G. Laquindanum, α/ω-dihexylquaterthiophene: A second thin film single-crystal organic semiconductor, *Chem. Mater.* **10**, 457 (1998a).

Katz, H. E., J. G. Laquindanum, and A. J. Lovinger, Synthesis, solubility, and field-effect mobility of elongated and oxa-substituted alpha,omega-dialkyl thiophene oligomers. Extension of "polar intermediate" synthetic strategy and solution deposition on transistor substrates, *Chem. Mater.* **10**, 633 (1998b).

Katz, H. E., W. Li, A. J. Lovinger, and J. G. Laquindanum, Solution-phase deposition of oligomeric TFT semiconductors, *Synth. Met.* **102**, 897 (1999).

Katz, H. E., A. J. Lovinger, J. Johnson, C. Kloc, T. Siegrist, W. Li, Y.-Y. Lin, and A. Dodabalapur, A soluble and air-stable organic semiconductor with high electron mobility, *Nature* **404**, 478 (2000a).

Katz, H. E., J. Johnson, A. J. Lovinger, and W. Li, Naphthalenetetracarboxylic diimide-based n-channel transistor semiconductors: Structural variation and thiol-enhanced gold contacts, *J. Am. Chem. Soc.* **122**, 7787 (2000b).

Katz, H. E., T. Siegrist, J. H. Schön, C. Kloc, B. Batlogg, A. J. Lovinger, and J. Johnson, Solid-state structural and electrical characterization of N-benzyl and N-alkyl naphthalene 1,4,5,8-tetracarboxylic diimides, *Chem. Phys. Chem.* **2**, 167 (2001).

Kloc, Ch., P. G. Simpkins, T. Siegrist, and R. A. Laudise, Physical vapor growth of centimeter-sized crystals of alpha-hexathiophene, *J. Crystal Growth* **182**, 416 (1997).

Kosbar, L., C. D. Dimitrakopoulos, and D. J. Mascaro, The effect of surface preparation on the structure and electrical transport in an organic semiconductor, in *Electronic, Optical, and Optoelectronic Polymers and Oligomers*, edited by G. E. Jabbour, B. Meijer, N. S. Sariciftci, and T. N. Swader, Materials Research Society Symposium Proceedings Series, Vol. 665 (2001).

Kudo, K., M. Yamashina, and T. Moriizumi, Field effect measurement of organic dye films, *Jpn. J. Appl. Phys. (Short Notes)* **23**, 130 (1984).

Kymissis, J., MS thesis, Massachusetts Institute of Technology, Cambridge, MA (1999).

Kymissis, J., C. D. Dimitrakopoulos, and S. Purushothaman, High performance bottom electrode organic thin-film transistors, *IEEE Trans. Electron Devices* **48**, 1060 (2001) [See also: correction to this paper, *IEEE Trans. Electron Devices* **48**, 1750 (2001).]

Laquindanum, J. G., H. E. Katz, A. Dodabalapur, and A. J. Lovinger, N-channel organic transistor materials based on naphthalene frameworks, *J. Am. Chem. Soc.* **118**, 11331 (1996).

Laquindanum, J. G., H. E. Katz, A. J. Lovinger, and A. Dodabalapur, Benzodithiophene rings as semiconductor building blocks, *Adv. Mater.* **9**, 36 (1997).

Laquindanum, J. G., H. E. Katz, and A. J. Lovinger, Synthesis, morphology, and field-effect mobility of anthradithiophenes, *J. Am. Chem. Soc.* **120**, 664 (1998).

Laudise, R. A., Ch. Kloc, P. G. Simpkins, and T. Siegrist, Physical vapor growth of organic semiconductors, *J. Crystal Growth* **187**, 449 (1998).

Le Comber, P. G., and W. E. Spear, Electronic transport in amorphous silicon films, *Phys. Rev. Lett.* **25**, 509 (1970).

Lin, Y.-Y., D. J. Gundlach, and T. N. Jackson, High mobility pentacene organic thin film transistors, in *54th Annual Device Research Conference Digest*, p. 80 (1996).

Lin, Y.-Y., D. J. Gundlach, S. Nelson, and T. N. Jackson, Pentacene-based organic thin-film transistors, *IEEE Trans. Electron Devices* **44**, 1325 (1997a).

Lin, Y.-Y., D. J. Gundlach, S. Nelson, and T. N. Jackson, Stacked pentacene layer organic thin-film transistors with improved characteristics, *IEEE Electron Device Lett.* **18**, 606 (1997b).

Lovinger, A. J., and L. J. Rothberg, Electrically active organic and polymeric materials for thin film transistor technologies, *J. Mater. Res.* **11**, 1581 (1996).

Malenfant, P. R. L., C. D. Dimitrakopoulos, J. D. Gelorme, L. L Kosbar, T. O. Graham, A. Curioni, and W. Andreoni, N-type organic thin film transistors based on a N,N'-dialkyl-3,4,9,10-perylene tetracarboxylic diimide derivatives, presented at the *2001 Fall MRS Meeting*, Symposium BB, Boston, MA (2001).

Malenfant, P. R. L., C. D. Dimitrakopoulos, J. D. Gelorme, L. L. Kosbar, T. O. Graham, A. Curioni, and W. Andreoni, N-type organic thin film transistor with high field effect mobility based on a N,N'-dialkyl-3,4,9,10-perylene tetracarboxylic diimide derivative, *Appl. Phys. Lett.* **80**, 2517 (2002).

March, J., *Advanced Organic Chemistry—Reaction Mechanisms and Structure*, 3rd edn, Wiley-Interscience, New York (1985).

Martel, R., T. Schmidt, H. R. Shea, T. Hertel, and Ph. Avouris, Single- and multi-wall carbon nanotube field-effect transistors, *Appl. Phys. Lett.* **73**, 2447 (1998).

McCullough, R. D., The chemistry of conducting polythiophenes, *Adv. Mater.* **10**, 93 (1998).

McCullough, R. D., and R. D. Lowe, *J. Chem. Soc., Chem. Commun.* 70 (1992).

Meyer zu Heringdorf, F.-J., M. C. Reuter, and R. M. Tromp, Growth dynamics of pentacene thin films, *Nature* **412**, 517 (2001).

Minakata, T., H. Imai, and M. Ozaki, Electrical properties of highly ordered and amorphous thin films of pentacene doped with iodine, *J. Appl. Phys.* **72**, 4178 (1992).

Minakata, T., M. Ozaki, and H. Imai, Conducting thin films of pentacene doped with alkaline metals., *J. Appl. Phys.* **74**, 1079 (1993).

Muzicante, I., and E. A. Silinsh, Investigation of local trapping states in organic molecular crystals by method of thermally modulated space-charge limited current, *Acta Phys. Pol. A (Poland)* **88**, 389 (1995).

Nariola, S., H. Ishii, D. Yoshimura, M. Sei, Y. Ouchi, K. Seki, S. Hasegawa, T. Miyazaki, Y. Harima, and K. Yamashita, The electronic structure and energy level alignment of porphyrin/metal interfaces studied by ultraviolet photoelectron spectroscopy, *Appl. Phys. Lett.* **67**, 1899 (1995).

Nelson, S. F., Y.-Y. Lin, D. J. Gundlach, and T. N. Jackson, Temperature-independent transport in high-mobility pentacene transistors, *Appl. Phys. Lett.* **72**, 1854 (1998).

Ostoja, P., S. Guerri, S. Rossini, M. Servidori, C. Taliani, and R. Zamboni, Electrical characteristics of field-effect trasistors formed with ordered α-sexithienyl, *Synth. Met.* **54**, 447 (1993).

Ostrick, J. R., A. Dodabalapur, L. Torsi, A. J. Lovinger, E. W. Kwok, T. M. Miller, M. Galvin, M. Berggren, and H. E. Katz, Conductivity type anisotropy in molecular solids, *J. Appl. Phys.* **81**, 6804 (1997).

Paloheimo, J., E. Punkka, H. Stubb, and P. Kuivalainen, in *Lower Dimensional Systems and Molecular Devices, Proceedings of NATO ASI*, Spetses, Greece, edited by R. M. Mertzger, Plenum, New York (1989).

Paloheimo, J. P. Kuivalainen, H. Stubb, E. Vuorimaa, and P. Yli-Lahti, Molecular field-effect transistors using conducting Langmuir-Blodgett films, *Appl. Phys. Lett.* **56**, 1157 (1990).

Paloheimo, J., H. Stubb, P. Yli-Lahti, and P. Kuivalainen, Field-effect conduction in Polyalkylthiophenes, *Synth. Met.* **41–43**, 563 (1991).

Paloheimo, J., H. Stubb, P. Yli-Lahti, P. Dyreklev, and O. Inganäs, Electronic and optical studies with Langmuir-lodgett transistors, *Thin Solid Films* **210/211**, 283 (1992).

Pope, M., and C. E. Swenberg, *Electronic Processes in Organic Crystals and Polymers*, 2nd edn, Oxford University Press, Oxford, p. 337 (1999).

Schön, J. H., and Z. Bao, Influence of disorder on the electron transport properties in fluorinated copper-phthalocyanine thin films, *J. Appl. Phys.* **89**, 3526 (2001).

Schön, J. H., and B. Batlogg, Trapping in organic field-effect transistors, *J. Appl. Phys.* **89**, 336 (2001).

Schön, J. H., and Ch. Kloc, Charge transport through a single tetracene grain boundary, *Appl. Phys. Lett.* **78**, 3821 (2001).

Schön, J. H., S. Berg, Ch. Kloc, and B. Batlogg, Ambipolar pentacene field effect transistors and inverters, *Science* **287**, 1022 (2000a).

Schön, J. H., Ch. Kloc, and B. Batlogg, Fractional quantum Hall effect in organic molecular semiconductors, *Science* **288**, 2338 (2000b).

Schön, J. H., Ch. Kloc, and B. Batlogg, On the intrinsic limits of pentacene field effect transistors, *Org. Electron.* **1**, 57 (2000c).

Schön, J. H., C. Kloc, Z. Bao, and B. Batlogg, Electron transport in fluorinated copper-phthalocyanine, *Adv. Mater.* **12**, 1539 (2000d).

Sheridon, N. K., Twisting ball panel display, *US Patent* No. 4,126,854 (1978).

Silinsh, E. A., and V. Čàpek, *Organic Molecular Crystals: Interaction Localization and Transport Phenomena*, Chapter 7, American Institute of Physics Press, New York (1994).

Sirringhaus, H., R. H. Friend, X. C. Li, S. C. Moratti, A. B. Holmes, and N. Feeder, Bis(dithienothiophene) organic field-effect transistors with a high ON/OFF ratio, *Appl. Phys. Lett.* **71**, 3871 (1997).

Sirringhaus, H., N. Tessler, and R. H. Friend, Integrated optoelectronic devices based on conjugated polymers, *Science* **280**, 1741 (1998).

Sirringhaus, H., P. J. Brown, R. H. Friend, M. M. Nielsen, K. Bechgaard, B. M. W. Langeveld-Voss, A. J. H. Spiering, R. A. J. Janssen, E. W. Meijer, P. T. Herwig, and D. M. de Leeuw, Two-dimensional charge transport in self-organized, high-mobility conjugated polymers, *Nature* **401**, 685 (1999a).

Sirringhaus, H., N. Tessler, and R. H. Friend, Integrated, high-mobility polymer field-effect transistors driving polymer light-emitting diodes, *Synth. Met.* **102**, 857 (1999b).

Sirringhaus, H., R. J. Wilson, R. H. Friend, M. Inbasekaran, W. Wu, E. P. Woo, M. Grell, and D. D. C. Bradley, Mobility enhancement in conjugated polymer field-effect transistors through chain alignment in a liquid-crystalline phase, *Appl. Phys. Lett.* **77**, 406 (2000a).

Sirringhaus, H., T. Kawase, R. H. Friend, T. Shimoda, M. Inbasekaran, W. Wu, and E. P. Woo, High-resolution inkjet printing of all-polymer transistor circuits, *Nature* **290**, 2123 (2000b).

Struijk, C. W., A. B. Sieval, J. E. J. Dakhorst, M. van Dijk, P. Kimkes, R. B. M. Koehorst, H. Donker, T. J. Schaafsma, S. J. Picken, A. M. van de Craats, J. M. Warman, H. Zuilhof, and E. J. R. Sudhölter, Liquid crystalline perylene diimides: Architecture and charge carrier mobilities, *J. Am. Chem. Soc.* **122**, 11057 (2000).

Swiggers, M. L., G. Xia, J. D. Slinker, A. A. Gorodetsky, and G. G. Malliaras, Orientation of pentacene films using surface alignment layers and its influence on thin film transistor characteristics, *Appl. Phys. Lett.* **79**, 1300 (2001).

Sze, S. M., *Physics of Semiconductor Devices*, 2nd edn, Wiley-Interscience, New York, p. 438 (1981).

Tang, C. W., and S. A. Van Slyke, Organic electroluminescent diodes, *Appl. Phys. Lett.* **51**, 913 (1987).

Taur, Y., and T. H. Ning, *Fundamentals of Modern VLSI Devices*, Cambridge University Press, Cambridge, p. 11 (1998).

Torsi, L., A. Dodabalapur, and H. E. Katz, An analytical model for short-channel organic thin-film transistors, *J. Appl. Phys.* **78**, 1088 (1995).

Tsumura, A., H. Koezuka, and T. Ando, Macromolecular electronic device: field-effect transistor with a polythiophene thin film, *Appl. Phys. Lett.* **49**, 1210 (1986).

Tsumura, A., H. Fuchigami, and H. Koezuka, Field-effect transistor with a conducting polymer film, *Synth. Met.* **41**, 1181 (1991).

Vissenberg, M. C. J. M., and M. Matters, Theory of the field-effect mobility in amorphous organic transistors, *Phys. Rev. B* **57**, 12964 (1998).

Warta, W., R. Stehle, and N. Karl, Ultrapure, high mobility organic photoconductors, *Appl. Phys. A* **36**, 163 (1985).

Wisnieff, R., Printing screens, *Nature* **394**, 225 (1998).

Xu, G., Z. Bao, and J. T. Groves, Langmuir-Blodgett films of regioregular poly(3-hexylthiophene) as field-effect transistors, *Langmuir* **16**, 1834 (2000).

Yu, G., and A. J. Heeger, Semiconducting polymers as materials for device applications, in *Proceedings of 23rd International Conference on the Physics of Semiconductors*, edited by M. Scheffler and R. Zimmerman, Vol. 1, World Scientific, Singapore, p. 35 (1996).

Yu, Z. G., D. L. Smith, A. Saxena, R. L. Martin, and A. R. Bishop, Molecular geometry fluctuation model for the mobility of conjugated polymers, *Phys. Rev. Lett.* **84**, 721 (2000).

Zang, D. Y., F. So, and S. R. Forrest, Giant anisotropies in dielectric properties of quasi-epitaxial crystalline organic semiconductor thin films, *Appl. Phys. Lett.* **59**, 823 (1991).

Advanced Semiconductor and Organic Nano-Techniques (Part II)
H. Morkoç (Ed.)

CHAPTER 5

Organic Optoelectronics: The Case of Oligothiophenes

G. Gigli, M. Anni, and R. Cingolani

Università di Lecce, Lecce

G. Barbarella

ISOF, Bologna

I. Introduction

Over the last few years, conjugated molecular systems, both polymers and oligomers, have attracted considerable attention by virtue of their techno-logical application to optoelectronic devices such as light emitting diodes (LEDs) (Sheats *et al.* 1996; Friend *et al.* 1999), field effect transistors (FETs) (Garnier *et al.* 1994; Dodabalapur *et al.* 1995; Horowitz 1998), lasers (Tessler *et al.* 1996; Schön *et al.* 2000) and photovoltaic cells (PCs) (Simon and André 1985; Roman *et al.* 1998).

The advantages of using organic materials with respect to inorganics stand in the easy synthesis, the high processability, and the easy tuning of their electronic properties. All these characteristics make them excellent candi-dates for low-cost, large-area devices eventually employing flexible substrates (Kallinger *et al.* 1998).

Since the pioneering works in the 1970s with *trans*-polyacetylene (Heeger 1986), great effort has been done in the characterization of other conjugated polymers with enhanced environmental stability both in the neutral and in the charged state. In this context, thiophene derivatives (Fig. 1) have received

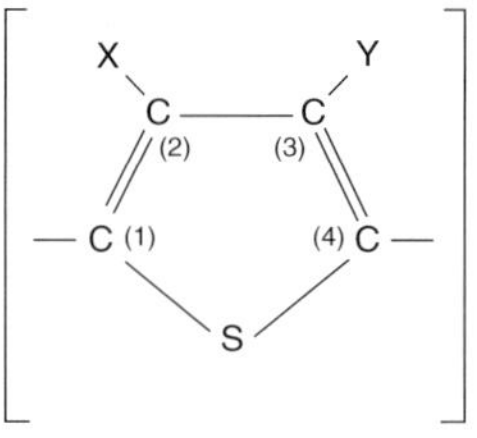

FIG. 1.   The thiophene molecule.

particular attention, due to their very high chemical stability and to the possibility of easily modulating their electronic properties by grafting different groups to the conjugated backbone. The interest has been focused both on thiophene oligomers and the corresponding polymers. In this chapter, we focus on thiophene oligomers which offer the advantage of presenting a well defined chemical structure and a higher processability/solubility relative to those of the parent polymers. Moreover from a theory standpoint, finite-size systems allow for sophisticated theoretical treatments, taking into account electron correlation effects. For these properties, well-defined oligomers are often used as model compounds to rationalize or predict the structures and properties of polymers.

Early studies of conjugated oligothiophenes were concentrated on the investigation of their high electrical conductivity upon chemical or electrochemical doping. A pioneering study by Schoeler *et al.* (1974) describes photocurrent measurements on Langmuir–Blodgett films of α-quinquethiophene. In the mid-1980s, oligothiophenes were used as model compounds as well as building blocks for electrically conducting polythiophenes. However, the milestone that originated the second generation of oligothiophenes was the discovery in 1988 of charge transport properties of α-sexithiophenes (Horowitz *et al.* 1989), an α-conjugated hexamer of thiophene. One year later, evaporated thin films of α-sexithiophenes were used as the p-type semiconductor of an organic FET. The mobility of the majority carriers measured in these devices was in the range $\mu = 10^{-3}$–10^{-4} cm^2/V/s, that is, one or two orders of magnitude higher than that of polythiophene-based FETs (Assadi *et al.* 1988), and substantially lower than that of conventional a-SiH based Metal Insulator Semiconductor Field Effect Transistor (MISFETs) (Sze 1981). In the same years, it was also demonstrated that α-6T FETs could be fabricated on flexible substrates, thus opening the era of "plastic electronics" (Garnier *et al.* 1990). The search for efficient organic FETs has recently involved a growing number of laboratories world-wide. Great progress has been accomplished in the understanding and fabrication of these devices resulting in higher mobilities, reduced size, and easy processability. Carrier mobility comparable to that of silicon based devices, up to $\mu_{FE} = 0.04$ cm^2/V/s in FETs using highly oriented α-6T thin films and up to $\mu_{FE} = 0.5$ cm^2/V/s in α-6T ultrapure single crystals (Schön *et al.* 1998), has been reported.

More recently, the research has been oriented towards the optical properties of conjugated materials in view of their application as active layers in electro-optic devices. Among these, organic LEDs constitute the most promising application, since nearly all the requirements for pre-industrialization seem to be fulfilled. Concerning thiophene oligomers, several studies have been reported about the correlation between optical and conformational properties, both for the isolated molecules in solution and for interacting molecules in the solid state. Full color tunability in the entire visible region has been demonstrated by varying the molecular distortion with proper substituent groups and changing the chain length (Grebner *et al.* 1995; Gigli *et al.* 1998). In this context, the advantages in studying oligomers rather than polymers is quite obvious: in small and well-defined molecules, the variation of the chain length and the molecular conformation can be controlled very carefully as opposed to the polymers, whose conformational parameters can hardly be controlled. Moreover, the availability of high-quality oligomers single crystals allows a fine control of the supramolecular structure in the solid state, which is not possible in polymer aggregates due to their high degree of disorder. However, even though a large effort has been done in the study of fundamental properties of oligothiophenes, their application as active layers in optoelectronic devices such as LEDs, has not been fully exploited due to their low photoluminescence (PL) quantum efficiency (Ziegler 1997b). Only very recently, a third generation of thiophene oligomers (Barbarella *et al.* 1999a, 2000; Gigli *et al.* 1999, 2001) has overcome this drawback, thanks to a new functionalization of the thienyl sulfur atom which determines an increase in emission efficiency up to values comparable to those of the best light emitting organic compounds.

Finally, α-oligothiophenes have also been used in other devices such as photovoltaic cells (Noma *et al.* 1995), spatial light modulators (Fichou *et al.* 1994), electro-optical modulators (Charra *et al.* 1994), and have been analyzed as biologically active compounds (Gommers 1972).

The present chapter is organized as follows: in Section II, we analyze the structural, electronic, and optical properties of thiophene-based oligomers, as well as the electronic excitation and recombination processes. In Section III, we discuss the basic properties of differently alkyl-substituted quaterthiophene crystals and oligothiophene-*S,S*-dioxides, and the relationship between their structural and optical properties. Photonic devices such as laser and LEDs based on oligothiophene-*S,S*-dioxides are discussed at the end of the chapter (Van Bolhuis *et al.* 1989; Chaloner *et al.* 1994; Pelettier and Brisse 1994; Horowitz *et al.* 1995; Siegrist *et al.* 1995, 1998; Fichou *et al.* 1996; Antolini *et al.* 1998a).

II.   Thiophene Oligomers

1.   STRUCTURAL CHARACTERISTICS

Since oligothiophenes have gained importance as molecular electronics materials, a large number of thiophene derivatives has been investigated.

With the exception of pentathiophene (α-5T) and septithiophene (α-7T), single crystals of all non-substituted α-oligothiophenes up to the α-8T have been grown and characterized (Van Bolhuis *et al.* 1989; Chaloner *et al.* 1994; Pelettier and Brisse 1994; Horowitz *et al.* 1995; Siegrist *et al.* 1995, 1998; Fichou *et al.* 1996; Antolini *et al.* 1998a). Due to the low solubility, single crystals of the longer unsubstituted oligothiophene can be successfully grown only by using a sublimation technique [Lipsett technique (Lipsett 1957)]. By this technique, α-6T and α-8T single crystals of macroscopic size were obtained, with a length of few millimeters and a thickness of some tens of micrometers. The problem of insufficient solubility and the related difficulty to crystallize the insoluble long oligothiophenes, have been overcome by substitution with alkyl side-chains, which makes them soluble in common organic solvents, and left them crystallize from solution. Recently the single crystal structure of a large number of soluble substituted oligothiophenes, such as di- and tetramethyl α-conjugated quaterthiophenes has been reported (Barbarella *et al.* 1993).

In the crystalline form, all non-substituted oligothiophenes are quasi-planar with a fully conjugated carbon skeleton as evidenced by a pronounced bond length alternation. X-ray structure determinations gave evidence for planar molecules with adjacent thiophene rings in the *trans* position and torsion angles $<10°$. Introduction of one or several side substituents in the β-positions of the thiophene rings, induces instead loss of planarity (Barbarella *et al.* 1993). In most cases, the twist angles are comparatively smaller in the center of the molecule and increase at the end of the chain.

Concerning the supramolecular packing, in unsubstituted and alkyl-substituted oligothiophenes, it is governed by weak van der Waals forces, which determine densely packed and highly ordered structures (Marseglia *et al.* 2000). All non-substituted α-nTs crystallize in the monoclinic system with a $P2_1$ space group. The number of molecules per unit cell is $Z = 2$ for α-2T and $Z = 8$ for α-3T, while it is $Z = 4$ for longer oligomers α-4T, α-5T, α-6T, and α-8T. The α-3T and α-8T single crystals exhibit a unique crystallographic phase, whereas two different structures have been identified for the α-4T and α-6T compounds (one with $Z = 2$ and the other with $Z = 4$) (Horowitz *et al.* 1995; Siegrist *et al.* 1995, 1998; Antolini *et al.* 1998a).

The predominant packing of the unsubstituted thiophene oligomers is of herringbone (HB) type (Fig. 2), with a parallel arrangement of the long molecular axis and an angle of about 66° between molecular planes (Van Bolhuis *et al.* 1989; Chaloner *et al.* 1994; Pelettier and Brisse 1994; Horowitz *et al.* 1995; Siegrist *et al.* 1995, 1998; Fichou *et al.* 1996; Antolini *et al.* 1998a). It is important to note that HB packing is a common scheme both in single crystals and in polycrystalline films, even though some differences in the structures may occur, primarily depending on the preparation conditions, for example, growth rate or temperature.

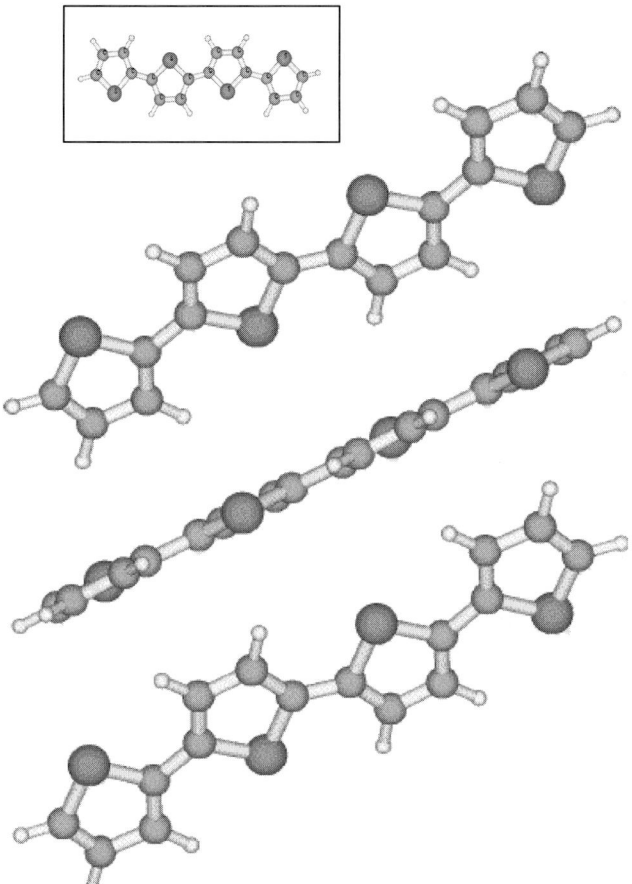

FIG. 2.   The HB supramolecular structure typical of unsubstituted oligothiophenes in the case of a quaterthiophene.

Substitution, or chemical modification of the molecule by appropriate substituents, may create strong intermolecular interactions, thus breaking the HB packing (Fig. 3). Alternatively, chemical or electrochemical oxidative doping of oligothiophenes can produce radical cation forms of the molecules whose fully coplanar quinoid structure may crystallize in a π-stack mode.

Besides the experimental studies, computational models have been used to confirm and predict the crystal structure of α-nTs. In particular mention the first work of Gavezzotti and Filippini, who calculated the crystal packing of unsubstituted α-nTs oligomers up to α-6T, and demonstrated that these molecules predominantly pack in a parallel fashion, with angles of 40–60° between molecular planes of side-by-side molecules and S $\cdots$ S contacts between 3.6 and 3.9 Å (Gavezzotti and Filippini 1991).

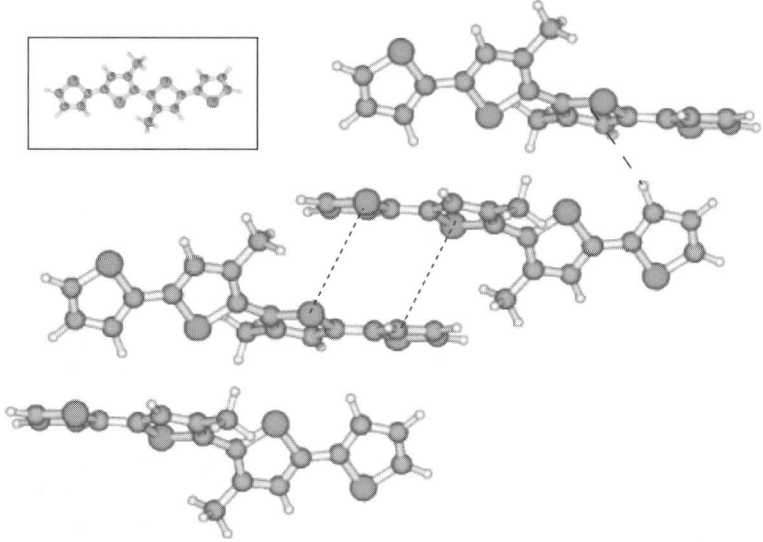

FIG. 3. Substitution with chemical groups breaks the HB packing. In 4′,3″-dimethyl-2,2′:5′,2″:5″,2‴-tetrathiophene, a combination of HB (dashed line) and π–π (dot lines) interactions occurs.

2. ELECTRONIC STRUCTURE AND OPTICAL PROPERTIES

The optical properties of conjugated molecules are correlated to the electronic structure and in particular to the frontier orbitals HOMO and LUMO (highest occupied molecular orbital and lowest unoccupied molecular orbital, respectively). *Ab initio* as well as semi-empirical calculations attribute the HOMO and LUMO to the bonding π and anti-bonding π* molecular orbitals determined by the overlapping of the p_z orbitals of carbon atoms in the chain (Brédas *et al.* 1983; Champagne *et al.* 1994), respectively. Any variation of such overlap, such as geometrical modifications induced by the presence of substituent groups, interactions with solvents, thermal effects leads to changes of the electronic structure. In particular, the torsion of the chain reduces the overlap of p_z orbitals and increases the gap between π and π* orbitals, thus causing shorter conjugation lengths and blue-shifted optical absorption (Gigli *et al.* 1998). Thermochromism and solvatochromism (Dossantos *et al.* 1992; Grastrom and Inganäs 1992; Pei *et al.* 1993; Tashiro *et al.* 1993) are clear demonstrations of these properties. Dilute solutions of oligothiophene based compounds at low temperatures exhibit a red-shift of the absorption spectrum (Chsrovin *et al.* 1993; Becker *et al.* 1995), indicating that at low temperatures the ground state is more planar.

The electronic structure of most oligothiophenes has been obtained by molecular orbital (MO) calculations, ultraviolet photoemission spectroscopy

(UPS) and X-ray photoemission spectroscopy (XPS) (Fichou 1999) for the determination of the occupied levels, and by a variety of optical techniques probing the transition between occupied and unoccupied levels (Horowitz *et al.* 1999; Schön *et al.* 1999). Hereafter, we report the main results concerning isolated unsubstituted oligothiophenes. A first interesting feature is the crossing of states belonging to different symmetries as a function of the chain length. The states of interest within the C_{2h} symmetry have $1B_u$ and $2A_g$ symmetry. As the chain length and the number of double bonds is increased, the lowest transition changes from an allowed (B_u) into a forbidden transition (A_g). The Calculation of the optical transition energies is a complex task, since electron correlation plays an important role and the simple one-electron pictures, such as Huckel and Hartree–Fock theories, do not describe the excited states properly (Schulten *et al.* 1976; Klessingner and Michl 1989). However, recent photophysical measurements on oligomers of different length (Becker *et al.* 1995) have shown that the $2A_g$ state lies below the $1B_u$ state for oligothiophenes constituted by more than eight rings, resulting in the inhibition of fluorescence for long oligothiophenes.

A second interesting feature of the unsubstituted oligothiophenes is the linear dependence of the $S_0 \rightarrow S_1$ transition energy on the inverse number of rings constituting the chain (Belijonne *et al.* 1996) (Fig. 4). A significant redshift of the lowest electronic transition occurs as the chain length is increased. This is related to the progressive extension of the π-delocalized states. The evolution with the chain size of the singlet triplet $S_0 \rightarrow T_1$ transition energies is instead much weaker than for the $S_0 \rightarrow S_1$ excitation (Belijonne *et al.* 1996) (Fig. 4). The singlet–triplet excitation is only lowered by ∼0.2 eV when going

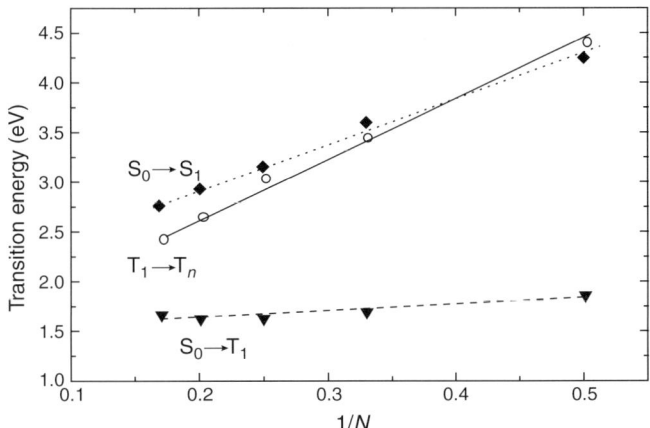

FIG. 4.   Chain size dependence of the $S_0 \rightarrow S_1$ transition energy between the ground state (S_0) and the first excited singlet state (S_1), of the singlet triplet transition energy $S_0 \rightarrow T_1$, and of the triplet–triplet transition energy $T_1 \rightarrow T_n$.

from the dimer to the hexamer, whereas a bathochromatic shift of $\sim1.4\,eV$ is observed for the singlet–singlet transition $S_0 \rightarrow S_1$. Such a behavior actually reflects the stronger confinement of the triplet exciton with respect to the singlet. This trend is consistent with optically detected magnetic resonance (ODMR) data for polythiophenes, indicating that the T_1 triplet state hardly extends over more than a single thiophene unit (Swanson *et al.* 1990). A larger delocalization of the higher-lying triplet state T_n is instead inferred by the more significant red-shift of the $T_1 \rightarrow T_n$ transition vs $1/N$ shown in Fig. 4.

So far we have discussed the electronic structure of isolated molecules. However in the solid state oligothiophenes interact by means of van der Waals forces and form densely packed and highly ordered molecular crystals. Optical excitations in such crystals are well known as Frenkel excitons (Craig and Walmsley 1968; Davydov 1971; Pope and Swenberg 1982), whose optical properties resemble very much those of the isolated molecules, since they are strongly confined in the molecule and only the weak interaction with the surrounding molecules leads to the formation of a collective excitation. This is opposed to the extended Mott–Wannier (MW) excitons in conventional semiconductors, where the electron and the hole are weakly bound (Pankove 1971). The main effect of intermolecular interactions is to split each molecular state in different crystalline states [Davydov splitting (Craig and Walmsley 1968; Davydov 1971; Pope and Swenberg 1982)] characterized by different symmetry properties (Fig. 5). In cases where there is more than one molecule per unit cell and the molecules are related by symmetry operations, the crystal wavefunctions are constructed out of the subsets of non-equivalent molecules which leads to representations of the crystal states which are symmetric or antisymmetric combinations of the subset wavefunctions (Pope and Swenberg 1982). We will see that this spitting of molecular states in crystalline states of well-defined symmetry, strongly affects the relaxation processes of the molecular excitations.

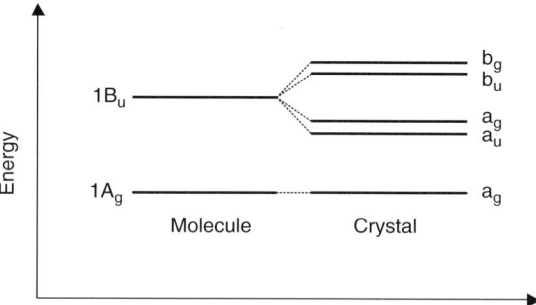

FIG. 5. Energy diagram (calculated by semi-empirical methods) of the electronic states for an isolated molecule (left) and in solid state (right). The presence of intermolecular interactions determines a splitting of the molecular states in a number of crystalline states depending on the number of equivalent molecules in the unit cell.

a.   Recombination Processes and Photoluminescence Efficiency

Excitons in oligomers and polymers can be produced by photo-excitation or charge recombination after electrical injection. Depending on the excitation process, optical or electrical, the radiative decay gives rise to PL or electroluminescence (EL), respectively. Competitive non-radiative processes provide additional paths of decay, thus reducing the quantum efficiency. Non-radiative decay channels can be intramolecular, such as internal conversion (IC), intersystem crossing (ISC), and singlet fission (SF), or intermolecular, such as the formation of aggregates and charge transfer excitons (CTEs). In addition, in the solid state quenching of the singlet excitons may occur due to extrinsic or conformational defects (Yang *et al.* 1998). In what follows, we describe the main intra- and intermolecular non-radiative decay channels in thiophene compounds, that is, IC, ISC, SF, formation of *H*-aggregates, and CTEs.

Internal conversion refers to relaxation from a highly excited state into a state of lower energy, within the same manifold, either singlet or triplet, by emission of phonons. This process is quite fast, on the femtosecond timescale. IC can provide an efficient non-radiative decay channel if the transition from the high energy state into the ground state is forbidden by the symmetry of the wavefunction of the relaxing state. If the lowest singlet excited state is for instance a $2A_g$ singlet state, radiative transitions to the ground state are symmetry forbidden and the coupling with the ground state is possible only via phonon emission.

Intersystem crossing refers to energy transfer between the singlet and triplet manifolds. Subsequent IC usually leads to the decay into the lowest triplet state (T_1), which is not fluorescent (Becker *et al.* 1995). The rate of ISC is governed by the spin–orbit coupling term (Bixon and Jortner 1968), which is enhanced in the presence of heavy atoms, such as the sulfur atoms in oligothiophenes. According to the calculation of Belijonne *et al.* (1996), in unsubstituted oligothiophenes the energy difference between the singlet S_1 and T_1 is too large to give efficient singlet–triplet overlap. However, INDO/MRD-CI calculations predict that there is one triplet excited state (T_4) of energy close to the lowest singlet excited state (S_1). For bithiophene, T_4 lies below S_1, so that ISC occurs readily, resulting in a very low fluorescence yield for short chains. With increasing the oligomer length, S_1 falls below T_4. Hence, the ISC becomes increasingly unlikely to occur, resulting in higher PL quantum efficiencies for longer oligomers (Belijonne *et al.* 1996). PL efficiencies of 0.07, 0.2, 0.28, and 0.42 have indeed been measured for dilute solutions of α-3T, α-4T, α-5T, and α-6T, respectively (Chsrovin *et al.* 1993). Time-resolved luminescence studies have demonstrated that with increasing the length of the oligomer and the extent of the π-conjugation, a rapid decrease of the non-radiative decay rate and of the yield of triplets (Becker *et al.* 1995) occurs (Chsrovin *et al.* 1993; Becker *et al.* 1995). A systematic decrease of the intersystem crossing rate as the rigidity and planarity of the molecules increase, has been also observed in dilute solutions of various oligomers (Nijegorodov and Downey 1994).

Singlet fission consists in the generation of two triplet excitons from the fission of one singlet excited state. The excited singlet state may be the lowest excited singlet state S_1 or a higher excited singlet, produced by singlet–singlet fusion under intense irradiation. In order to be singlet fission an energetically allowed process, the energy of the singlet excited state undergoing fission must be at least twice that of the triplet excited state. Considering the case of the lowest singlet state S_1, which may play a role in the non-radiative decay of oligomers, it must be: $E(S_0–S_1) \geq 2E(S_0–T_1)$. Belijonne *et al.* (1996) showed that this condition is satisfied for short oligomers, whereas for longer oligomers this condition is not fulfilled.

Aggregation and Davydov splitting. The quantum yield of fluorescence for conventional oligothiophenes in the solid state is generally lower than in solution by one or two orders of magnitude. This is attributed to quenching induced by aggregation effects. The fluorescence quantum yields for thin films of unsubstituted oligothiophenes are quoted in Ziegler (1997a). α-5T–α-8T exhibit fluorescence quantum yield of $\Phi_F \leq 10^{-4}$ in highly ordered thin films ($d = 3$–5 nm) and about one order of magnitude larger values in thicker films (Egelhaaf and Oelkrug 1995; Oelkrug *et al.* 1995). For the same thickness, the yield is smaller for more ordered films, as expected for molecules with strong excitonic coupling. Microcrystalline α-4T has a yield of $\Phi_F = 7 \times 10^{-3}$ even in thin films, and of $\Phi_F = 2 \times 10^{-2}$ in thick films. The yield of end-capped oligothiophenes is around 3×10^{-3} regardless of the layer thickness, due to the reduced tendency to form ordered films. Furthermore, the fluorescence yields of α-substituted (highly ordered) films are much lower than β-substituted (disordered) films (Yassar *et al.* 1994, 1995). These low PL efficiencies have been correlated to the supramolecular arrangement of the molecules in the solid state. In the HB packing, typical of both unsubstituted and most α-substituted oligothiophenes, the coupling of the dipole moments between translationally inequivalent molecules in the unit cell, gives rise to a splitting (Davydov splitting) of the single molecule excited states into a set of crystalline levels, with the lowest energy one optically forbidden. The effects of different crystalline structures on the electronic states can be explained by considering a simple monodimensional model in which infinite molecules are orderly located at distance a (Fig. 6). If only first neighbors interactions are taken in account, the energy of an excited state can be expressed as:

$$E^e(0) \approx \Delta E^e + D^e + L^e\left(\vec{k}\right)$$

$$= \Delta E^e + D^e + 2d^2 \frac{1 - 3\cos^2(\alpha)}{a^3} \cos\left(\vec{k}a\right) \qquad (1)$$

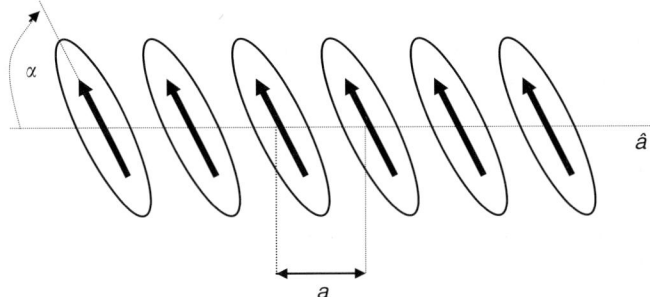

FIG. 6.   Crystal monodimensional model. An infinite number of molecules, forming an angle α with the $\hat{a}$ axis, are periodically located at distance a.

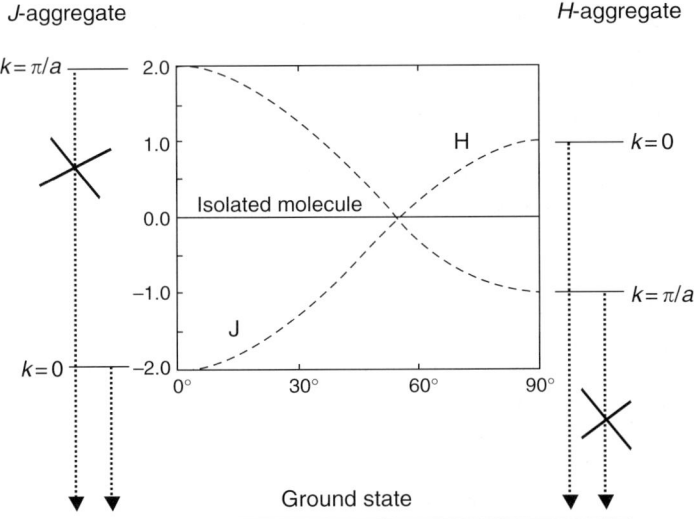

FIG. 7.   Evolution of the $k = 0$ excitonic state energy with the angle α. For values of $α < 54.7°$ (*H*-aggregates) the resonance interaction is positive and the excitonic state energy is larger than that of the single molecule. For values of $α > 54.7°$ (*J*-aggregates) the resonance interaction is negative and the excitonic state energy is lower than the single molecule one. The energy of the $k = π/a$ excitonic state (as all the other $k \neq 0$ excitonic states are not optically allowed) is also shown. If the supramolecular structure is that of an *H*-aggregate the lowest energy state is optically forbidden.

where d is the yield of the dipole transition and α is the angle between the dipole and the straight line a. In Fig. 7 the plot of $k = 0$ exciton energy vs the angle α is shown. For values of $α < 54.7°$ (*H*-aggregates) the resonance interaction $L^e(\vec{k})$ is positive and the excitonic state energy is larger than that of the single molecule. For values of $α > 54.7°$ (*J*-aggregates) the resonance interaction is instead negative and the excitonic state energy is lower than the single molecule one. The energy of the $k = \pi/a$ excitonic state (as all the other $k \neq 0$ excitonic states are not optically allowed) is also reported. If the supramolecular structure is that of an *H*-aggregate, the lowest energy state where termalization occurs is optically forbidden, determining a quenching of the PL efficiency. Actually, this selection rule is strictly valid only for an ideal infinite single crystal. The dipole selection rules are relaxed when the crystalline size is reduced or in the presence of disorder.

Charge transfer excitons. Up to now, we focused on neutral excitations where the charges are not separated, but localized on the same oligomer unit (Frenkel excitons). Excitations where a separation of charges occurs are termed as charge transfer excitons or CTEs and consist of a weakly bound electron–hole pair localized over two or more adjacent oligomers (Fig. 8). CTE represent the intermediate step between excitons and carrier separation. If the weakly bound electron–hole pair can recombine quickly, it can form in fact the intramolecular singlet exciton before charge separation occur, thus originating a new non-radiative channel. In order to establish how efficient this non-radiative channel is, the energy separation between intramolecular Frenkel excitons and CTE has to be taken in account. Such separation depends on the oligomer chain-length, on the first ionization energy, on the electron affinity, and on the intermolecular distance. An increase of the oligomer chain length determines a larger π-electron delocalization and a stronger π–π van der Waals force, which increase the electron affinity and reduce the intermolecular distances, as well as the first ionization potentials. All of these factors tend to reduce the energy separation between the intermolecular CTEs and the intramolecular Frenkel excitons, determining a competing non-radiative decay channel.

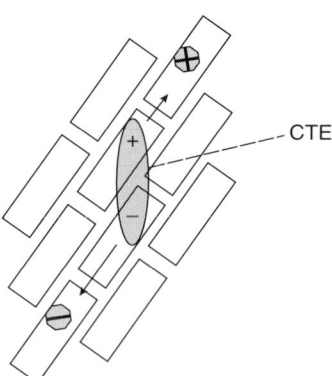

Fig. 8. Excitations where a separation of charges occurs are termed as charge transfer excitons (CTEs). They consist of a weakly bound electron–hole pair localized over two or more neighboring oligomers. CTE is the intermediate step towards carrier separation and transport.

CTEs are not easily observed in the linear absorption spectrum due to the poor overlap between the wavefunction of the ground state, localized on a single oligomer, and excited states spread over two or more adjacent oligomers. However, they can be detected in electric field modulated spectroscopy (Sebastian *et al.* 1981), exploiting the large dipole moment of the excited states which generates an electroabsorption signal proportional to the second derivative of the absorption spectrum. Nevertheless, CTEs have been identified for many conjugated molecules and the corresponding energies are much higher than those of the respective first singlet absorption bands (Sillinsh and Capek 1994).

For thin films of α-sexithiophene, crystallographic studies have demonstrated that the oligomers are arranged in a stacked layer structure, in which conductivity parallel to the stacks (perpendicular to the long oligomer axis) is much greater than that between the layers (Garnier *et al.* 1993). The photoconductivity has therefore been explained as due to to the rapid charge separation of longitudinal CTEs extending along the stack (Dippel *et al.* 1993). Therefore, close packing of oligomers within the layers should favor efficient charge separation and photoconductivity rather than fluorescence. In order to favor fluorescence, efficient charge transport can be suppressed by reducing the degree of ordering within the film, either by using polycrystalline films with small crystallite size, or by increasing the separation of oligomers by the addition of bulk substituent groups.

In reality, all the above effects occur simultaneously, with a relative efficiency ultimately depending on the microscopic properties of the sample. In Fig. 9, a simple scheme of all the processes described is shown.

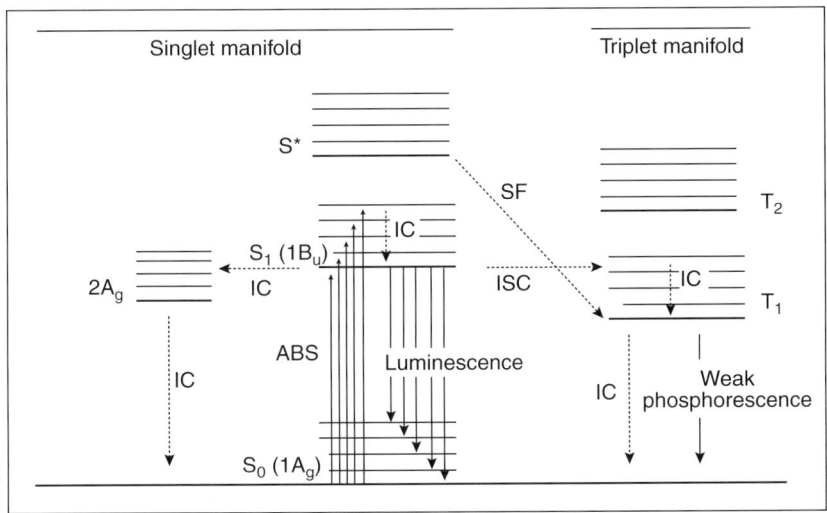

FIG. 9.   Scheme of the radiative and non-radiative recombination processes.

III. Basic Studies

1. Alkyl-substituted Oligothiophenes

a. Relationship between Optical and Structural Properties in Methyl-substituted Quaterthiophenes

The correlation between the molecular structure and electro-optical properties is a powerful tool to control and optimize the optical and electrical characteristics of oligomer-based devices. In this context, the analysis of molecular crystals is particularly useful for determining the exact conformational structure of the molecules. For the oligomers, a structural parameter that deserves particular attention is the molecular distortion, as the backbone distortion of the molecules is strongly correlated to the π–π conjugation between adjacent thiophene rings and, consequently, it strongly affects the optical properties.

The methyl-substituted quaterthiophene crystals are prototype materials where the effects of the molecular distortion are quite evident. By a proper β-functionalization of the thienyls, it is in fact possible to tune the torsion angles between adjacent rings, that is, of the degree of molecular distortion. In what follows, we discuss four representative compounds of this class of crystals: a unsubstituted quaterthiophene crystal (**A**), two different tetramethyl quaterthiophene crystals (**B** and **D**) and a dimethyl quaterthiophene crystal (**C**) (Fig. 10). Details on the synthesis and on the X-ray characterization of these compounds can be found in Van Pham *et al.* (1989), Barbarella *et al.* (1993), and Antolini *et al.* (1998a).

According to single crystal X-ray diffraction data, the unsubstituted quaterthiophene is conformationally homogeneous and it exists in the all-anti-planar conformation (Siegrist *et al.* 1998) shown in Fig. 10(a). Instead, the methyl-substituted quaterthiophenes deviate significantly from coplanarity and exhibit an increasing degree of molecular distortion (Van Bolhuis *et al.* 1989; Chaloner *et al.* 1994; Pelettier and Brisse 1994; Horowitz *et al.* 1995; Siegrist *et al.* 1995, 1998; Fichou *et al.* 1996; Antolini *et al.* 1998a). Compound **B** is characterized by statistical disorder arising from two alternative *anti* (84% probability) and *syn* (16% probability) orientations of one of the terminal rings. Therefore, two conformations exist in crystals **B**, namely the prevailing *anti–anti–anti* and the *anti–anti–syn*. The dihedral angles of the prevailing conformation [Fig. 10(b)] are $\omega_1 = 11.8°$, $\omega_2 = 28.8°$, and $\omega_3 = 21.3°$, respectively. Compound **C** is characterized by a large torsion angle ($\omega_2 = 47°$) between the inner rings and by statistical disorder arising from two alternative *anti* (88% probability) and *syn* (12% probability) orientations of both terminal rings. As a consequence, both the *syn–anti–syn* and the mixed and indistinguishable *syn/anti–anti–anti/syn* conformations are present in the crystal. In the prevailing conformation of compound **C**

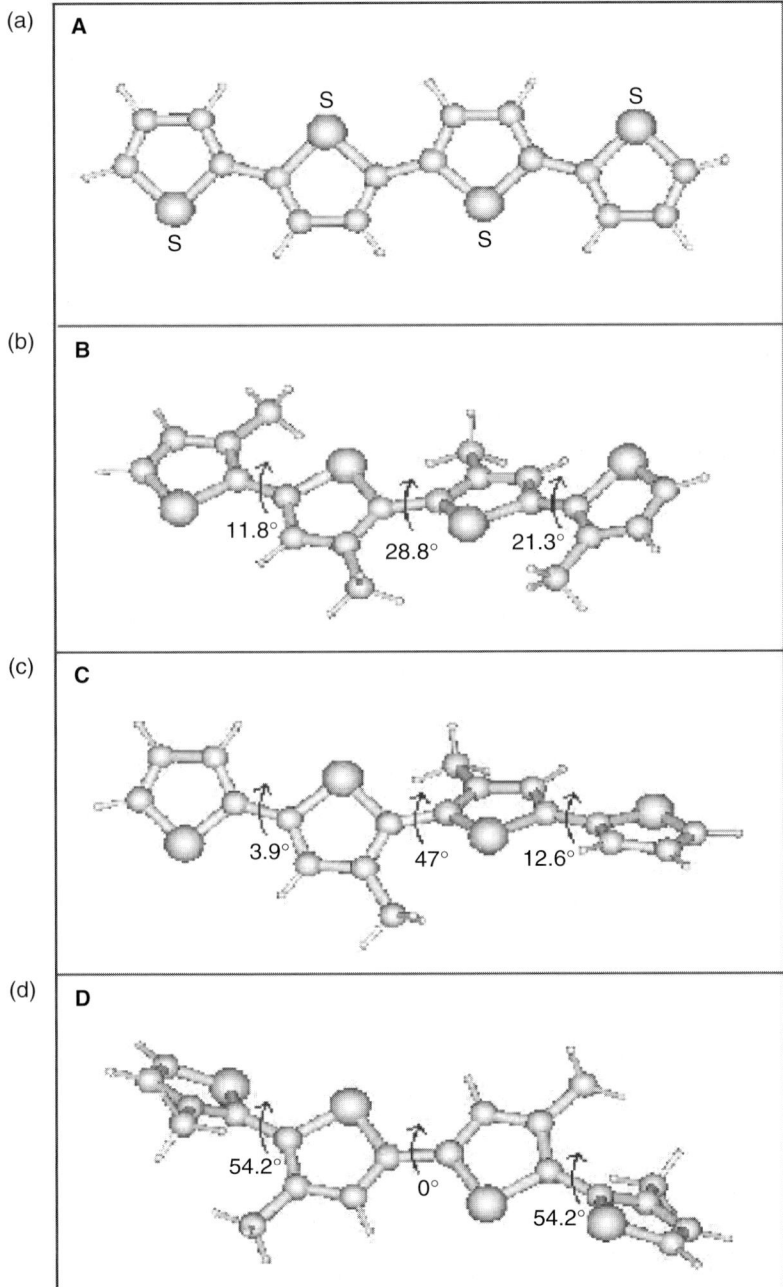

FIG. 10. Molecular structure of the unsubstituted oligothiophenes compounds: **A**, **B**, **C**, and **D**.

[Fig. 10(c)], the terminal rings are twisted by $\omega_3 = 12.6°$ and $\omega_1 = 3.9°$ with respect to the inner ones. Finally, compound **D** [Fig. 10(d)] is conformationally homogeneous and it exhibits a *syn–anti–syn* conformation characterized by two exactly coplanar inner rings, and by an extremely large torsion angle (54.2°) of the external rings.

In order to study the effect of the molecular conformation on the optical properties and electronic states, it is useful to introduce a macroscopic parameter, defined as the average inter-ring angle $\theta = (\omega_1 + \omega_2 + \omega_3)/3$, which takes into account, in a simple way, the level of molecular distortion.

In Fig. 11, we plot the PL and excitation photoluminescence (PLE) spectra of the compounds ordered according to the average inter-ring angle $\theta = (\omega_1 + \omega_2 + \omega_3)/3$. The PLE spectra show evident singlet Frenkel exciton resonances shifting to high energy as the structure of the molecules deviates from the planar conformation, that is, with increasing θ. The blue-shift is attributed to the loss of π-conjugation induced by the steric effect of the methyl groups. The luminescence spectra are Stokes shifted by about 250 meV, also showing well-resolved vibronic structures with equally spaced replicas ($\Delta E = 140$ meV) in the reference unsubstituted crystal. These features, which

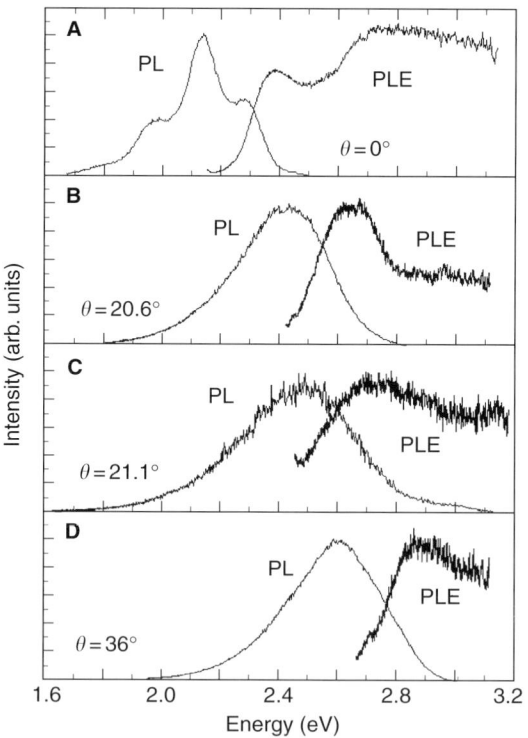

Fig. 11.   CW-PL and PLE of compounds **A–D**.

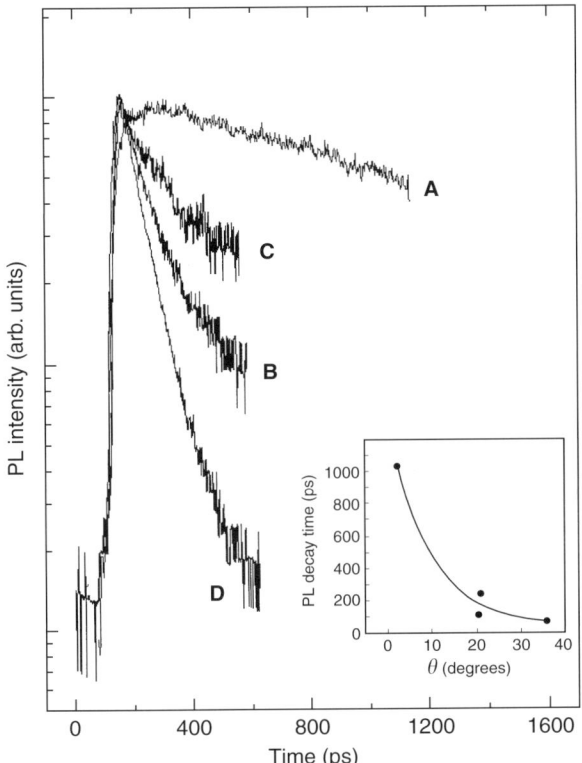

FIG. 12. Temporal evolution of the PL peaks of compounds **A–D** at room temperature. Inset: PL decay time vs average molecular distorsion angle $\theta = (\omega_1 + \omega_2 + \omega_3)/3$. The line is a guide for the eye.

are inherent to the planar structure of sample **A**, are completely absent in the more distorted compounds **B**, **C**, and **D**. In these compounds, the strong molecular distortion due to the substituents groups and the crystalline disorder introduced by the statistical distribution of external rings, probably prevent the formation of collective vibrational modes determining broad and featureless PL bands.

The life-time measurements (Fig. 12) show that the decay time of the luminescence strongly decreases as the molecular distortion increases (inset Fig. 12). The radiative life-time is expected to be in the order of nanoseconds for all the compounds, as estimated from the calculated transition dipole: this value also agrees with the one measured for the molecules in solution (Kanemitsu *et al.* 1994). Such a decrease of exciton life-time is attributed to the increasing number of vibrational degrees of freedom in the molecules distorted by the substituents groups, that is, on the increasing probability of scattering processes between exciton and vibrational excitations.

Quasi-monoexponential decays with time constant of 1020, 110, 240, and 70 ps are found for samples **A**, **B**, **C**, and **D**, respectively. We note that samples **A** and **D**, having widely different θ values, exhibit a life-time variation of more than one order of magnitude. Conversely, samples **B** and **C** having almost the same average distortion θ, exhibit a comparable luminescence decay time (within a factor 2). This small difference is probably due to the statistical distribution of the external rings which introduces conformational disorder and an aperiodic crystalline structure depending on the percentage of the less-probable molecular conformations. These data suggest that the effect of conformational disorder on the emission energy and on the life-times is less relevant than that of the molecular distortion exemplified by the average distortion angle θ. We recall that compounds **B** and **C** have the same emission energy and different degree of conformational disorder.

In order to provide a more quantitative description of the measured optical absorption spectra, one has to calculate the first singlet excited state of the functionalized molecule constituting the different crystals. Even though the calculations are performed for the individual molecule, the effect of the crystal bonding has to be taken into account directly by using the distortion and the atomic coordinates determined experimentally. The calculation can be performed by using spectroscopic intermediate neglect of diatomic overlap/spectroscopic parameterization (INDO/S) (Ridely and Zerner 1975, 1976; Zerner *et al.* 1980), a semi-empirical Hartree–Fock technique, followed by a configuration interaction (CIS), considering single excitations from the highest 20 occupied molecular orbitals to the lowest twenty unoccupied orbitals. Details on the theoretical method, which is commonly used for oligothiophenes, can be found in Belijonne *et al.* (1993, 1996) and Cornil *et al.* (1995).

The calculated transition energies overestimate the absolute value of the crystal gap by about 0.3 eV. However, the calculated values predict quite well the difference in energy between all the compounds and their dependence on the average distortion angle (Fig. 13). This indicates that the energy gap of the crystals, at first approximation, depends on the level of distortion of the single molecule and on the degree of π–π conjugation. The difference, in absolute value, between the transition energy of an individual molecule and the experimental results of the crystal is essentially due to the neglect of exciton delocalization (Craig and Walmsley 1968; Davydov 1971; Pope and Swenberg 1982), which is known to cause a red-shift.

b. The Case of Polymorphism

An intriguing effect influencing relationship between optical and structural properties is the conformational polymorphism. The phenomenon of polymorphism originates from the occurrence of different molecular packings for

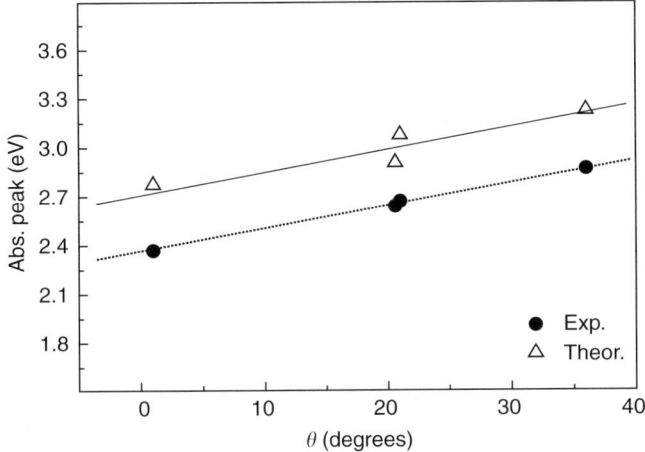

FIG. 13. Experimental (circles) and theoretical (triangles) absorption peak energy vs the average molecular distorsion angle $\theta = (\omega_1 + \omega_2 + \omega_3)/3$.

the same molecules. When different crystal packings are characterized by a different conformation of the molecules, the polymorphism is called *conformational*. A case in which a clear correlation between polymorphism (conformational) and optical properties has been demonstrated is that of the methylsulfanyl (SCH_3) polysubstituted α-conjugated quaterthiophene (Fig. 14) [3,3',4'',3'''-tetrakis(methylsulfanyl)-2,2':5',2'':5'',2'''-quaterthiophene]. This compound exists as amorphous and in two different crystalline forms: triclinic (**E**) and monoclinic (**F**), obtained at room temperature depending on the solvent used (Barbarella *et al.* 1996, 1997). For the amorphous compound, no structural data are available, whereas an accurate structural analysis has been carried out on the two crystalline forms. According to the X-ray diffraction measurements (Becker *et al.* 1996), in the crystal **E** the inner thiophene rings are exactly coplanar whereas the terminal ones are twisted by 28.9°; the crystal **F** instead exhibits a distorted conformation with the external rings twisted by 57° with respect to the inner ones. Another difference is that the SCH_3 groups are roughly coplanar in the triclinic form, whereas in structure **6** they are randomly distributed over two different orientations.

In Fig. 15, we show the PLE spectra of the two crystals. The HOMO–LUMO gap of sample **F** is blue-shifted by about 140 nm as compared to that of sample **E**. Such a blue-shift may be attributed to the more distorted conformation of 3,3',4'',3'''-tetrakis(methylsulfanyl)-2,2':5',2'':5'',2'''-quaterthiophene in the monoclinic single crystal. In this conformation, the loss of π-conjugation induced by the steric effect is not compensated by the delocalization of the electron pairs of the sulfur atoms of the *S*-methyl group onto the backbone. This mesomeric effect depends on the inter-ring twist

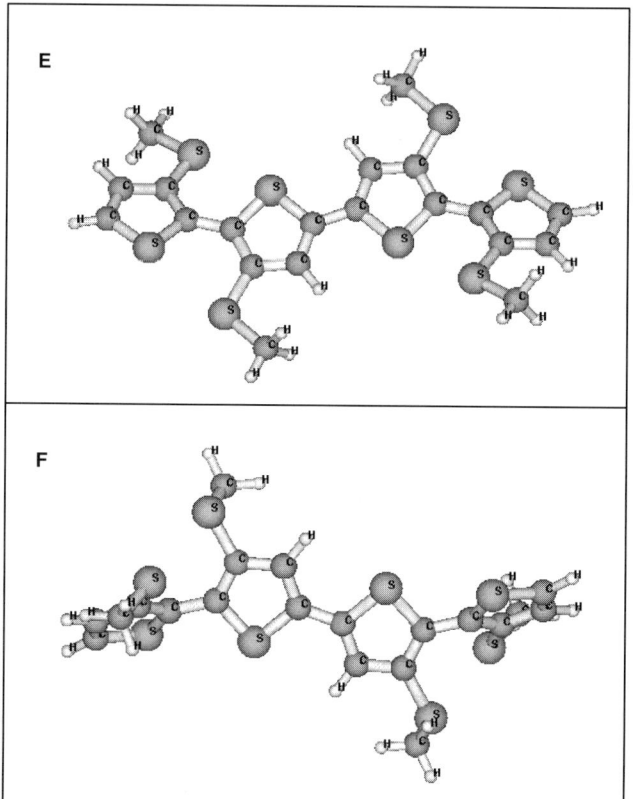

FIG. 14.   Molecular structure of substituted oligothiophene compounds **E** and **F**.

angle and on the orientation of the SCH_3 with respect to the thiophene ring to which it is bound. A large value of these angles, just like in sample **F**, results in a stronger electron localization and in a larger HOMO–LUMO gap. Furthermore, the broad low-energy tail in the PLE spectrum of crystal **F** is reminiscent of the inhomogeneous spectral broadening occuring in disordered crystal induced by the density of state tails in the forbidden gap.

The PL spectra (Fig. 15) reflect the shift and the broadening of the absorption edge, showing also a marked vibronic structure with a secondary peak red-shifted by 160 meV in sample **E**. This feature, which is inherent to the crystalline order of structure **E**, is completely absent in the **F** form, which exhibits a broad and featureless PL band. The different vibronic structure can be reasonably understood by considering the distorted structure of sample **F** and, in particular, by considering the high level of disorder introduced by the two equiprobable orientations of the SCH_3 group.

Figure 16 shows the temporal evolution of the PL peak at 624 nm for sample **E** and at 550 nm for sample **F**. Distinct monoexponential decays with

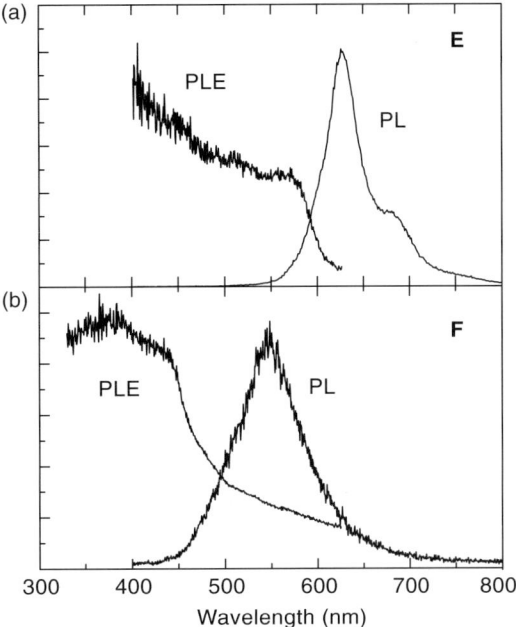

FIG. 15.   CW-PL and PLE of compounds **E** and **F**.

time constant (τ_R) of 190 and 560 ps are found for **F** and **E**, respectively. A slightly longer τ_R is expected for the triclinic crystal (**E**) since the spontaneous emission rate is proportional to the cube of the frequency of the transition. However, this cannot account for the large difference measured in the decay time of the monoclinic and triclinic crystals. This is most probably due to the effect of disorder on the Frenkel exciton life time, as also inferred from the PL line-width broadening (FWHM $\sim 1/\tau$). The distorted structure and the statistical distribution of the SCH_3 group orientation determines a strong interaction between electronic and vibrational excitations through the increase of the exciton–lattice scattering processes into lower energy states. These processes cause a loss of coherence of the exciton wavefunction and a consequent decrease of the excitonic life-time.

These experiments clearly show the strong correlation between the conformational structure of a single molecule and its photoluminescence, showing that a proper molecular engineering allows one to tune and control the optical properties.

c.   Photoluminescence Efficiency

A key issue for the improvement of the LEDs and laser performances is the PL quantum yield of the active material, that is the ratio between emitted and

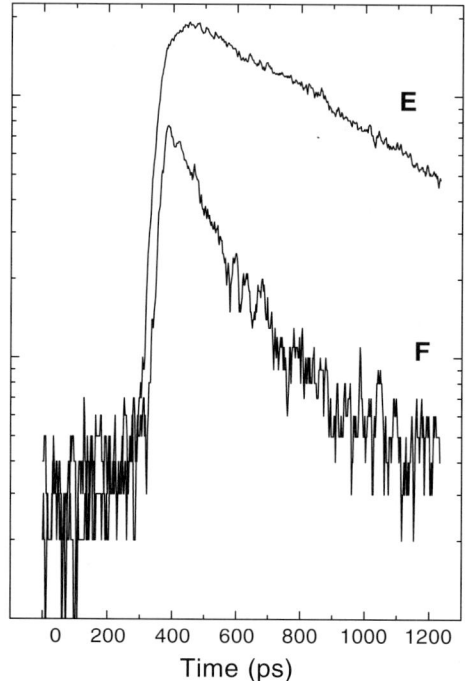

0   200   400   600   800   1000   1200

Time (ps)

FIG. 16.   Temporal evolution of the PL peaks at room temperature: at 624 nm for compound **E**, at 550 nm for compound **F**.

absorbed photons. We have seen in the previous section that this ratio can be affected by several processes: in isolated oligomers and polymers neutral excited states produced by photoexcitation can decay radiatively or through non-radiative processes such as IC and ISC. In the solid state, intermolecular interactions can increase the probability of non-radiative decay of exciton due to the formation of *H*-aggregates with by lower-energy optically forbidden states or the formation of CTEs. In addition, extrinsic processes, such as exciton migration to structural defects and impurities, influence the PL efficiency in a substantial way. All these processes are strictly correlated to the molecular geometry and the supramolecular structure, so that a deep knowledge of the structural parameters is fundamental for a clear understanding and control of molecular devices. In molecular crystals, the contribution of each non-radiative process can be evaluated and correlated to the structural properties, differently from disordered systems such as polymers and amorphous compounds. In what follows we discuss representative results taken on microcrystals constituted by the compounds **A–F** described in the previous sections. The microcrystals must be selected by optical microscopy to make sure that structural defects as dislocations (one of the

TABLE I

Measured PL Efficiency (η), Intermolecular Resonance Interaction for the Two *H*-Aggregate Directions (γ_{H1} and γ_{H2}) and for the *J*-Aggregate Direction (γ_J), Hole Transfer (t_h), and Electron Transfer (t_e). All the Calculated Values are in Millielectron Volts

Sample	η (%)	γ_{H1}	γ_{H2}	γ_J	$|t_h|$	$|t_e|$
A	12	115	103	−47	42	40
B	1.5	90	74	−47	43	26
C	1.6	119	61	−47	45	13
D	1.8	60	25	−38	17	3
E	14	59	58	−59	24	7
F	1.1	52	51	−25	8	1

main causes of PL quenching) are reduced as far as possible. The high purity of the materials for all the compounds must be guaranteed by the state-of-the-art purification methods, such as silica gel chromatography, applicable to soluble compounds.

The measurements of the PL efficiency are normally carried out by a carefully calibrated integrating sphere. Values are quoted in Table I. Crystal **A** constituted by planar non-substituted quaterthiophene molecules arranged in a strongly packed HB structure, shows a high PL efficiency of 12%. Comparable quantum yield has been recently reported in crystalline thin film (Yang *et al.* 1998). Crystals **B–D**, where the functionalization with methyl groups distorts the molecules in a less-compressed quasi-HB structure, exhibit a PL efficiency reduced to 1–2%. Compound **E**, which is characterized by a slipped π–π stacking structure, shows PL efficiency of 14%. Finally, compound **F** presents a packing structure similar to compound **4** and a low PL efficiency.

Noticeably, these efficiency values cannot be directly compared to the efficiency measured in solution, frozen solution or inclusion compound, due to the different conformation taken by the single molecules in these environments. In fact, the molecular distortion in crystal samples is not only an intrinsic molecular property (induced by substituents), but is also strongly affected by the crystal packing. We should also mention that, in solution, the luminescence is expected to come from a more planarized conformation (Becker *et al.* 1996), whereas a geometrical relaxation is much more difficult to occur in the crystalline form. In order to elucidate whether the intermolecular interactions can be responsible for the strong PL efficiency variations, one has to take in account the exciton resonance interaction, and then the role of CTEs (Wu and Conwell 1997).

The exciton resonance interaction (γ) is proportional to the rate of excitation transfer between molecules: thus, the optical properties of single molecules are fully retained in a crystal with small γ, whereas for high values of γ the crystalline supramolecular structure should play an important role. We have calculated the exciton resonance interaction (Silbey *et al.* 1963) of crystals **A–F**, for each couple of molecules, along all the directions, using the X-ray structural data (Barbarella *et al.* 1993; Antolini *et al.* 1998a):

$$\gamma_{AB} = \int \int d\,r_1\,d\,r_2\,P_A(r_1)\frac{1}{r_1 - r_2}P_B(r_2) \qquad (2)$$

The transition density $P(r_i)$ and the two-electron integrals have been calculated within the intermediate neglect of differential overlap/single configuration interaction (INDO/SCI) scheme (Ridley and Zerner 1973, 1979). The evaluation of the exact resonance interaction overcomes all the limitation of the dipole–dipole approximations (Silbey *et al.* 1963) and, even though the absolute values may be improved (Chandross and Mazumdar 1997), the method well reproduces the relative values of the different compounds. We also want to point out that the molecular distortion and the presence of substituents have only a minor effect on the transition density, that is, the intermolecular resonance interaction depends mainly on the relative distance and position of the molecules in the crystal.

In Table I, the values of γ are reported for the first couples of neighbor molecules along three almost orthogonal directions. In all our crystals along two of these directions, the molecules form closely packed *H*-aggregates ($\gamma > 0$) (Kanemitsu *et al.* 1994) whereas along the third direction, a *J*-aggregate-type interaction ($\gamma < 0$) (Kirstein and Möwald 1995) is present. The resonance interaction is, however, always maximum along the *H*-aggregates indicating a clear predominance of these structures.

We found that γ is very high for compounds **A** and **C**, but sample **A** shows high PL efficiency. This is clearly in contrast with the general argument that the PL efficiency must be strongly reduced in *H*-aggregates, whose lowest exciton state is optically forbidden (Kanemitsu *et al.* 1994; Yassar *et al.* 1995; Cornil *et al.* 1998; di Cesare *et al.* 1999). Furthermore, we note that compound **E**, having different crystalline structure and lower γ compared to compound **A**, has instead a comparable PL efficiency. Hence, in these samples no direct correlation can be found between the resonance interactions and the measured values of PL efficiency. Even if the description of the absorption spectra in terms of excitonic band structure is well established (Craig and Walmsley 1968; Davydov 1971; Pope and Swenberg 1982), the PL process cannot be directly explained by the exciton relaxation into low-energy optically forbidden states.

The role of CTEs, in PL quenching, must be evaluated by considering the interchain transfer integrals (Wu and Conwell 1997). The interchain hopping process is in fact proportional to the transfer integrals, which describe the coupling between Frenkel excitons and CTEs (Hennessy *et al.* 1999). We have calculated the electron and the hole transfer integral by using the INDO/SCI wavefunctions, for each couple of molecules in their crystalline positions (Hennessy *et al.* 1999). The maximum values obtained (see Table I) are quite small, due to the absence of a complete π–π stacking in all these samples. We found that the charge-transfer contribution is strongly suppressed in samples **F** and **D**, but again we do not find any direct correlation with the PL efficiency. Assuming that extrinsic process are not dominant in these crystals due to the comparable and high degree of purity, the variations in the PL efficiency must therefore be caused by intramolecular processes.

A non-radiative channel that deserves special attention is the crossing of the excitations from the singlet to the triplet manifold (ISC), which in turn decays non-radiatively. Actually, ISC is known to be the main cause of PL quenching in oligothiophenes (Grebner *et al.* 1995; Belijonne *et al.* 1996; Benincori *et al.* 1998) compared to other non-radiative processes, such as IC. Using the theory of the radiationless transitions (Bixon and Jortner 1968), the ISC rate from the first singlet excited state S_{1i} to the triplets manifolds T_{nj} (*i* and *j* are vibrational quantum numbers) can be written as:

$$k_{ISC} = \sum_n |C_{1n}^{ISC}|^2 \sum_{ji} p_{1i} |S_{1i,nj}|^2 \delta(E_{S1i} - E_{Tnj}) \tag{3}$$

where C_{1n}^{ISC} contains the spin–orbit coupling matrix element, p_{1i} is the occupancy probability and $S_{1i,nj}$ is the Franck–Condon overlap factor.

For large organic molecules, it can be theoretically shown that a good approximation to Eq. (3) is the phenomenological energy-gap law:

$$k_{ISC} \approx A \exp\left[-\frac{\Delta E}{\hbar\Omega}\right] \tag{4}$$

where $\Delta E = E_{S1} - E_{Tn}$ is the singlet–triplet energy splitting for the most coupled triplet T_n, A is the pre-exponential factor proportional to $|C^{ISC}|^2$ and $\hbar\Omega$ is related to the highest vibrational frequency. According to the principle of detailed balance, if $\Delta E < 0$ a Boltzmann factor

$$\exp\left[-\frac{|\Delta E|}{k_B T}\right] \tag{5}$$

must be inserted into Eq. (4) (Burin and Ratner 1998).

In Fig. 17, we report the energy position, calculated with the INDO/SCI (Ridley and Zerner 1973, 1979) method, of the first singlet excited S_1 state and of the triplet state T_4 closest in energy to the singlet state. The state T_4 has been already identified (Grebner *et al.* 1995; Belijonne *et al.* 1996; Benincori *et al.* 1998) as the main responsible for the ISC in oligothiophenes.

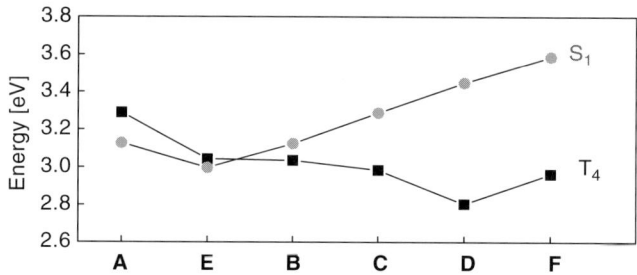

Fig. 17. INDO/SCI energies of the first singlet excited state (S_1) and the fourth triplet state (T_4) of compounds **A**–**F**, calculated by using X-ray determined crystal structure. Other triplets are well separated in energy.

It is worth noting that the energy splitting $\Delta E = S_1 - T_4$ increases with increasing the molecular distortion, and that for compounds **A** and **E** it takes a negative value which implies a thermal activation of the ISC process. The simple exponential law in Eq. (4) cannot, however, account for significant differences in the ISC rate because $\hbar\Omega$ can be quite large (Burin and Ratner 1998). An important contribution is instead given by the pre-exponential factor A in Eq. (4). In fact the calculated spin–orbit matrix elements (Ellis *et al.* 1971) are found to be small in the more planar compounds, whereas they increase by several orders of magnitude in the distorted ones. Thus, the ISC rate is higher in the distorted compound mainly due to the higher spin–orbit coupling. Even if a more complete treatment should include vibrational coupling (Bixon and Jortner 1968) to all triplets, the differences of the experimental PL efficiency between samples **A** and **E** and the others can be well explained by the different ISC rate.

These findings are consistent with the recent studies on bridged quater-thiophenes in solution (Benincori *et al.* 1998), where higher quantum yields have been obtained for the more planarized compounds. It is interesting to note that the PL efficiency at 20 K can be estimated by normalizing the room temperature values to the quenching factor of the PL and absorption intensity with temperature. The efficiencies estimated for compounds **A** and **E** are 22 and 17%, respectively, consistent with the idea that in the more planar compound **A** the spin–orbit coupling is smaller and $|\Delta E|$ is larger. Further support to this interpretation comes from the temperature dependence of the PL decay times. Assuming that the radiative decay time is temperature independent, these measurements provide directly the temperature dependence of the non-radiative decay rate k_{NR}, which is generally expressed as (Rossi *et al.* 1993)

$$k_{NR}(T) = k_1 + k_2(T) = k_1 + A \exp\left[-\frac{E_a}{k_B T}\right] \qquad (6)$$

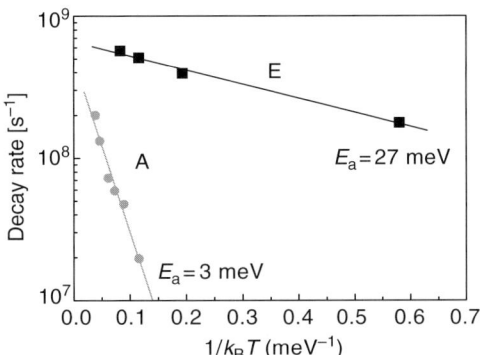

FIG. 18.   Temperature dependence of PL decay rate. $k_2(T)$ is null for compounds **B, C, D,** and **F**.

where k_1 accounts for the temperature-independent process decay rate (Belijone *et al.* 1996), and $k_2(T)$ accounts for the temperature-dependent decay processes, characterized by the activation energy E_a. The main result is that the measured activation energies E_a are in agreement with the calculated singlet–triplet energy splitting, showing that the measured $k_2(T)$ directly reflects the temperature activated ISC rate. In fact, samples **B, C, D,** and **F** do not show any measurable temperature dependence of the decay times, whereas samples **A** and **E**, with $\Delta E < 0$ show the temperature dependence reported in Fig. 18. From these data, we estimate activation energies of 27 and 3 meV for compounds **A** and **E**, respectively. We note also that the extrapolated values of k_2 is 6×10^8. This value is comparable to that measured in solution (Rossi *et al.* 1993), confirming the ISC origin of the non-radiative decay. For sample **E** the measured E_a is very low, in agreement with the small ΔE value resulting from the calculations. We should mention that though the qualitative agreement between the data is very good, exact quantitative agreement cannot be obtained due to the limited accuracy of the INDO/SCI scheme and to the neglect of the effects of the surface potential.

In summary, for compounds **A** and **E** the main non-radiative channel is the activated ISC process, whereas for the other compounds the non-radiative decay is a balance of non-activated ISC and exciton–phonon scattering process, which increases with increasing the molecular distortion induced by the intermolecular interactions (Craig and Walmsley 1968; Davydov 1971; Pope and Swenberg 1982). Thus, the luminescence of these quaterthiophene crystals is more correlated to the single-molecule electronic levels than to crystal packing. Obviously, the general validity of the results might be influenced by the possible occurrence of defects and impurity which might contribute in a substantial way to the non-radiative processes, depending on sample quality.

2. Oligothiophene-*S*,*S*-dioxides

a. Introduction

We have already mentioned that due to their typical low PL efficiency (Wu and Conwell 1997), unsubstituted and alkyl-substituted oligothiophenes have not be fully exploited as active layers in opto-electronic devices such as LEDs. Recently, a third generation of high PL efficiency thiophene-based oligomers, characterized by the functionalization of the thienyl sulfur atom with two oxygens has overcome this drawback (Barbarella *et al.* 1999a, 2000; Gigli *et al.* 1999, 2001). The Compound specifications are quoted in Table II. Through the chapter, these will be labeled by numbers to differentiate from the first-generation thiophenes discussed so far (labeled by letters).

The oligothiophene-dioxides have PL efficiencies increased up to two orders of magnitude with respect to the typical values of unsubstituted oligothiophenes. Such a strong increase of PL efficiency is correlated to the effects to the sulfur functionalization with oxygen atoms both on the intra- and intermolecular properties. In the following, the effects of oxygenation on molecular structure, crystal packing, and optical properties will be analyzed.

b. Structural Properties

In conventional thiophene oligomers, we have seen that the typical HB structure taken by the molecules in the solid state can lead to a quenching of photoluminescence (Belijonne *et al.* 1996; Brédas *et al.* 1999; Garnier *et al.* 1999). In oligothiophene-*S*,*S*-dioxides the presence of the strongly dipolar sulfonyl group introduces novel electrostatic intermolecular interactions, which counteract the tendency of the molecules to organize in parallel layers. The impact of these interactions on the supramolecular arrangements of the molecules, has been evaluated by studying the structural properties of oligothiophene-*S*,*S*-dioxides with different chain lengths, respectively a trimer (**8**), a pentamer (**13**), and a heptamer (**17**) (Antolini *et al.* 2000).

The X-ray structures of compounds **8**, **13**, and **17** are shown in Fig. 19. All the compounds show conformational disorder with two different orientations of the outer thienyl rings (*anti* or *syn* with respect to the S atoms of adjacent rings) and different orientations of the two hexyl chains.

The conformation of trimer **8** is characterized by S–C–C–S inter-ring torsional angles of −168.4 and 28.5° or, due to disorder, −151.6°. The slightly preferred conformation (51.6%) is the *anti–anti* one. The deviations from coplanarity between rings become considerably smaller in pentamer **13**, whose inter-ring torsion angles range from −165.3 to −171.9°. All ring junctions are of the *anti–anti* type in the prevailing orientation (66.2%) for the disordered outer ring. In heptamer **17**, the increase in the number of

TABLE II

Molecular Structure, Maximum Absorption Wavelengths for Chloroform Solution (λ_{max}, nm) and Crystalline Powder (λ_A, nm), and Solid State PL Wavelength (λ_{PL}, nm), and PL Efficiencies (η, %) of Selected Oligothiophene-S,S-dioxides and Polymers

Compound	R	R$_1$	λ_{max} (nm)	λ_A (nm)	λ_{PL} (nm)	η (%)
1			314	307	405	29
2	Methyl		344	339	470	42
3	n-Hexyl		336	352	485	63
4	neo-Pentyl		348	344	500	7
5	H		383		500	24
6	Phenyl		375	385	510	42
6b	neo-Pentyl				491	64
7	n-Hexyl		356	397	500	70
8	n-Hexyl		412	400	525	45
9	Methyl		402		535	6
10	Phenyl		437	427	565	8

269

TABLE II
(Continued)

Compound	R	R₁	λ_{max} (nm)	λ_A (nm)	λ_{PL} (nm)	η (%)
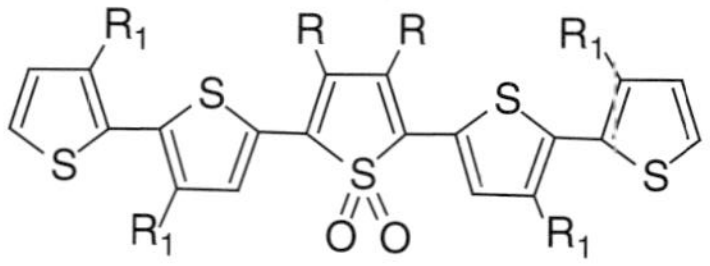 **11**	n-Hexyl	Methyl	454	461	600	37
12	neo-Pentyl	neo-Pentyl	442		605	6
13	n-Hexyl	H	469	519	625	12
14	Methyl	Methyl	445	429	632	12
15	Phenyl	Methyl	482		660	4
16	n-Hexyl	Cyclohexyl		448	608	22
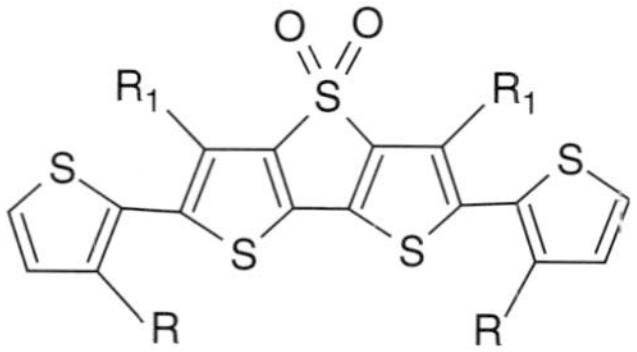 **16b**	Methyl	Methyl	400	410	564	48
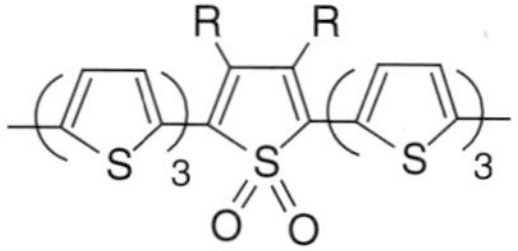 **17**	n-Hexyl		495		725	
P5 Poly(13)			585		801	
P3 Poly(9)					910	

270

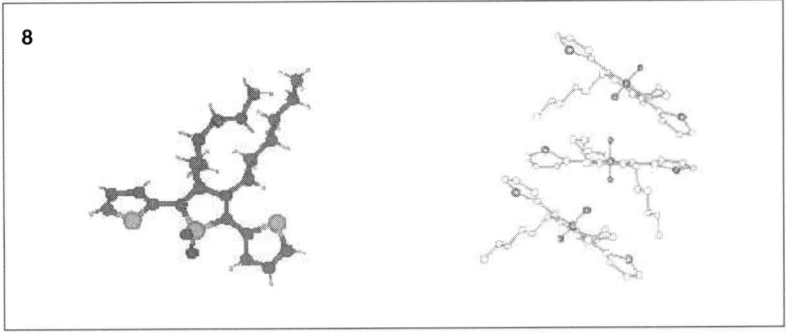

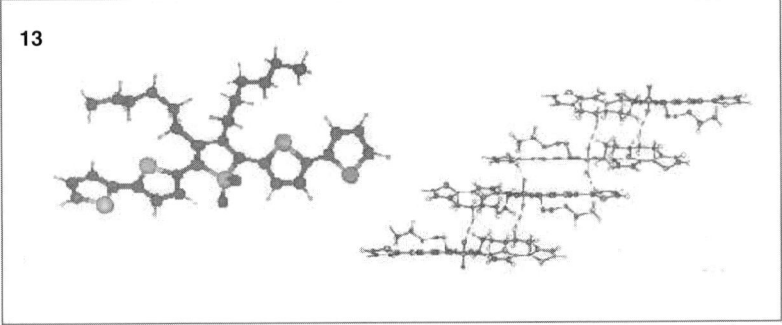

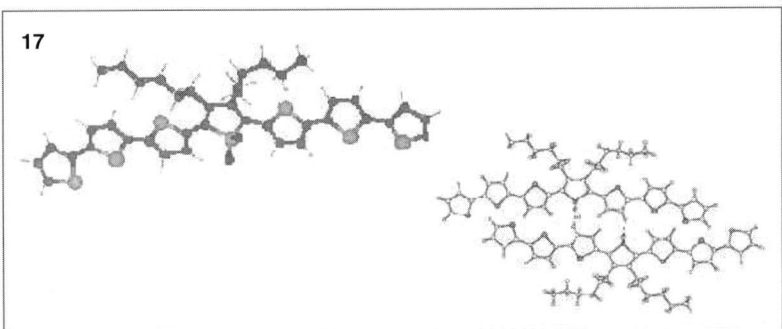

FIG. 19.   Molecular conformation and supramolecular packing of compounds **8**, **13**, and **17**.

thienyl units leads to a much more planar structure in which the dihedral angles between ring planes range between 1.4 and 7.3°.

The crystal packing of **8** (Fig. 19) leads to marked conformational distortion from coplanarity and to the lack of short intermolecular contacts. In pentamer **13**, in spite of the strong out-of-plane displacement of the oxygen atoms of the sulfonyl group and of the hexyl chains, the molecular packing (Fig. 19) is close to the HB packing motif typical of planar or quasi-planar oligothienyls. Nonetheless, the dihedral angle between the mean planes through aromatic moieties related by glide plane ("HB" angle) is 38.51°, which is far from the 55–70° range normally observed in oligothiophenes (Porzio *et al.* 1993; Antolini *et al.* 1998b). Unlike compound **8**, the molecular packing of compound **13** is characterized by a number of very short van der Waals contacts, quite rarely observed in oligothiophenes (Barbarella *et al.* 1993, 1994, 1999b; Antolini *et al.* 1998c).

In Fig. 19, we also show heptamer **17** whose longer and more planar molecules stack roughly perpendicular to the *b* cell axis (Fig. 19), forming parallel layers with crystallographic inversion center. The long interlayer distance and, more important, the scarce molecular overlap do not allow π–π interactions among the stacked thienyl rings. Nevertheless, a lot of short S · · · O, S · · · C, C · · · O, and C · · · C intermolecular contacts, stronger than those observed in compound **13**, characterize the crystal packing. This holds for the S · · · S interactions as well, three of them being shorter than 3.8 Å.

The most relevant structural features of compounds **8**, **13**, and **17** are summarized in Table III, where we quote the torsion angles (φ), the packing coefficients (PC) and the packing potential energies (ppe) estimated by the atom–atom potential energy method (Gavezzotti 1983; Persin and Kitaigorodsky 1987). Moreover, in Table III we quote the short intra- and intermolecular contacts of the type C–H · · · O, C–H · · · S and S · · · S observed in compounds **8**, **13**, and **17**. These data show that the hydrogen bonding interactions between donor groups (oxygen atoms, thienyl sulfurs, and thienyl π-systems) and acceptor neighboring C–H groups as well as intramolecular S · · · S interactions, play a dominant role in compounds **8**, **13**, and **17**. Apparently, the balance between all these interactions and the

TABLE III

Torsion Angles (φ), Packing Coefficient (PC), Packing Potential Energies (ppe) and Short Intermolecular Contacts of Compounds **8**, **11**, and **17**

	8	13	17
φ_1 (°)			−7.8
φ_2 (°)		−165.3	178.0
φ_3 (°)	−168.4	−171.9	−178.5
φ_4 (°)	28.5	−168.5	−174.9
φ_5 (°)	−151.6	−171.2	178.9
φ_6 (°)			−3.4
PC	0.636	0.676	0.681
ppe (kcal/mol)	−87.6	−124.6	−165
H · · · O (Å)	2.638	2.474	2.318
C–H · · · O (°)	164.2	152.01/166.75/152.55	141.475/161.75
S · · · S (Å)			3.253 (intra)

dipole–dipole intermolecular interactions due to the presence of the SO_2 groups, defines the conformation and packing of the solid state ensemble. The introduction of the oxygen atoms breaks the $HB/\pi–\pi$ stacking duality generally found in the solid state for unsubstituted oligothiophenes (Van Bolhuis *et al.* 1989; Chaloner *et al.* 1994; Pelettier and Brisse 1994; Horowitz *et al.* 1995; Siegrist *et al.* 1995, 1998; Fichou *et al.* 1996; Antolini *et al.* 1998a).

Figure 19 shows that when going from the trimer **8** (one SO_2 for two thienyl rings) to the heptamer **17** (one SO_2 for six thienyl rings), the strength the weak hydrogen bonding and sulfur–sulfur interactions overcomes the dipolar interactions, leading to more planar conformations, greater packing energies, and alignment of molecular long axes. These structural character- istics are accompanied by a strong decrease of the PL efficiency with decreasing the electrostatic interactions.

Table II shows that the PL efficiency in the solid state decreases with increasing the oligomer size and varies from 45% in trimer **8** to 12% in pentamer **13**, and 2% in heptamer **17**. These data indicate that there is a correlation between packing arrangements and fluorescence efficiency. The quite remarkable 45% PL efficiency of trimer **8** is presumably due to the lack of alignment of molecular long axes in the crystalline packing, whereas the low 2% efficiency of heptamer **17**, is associated to the parallel arrangement of molecular long axes. The 12% efficiency of pentamer **13** is related to an intermediate type of packing. Therefore, the more the packing arrangement resembles that of conventional oligomers, the lower is the fluorescence yield.

We can thus summarize this section, by saying that the self-assembly modality in the solid state of thiophene oligomers modified by the intro- duction of one thienyl-*S,S*-dioxide unit, depends on the relative number of conventional and modified thienyl units present in the molecule. When the oligomer is short, the packing is dominated by the dipolar intermolecular interactions due to the presence of the sulfonyl groups, and a high fluores- cence efficiency is observed. On the contrary, when the packing is dominated by the intra- and intermolecular interactions due to the presence of numerous unmodified thienyl rings, the PL efficiency drops to a few percent. The data confirm that the introduction of a thienyl *S,S*-dioxide unit into the backbone of a thiophene oligomer is a useful strategy to orient the packing forces and to control the fluorescence efficiency in the solid state.

c.   *Theoretical Modeling of Oligothiophenes*

A theoretical description of the optical properties of thiophenes needs an accurate treatment of the excited states. Unfortunately, for large molecules like oligothiophenes, accurate *ab initio* methods such as complete active space perturbation theory (Roos *et al.* 1995), require too much computational effort and cannot be applied. Therefore, most theoretical investigations of the optical properties of oligothiophenes have been carried out with semi-empirical methods (Belijonne *et al.* 1993, 1996; Colditz *et al.* 1995; Cornil *et al.* 1995). Recently, methods based on time-dependent density functional theory (TD-DFT) were introduced to treat the optical properties by a first-principles approach which requires only moderate computational effort and thus can be applied to large systems (Gross *et al.* 1996).

The effects of sulfur functionalization with oxygens on the excitation energies of prototype terthiophene-*S*,*S*-dioxide (**8**) and unsubstituted terthiophene have been analysed in detail by Della Sala *et al.* (2001). As opposed to semi-empirical calculations (Belijonne *et al.* 1996; Belletête *et al.* 1996) (usually optimized for singlet–singlet transitions and not for singlet–triplet transitions), the T_2 excitation energy in terthiophene is correctly predicted by TD-DFT to occur at higher energy than the S_1 transition, consistently with recent experimental results (Rentsch *et al.* 1999). In terthiophene-*S*,*S*-dioxide, the energy difference between the T_2 and S_1 transitions is found to be about 0.2 eV larger than in terthiophene. Thus, the ISC rate and the non radiative de-excitation processes are expected to be less efficient in the oxygenated molecule, in agreement with the increased quantum efficiency of these compounds (Assadi *et al.* 1988; Garnier *et al.* 1990).

Another important result is the account of the strong red-shift of the transitions occurring in terthiophene-*S*,*S*-dioxide with respect to the unsubstituted terthiophene. For the S_1 transition the TD-ALDA calculations yield a red-shift of 0.41 eV, which agrees reasonably well with the difference of 0.58 eV obtained by absorption experiments in solution. Semi-empirical calculations predict a similar red-shift for the S_1 transition energy in compound **8** (Barbarella *et al.* 1999c). Such a red-shift is attributed to the effects of the central sulfur functionalization on the HOMO and LUMO orbitals. The HOMO wavefunction is slightly affected by the oxidation, which induces only a small decrease of the HOMO energy of 0.42 eV related to the increased nuclear attraction due to the two oxygens. Instead, significant differences occur upon oxidation in the LUMO wavefunction, which is much more localized on the central ring and it is characterized by the vanishing of the sulfur p_z contribution. The strong sulfur p_z suppression induced by the oxidation allows the formation of a new bond which comprises the carbon atoms C_1 and C_4 (Fig. 1). The formation of this new bonding interaction determines the lowering of the LUMO by 0.9 eV, resulting in a overall decrease of the HOMO–LUMO gap of 0.48 eV, with respect the unsubstituted terthiophene (Della Sala *et al.* 2001).

An important point to be taken into account in the discussion of the optical properties of oligothiophenes is the very low rotational barrier for inter-ring torsion (Barbarella *et al.* 1998). This means that the molecular configuration, and consequently the optical properties, can be strongly affected by environmental effects. The increase of the transition energies due to molecular distortion is well known for unsubstituted oligothiophenes

(Belletête *et al.* 1996; Gigli *et al.* 1998). The behavior of the excitation energies of the terthiophene-*S,S*-dioxide (**8** in Fig. 19) induced by the change of the interring torsion angle θ, is found to be completely different. The S_1 and T_2 transitions are almost independent of θ and the T_1 transition increases only slightly, much less than the T_1 transition in the unsubstituted terthiophene. This means that the ISC in compound **8** should be much less affected by distortion configuration than in unsubstituted terthiophene.

The different sensitivity of the excitation energies on the configuration in unsubstituted terthiophene and in sample **8** is primarily due to the change of the LUMO upon change of θ. In compound **8**, the LUMO energy is not affected by the change of θ, whereas it increases in unsubstituted terthiophene. This is a consequence of the enhanced localization of the orbital on the central ring mentioned above. The HOMO energy instead behaves similarly in unsubstituted terthiophene and in compound **8**, showing a slight decrease with θ. This is due to the very similar HOMO wavefunctions. As a result, the HOMO–LUMO gap in compound **8** increases much less with θ than in the unsubstituted terthiophene. The fact that in **8** the S_1 transition energy does not change at all, that is, it does not reflect the slight increase of the HOMO–LUMO gap, is attributed to the additional change of energy caused by the coupling corrections, which includes the response effects of the Coulomb and exchange-correlation potential (Della Sala *et al.* 2001).

d. *Optical Properties*

Similarly to unsubstituted oligothiophenes, photoluminescence emission of the thiophene-*S,S*-dioxide molecules can be finely tuned across the entire visible range with a great variety of tones. Given the richness of substituents that can be grafted to the oxygenated unit and to the different positions of the aromatic rings, a very fine color tuning may be achieved with these materials. Normalized PL spectra of a few selected oligothiophene-*S,S*-dioxides, spanning from blue to near-IR, are reported in Fig. 20 (Barbarella *et al.* 1999a, 2000; Gigli *et al.* 1999, 2001).

The PL and absorption wavelengths in solution as well as in the solid state are strongly correlated to the degree of π-electron delocalization between the inner oxygenated unit and the α-linked aromatic rings. This delocalization depends on the chain length, on the molecular distortion and on the nature of the substituent groups. As the number of molecular units increases and/or the degree of molecular distortion decreases, a major π-electron delocalization is induced. This determines a shift of PL emission towards lower energies. In this context the effect of the oxygenated moiety is to determine a further red-shift of the optical spectra with respect to the values of the corresponding unsubstituted oligothiophenes. Such a property allows us to obtain PL emission in the near-IR region, a spectral range hardly achieved by

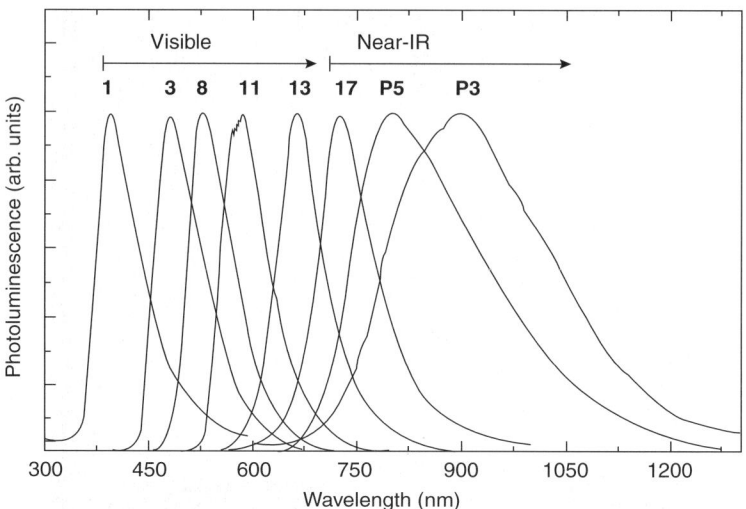

Fig. 20.   PL spectra of selected oligothiophene-*S,S*-dioxides. Compounds **P3** and **P5** have been obtained by polymerization of **8** and **13**, respectively. Full color tenability in the visible and near-IR is obtained.

most organic materials. PL emission up to more than 900 nm have been obtained by polymers having as repetitive unit a thiophene-*S,S*-dioxide. On the contrary, for the same reasons, emission in the blue region is difficult to obtain. The attempt to obtain blue PL by causing very large backbone distortions through the introduction of bulky *t*-butyl groups or through the distribution of the degree of distortion over the entire molecule by grafting head-to-head *neo*-pentyl groups, failed (Barbarella *et al.* 1999a, 2000; Gigli *et al.* 1999, 2001). Despite the many bulky substituents, π-electron conjugation is still very efficient and no blue emission was obtained. The only materials emitting in the blue region are the short monomer **1**, and the phenyl–thienyl compounds **2** and **3** where the red-shift of the emission energy induced by the oxygenated moiety is compensated by the blue-shift due to the π-electron delocalization introduced by the phenyl groups.

3.   Device Applications

a.   Towards the White Emission

A finer color tunability across the entire visible range can be obtained by making blends of different compounds. Blends of oligothiophene-*S,S*-dioxides can nicely extend the colors of light emission (see, e.g., the magenta colors in Fig. 21) and even give rise to white photoluminescence (Anni *et al.* 2000).

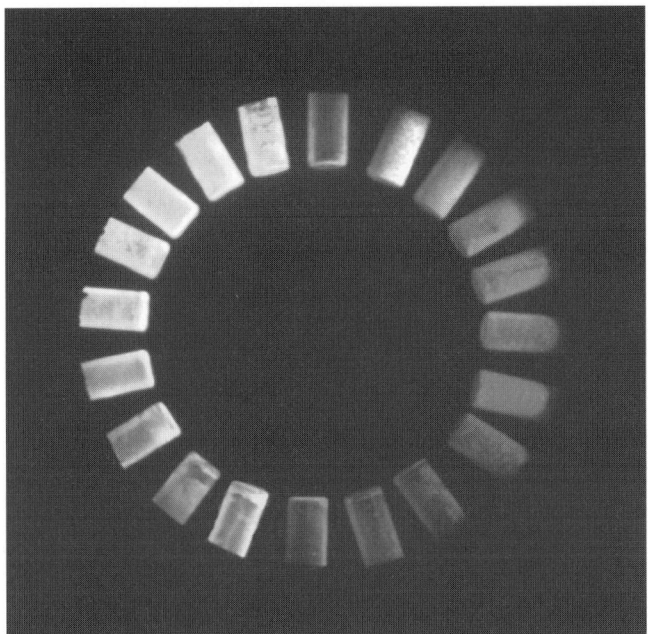

FIG. 21. Photograph of UV-illuminated cast films of selected pure oligothiophene-S,S-dioxides and blends of different compounds.

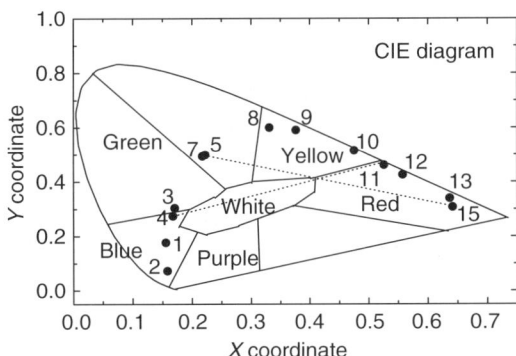

FIG. 22. Spectral distribution of the oligothiophene-S,S-dioxides (dots labeled by numbers) according to the CIE standard.

Figure 21 shows the PL emission of cast films of several pure and binary blends of oligothiophene-S,S-dioxides under UV excitation at 363 nm. The color coordinates of the constituting molecules according to the Commission International de l'Enclairage (CIE) standards are displayed in Fig. 22, showing the complete range of colors covered by these modified oligothiophenes. In particular, binary blends emitting in the white region can be obtained.

The color tuning allowed by the blending of the different compounds is strictly connected to the energy separation between the HOMO–LUMO energy gap of the constituting molecules. For instance, the emission of the purple blend [Fig. 23(a)] is given by the bare superposition of the spectra of the constituting molecules (compound **1** emitting in the blue, and compound **11** emitting in the red) which are separated by $\Delta E = 0.745$ eV. Conversely, the red blend [Fig. 23(b)] constituted by two compounds separated by $\Delta E = 0.542$ eV (compound **15** emitting in the red and compound **7** emitting in the green) shows a strong transfer of oscillator strength from the high-energy

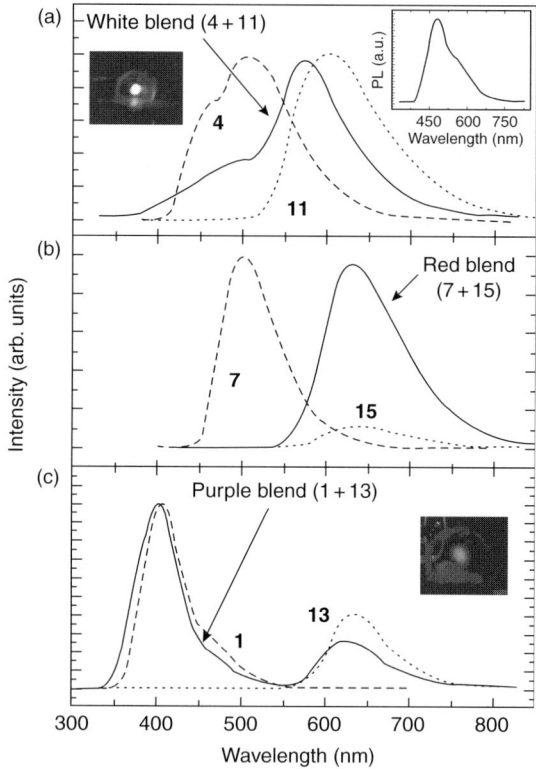

FIG. 23. (a) Emission spectrum of the white blend and of the constituent molecules. The dashed line and the dotted line represent the emission spectra of the constituting molecules emitting at high energy (HE) and at low energy (LE), respectively. The white color obtained in this spectrum is predominantly reddish. Right inset: same white blend with predominant greenish tone, obtained by using a 7:1 ratio of compounds **4** and **11**. Left inset: photograph of the emission spot under CW excitation. (b) Emission spectrum of the red blend and of the constituent molecules. (c) Emission spectrum of the purple blend at room temperature. Inset: photograph of the emission spot under CW excitation.

component to the low energy one, which is typical of the Förster transfer (Gupta *et al.* 1999). Another important parameter which has strong influence on the final emission spectrum of the blend is the relative PL efficiency of the constituting molecules. The different PL efficiency of the molecules must be balanced by varying proportionally the relative amounts of the two materials. This is exemplified in the inset of Fig. 23(c), where the white color of the blend is tuned by varying the relative amount of the constituent compounds, from a reddish white to a greenish white along the dashed line connecting compounds **4** and **11** in Fig. 22.

In order to have a deeper understanding of the Förster transfer, in oligothiophenes, in Fig. 24 we plot the ratio R_η between the PL efficiency of the high-energy (HE) and low-energy (LE) compounds constituting the blend, and the ratio R_I between the intensity of the high- and low-energy bands of the blend emission spectrum (obtained by the Gaussian deconvolution of the experimental spectra) vs ΔE. It appears quite clearly that the blend emission is given by the bare superposition of the emission spectra of the constituting molecules ($R_\eta \cong R_I$) only when the energy separation ΔE is rather large (above 0.56 eV). In contrast, the low-energy component is strongly enhanced by the Förster transfer, almost independently of the efficiency of the high-energy compound, in blends with $\Delta E < 0.56$ eV ($R_\eta \gg R_I$).

The rate of Förster transfer can be evaluated by considering the expression (Kozlov *et al.* 1998)

$$k \propto \frac{1}{\tau}\frac{1}{r^6} \int\limits_{0}^{+\infty} \frac{1}{E^4} \alpha(E)\sigma(E)\,\mathrm{d}E \tag{7}$$

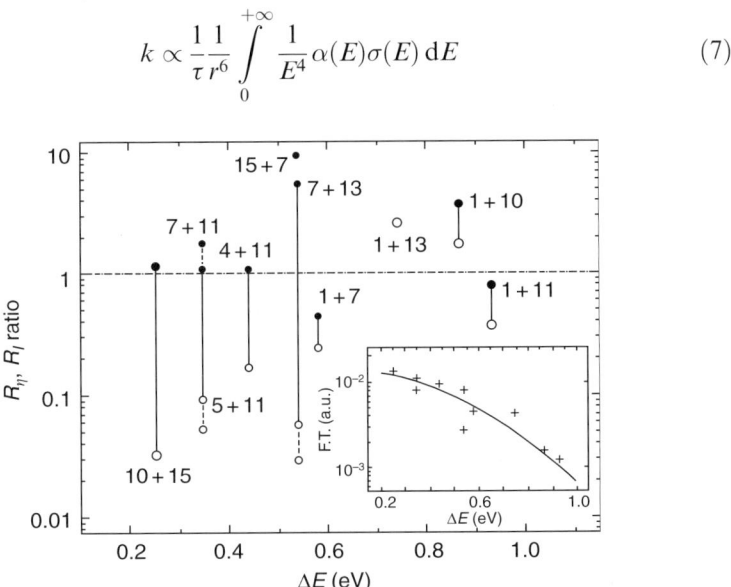

FIG. 24. R_η (full dots) and R_I (empty dots) ratio vs the energy difference between the HOMO–LUMO gap of the molecules constituting the blends (ΔE). Inset: Förster transfer rate vs ΔE (symbols). The solid line is the theoretical curve obtained by our approximated model [Eq. (8)].

where $\alpha(E)$ and $\sigma(E)$ are the normalized absorption and emission spectra of the two species, τ the radiative life-time of the emitting specie, and r the average distance between the different molecules. Assuming that r and τ do not vary strongly in the investigated blends, which were realized by using equal amounts of HE and LE compounds, the rate of Förster transfer can be calculated by a numerical integration of the overlap integral of the PL spectrum of the HE molecule and the absorption spectrum of the LE one. The rate of Förster transfer thus turns out to be:

$$k \propto A \, e^{-[(\Delta E)^2 / 4\sigma^2]} \tag{8}$$

where $\Delta E = E - E_{gap}$, E_{gap} is the HOMO–LUMO gap of the molecule, and A is a term only slowly depending on ΔE. Equation (8) is displayed by the continuous curve in the inset of Fig. 24, and it is found to agree quite well with the experimental data, demonstrating the importance of the Förster transfer in the analyzed blends.

b. Tunable Optical Gain and Lasing

Since the demonstration of optically pumped lasing from a poly(p-phenylene vinylene) (PPV) layer incorporated in a optical microcavity (Tessler et al. 1996; Schön et al. 2000), great attention has been paid to the research of good active materials for solid state organic lasers. This has led to the demonstration of optical gain in different conjugated materials (Frolov et al. 1996, 1997b; Graupner et al. 1996; Tessler et al. 1996; Denton et al. 1997; Gelink et al. 1997; Schweitzer et al. 1998; Stagira et al. 1998a), to the achievement of optically pumped laser built using many different cavity geometry (Berggren et al. 1997a,b, 1998; Frolov et al. 1997a; Kozlov et al. 1997; Brouwer et al. 1998; Stagira et al. 1998b; Spiegelberg et al. 1999), and recently to the first evidence of lasing in an electrically injected device (Schön et al. 2000), based on organic single crystals.

Compared to conventional laser materials, soluble conjugated molecules offer important advantages. One of the main advantages stands in low-cost deposition techniques, such as spin coating or direct printing. Moreover, the usually broad density of states and the strong electron–phonon coupling give rise to broad gain spectra, which can be exploited to obtain spectral tuning of the laser emission (Wegmann et al. 1998). All these factors make soluble conjugated molecules appealing compounds for applications in flexible laser devices. In particular, the class of thienyl-S,S-dioxide oligothiophenes is remarkable for the excellent solubility in common organic solvents and the tunability of the optical gain in a broad part of the visible region. Recent pump–probe measurements by Anni et al. (2001) have shown that it is possible to obtain optical gain from 470 to 660 nm in differently substituted

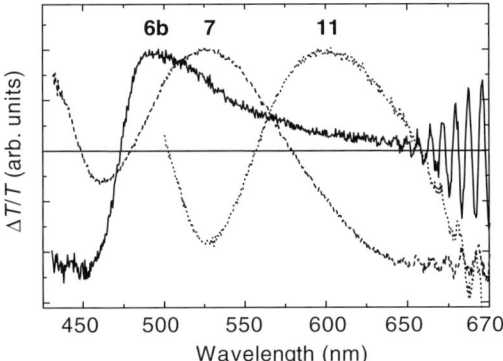

FIG. 25.   Normalized differential transmission spectra for a pump–probe delay of 300 fs. The negative signal is due to photoinduced absorption, while the positive signal is due to absorption bleaching at shorter wavelengths and to stimulated emission at longer wavelengths.

soluble thienyl-S,S-dioxides (Fig. 25). Noticeably, the other studies of optical amplification in oligothiophenes are limited to single crystals of short insoluble molecules, not useful for device applications (Garnier *et al.* 1998).

The maximum value of the gain cross-section σ_g has been measured in sample **7**. Anni *et al.* (2001) estimated for this compound a value of $\sigma_g = 9 \times 10^{-18}\,\mathrm{cm}^2$ at 525 nm using the relation, valid in the small signal limit:

$$\sigma_g \approx \left(\frac{\Delta T}{T}\right)_{\mathrm{M}} \frac{1}{Nd} \tag{9}$$

where $(\Delta T/T)_{\mathrm{M}}$ is the maximum value of the differential transmission in the gain region, N is the density of photo-excited states involved in the transition, and d is the sample thickness.

We recall that in order to be a good material for laser application, an organic molecule has to show not only large gain values, but also large gain bandwidth. This can be defined as (Kretsch *et al.* 1999):

$$F = \frac{f \cdot g_{\mathrm{max}}}{P_{\mathrm{abs}}} \tag{10}$$

where f is the gain resonance FWHM in nanometers, P_{abs} is the absorbed pump energy density in Joules per square centimeter and g_{max} is the maximum gain value in centimeter inverse.

As shown in Fig. 26, F values measured in thienyl-S,S-dioxide oligothiophenes are better than those of phenyl-substituted PPV (PPPV) deduced from (Wegmann *et al.* 1998) and of methyl substituted ladder type poly(*para*-phenylene) (Me-LPPP) deduced from (Schweitzer *et al.* 1998), whereas they are worse than those of LPPP, BuEH-PPV, and polystyrene-doped waveguides (Kretsch *et al.* 1999) (G33 and SP35) (Fig. 26).

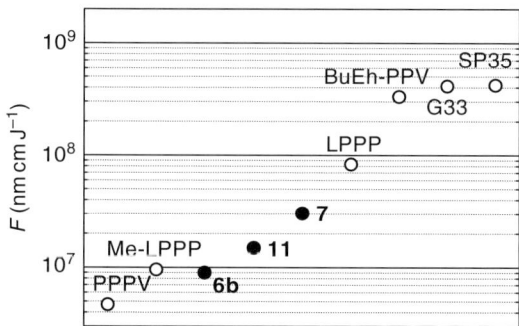

FIG. 26.   Comparison between the gain bandwidth of molecules **6b**, **11**, and **7** (full dots) and that of some polymers showing optical gain (empty dots). The figure of merit of oligothiophene-S,S-dioxides is better than that of some PPV or PPP derivatives, like PPPV or Me-LPPP, while it is about one order of magnitude smaller than that of BuEh-PPV and of doped poly(styrene) waveguides.

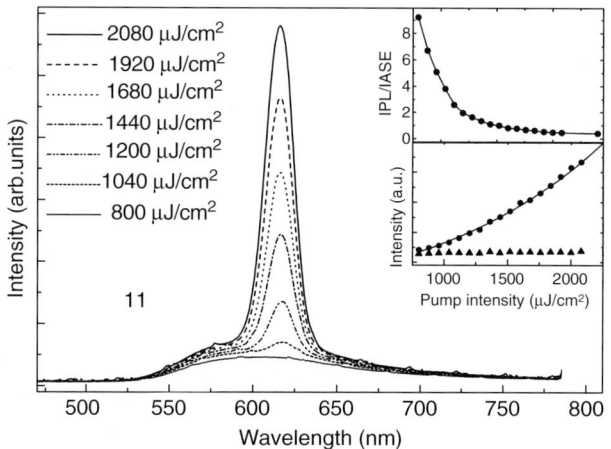

FIG. 27.   PL spectra of compound **11** for different pump energy densities below and above the threshold for line narrowing. The top inset shows the pump intensity dependence of the broad spontaneous emission band relative to the narrow peak intensity. The continuous line is a guide for the eyes. The bottom inset shows the evolution of the broad emission band (full triangles) and of the PL integrated area (full circles) as a function of the pump intensity. The continuous line is a fit of the PL intensity with the analytic expression for ASE.

In Fig. 27, we show the photo-pumped emission spectra of a 700 nm thick film of compound **11**. A spectrally narrow peak appears for optical excitation powers above 960 μJ/cm^2 and the total intensity grows almost exponentially with the pump intensity without a clear threshold for the line narrowing. Concomitantly, suppression of the broad spontaneous emission relative to the narrow band and decay time shortening of the gain are observed, indicating the occurrence of optical pumped amplified stimulated emission (ASE).

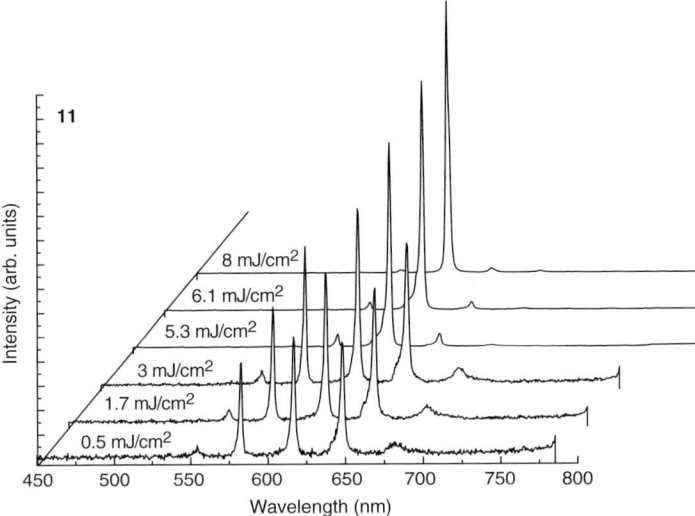

FIG. 28. PL spectra of compound **11** for various pump pulse energies of the microcavity laser. The progressive narrowing of the PL spectrum in a single mode laser emission is evident.

For a film of thickness 50 nm, no line narrowing was observed, pointing out the importance of waveguiding in the plane of the film. Hence, the line narrowing has been attributed to ASE assisted by waveguiding.

Lasing was obtained from thienyl-S,S-dioxide quinquethiophene by realizing a cavity with a spin-coated 600 nm thick layer on a planar distributed Bragg reflector (DBR), used as the bottom mirror, and a spherical dielectric output coupler. In this case, with increasing the pump intensity the PL spectrum evolved from the broad PL spectrum of the active material, filtered by the cavity modes, into a single mode emission (FWHM < 1 nm) (Fig. 28). A clear threshold at an excitation density of about 4 mJ/cm^2 was observed, followed by a linear increase of the emitted intensity (Fig. 29).

c. Electroluminescent Devices

The use of organic materials in opto-electronic devices like LEDs, is motivated by the possibility to realize low-cost, large-area devices. Organic EL devices are characterized by a semiconductor layer sandwiched between two electrodes, a high work function anode (usually gold or indium tin oxide) for injection of positive charges, and a low work function cathode (usually aluminum, calcium, or magnesium) for injection of negative electrons [Fig. 30(a)]. Once the charges are injected, electrons and holes form singlet or triplet excitons, depending on the relative alignment of the spins. EL is due to the radiative decay of singlet excitons from the first excited state (S_1) into the

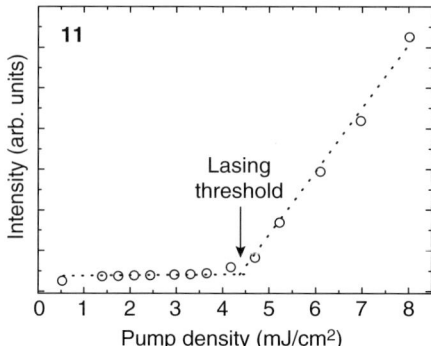

FIG. 29.   Input–output characteristic of the microcavity laser based on compound **11**. The threshold for lasing and the linear dependence of the emission intensity on the pump intensity above threshold are clearly visible.

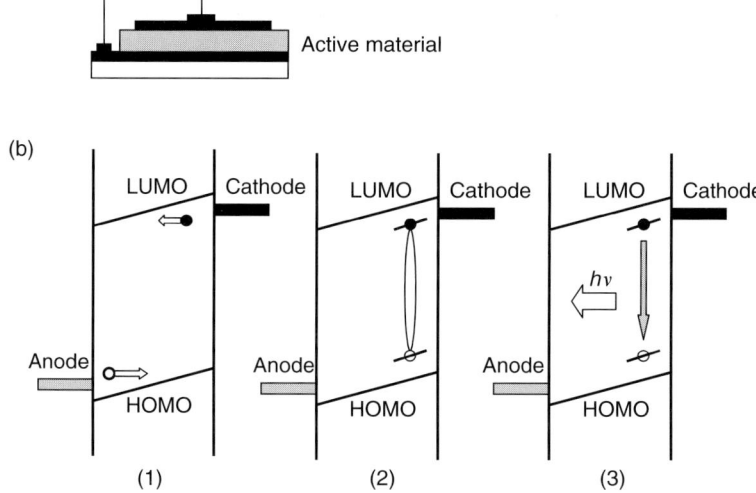

FIG. 30.   (a) Scheme of a single layer LED. (b) Schematic representation of the charge carrier injection (1), exciton formation (2), and radiative recombination in organic LEDs.

ground state (S_0) [Fig. 30(b)]. However, as we have seen in the Section II.2.a, several non-radiative processes can reduce the EL efficiency. The internal efficiency of LEDs η_{int} can be expressed as the product of three term: $\eta_{int} = \chi r_{st} q$, where χ is the ratio between the number of exciton formation and the electrons flowing in the external circuit, r_{st} is the fraction of singlet excitons (this value is generally taken to be 0.25 for statistical reasons) and q is the efficiency of singlet exciton radiative decay [related to the quantum efficiency of the active material and to the optical design of the device

(Greenham *et al.* 1994)]. In order to have good device performances, the choice of the active materials must take in account all these elements. In this context, standard thiophene derivatives are remarkable for their chemical stability and for the wide color tunability but they have relatively poor quantum efficiency, thus preventing their use as active materials of LEDs. However, the thienyl-*S,S*-dioxides discussed in Section III.2 has overcome this problem, offering the possibility to combine the well-known stability properties of thiophene derivatives with high PL efficiency. Furthermore, the high electron affinities induced by the oxygenated moiety strongly increases the electron injection capability from low workfunction metals into the active material (Barbarella *et al.* 1999a, 2000; Gigli *et al.* 1999, 2001). These characteristics have been exploited to realize high efficiency multicolor LEDs with a hole transporting material, namely poly(3,4-ethylene dioxythiophene) (PEDOT) doped with poly(styrene sulfonate) (PSS), inserted between the indium tin oxide (ITO) anode and the pure active material (Barbarella *et al.* 1999a, 2000; Gigli *et al.* 1999, 2001). The deposition of the PEDOT-PSS layer is a general route to increase the hole injection from the anode into the active material (to have a better balancing of charge carriers and hence a higher χ), as well as to improve the forming properties of the films. As the cathode, we used calcium followed by an Al cap.

In Fig. 31, we display a few representative room temperature EL spectra of these LEDs. EL emission from green to near-IR region can be obtained by using as active materials differently substituted thienyl-*S,S*-dioxide oligo-thiophenes. For all the LEDs, the turn-on voltages for luminance at $0.01 \, \text{cd/m}^2$ are between 2 and 4 V, strongly reduced as compared to the values

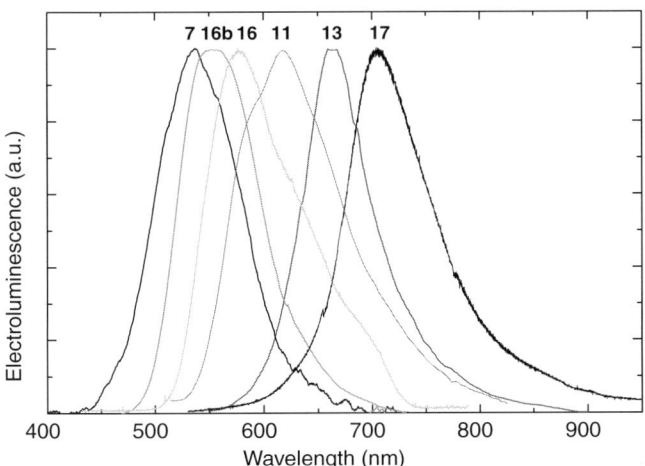

FIG. 31.   EL spectra of selected oligothiophene-*S,S*-dioxides. Color tunability from green to near-IR is obtained.

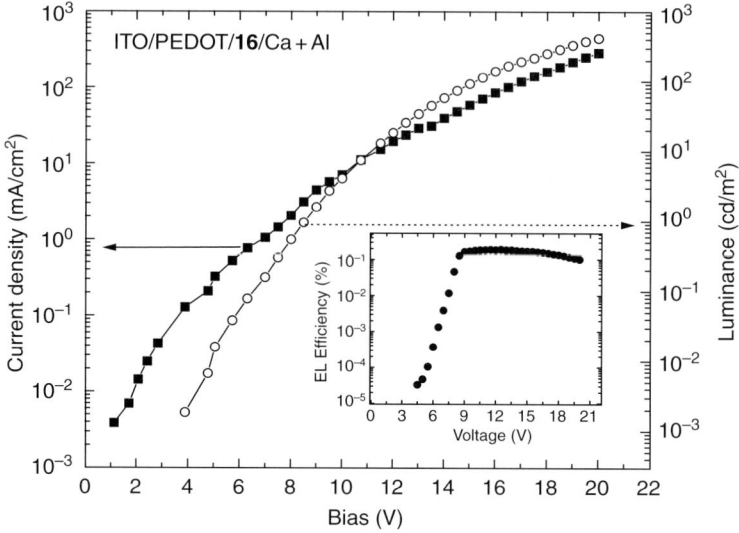

Fig. 32. Current–voltage (*I–V*) and luminance–voltage (*L–V*) characteristics of a device based on compound **16**. Inset: EL efficiency vs voltage of the same device. The device characterization was carried out in air atmosphere.

reported for poly(alkylthiophenes)-based devices (Barta *et al.* 1998). This is correlated to the increase of electron affinity induced by the dioxide functionalization and to the consequent reduction of the electron injection barrier.

The values measured for luminance and EL efficiency are strongly increased as well. In particular, for the device made by compound **16** a luminance of $400 \, \text{cd/m}^2$ at 20 V and an EL efficiency of 0.2% are obtained (Fig. 32), more than two order of magnitude larger than the typical values reported for non-substituted oligothiophenes (Ziegler 1997b).

A possible route to further improve the film forming properties of the oligomer in these LEDs is to blend the active material in polymeric matrices. By choosing a proper polymer with electron and/or hole transporting properties it is also possible to increase the carrier injection in the active material. Following this method, LEDs with a spin-coated blend of compound **11** and a polyfluorene-based material (Kim *et al.* 1999) as active medium were prepared. For the single layer device, the maximum values measured for luminance and EL efficiency were found to be, respectively, $\sim 100 \, \text{cd/m}^2$ at 7 V and $0.03 \, \text{cd/A}$ at $\sim 180 \, \text{mA/cm}^2$. Turn-on voltages of 2.8 V for the luminance at $0.01 \, \text{cd/m}^2$ confirm, also in the case of a blend as active medium, the good charge carrier balancing connected to the high electron affinity of the thienyl-*S,S*-dioxides.

In the double layer devices, the insertion of a PEDOT layer between anode and active material determines a further reduction of the turn-on voltage

$(0.01\,cd/m^2)$ down to $\sim 1.9\,V$ and an increase of the maximum value for the luminance and EL efficiency to $\sim 200\,cd/m^2$ at 7 V and at $\sim 1.2\,cd/A$ at 3.5 V $(44\,cd/m^2$ and $3.6\,mA/cm^2)$, respectively. The large efficiency improvement is consistent with the effect of PEDOT in cells employing other semi-conductors, and is due to the increased distance of the recombination region from the ITO interface (Carter *et al.* 1997), and, therefore, to the reduction of quenching effects induced by physical and chemical interactions between the EL species and the electrode.

IV. Conclusions

In this review, we have illustrated the relationships between structural and optical properties in oligothiophenes. In particular, we have shown that, thanks to a proper functionalization of the thienyl units with oxygen atoms, it is possible to increase the photoluminescence efficiency up to values comparable and even better than those of the best organic materials commonly used as active compounds in light emitting devices. Such a characteristic, together with the high chemical stability and the wide color tunability, makes oligothiophene-*S,S*-dioxides excellent candidates not only for applications to electronic devices, but also to optoelectronic devices, such as multi-color LEDs and lasers. Recent achievement of high-performances thiophene-based devices have also been presented.

Acknowledgments

The authors wish to acknowledge the collaboration of Dr G. Sotgiu, L. Favaretto, M. Zambianchi, Dr F. Della Sala, Dr F. Cacialli, Prof. G. Lanzani, and Prof. O. Inganäs.

References

Anni, M., G. Gigli, V. Paladini, R. Cingolani, G. Barbarella, L. Favaretto, G. Sotgiu, and M. Zambianchi, *Appl. Phys. Lett.* **77**, 2458 (2000).
Anni, M., G. Gigli, M. Zavelani-Rossi, C. Gadermaier, G. lanzani, G. Barbarella, L. Favaretto, and R. Cingolani, *Appl. Phys. Lett.* **78**, 2679 (2001).
Antolini, L., G. Horowitz, F. Kouki, and F. Garnier, *Adv. Mater.* **10**, 382 (1998a).
Antolini, L., G. Horowitz, F. Kouki, and F. Garnier, *Adv. Mater.* **10**, 385 (1998b).
Antolini, L., U. Folli, F. Goldoni, A. Mucci, and L. Schenetti, *Acta Polym.* **49**, 248 (1998c).
Antolini, L., E. Tedesco, G. Barbarella, L. Favaretto, G. Sotgiu, M. Zambianchi, D. Casarini, G. Gigli, and R. Cingolani, *J. Am. Chem. Soc.* **122**, 9006 (2000).
Assadi, A., S. Svensson, M. Wilader, and O. Inganäs, *Appl. Phys. Lett.* **53**, 195 (1988).

Barbarella, G., M. Zambianchi, A. Bongini, and L. Antolini, *Adv. Mater.* **5**, 834 (1993).
Barbarella, G., M. Zambianchi, A. Bongini, and L. Antolini, *Adv. Mater.* **6**, 561 (1994).
Barbarella, G., M. Zambianchi, R. DiToro, M. Colonna, L. Antolini, and A. Bongini, *Adv. Mater.* **8**, 327 (1996).
Barbarella, G., M. Zambianchi, Montserrat del Fresno I Marimom, L. Antolini, and A. Bongini, *Adv. Mater.* **9**, 484 (1997).
Barbarella, G., O. Pudova, C. Arbizzani, M. Mastragostino, and A. Bongini, *J. Org. Chem.* **63**, 5479 (1998).
Barbarella, G., L. Favaretto, G. Sotgiu, M. Zambianchi, V. Fattori, M. Cocchi, F. Cacialli, G. Gigli, and R. Cingolani, *Adv. Mater.* **11**, 1375 (1999a).
Barbarella, G., M. Zambianchi, L. Antolini, P. Ostoja, P. Maccagnani, A. Bongini, E. A. Marseglia, E. Tedesco, G. Gigli, and R. Cingolani, *J. Am. Chem. Soc.* **121**, 8920 (1999b).
Barbarella, G., L. Favaretto, G. Sotgiu, M. Zambianchi, C. Arbizzani, A. Bongini, and M. Mastragostino, *Chem. Mater.* **11**, 2533 (1999c).
Barbarella, G., L. Favaretto, G. Sotgiu, M. Zambianchi, A. Bongini, C. Arbizzani, M. Mastragostino, M. Anni, G. Gigli, and R. Cingolani, *J. Am. Chem. Soc.* **122**, 11971 (2000).
Barta, P., F. Cacialli, R. H. Friend, and M. Zagorska, *J. Appl. Phys.* **84**, 6279 (1998).
Becker, R. S., J. S. de Melo, A. L. Macanita, and F. Elisei, *Pure Appl. Chem.*, **67** 9 (1995).
Becker, R. S., *et al.*, *J. Phys. Chem.* **100**, 18683 (1996).
Belijonne, D., Z. Shuai, and J. L. Brédas, *J. Chem. Phys.* **98**, 8819 (1993).
Belijonne, D., J. Cornil, R. H. Friend, R. A. Janssen, and J. L. Brédas, *J. Am. Chem. Soc.* **118**, 6453 (1996).
Belletête, M., N. Di Cesare, M. Leclerc, and G. Durocher, *Chem. Phys. Lett.* **250**, 31 (1996).
Benincori, T., *et al. Phys. Rev. B* **58**, 9082 (1998).
Berggren, M., A. Dodabalapur, and R. E. Slusher, *Appl. Phys. Lett.* **71**, 2230 (1997a).
Berggren, M., A. Dodabalapur, R. E. Slusher, and Z. Bao, *Nature* **389**, 466 (1997b).
Berggren, M., A. Dodabalapur, R. E. Slusher, A. Timko, and O. Nalamasu, *Appl. Phys. Lett.* **72**, 410 (1998).
Bixon, M., and J. Jortner, *J. Chem. Phys.* **48**, 715 (1968).
Brédas, J. L., R. L. Elsembaumer, R. R. Chance, and R. Silbey, *J. Chem. Phys.* **78**, 5656 (1983).
Brédas, J. L., J. Cornil, D. Beljonne, D. A. Dos Santos, Z. Shuai, *Acc. Chem. Res.* **32**, 267 (1999).
Brouwer, H. J., V. V. Krasnikov, T. A. Pham, R. E. Gill, and G. Hadziioannou, *Appl. Phys. Lett.* **73**, 708 (1998).
Burin, A. L., and M. A. Ratner, *J. Chem. Phys.* **109**, 6092 (1998).
Carter, S. A., M. Angelopoulos, S. Karg, P. J. Brock, and J. C. Scott, *Appl. Phys. Lett.* **70**, 2067 (1997).
Chaloner, P. A., S. R. Gunatunga, and P. B. Hitchcock, *Acta Cryst.* **C50**, 194 (1994).
Champagne, B., D. H. Mosley, and J.-M. Anré, *J. Chem. Phys.* **100**, 2034 (1994).
Chandross, M., and S. Mazumdar, *Phys. Rev. B* **55**, 1497 (1997).
Charra, F., M. P. Lavie, and D. Fichou, *Synth. Met.* **65**, 13 (1994).
Chsrovin, H., S. Rentsh, D. Grebner, D. U. Dahm, E. Birckener, and H. Naarmann, *Synth. Met.* **60**, 23 (1993).
Colditz, R., D. Grebner, M. Helbig, and S. Rentsch, *Chem. Phys.* **201**, 309 (1995).
Cornil, J., D. Beljonne, and J. L. Brédas, *J. Chem. Phys.* **103**, 842 (1995).
Cornil, J., *et al.*, *J. Am. Chem. Soc.* **120**, 1289 (1998).
Craig, D. P., and S. H. W. Walmsley, *Excitons in Molecular Crystal*, W.A. Benjamin, Inc., New York (1968).
Davydov, A. S., *Theory of Molecular Excitons*, Plenum Press, New York (1971).
Della Sala, F., H. H. Heinze, and A. Görling, *Chem. Phys. Lett.* **339**, 343 (2001).

Denton, G. J., N. Tessler, M. A. Stevens, and R. Friend, *Adv. Mater.* **9**, 547 (1997).

di Cesare, N., *et al.*, *J. Phys. Chem. A* **103**, 3864 (1999).

Dippel, O., V. Brandl, H. Bassler, R. Danieli, R. Zamboni, and C. Taliani, *Chem. Phys. Lett.* **216**, 418 (1993).

Dodabalapur, A., L. Torsi, and H. E. Katz, *Science* **268**, 270 (1995).

Dossantos, D. A., S. S. Galvao, B. Lacks, and M. C. Dossantos, *Synth. Met.* **51**, 203 (1992).

Egelhaaf, H.-J., and D. Oelkrug, *SPIE* **2362**, 398 (1995).

Ellis, R. L., R. Squire, and H. H. Jaffe, *J. Chem. Phys.* **55**, 3499 (1971).

Englman, R. and J. Jortner, *Mol. Phys.* **18**, 145 (1970).

Fichou, D., editor, *Hanbook of Oligo- and Polythiophenes*, Wiley, New York (1999).

Fichou, D., J. M. Nunzi, F. Charra, and N. Pfeffer, *Adv. Mater.* **6**, 64 (1994).

Fichou, D., B. Bachet, F. Demanze, I. Billy, G. Horowitz, and F. Garnier, *Adv. Mater.* **6**, 500 (1996).

Friend, R. H., *et al.*, *Nature* **397**, 121 (1999).

Frolov, S. V., M. Ozaki, W. Gellerman, V. Z. K. Yoshino, and Z. V. Vardeny, *Jpn. J. Appl. Phys.* **35**, L1371 (1996).

Frolov, S. V., M. Shkunov, Z. V. Vardeny, and K. Yoshino, *Phys. Rev. B* **56**, 4363 (1997a).

Frolov, S. V., W. Gellerman, M. Ozaki, K. Yoshino, and Z. V. Vardeny, *Phys. Rev. Lett.* **78**, 729 (1997b).

Garnier, F., G. Horowitz, D. Fichou, and X. Peng, *Adv. Mater.* **2**, 592 (1990).

Garnier, F., *et al.*, *J. Am. Chem. Soc.* **115**, 8716 (1993).

Garnier, F., R. Hajlaoui, A. Yassar, and P. Srivastava, *Science* **265**, 1684 (1994).

Garnier, F., G. Horowitz, P. Valat, F. Kouki, and V. Wintgens, *Appl. Phys. Lett.* **72**, 2087 (1998).

Garnier, F., *Acc. Chem. Res.* **32**, 209 (1999).

Gavezzotti, A., *J. Am. Chem. Soc.* **105**, 5220 (1983).

Gavezzotti, A., and G. Filippini, *Synth. Met.* **40**, 257 (1991).

Gelink, G. H., J. W. Warman, M. Remmers, and D. Neher, *Chem. Phys. Lett.* **265**, 320 (1997).

Gigli, G., M. Lomascolo, R. Cingolani, G. Barbarella, M. Zambianchi, L. Antolini, F. Della Sala, A. Di Carlo, and P. Lugli, *Appl. Phys. Lett.* **73**, 2414 (1998).

Gigli, G., G. Barbarella, L. Favaretto, F. Cacialli, and R. Cingolani, *Appl. Phys. Lett.* **75**, 439 (1999).

Gigli, G., O. Inganäs, M. Anni, M. De Vittorio, R. Cingolani, G. Barbarella, and L. Favaretto, *Appl. Phys. Lett.* **78**, 1493 (2001).

Gommers, F. J., *Nematologica* **18**, 458 (1972).

Grastrom, M., and O. Inganäs, *Synth. Met.* **48**, 21 (1992).

Graupner, W., G. Leising, G. Lanzani, M. Nisoli, S. De Silvestri, and U. Scherf, *Phys. Rev. Lett.* **76**, 847 (1996).

Grebner, D., M. Helbig, and S. Rentsch, *J. Phys. Chem.* **99**, 1699 (1995).

Greenham, N. C., R. H. Friend, and D. D. Bradley, *Adv. Mater.* **6**, 491 (1994).

Gross, E. K. U., J. F. Dobson, M. Petersilka, in *Density Functional Theory II*, edited by R. F. Nalewajski, Springer Series in Topics in Current Chemistry, Vol. 181, p. 81, Springer, Heidelberg (1996).

Gupta, R., M. Stevenson, M. D. McGehee, A. Dogariu, V. Srdanov, J. Y. Park, and A. J. Heeger, *Synth. Met.* **102**, 875 (1999).

Heeger, A. J., in *Handbook of Conducting Polymers*, edited by T.A. Skotheim, Chapter 21, Marcel Dekker, New York (1986).

Hennessy, M. H., *et al.*, *Chem. Phys.* **245**, 199 (1999).

Horowitz, G., *Adv. Mater.* **10**, 365 (1998).

Horowitz, G., D. Fichou, and G. Garnier, *Solid State Commun.* **70**, 385 (1989).

Horowitz, G., B. Bachet, A. Yassar, P. Lang, F. Demanze, J. L. Fave, and F. Garnier, *Chem. Mater.* **7**, 1337 (1995).

Horowitz, G., F. Kouki, A. El Kassmi, P. Valat, V. Wintgens, and F. Garnier, *Adv. Mater.* **11**, 234 (1999).

Kallinger, C., *et al.*, *Adv. Mater.* **10**, 920 (1998).

Kanemitsu, Y., K. Suzuki, Y. Masumoto, Y. Tomiuchi, Y. Shiraishi, and M. Kuroda, *Phys. Rev. B* **50**, 2301 (1994).

Kim, J. S., R. H. Friend, and F. Cacialli, *Appl. Phys. Lett.* **74**, 3084 (1999).

Kirstein, S., and H. Möwald, *J. Chem. Phys.* **103**, 826 (1995).

Klessingner, M., and J. Michl, *Light Absorption and Photochemistry of Organic Molecules*, VCH, Weinheim (1989).

Kozlov, V. G., V. Bulovic, P. E. Burrows, and S. R. Forrest, *Nature* **389**, 362 (1997).

Kozlov, V. G., V. Bulovic, P. E. Burrows, M. Baldo, V. B. Khalfin, G. Parthasarathy, and S. R. Forrest, *J. Appl. Phys.* **84**, 4096 (1998).

Kretsch, K., C. Belton, S. Lipson, W. J. Blau, F. Z. Henari, H. Rost, S. Pfeiffer, A. Teschel, H. Tillmann, and H. H. Hörhold, *J. Appl. Phys.* **86**, 6155 (1999).

Lipsett, F. R., *Can. J. Phys.* **5**, 284 (1957).

Marseglia, E. A., F. Grepioni, E. Tedesco, and D. Braga, *Mol. Cryst. Liq. Cryst.* **348**, 137 (2000).

Nijegorodov, N. I., and W. S. Downey, *J. Phys. Chem.* **98**, 5639 (1994).

Noma, N., T. Tsuzuki, and Y. Shirota, *Adv. Mater.* **7**, 647 (1995).

Oelkrug, D., H.-J. Egelhaaf, D. R. Worral, and F. Wilkinson, *J. Fluorescence* **5**, 165 (1995).

Pankove, J. I., *Optical Process in Semiconductors*, Prentice-Hall, Englewood Cliffs, NJ (1971).

Pei, Q., *et al.*, *Synth. Met.* **55**, 1221 (1993).

Pelettier, M., and F. Brisse, *Acta Cryst.* **C50**, 1942 (1994).

Persin, A. J., and A. I. Kitaigorodsky, *The Atom–Atom Potential Method*, Springer-Verlag, Berlin (1987).

Pope, M., and C. E. Swenberg, *Electronic Processes in Organic Crystal*, Clarendon Press, Oxford (1982).

Porzio, W., S. Destri, M. Mascherpa, and S. Bruckner, *Acta Polym.* **44**, 266 (1993).

Rentsch, S., J. P. Yang, W. Paa, E. Birckner, J. Schiedt, and R. Weinkauf, *Phys. Chem.* **1**, 1707 (1999).

Ridley, J. E., and M. C. Zerner, *Theor. Chim. Acta* **32**, 111 (1973).

Ridley, J. E., and M. C. Zerner, *Theor. Chim. Acta* **42**, 223 (1976).

Ridley, J. E., and M. C. Zerner, *Theor. Chim. Acta* **53**, 21 (1979).

Roman, L. S., W. Mammo, L. A. A. Petersson, M. R. Andersson, and O. Inganäs, *Adv. Mater.* **10**, 774 (1998).

Roos, B. O., M. Fulsher, P. A. Malmqvist, M. Merchan, and L. Serano-Andres, in *Quantum Mechanical Electronic Structure Calculations with Chemical Accuracy*, edited by S. R. Langjoff, Kluwer Academic, Dordrecht (1995).

Rossi, R., *et al.*, *J. Photochem. Photobiol. A* **70**, 59 (1993).

Schoeler, U., K. H. Tews, and H. Kuhn, *J. Chem. Phys.* **61**, 5009 (1974).

Schön, J. H., Ch. Kloch, R. A. Laudise, and B. Batlogg, *Appl. Phys. Lett.* **73**, 3574 (1998).

Schön, J. H., Ch. Kloch, R. A. Laudise, and B. Batlogg, *J. Appl. Phys.* **85**, 2844 (1999).

Schön, J. H., Ch. Kloch, A. Dodabalapur, and B. Batlogg, *Science* **289**, 599 (2000).

Schulten, K., I. Ohmine, and M. Karplus, *J. Chem. Phys.* **64**, 4422 (1976).

Schweitzer, B., G. Wegmann, H. Giessen, D. Hertel, H. Bässler, R. F. Mahrt, U. Scherf, and K. Müllen, *Appl. Phys. Lett.* **72**, 2933 (1998).

Sebastian, L., G. Weiser, and H. Bassler, *J. Chem. Phys.* **61**, 125 (1981).

Sheats, J. R., H. Antoniadis, M. Hueschen, W. Leonard, J. Miller, R. Moon, D. Roitman, and A. Stocking, *Science* **273**, 884 (1996).

Siegrist, T., *et al. J. Mater. Res.* **10**, 2170 (1995).

Siegrist, T., Ch. Kloch, R. A. Laudise, H. E. Katz, and R. C. Haddon, *Adv. Mater.* **10**, 379 (1998).

Silbey, R., J. Jortner, and S. A. Rice, *J. Chem. Phys.* **42**, 1515 (1963).

Sillinsh, E. A., and V. Capek, *Organic Molecular Crystals, Interaction, Localization and Transport Phenomena*, AIP Press, Woodbury (1994).

Simon, J., and J. J. André, *Molecular Semiconductors*, Springer, Berlin (1985).

Spiegelberg, C., N. Peyghambarian, and B. Kippelen, *Appl. Phys. Lett.* **75**, 748 (1999).

Stagira, S., M. Nisoli, G. Cerullo, M. Zavelani-Rossi, S. De Silvestri, G. Lanzani, W. Graupner, and G. Leising, *Chem. Phys. Lett.* **289**, 205 (1998a).

Stagira, S., M. Zavelani-Rossi, M. Nisoli, S. De Silvestri, G. Lanzani, C. Zenz, P. Mataloni, and G. Leising, *Appl. Phys. Lett.* **73**, 2860 (1998b).

Swanson, L. S., J. Shinarand, and K. Yoshino, *Phys. Rev. Lett.* **65**, 1140 (1990).

Sze, S. M., in *Semiconductor Devices*, edited by Hoepli, Milan, J. Wiley & Sons, New York (1981).

Tashiro, K., Y. Mingaua, M. Kobayashi, S. Morita, T. Kawai, and K. Yoshino, *Synth. Met.* **55**, 321 (1993).

Tessler, N., G. J. Denton, and R. Friend, *Nature* **382**, 695 (1996).

Van Bolhuis, F., H. Winberg, E. E. Havinga, E. W. Meijer, and E. G. J. Staring, *Synth. Met.* **30**, 381 (1989).

Van Pham, C., A. Burkhardt, A. Nkansah, R. Shabana, D. D. Cunnigham, H. B. Mark, Jr., and H. Zimmer, *Phosphorus, Sulfur, Silicon Relat. Elem.* **46** 153 (1989).

Wegmann, G., H. Giessen, A. Greiner, and R. F. Mahrt, *Phys. Rev. B* **57**, R4218 (1998).

Wu, M. W., and E. M. Conwell, *Phys. Rev. B* **56**, R10060 (1997).

Yang, A., M. Kuroda, Y. Shiraishi, and T. Kobayashi, *J. Chem. Phys.* **109**, 8442 (1998).

Yang, A., *et al.*, *J. Chem. Phys.* **109**, 8442 (1998).

Yassar, A., *et al.*, *Synth. Met.* **67**, 277 (1994).

Yassar, A., *et al.*, *J. Phys. Chem.* **99**, 9155 (1995).

Zerner, M. C., G. H. Loew, R. Kichner, and U. T. Mueller-Westerhoff, *J. Am. Chem. Soc.* **102**, 589 (1980).

Ziegler, C., in *Handbook of Organic Conductive Molecules and Polymers*, edited by H. Nalwa, Vol. 3, p. 667, Wiley, New York (1997a).

Ziegler, C., in *Handbook of Organic Conductive Molecules and Polymers*, edited by H. S. Nalwa, Vol. 3, p. 678, Wiley, Chichester (1997).

Advanced Semiconductor and Organic Nano-Techniques (Part II)
H. Morkoç (Ed.)

CHAPTER 6

Single-Walled Carbon Nanotubes for Nanoelectronics

M. S. Fuhrer

UNIVERSITY OF MARYLAND, COLLEGE PARK, MARYLAND

I. Introduction

Around 1991, inspired by the discovery and mass production of cage-like fullerene molecules, several research groups began to consider the properties of a hypothetical carbon structure: a single layer of graphite wrapped into a seamless cylinder—the carbon nanotube (Hamada *et al.* 1992; Mintmire *et al.* 1992; Saito *et al.* 1992a). It was soon realized that carbon nanotubes may be metallic or semiconducting, and should be excellent one-dimensional (1D) conductors at room temperature. About the same time, carbon deposits resulting from the arcing of graphite rods to obtain fullerenes were investigated via transmission electron microscope (TEM) and revealed to contain concentrically nested carbon nanotubes (Iijima 1991) ("multi-walled carbon nanotubes" or MWNTs). Soon thereafter, single-walled carbon nanotubes (SWNTs) were synthesized (Bethune *et al.* 1993; Iijima and Ichihashi 1993). Measurments soon revealed a material with an extraordinary convergence of exceptional thermal, mechanical, and electrical properties: the thermal conductivity of individual MWNTs exceeds that of diamond at room temperature (Kim *et al.* 2001); the elastic modulus of carbon nanotubes may exceed 1 TPa, making them the strongest known fibers (Treacy *et al.* 1996; Wong *et al.* 1997; Poncharal *et al.* 1999); electrons are transported ballistically through SWNTs over distances greater than 1 μm at room temperature (Bachtold *et al.* 2000).

The first electrical experiments on individual metallic SWNTs were reported in 1997 (Bockrath *et al.* 1997; Tans *et al.* 1997), with experiments on individual semiconducting SWNTs in 1998 (Tans *et al.* 1998b). In the following 5 years, research into the electrical properties of nanotube devices has exploded. A picture has emerged of a material with superlative electrical properties: metallic SWNTs have conductivities comparable to the best metals, and can carry current densities exceeding $10^9 \, A/cm^2$; semiconducting SWNTs have mobilities exceeding the best silicon MOSFETs. Still, enormous challenges remain to incorporating this material into a useful device technology: nanotubes are still expensive to manufacture, currently nanotubes cannot be sorted according to electronic property (metallic and semiconducting nanotubes are randomly mixed), and methods for placing nanotubes with precision onto substrates are in their infancy.

This chapter will serve to review the current status of research on carbon nanotubes for nanoelectronics applications. The focus will be on single-walled carbon nanotubes, whose properties are closest to ideal, though some of the conclusions will also apply to multi-walled nanotubes. It should be noted that there exist a number of excellent reviews of the electronic transport properties of individual single-walled (Dekker 1999; Nygard *et al.* 1999; McEuen 2000; Louie 2001; Yao *et al.* 2001) and multi-walled carbon nanotubes (Schonenberger *et al.* 1999; Schonenberger and Forró 2000; Forró and Schonenberger 2001).

This chapter is structured as follows. Sections II–IV will serve as a review of the fundamental theoretical and experimental work on the electronic properties of carbon nanotubes. Section II will introduce the theory of the electronic structure of carbon nanotubes, including the remarkable dependence of their electronic properties on structure. Section III will discuss the synthesis of carbon nanotubes, as well as the techniques used to place individual nanotubes into electronic circuits. Section IV will discuss the room temperature and cryogenic electronic transport properties of individual metallic and semiconducting SWNTs.

Section V will give an overview of the current state of research on nanoscale electronic devices incorporating carbon nanotubes. These devices range from field-effect and single-electron transistors to more exotic junction devices and electromechanical devices. Some significant advantages of carbon nanotubes for nanoelectronics become obvious: the exposed channel of the semiconducting nanotube transistor makes it an excellent candidate for chemical and biological sensing, the geometry of the carbon nanotube automatically guarantees small junction capacitance and high transconductance in nanotube single-electron transistors, the long electron mean free paths and high thermal conductivity in nanotubes suggests their use as interconnects to devices located at nanotube junctions, and the high stiffness and robustness of nanotubes makes them attractive for high-speed mechanical devices.

Sections VI–VII will attempt to look to the future of nanotube research. Section VI discusses the challenges that stand in the way of developing nanotube devices into useful technologies, such as the production of electronically uniform material, and the development of techniques for precise placement of nanotubes within circuits. Section VII will move beyond carbon nanotubes to pose the question: Are there other materials that have some of the advantageous properties of carbon nanotubes, but avoid some of the difficulties? Indeed a rapidly growing number of non-carbon nanotubes and nanowires have been synthesized, some with very attractive properties.

II. Electronic Properties of Carbon Nanotubes

Soon after the discovery of fullerene synthesis by the arc-discharge technique, it was found that, after arcing, the carbon cathode held a deposit which was dense in nanotubes (Iijima 1991). TEM imaging of these tubules showed that they consisted of concentric shells of seamless graphene sheets. These carbon nanotubes were found to have between two and over 50 walls, and be many micrometers in length. It was later found that the presence of certain transition metals in the arc catalyzed the production of carbon nanotubes with only one layer (Bethune *et al.* 1993; Iijima and Ichihashi 1993). This divided nanotubes into two classes: single-SWNTs and MWNTs.

Even before the discovery of carbon nanotubes by Iijima (1991), several research groups became interested in the possibility of extending the fullerene cage structure into a one-dimensional wire. Soon after the publication of Iijima's discovery, three groups published results on the electronic structure of SWNTs (Hamada *et al.* 1992; Mintmire *et al.* 1992; Saito *et al.* 1992a). It was realized that this new form of carbon could be metallic or semiconducting depending sensitively on its atomic structure (Hamada *et al.* 1992; Saito *et al.* 1992a,b), and that the Peierls distortion, to which all 1D metals are unstable, should occur only at very low temperatures in metallic nanotubes (Mintmire *et al.* 1992).

1. Band Structure of Carbon Nanotubes

Graphite is the starting point for considering the electronic structure of the carbon nanotube. Graphite is a solid form of carbon in which the atoms are covalently bonded in a hexagonal network to form 2D planes, called graphene, with the planes weakly bonded together through the van der Waals interaction. Figure 1(a) shows the graphene atomic structure. The low-energy band structure of graphene is unusual [see Fig. 1(b)]; the bands are cones, meeting in points at the Fermi surface at the K points in the Brillouin zone. In most directions in k-space, graphene is a semiconductor, only for special directions [e.g., k_y, in Fig. 1(b)] is graphene metallic.

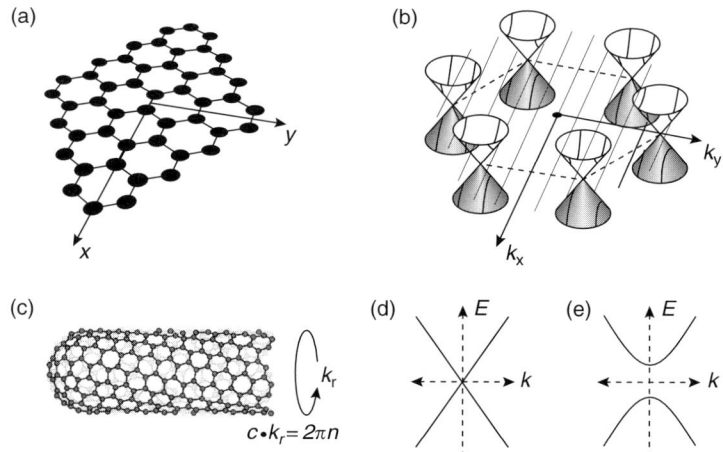

Fig. 1. Band structure of the single-walled carbon nanotube. The atomic structure of the graphene sheet is shown in (a). The band structure for low-energy excitations is shown schematically in (b), with k_x and k_y the wavevectors in the x and y directions in (a). The bands are cones, which meet the Fermi surface at points located at the corners of the hexagonal Brillouin zone (shown by the solid hexagon). When the graphene sheet is rolled up to form a nanotube (c), the circumferential wavevector k_r is quantized, such that $c \cdot k_r = 2\pi n$, where c is the circumference and n is an integer. The 1D bands are slices of the 2D band structure in (b) taken at the allowed values of k_r. The thin lines in B show the slices for a particular semiconducting SWNT. D and E show the resulting 1D band structures at low energies for the metallic and semiconducting nanotubes respectively. The metallic nanotube (d) has two linear bands which cross at the Fermi level. The semiconducting nanotube (e) has two hyperbolic bands, with an energy gap $E_g \approx 600$ meV for a 1.5-nm diameter nanotube.

Conceptually, an SWNT may be formed by cutting a strip of the graphene sheet and rolling it into a tube (Saito *et al.* 1992b). The lattice vector on the graphene sheet that connects the two points that will roll into each other is called the circumferential vector c, and completely defines the type of SWNT. SWNTs are then denoted by their circumferential vector in terms of the graphene lattice vectors a_1 and a_2; thus, an SWNT with a circumferential vector $n_1 a_1 + n_2 a_2$ would be denoted (n_1, n_2). (Two non-chiral types of tubes exist and are given special names in the literature: (n, n) tubes are called armchair and $(n, 0)$ tubes are called zig-zag, after the shapes of the bands of carbon atoms which encircle each nanotube.)

The electronic structure of an SWNT may be approximated by simply taking the electronic structure of the graphene sheet and quantizing it in the circumferential direction such that $c \cdot k_r = 2\pi n$, where n is an integer (Saito *et al.* 1992b). [Other calculations give similar results (Hamada *et al.* 1992; Mintmire *et al.* 1992).] This condition is equivalent to stating that the electronic wavefunction on the tube be single-valued around the circumference. Thus the graphene sheet is sliced at regular intervals in k-space [see Fig. 1(b)]. The result is a 1D dispersion relation with many sub-bands corresponding to the various slices.

The bands of the graphene sheet only cross the Fermi surface at the K point. This means only those tubes whose band structure slices cross the K point will be metallic. For a tube (n_1, n_2) this can be expressed by the condition: $n_1 - n_2 = 3q$ where q is an integer. A more careful treatment (Hamada *et al.* 1992) which takes into account the curvature of the graphene sheet finds that only the tubes with $n_1 = n_2$ will be metallic, the other tubes satisfying $n_1 - n_2 = 3q$, with non-zero q will have a small gap on the order of milli electron-Volts, considerably smaller than room temperature.

The low-energy band structure of metallic and semiconducting SWNTs is depicted in Fig. 1(d) and (e). Metallic SWNTs have two linear bands which cross at the Fermi level. Semiconducting SWNTs have two hyperbolic bands, separated by an energy gap, with $E_g \approx 900\,\text{meV}/d$, where d is the diameter in nanometers. In both cases, the next closest bands are on order $500\,\text{meV}$ away for typical diameter SWNTs; thus SWNTs are extremely 1D at room temperature, where the thermal energy is $25\,\text{meV}$.

This result is quite striking. It is an extreme example of a size effect in small structures, and manifests because of the unusual nature of the graphene Fermi surface—metallic for some directions in k-space, and semiconducting in others. This result has been verified by simultaneous measurement of the geometry and electronic structure of individual SWNTs by scanning tunneling microscope (STM) (Odom *et al.* 1998; Wildoer *et al.* 1998). In these experiments, the atomic structure of individual SWNTs was measured topographically by the STM, and then the STM tip was placed at a fixed distance above the SWNT to perform tunneling spectroscopy measurements of the band structure. The results were in excellent agreement with the model presented above.

The large number of possible band structures for SWNTs would seem to speak against their possible usefulness as an electronic material, unless the atomic structure of nanotubes could be controlled exactly. However, the picture is less complicated than is immediately obvious. It was soon pointed out that SWNTs fall into three broad classes: metallic SWNTs with $n_1 = n_2$, small-gap semiconducting SWNTs with $n_1 - n_2 = 3q$ but $q \leq 0$, and moderate-gap semiconducting SWNTs, with $n_1 - n_2 \leq 3q$. Furthermore, the typical energy gaps of the semiconducting SWNTs were found to depend almost entirely on diameter d, with the moderate semiconducting gaps scaling as $1/d$ and the small semiconducting gaps scaling as $1/d^2$ (Kane and Mele 1997). In fact, all the features in the density of states are determined to good approximation by diameter alone (Mintmire and White 1998). Since the energy gaps in small-gap semiconducting SWNTs are smaller than room temperature for $d > 1.4$ nm (Kane and Mele 1997), these SWNTs may be grouped with the metallic SWNTs for the purposes of room temperature applications. SWNTs may be produced with narrow diameter distributions, so the problem of obtaining an electronically uniform material is reduced to the separation of SWNTs into two classes, metallic (including small band-gap semiconductor) and large-gap semiconductor. More will be said on this issue below in Section VI.

2. PEIERLS DISTORTION

All 1D metals are unstable to a Peierls distortion, a periodic modulation of the atomic positions which opens a gap at the Fermi surface (Peierls 1955). For instance, in polyacetylene, a 1D chain of carbon atoms, the symmetry of the chain is broken by alternating single and double bonds. The Peierls transition temperature T_c, above which the system would be metallic, is given by $k_B T_c \approx 0.28 E_g$, where k_B is Boltzmann's constant, and E_g is the Peierls energy gap. For polyacetylene, the Peierls energy gap is ~ 1.4 eV (Su et al. 1980) therefore T_c is ~ 5000 K, significantly greater than room temperature, and in fact significantly greater than the decomposition temperature of the polymer.

Mintmire, Dunlap and White estimated the Peierls temperature for a (5,5) metallic SWNT to be of order 1 K, orders of magnitude below that of polyacetylene. The reason for the much lower Peierls transition temperature in SWNT is a much smaller gain in electronic energy *per carbon atom*, since the one dimensional state is shared over, for example, 10 carbon atoms around the circumference of the (5,5) SWNT, compared with a slightly greater (1.5×) elastic energy cost per carbon atom to distort the carbon–carbon bonds in the three-fold-coordinated SWNT lattice compared to the two-fold-coordinated polyacetylene.

More careful analyses of the particular phonon modes and electron–phonon coupling involved in Peierls distortion of a metallic SWNT lead to estimates of the Peierls transition temperature as below 9 K (Huang *et al.* 1996) and 15 K (Sedeki *et al.* 2000), still well below room temperature.

3.   SUMMARY OF ELECTRONIC STRUCTURE

SWNTs may be metallic or semiconducting, depending very sensitively on how the graphene sheet is wrapped up to form the nanotube. However, the gross features in the density of states are determined by diameter alone, so nanotubes of a given diameter fall into three classes: metallic, small band-gap semiconductor, and moderate band-gap semiconductor. For room temperature purposes, small band-gap semiconductor SWNTs may be treated as metallic. For example, SWNTs of diameter 1.5 nm will be either metallic, or semiconducting with a bangap $E_g \approx 0.6$ eV. Nanotubes are highly 1D: a 1.5-nm SWNT has a sub-band spacing of order 1 eV, much greater than the room temperature thermal energy of 25 meV.

Because of the high stiffness and small number of 1D modes of the nanotube (low density of states) nanotubes are relatively immune to the Peierls distortion, which destroys metallicity. The Peierls transition is estimated to occur only well below room temperature in metallic SWNTs.

III.   Nanotube Synthesis and Device Fabrication

1.   SYNTHESIS OF SINGLE-WALLED CARBON NANOTUBES

It is now known that SWNTs may be synthesized by a variety of methods. These methods have in common a source of carbon and a catalyst, typically a transition metal, such as iron, nickel, cobalt, yttrium, or molybdenum, in the form of nanoparticles. The basic picture that arises is that of a vapor–liquid–solid reaction; atomic carbon dissolves in a small metal particle, and, once supersaturated, the particle extrudes solid carbon. A graphite tube satisfies the need to extrude a narrow carbon shape with no dangling bonds.

The first SWNTs were synthesized in a carbon plasma obtained by striking an electric arc between carbon rods containing small amounts of catalyst, typically iron, nickel, cobalt, and/or yttrium (Bethune *et al.* 1993; Iijima and Ichihashi 1993; Journet *et al.* 1997). Higher yields were demonstrated by vaporizing a similar carbon target with intense laser pulses (Thess *et al.* 1996).

In both cases, the plasma is surrounded by an inert atmosphere, such as helium. Unlike MWNTs, which form as a growth on the cathode during arc synthesis (without a catalyst), SWNTs deposit on the synthesis chamber walls.

SWNTs synthesized by these methods are found by TEM to form bundles of tens or hundreds of nanotubes arranged in a triangular lattice. Along with the nanotube bundles, fullerenes, amorphous carbon, and metal catalyst particles are also seen in the as-formed material.

Several techniques have been employed to purify the as-grown SWNT material. Baking in a vacuum oven at 1000°C removes much of the metal catalyst and fullerenes. Alternatively, soluble fullerenes may be removed with toluene or CS_2, and the insoluble material may be suspended in surfactant-containing H_2O via ultrasonication and then filtered (Bandow et al. 1997; Rao et al. 1997). Refluxing in boiling nitric acid before washing and filtration has been found to be effective at removing the catalyst particles and amorphous carbon (Rinzler et al. 1998).

Various methods have been derived to separate nanotubes from bundles in suspension or solution. Ultrasonication of SWNT material that has been purified by baking in a vacuum oven in dichloroethane (DCE) results in a suspension of small bundles and some individual SWNTs. More recent efforts to solubilize nanotubes will be discussed in Section VI.

Chemical vapor deposition (CVD) offers an alternative synthesis method. SWNTs have been prepared via CVD from a carbon-containing feedstock gas, such as carbon monoxide (Dai et al. 1996), methane (Kong et al. 1998), or ethylene (Hafner et al. 1998). In general, a nanostructured transition metal catalyst is prepared, and the carbon feedstock gas is flowed over the catalyst at high temperature. The catalyst catalyzes the decomposition of the feedstock gas, and carbon dissolves in the catalyst particles. Once supersaturated, the catalyst particles extrude excess carbon in the form of nanotubes.

The CVD synthesis technique has been used to successfully prepare carbon nanotubes directly on flat substrates which are suitable for electronic device fabrication (Kong et al. 1998; Hafner et al. 2001). Kong et al. (1998) used a catalyst of alumina nanoparticles coated with iron and molybdenum. This catalyst could be patterned into islands on a flat SiO_2-capped silicon wafer substrate using a conventional lift-off lithography technique. The substrate supporting the catalyst was exposed to flowing methane at 900°C. After CVD, SWNTs were found lying on the flat substrate anchored to the catalyst islands. Hafner et al. (2001) prepared a catalyst of iron nanoparticles by depositing a thin film of ferric nitrate deposited on the substrate simply by dipping the substrate into a ferric nitrate solution in 2-propanol followed by rinsing in hexane. The ferric nitrate was reduced to iron in a hot hydrogen-containing atmosphere, followed by CVD growth using ethylene as the feedstock gas at 700°C. This method produced SWNTs dispersed about the surface of the substrate. Some of the SWNTs were found to be projecting upward from the substrate after growth, and could be plucked from the surface by a passing atomic force microscope (AFM) tip. Unlike the arc- and laser-ablation derived SWNT material, most CVD-grown SWNTs are found as individual nanotubes, with a few small bundles are formed. However, the size distribution of CVD-grown SWNTs is typically much larger.

2. FABRICATION OF ELECTRONIC DEVICES

The first attempts to attach electrodes to individual SWNTs were made using material derived from the laser ablation process (Bockrath *et al.* 1997; Tans *et al.* 1997). These SWNTs were suspended in DCE via ultrasonication and then deposited on SiO_2-capped silicon substrates by spinning or simply applying a drop of suspension to the substrate for a short time and then washing off the excess solution with 2-propanol.

Two schemes for contacting SWNTs with electrodes were initially investigated: deposition of SWNTs on top of prepatterned electrodes ("tube-on-top") and evaporating metal electrodes on top of SWNTs after deposition ("metal-on-top"). Figure 2 shows AFM images of examples of each top of device, and illustrates each schematically. The prototypical SWNT device has two electrodes attached to the SWNT ("source" and "drain") and a third "gate" electrode (either the conducting silicon substrate beneath the oxide, or a nearby metallic electrode).

Tube-on-top devices were fabricated by prepatterning gold or platinum electrodes and then depositing SWNTs from suspension by spin coating or

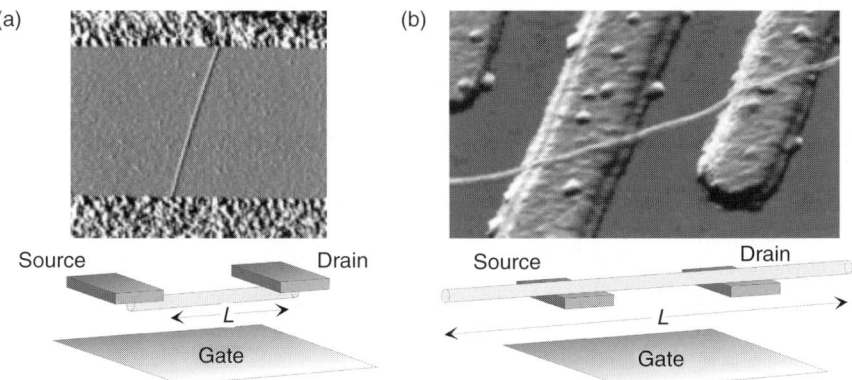

FIG. 2. An AFM image of a "metal-on-top" device is shown in (a). The rough areas at top and bottom are Cr/Au electrodes, and the narrow line is the nanotube, lying on an SiO_2-capped Si substrate. A schematic of the device is shown below. A "tube-on-top" device is shown in (b) (image courtesy Cees Dekker). Here, a nanotube (thin line) was deposited from solution over pre-patterned Pt leads on SiO_2-capped Si. The metal leads are spaced approximately 140 nm apart. The lower panel shows a schematic of the device.

immersion in solution followed by rinsing. In general, tube-on-top devices were found to have higher contact resistances than metal-on-top devices. It was speculated that this was due to the sharp bending of the nanotube over the contacts (Bezryadin *et al.* 1998), which could open an electronic energy gap in a nominally metallic SWNT. This theory was born out by the fact that the contact resistance of tube-on-top devices was improved by planarizing the contacts before deposition (Yao *et al.* 2000).

Metal-on-top devices were fabricated by first depositing SWNTs on substrates, and then patterning electrodes via a liftoff process. Bockrath *et al.* (1997) developed a technique to contact specific individual SWNT that had been located with AFM. First a pattern of metal (Cr/Au) alignment marks was defined on the substrate using conventional electron-beam lithography and lift-off. An AFM was then used to image the SWNTs relative to the alignment marks. Resist was again spun on the substrate, and a second electron-beam lithography step established electrical contacts to the desired nanotubes.

Kong *et al.* (1998) exploited the ability to pattern the catalyst for CVD growth of SWNTs to fabricate devices with high yield. Islands of catalyst were patterned on substrates with separations of a few micrometers using electron beam lithography and lift-off. After CVD growth of SWNTs, a second lithography step was used to cover the areas containing the catalyst islands with metal electrodes. Often SWNTs would span the gaps between catalyst islands, and would thus be contact by the electrodes.

IV.   Electronic Transport Properties

In general, the device behavior of individual two-terminal SWNT devices was found to fall into two classes. One class, identified as metallic, has a source–drain conductance that is nearly independent of voltage applied to the gate electrode (see Fig. 3). The other class, identified as semiconducting, has a source–drain conductance that depends strongly on gate voltage V_g, conducting at negative V_g, and becoming insulating at positive V_g (see Fig. 3). This behavior is reminiscent of a p-type field-effect transistor (FET), and hence the semiconducting SWNT device has been termed the tube-FET or NT-FET. More will be said below in Section V.1 about the theory of operation of the semiconducting SWNT FET.

1.   Conductance Quantization in One Dimension

The starting point for the discussion of electronic transport properties of highly 1D SWNTs is the Landauer conductance formalism. Landauer (1958)

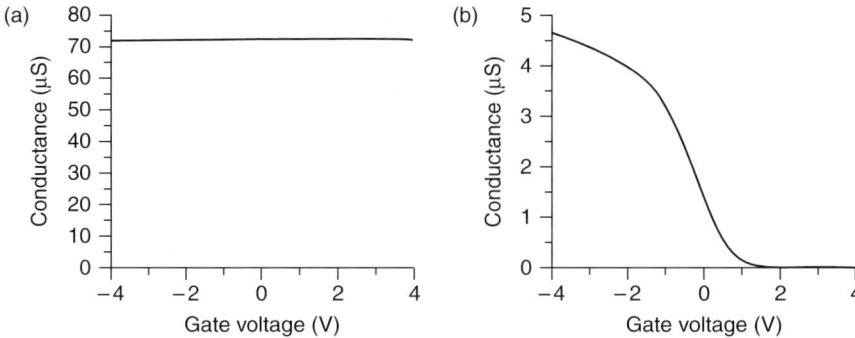

FIG. 3.   The conductance G as a function of gate voltage V_g for metallic and semiconducting SWNT devices at room temperature. The conductance of the metallic nanotube (a) is nearly independent of gate voltage, while the conductance of the semiconductaing nanotube (b) varies strongly with gate voltage, conducting at negative gate voltage, and becoming insulating at positive gate voltage. Both devices are "metal-on-top", fabricated using nanotubes prepared by CVD.

first derived the conductance of an ideal (zero scattering, or ballistic) 1D channel, and found the surprising result that the conductance was finite, and quantized: the conductance $G = e^2/h$ for each mode of the channel, where e is the electronic charge, and h is Planck's constant [for a general discussion, see, e.g., Datta (1995)]. The general result for a 1D channel with multiple modes and less than unity transmission probability is

$$G = \frac{e^2}{h}\sum_i T_i$$

where T_i is the transmission probability of the ith mode. For an individual metallic SWNT, there are two doubly-spin-degenerate bands at the Fermi level, and the conductance in the absence of scattering would be $4e^2/h$, or approximately $155\,\mu S$ (the result would be the same for a semiconducting nanotube, as long as the Fermi level lies with in the first sub-band, due to electrostatic or chemical doping). This corresponds to a minimum resistance of approximately $6.5\,k\Omega$ for a single SWNT. This quantized resistance appears as a contact resistance—a drop in the electrochemical potential occurs at each contact to the SWNT.

Using the Landauer formalism in the ohmic (classical transmission) limit (the phase relaxation length is less than the distance between scatterers), the two-terminal resistance of a nanotube device is

$$R = h/4e^2 + R_i + R_{c1} + R_{c2}$$

where $h/4e^2$ is the quantized contact resistance of the nanotube, R_i the intrinsic resistance from scattering processes within the tube arising from, for example, disorder or phonons, and the contact resistances $R_{c1,2}$ the resistances due to transport barriers formed at the metal electrode/nanotube junctions.

2. METALLIC SINGLE-WALLED NANOTUBES

a. Two-Terminal Conductance at Room Temperature

The first metallic SWNT devices fabricated and measured in 1997 had two-terminal resistances ranging from tens of kilo ohms to mega ohms (Bockrath *et al.* 1997; Tans *et al.* 1997) significantly greater than the minimum resistance of $\sim 6.5\,\text{k}\Omega$. It is difficult to separate the contact resistances $R_{c1,2}$ from the intrinsic resistance R_i in mesoscopic devices. In macroscopic devices, a four-terminal measurement may be made, but the placement of additional strongly interacting voltage probes onto a SWNT will significantly disturb the measurement. Electrostatic force microscopy (EFM) provides a way around this dilemma (Bachtold *et al.* 2000). In an EFM measurement, a conducting-tip AFM probe is used as a local voltmeter which interacts only weakly with the device under test.

Figure 4 shows an EFM measurement of the voltage in a current-carrying metallic SWNT device (Bachtold *et al.* 2000). The two-probe resistance of this 2.5 nm diameter bundle is 40 kΩ. (From the current-carrying capacity of the bundle, it was estimated that two metallic SWNTs were present.) The EFM image of this SWNT bundle, as well as a line trace along the backbone of the bundle is shown in the figure. The potential is flat over its length,

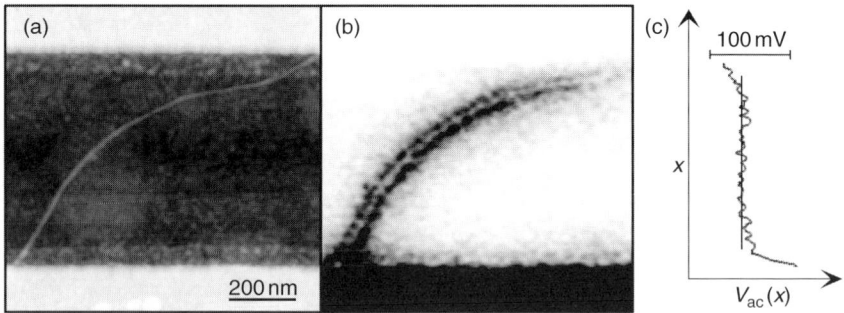

FIG. 4. EFM measurement of a metallic SWNT bundle at room temperature (Bachtold *et al.* 2000). A topographic AFM image is shown in (a); the white areas at top and bottom are the Cr/Au electrodes, and the faint white line is the nanotube. The EFM signal is shown in (b); 100 mV was applied to the bottom electrode, and the top electrode was grounded. The total device resistance is 40 kΩ. A trace of the local voltage along the backbone of the nanotube is shown in (c). The voltage drops at each contact, but is flat in between ($R < 3\,\text{k}\Omega$), indicating ballistic conduction over the length of the nanotube ($\sim 1.6\,\mu\text{m}$).

indicating no measurable intrinsic resistance; R_i of the bundle was estimated to be at most $3\,k\Omega$. The contact resistances are measured to be approximately 28 and $12\,k\Omega$ for the upper and lower contacts, respectively.

The measured resistance may be related to a transmission probability using the four-terminal Landauer formula: $R_i = (h/4e^2)(1 - T_i)/T_i$ per nanotube, where T_i is the transmission coefficient for electrons along the length of the nanotube. Here $T_i > \frac{1}{2}$: the majority of electrons traverse the bundle without scattering. Transport in metallic nanotubes is ballistic over a length of $>1\,\mu m$, even at room temperature. This result indicates that the resistance of metallic SWNT devices arises from imperfect contact resistances and not from intrinsic scattering within the nanotube.

A number of techniques have been employed to improve the contact resistance of SWNT devices. The best contacts have been obtained by evaporating Au or Pt over the tube, often followed by a subsequent anneal. A number of groups have seen conductances approaching the value $G = 4e^2/h$ (Kong *et al.* 2001; Liang *et al.* 2001). Figure 5 shows the differential conductance dI/dV as a function of V_{sd} for a 1 μm long SWNT with low contact resistance. At low V_{sd}, the conductance is $\sim 2e^2/h$ at room temperature, growing to $\sim 3.3\,e^2/h$ as the temperature is lowered, implying an intrinsic resistance $R_i < 3.3\,k\Omega$ at room temperature, and $<1.4\,k\Omega$ at low temperature. For a 1.5 nm diameter tube, this corresponds to a room temperature resistivity ρ less than $10^{-6}\,\Omega\,cm$. The conductivity of metallic nanotubes at room temperature can thus be equal to, or even exceed, the conductivity of the best metals at room temperature.

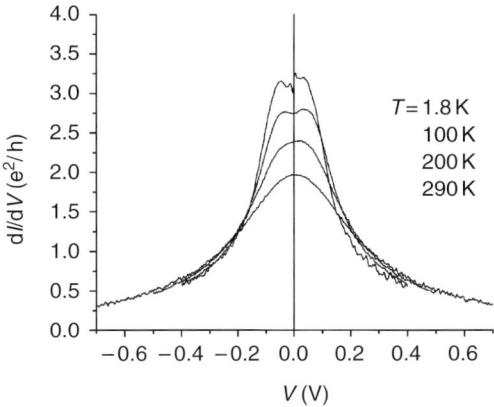

FIG. 5. Differential conductance as a function of bias voltage for a metallic SWNT device at temperatures 1.8, 100, 200, and 290 K. At zero bias, the conductance rises as the temperature is lowered, reaching a low-temperature value of $\sim 3.3e^2/h$, more than 80% of the maximum conductance $4e^2/h$, at 1.8 K. At higher bias, the conductance falls, due to electron scattering via emission of optical phonons.

The micrometer-scale scattering length for electrons at room temperature in metallic SWNTs is a striking result. The mean free path for electrons at room temperature in the best 3D metals, for example, copper, is on order tens of nanometers, due to phonon scattering. The difference arises from the 1D nature of the metallic SWNT. In one dimension, the phase space for scattering electrons is significantly reduced. Electrons may propagate only forward or backward in the nanotube, with a large momentum transfer required to backscatter the forward-moving electrons. This momentum is not present in any single acoustic phonon at room temperature. In three dimensions, electrons backscatter by undergoing a large number of small-angle collisions with acoustic phonons. Weak scattering has been predicted theoretically in metallic SWNTs (Anantram and Govindan 1998; White and Tudorov 1998; McEuen *et al.* 1999; Nakanishi and Ando 1999). McEuen *et al.* (1999) showed that the forward and backscattered electronic states in metallic SWNTs have opposite parity, and would require a short wavelength scattering potential to cause scattering. Nanotubes have also been shown to be relatively insensitive to point defect disorder: since the electronic states are extended over the circumference of the nanotube, point disorder is effectively averaged out (White and Tudorov 1998).

While the mean free path is orders of magnitude larger in metallic SWNTs than in traditional metals, the conductivity is only comparable or slightly better. This is due to the very low density of states in nanotubes, arising primarily from the semimetallic nature of graphene. There is a trade-off in 1D metals; the same low density of states allows the Peierls distortion to be pushed well below room temperature. A 1D system with significantly larger number of conducting modes per area would likely be a Peierls insulator at room temperature.

There are of course limits to the astounding conductivity of metallic SWNTs. While room temperature acoustic phonons cannot backscatter electrons, optic and zone-boundary phonons do have the necessary momentum. They are too high in energy ($hf \sim 150\,\text{meV}$) to be present at room tempererature and low V_{sd}. At high source–drain voltages, however, electrons can emit these phonons and efficiently backscatter. This leads to a dramatic reduction of the conductance at high biases, as was first reported by Yao *et al.* (2000). This can be readily seen in the data of Figure 5. The scattering rate grows linearly with V_{sd}, leading to a saturation of the total current through the SWNT. This saturation value is $\sim(4e^2/h)\,hf \sim 25\,\mu\text{A}$ for small diameter SWNT. This corresponds to a current density of $j = 2.5 \times 10^9\,\text{A/cm}^2$ for a 1 nm diameter tube—orders of magnitude larger than current densities found in present-day interconnects. Metal interconnects typically fail via electromigration. In covalently bonded carbon nanotubes, however, there are no low-energy defects or dislocations that can lead to migration of atoms. Nanotubes have been observed to carry current densities of $10^9\,\text{A/cm}^2$ for periods of weeks without failure (Wei *et al.* 2001).

b. Low Temperature: Coulomb Blockade

The low-temperature behavior of metallic SWNT devices depends strongly on the magnitude of the contact resistance, and a rich spectrum of behaviors has been observed. Electronic transport through devices with high resistance contacts will be dominated by Coulomb blockade at low temperature (Bockrath et al. 1997; Tans et al. 1997); the energy associated with adding an electron to the isolated nanotube island will exceed the thermal energy $k_B T$. If the contact resistance is very low, however, electrons will pass nearly freely into the nanotube from the leads. Small residual scattering at the two contacts will lead to interference and Fabry–Perot-like oscillations of the conductance of the nanotube (Liang et al. 2001). In the intermediate regime of contact resistance $R_c \leq h/e^2$, more exotic transport mechanisms, such as tunneling via the Kondo resonance, may be observed (Nygard et al. 2000). This section will focus primarily on the case of fairly resistive barriers, and the resulting Coulomb blockade, because of its application to single-electron tunneling devices.

A finite-sized metallic island has an associated charging energy $E_c = e^2/2C_\Sigma$ where C_Σ is the total capacitance of the island. The charging energy represents the electrostatic energy required to add a single electron to the island. If the island is well separated from its surroundings (i.e., the conductance of the contacts to the island $G_{\text{contact}} < e^2/h$) and the thermal energy $k_B T$ is much less than the charging energy E_c, then the charge on the island cannot fluctuate and hence current cannot be transported across the island. This is the Coulomb blockade.

In a single-electron transistor (SET), two leads are coupled to a small island through tunnel barriers, such that $G_{\text{contact}} < e^2/h$. A third gate electrode is coupled capacitively to the island. The total island capacitance is thus $C_\Sigma = C_l + C_r + C_g$, where C_l and C_r are the capacitances to the left and right leads, and C_g the capacitance to the gate electrode. In general, when $k_B T \ll E_c$ no current may pass from the left lead through the island to the right lead. However, if the electrochemical potential of the island can be tuned by a nearby gate electrode, the island will conduct whenever the n and $n + 1$ electron states are degenerate. As the voltage on the gate electrode is changed, a charge $q = C_g V_g$ is induced on the island. Of course this charge must be added discretely, so the charge state of the island changes by one whenever $C_g V_g$ changes by and amount e. This gives rise to periodic oscillations in the conductance through the island as a function of gate voltage, with a period $\Delta V_g = e/C_g$.

For an SWNT device, it is found that the capacitance is simply proportional to length L; $C = L$ in cgs units. The charging energy E_c of the SWNT is approximately $3\,\text{eV}/L$, where L is measured in nanometers. This points out the importance of charging effects even for room temperature nanodevices; the charging energy will exceed room temperature for nanotube devices of

$L < 50$ nm. Room temperature nanotube SETs will be discussed in greater detail below in Section V.2.

Experimentally, it is found that for metal-on-top devices L measures the length of the SWNT between electrodes (Bockrath *et al.* 1997) [see Fig. 2(a)]. In essence, although the SWNT extends underneath the electrode, its electronic structure is significantly perturbed by the metal electrode, and therefore it may be thought of as electronically "cut" at the edge of the metal electrode. For some tube-on-top devices, apparently the less strongly interacting contacts do not cut the SWNT, and L may represent the entire length of the SWNT (Tans *et al.* 1997) [see Fig. 2(b)].

An interesting property of SWNT SET devices is that even long SWNT islands may be *quantum dots* at low temperature; that is, the single-particle level spacing ΔE may exceed the thermal energy $k_B T$. For a 1D island, the single particle level spacing scales inversely with length: $\Delta E = \pi \hbar v_F / L$, where $\hbar$ is Planck's constant, and v_F is the Fermi velocity. For a metallic SWNT, $\Delta E \approx 0.5\,\text{eV}/L$ where L is measured in nanometers. Since the charging energy is also inversely proportional to length, the ratio of charging energy to level spacing $E_c / \Delta E$ is fixed. This is not true, in general, in one dimension; the inverse dependence on L comes from the linear dispersion at the Fermi surface, so only metallic SWNTs have a fixed $E_c / \Delta E$. The implication is that quantum effects will be important for any metallic SWNT SET device; because operation at temperatures $k_B T \ll E_c$, will necessitate $k_B T$ less than or approximately equal to ΔE.

The discreteness of the energy levels modifies the Coulomb blockade somewhat. The addition energy E_{add} for adding an electron to the quantum dot is now $E_{\text{add}} = e^2 / C + \Delta E$, and the period of oscillations in V_g becomes

$$\Delta V_g = \frac{C_\Sigma}{e C_g} \left(\Delta E + \frac{e^2}{C_\Sigma} \right)$$

which reduces to $\Delta V_g = e / C_g$ in the limit $\Delta E \to 0$.

The presence of discrete energy levels in the SWNT quantum dot gives rise to additional features in the electronic transport (Bockrath *et al.* 1997; Tans *et al.* 1997, 1998a; Cobden *et al.* 1998b). This may be seen clearly by performing "transport spectroscopy" measurement of the differential conductance dI/dV as a function of both the source–drain voltage and gate voltage (Kouwenhoven *et al.* 1997). Such a transport spectroscopy measurement on a SWNT quantum dot is shown in Fig. 6.

The observation of discrete energy levels with spacing consistent with the length of the entire SWNT (Tans *et al.* 1997; McEuen *et al.* 1999) (up to 8 μm) in metallic tube-on-top devices indicates that at low temperature, electronic transport is coherent across the entire length of the SWNT, that is, the coherence length (and the mean free path) is at least 8 μm.

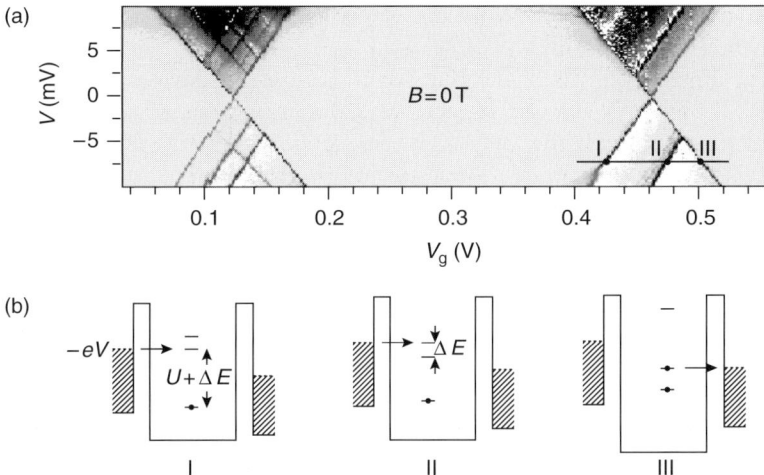

FIG. 6. Conductance spectroscopy of a metallic SWNT at 100 mK (Cobden *et al.* 1998b) (images courtesy of P. L. McEuen). The plots show the differential conductance (greyscale, dark is higher conductance) as a function of bias voltage V and gate voltage V_g for zero magnetic field (a) and a magnetic field of 5 T (b). The two points of finite conductance at zero bias voltage in (a) and (b) each represent the addition of one electron to the nanotube. The diagonal lines of finite conductance appear as states become available for tunneling within the source–drain bias window, as represented schematically in (c). As the gate voltage is increased [from points I to II to III in panel (a)], the single-particle states move down in energy as shown in (c). At point I, the addition of an electron from the left lead first becomes allowable, and a peak in the differential conductance occurs. At point II, an additional single-particle level becomes available for tunneling, and another peak in the differential conductance appears. At point III, the state corresponding to the added electron has moved below the potential of the right lead; beyond this point the current drops again, so another peak in differential conductance is seen. Application of a magnetic field (b) results in Zeeman splitting of some of the single-particle levels.

As discussed above, the two conditions necessary for Coulomb blockade are: (1) that the temperature be low enough such that $k_B T \ll E_c$; and (2) that the contacts are resistive such that $R_c \ll h/e^2$. The second condition is not necessarily true for nanotube devices—near ohmic contacts to nanotubes have been demonstrated (Kong *et al.* 2001; Liang *et al.* 2001). As the contact resistance is reduced, the charging energy is also reduced; E_c must go to zero as R_c goes to zero. This is indeed observed in nanotube devices (Liang *et al.* 2001). Suitable engineering of the contact resistance would allow for both ballistic and single-electron charging devices to be constructed from metallic SWNTs.

c. Temperature Dependence: Luttinger Liquid Tunneling

The conductance of high-contact-resistance metallic SWNT devices was observed to decrease as the temperature was lowered (Bockrath *et al.* 1999)

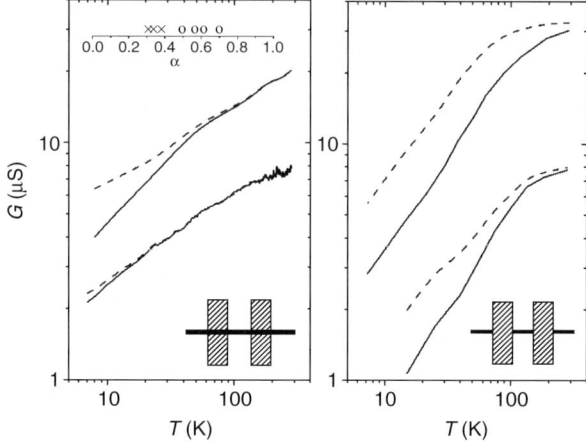

FIG. 7. The linear response (low-bias) conductance as a function of temperature for a tube-on-top device SWNT (left panel) and a metal-on-top SWNT device (right panel), showing power-law behavior due to Luttinger liquid tunneling (Bockrath *et al.* 1999). The solid lines are the measured data, the dashed lines are corrected for temperature dependence due to Coulomb blockade (Bockrath *et al.* 1999). The dashed lines show a power-law temperature dependence, in accordance with predictions for tunneling into a Luttinger liquid (see text). The inset in the left panel shows the exponents from the power-law dependence inferred from the data for several devices. Open circles denote metal-on-top devices, crosses denote tube-on-top devices.

(see Fig. 7). The zero bias conductance G varied as a power law in temperature T: $G(T) \sim T^\alpha$ (see Fig. 7). In addition, the differential conductance dI/dV varied as a power law in bias voltage V: $dI/dV \sim V^\alpha$. The exponent α was the same in each case, but depended on the geometry of the device (metal-on-top vs tube-on-top).

As shown above in Section IV.2.a, the two-terminal conductance of a metallic nanotube device largely measures the contact properties. The results were interpreted in terms of electrons tunneling from metallic leads into a Luttinger liquid, the interacting ground state of a 1D metal (Fisher and Glazman 1997). The Luttinger liquid is characterized by a single parameter g, which measures the strength of the Coulomb interaction between electrons. In the non-interacting limit $g = 1$, while for strongly repulsive interactions, $g \ll 1$. For any system with $g \neq 1$, the elementary excitations of the system cannot be described as single-particle-like, and thus Fermi liquid theory breaks down.

The metallic SWNT is expected to have a value of g moderately less than unity (Egger and Gogolin 1997; Kane *et al.* 1997). For a finite length SWNT, the Luttinger liquid parameter is given by:

$$g = \left[1 + \frac{2E_c}{\Delta E}\right]^{-1/2}$$

where E_c is the charging energy of the SWNT and ΔE is the single-particle energy level spacing. Experiments (Bockrath *et al.* 1997; Tans *et al.* 1997; Cobden *et al.* 1998a) give the ratio $E_c/\Delta E = 6\text{--}10$, which gives $g = 0.22\text{--}0.28$ (Bockrath *et al.* 1999); a theoretical estimate using a screening length of $1000\,\text{Å}$ gives $g = 0.3$ (Kane *et al.* 1997).

Tunneling from a Fermi liquid into a Luttinger liquid is suppressed at low energy, since the tunneling electron must excite a packet of excitations in the Luttinger liquid. The tunneling conductance of the Luttinger liquid is predicted to follow a power law in both temperature and bias voltage $G(T) \sim T^{\alpha}$ and $dI/dV \sim V^{\alpha}$ (Bockrath *et al.* 1999) (see Fig. 7). The power-law exponent α depends on the number of 1D modes in the Luttinger liquid, and on the geometry of the tunnel junctions (whether the electron tunnels into the end or the bulk of the Luttinger liquid). For an SWNT with four conducting modes, it was found that:

$$\alpha_{\text{bulk}} = \frac{g^{-1} + g - 2}{8}$$

$$\alpha_{\text{end}} = \frac{g^{-1} - 1}{4}$$

$$\alpha_{\text{bulk-bulk}} = 2\alpha_{\text{bulk}} = \frac{g^{-1} + g - 2}{4}$$

$$\alpha_{\text{end-end}} = 2\alpha_{\text{end}} = \frac{g^{-1} - 1}{2}$$

where α_{bulk}, α_{end}, $\alpha_{\text{bulk-bulk}}$, and $\alpha_{\text{end-end}}$ are the power-law exponents for tunneling from electrode into the SWNT bulk, from electrode into the SWNT end, from the bulk of one SWNT to the bulk of another SWNT, and from one SWNT end to another SWNT end, respectively.

These four geometries have been realized experimentally. Tube-on-top SWNT devices act as bulk-contacted Luttinger liquids, while metal-on-top SWNT devices act as end-contacted Luttinger liquids (see Fig. 7). The other geometries, the bulk-to-bulk and end-to-end Luttinger liquids, were realized by Postma *et al.* (2000) by manipulating SWNTs using an AFM tip. A sharp kink in a SWNT formed a tunnel barrier, and thus an end-to-end contact, while the bulk-to-bulk contact was formed by cutting a SWNT with the AFM tip and then pushing the ends back together so they crossed. Experimentally, these geometries give $\alpha_{\text{bulk}} = 0.3$, $\alpha_{\text{end}} = 0.6$, $\alpha_{\text{bulk-bulk}} = 0.5$, and $\alpha_{\text{end-end}} = 1.4$. The Luttinger interaction parameters extracted from these four experiments using the formulas above are 0.24, 0.29, 0.26, and 0.27, respectively, in very good agreement with the theoretical prediction $g = 0.22\text{--}0.28$.

3. SEMICONDUCTING SINGLE-WALLED NANOTUBES

a. Two-Terminal Conductance at Room Temperature

Semiconducting behavior in SWNTs was first reported by Tans *et al.* (1998b). Figure 3(b) shows a measurement of the conductance of a semi-conducting SWNT as the gate voltage applied to the conducting substrate is varied. The nanotube conducts at negative V_g and turns off with a positive V_g. The resistance change between the on and off state is many orders of magnitude. This device behavior is analogous to a p-type MOSFET, with the nanotube acting as the semiconductor channel. At large positive gate voltages, n-type conductance is sometimes observed, especially in larger-diameter nanotubes (Park and McEuen 2001; Javey *et al.* 2002). The conductance in the n-type region is typically less than in the p-type region because of the work function of the Au electrodes. The Au Fermi level aligns with the valence band of the SWNT, making a p-type contact with a barrier for the injection of electrons.

Semiconducting nanotubes are found to be doped slightly p-type; that is, the conductance is finite at $V_g = 0$. This arises because chemical species, particularly oxygen, adsorb on the tube and act as weak p-type dopants. Experiments have shown that changing a tube's chemical environment can change this doping level—shifting the voltage at which the device turns on by a significant amount (Bockrath *et al.* 2000; Kong *et al.* 2000a,b; Derycke *et al.* 2001). This has spurred interest in nanotubes as chemical sensors. The threshold voltage is very sensitive to the processing history of the device—for example, heating or exposure to UV radiation drives off oxygen, lowering the p-doping level of the device. Controlling adsorbate doping is an important challenge to be addressed. More will be said on this below in Section V.1.

In the data of Fig. 3(b), the conductance initially rises linearly with V_g as additional holes are added to the nanotube. At higher gate voltages, the conductance stops increasing and instead is nearly constant. This limiting conductance is presumably due to the contact resistance between the metallic electrodes and the nanotube. As with metallic SWNT devices, EFM measurements again may be used to determine the nature of the resistance (Fuhrer *et al.* 2001).

Figure 8 shows an EFM trace on a 4.5 μm long semiconducting SWNT grown by CVD. The voltage drop along this nanotube indicates a resistance of approximately $9.2 \, k\Omega/\mu m$. The mean free path l, the distance over which the transmission probability $T = \frac{1}{2}$, is given by the length over which the conductance is $4e^2/h$, or the resistance is $\sim 6.5 \, k\Omega$. This gives $l = 700$ nm for this semiconducting SWNT at room temperature and zero gate voltage (only lightly p-doped) (Fuhrer *et al.* 2001).

During the EFM measurement, there is a large potential difference between the tip and the sample. The tip may therefore locally modify the

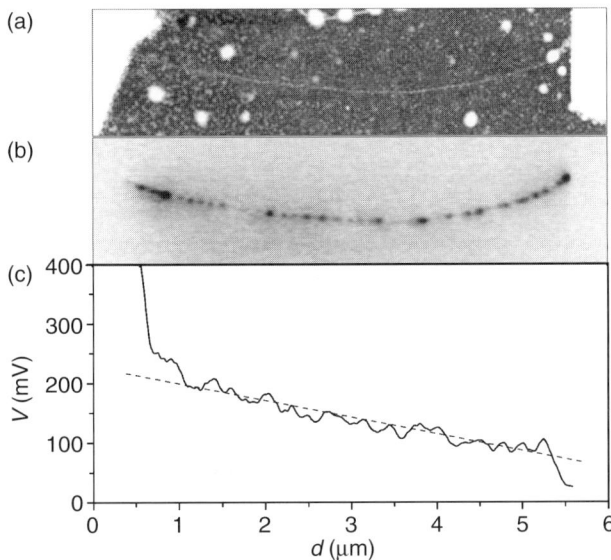

FIG. 8. EFM and SGM measurements of a semiconducting SWNT at room temperature (Fuhrer *et al.* 2001). A topographic AFM image is shown in (a); the white areas at left and right are the Cr/Au electrodes, and the faint gray line is the nanotube. An SGM image of the same device is shown in (b). The SGM image was taken with −1 V applied to the tip, and 1 V applied across the nanotube. Dark color indicates decreased resistance; black corresponds to a change in resistance of approximately 500 Ω. The local potential in the nanotube under an applied bias of 400 mV as measured by EFM is shown in (c). The dashed line is a guide to the eye, and represents a voltage drop corresponding to 9.2 kΩ/μm. The distance scale in (c) applies to all panels.

conducting properties of the sample as it scans over it; this is particularly likely in the case of the semiconducting nanotubes, since their conductance changes with shifts in the Fermi level. Scanned gate microscopy (SGM) images this perturbation by measuring the conductance of the sample as a function of tip position. The conductance changes when the tip locally depletes, or gates, the underlying electron system. This technique has been used in the past to study quantum point contacts (Eriksson *et al.* 1996) and quantum Hall conductors (Woodside *et al.* 2001).

Figure 8(b) shows an SGM image of a semiconducting SWNT device. The tip affects the conductance only at discrete locations along the nanotube. The resistance change at each point under application of a negative gate voltage is less than 1 kΩ, indicating each point is only weakly scattering the electrons; the intrinsic resistance arises from a large number of weakly scattering defects. These locations are likely areas of high local potential, caused by disorder in the nanotube or defects, for example, trapped charge, in the substrate (Fuhrer *et al.* 2001).

Unlike in metallic SWNTs, the intrinsic resistance of semiconducting SWNT is strongly dependent on the synthesis method and treatment of the nanotube material. Previous EFM measurements on semiconducting SWNTs derived from the laser ablation process and deposited on substrates after ultrasonication show that large barriers to transport ($R_i \sim M\Omega$) are present along the nanotube with a spacing of order ~ 100 nm (Bachtold *et al.* 2000).

The mean free path may be used to estimate the carrier mobility for the semiconducting nanotube. The mobility is given by $\mu = e\tau/m^* = el/m^*v_F$, where the τ is the scattering time and m^* is the effective mass of the holes. For a 1D conductor with two bands at the Fermi level, $m^*v_F = \hbar k = \pi\hbar n/4$ where $\hbar$ is Planck's constant and n the hole density per unit length. The hole density may be estimated from the capacitance per length c_g, using $n = c_g(V_t - V_g)$, where V_t is the threshold voltage. The capacitance per unit length of the tube can be obtained from low temperature measurements of metallic SWNT (Bockrath *et al.* 1997; Tans *et al.* 1997; Cobden *et al.* 1998b). Using $c_g = 6.0\,e^-/V \cdot \mu m$, and $V_t - V_g \approx 10$ V, a mobility $\mu = 23\,000\,\mathrm{cm}^2/V \cdot s$ is obtained.

The transport data $G(V_g)$ [see Fig. 3(b)] may also be analyzed to extract a carrier mobility. The Drude conductivity relation $\sigma = ne\mu$, where σ is the conductivity and n the linear carrier density, predicts that the conductance of the nanotube will grow linearly with carrier density (and hence gate voltage) if the mobility is constant: $G = c_g(V_t - V_g)\mu/L$. From the linear portion of the $G(V_g)$ curve a mobility of the tube μ may be inferred. For the 4.8 μm long semiconducting SWNT device in Fig. 3(b), the slope $dG/dV_g = 2.2\,\mu S/V$. Using a capacitance of $c_g = 6.0\,e^-/V \cdot \mu m$, a mobility $\mu = 11\,000\,\mathrm{cm}^2/V \cdot s$ is obtained.

Typical mobilities for CVD-grown SWNTs obtained from the $G(V_g)$ characteristics are in the range of a few thousand to occasionally as high as $20\,000\,\mathrm{cm}^2/V \cdot s$. Mobility determined from $G(V_g)$ may underestimate the true mobility if the contact resistance also varies with gate voltage, which is not unreasonable. The fact that the EFM measurements are consistent with the highest mobilities from $G(V_g)$ bears this out.

These mobilities significantly exceed the mobilities of laser-ablation nanotubes deposited from suspension, where μ is on order 10–$100\,\mathrm{cm}^2/V \cdot s$ (Martel *et al.* 1998). The mobilities of CVD-grown semiconducting SWNTs are extraordinary, exceeding the best room temperature Si MOSFETs, indicating that SWNTs are a remarkably high-quality semiconducting material.

b.   Low Temperature

Initial low-temperature experiments were performed on laser-ablation derived SWNT. Transport spectroscopy performed on such semiconducting

SWNT devices shows a large gap in the conductance as a function of source–drain voltage at positive V_g which narrows at negative V_g, but never closes. This behavior has been interpreted as follows: at positive V_g, the semiconducting SWNT is depleted of carriers and the barrier to transport is the intrinsic bandgap. At negative V_g, the SWNT contains hole carriers, but disorder causes the SWNT to break up into a string of quantum dots in series separated by large barriers (McEuen *et al.* 1999). Because the dots are in series, at any given V_g some of the dots will be blockaded and there will be a gap to transport. The size of the gap (25–50 meV) indicates that the smallest dots are on order 100 nm. This is consistent with the observation via EFM of large barriers to transport along the length of the nanotube (Bachtold *et al.* 2000).

More recently, devices have been made using semiconducting SWNTs grown via CVD. At low temperature, and at negative gate voltage, a single oscillation period of the conductance as a function of gate voltage is seen, similar to metallic SWNT quantum dots (Fuhrer *et al.* 2001). The period in V_g and charging energy are consistent with a quantum dot with the length of the entire semiconducting SWNT, in this case $\sim 1\,\mu m$. This indicates a mean free path of at least 1 μm at low temperature; implying that if semiconducting SWNTs are made relatively free of disorder, they may also have long mean free paths at low temperature.

No reports have to date been made of ohmically contacted semiconducting SWNTs. This may indicate the presence of an intrinsic barrier at the semiconducting SWNT–electrode interface, at least for the electrodes currently being employed. It is also not clear how to apply the model of Luttinger liquid tunneling to the semiconducting nanotubes. The contact resistance of semiconducting nanotubes does increase as the temperature is lowered, but connection with the Luttinger tunneling model is greatly complicated by the nonlinear band dispersion and the closeness of the Fermi level to the band edge.

4. SUMMARY OF ELECTRONIC TRANSPORT PROPERTIES

Nanotubes are extremely 1D; a 1.5 nm SWNT has a sub-band spacing of order 1 eV, much greater than the room temperature thermal energy of 25 meV. Because of the 1D nature of carbon nanotubes, electronic transport properties must be interpreted carefully. For instance, a perfect (zero scattering) SWNT with perfect contacts will still have a two-terminal resistance of $\sim 6.5\,k\Omega$, the quantized contact resistance of a 1D channel with four modes.

The interacting ground state of a 1D conductor is termed the Luttinger liquid. The excitations of the Luttinger liquid are no longer single-particle-like. This has consequences for the conductance of a tunnel junction contact to a SWNT; a power-law gap is observed in the tunnel conductance as a function of voltage and temperature.

The intrinsic resistance of the SWNT channel may be much less than the measured two-probe resistance. Metallic SWNTs have mean free paths of $>1\,\mu m$, while moderately doped semiconducting SWNTs have mean free paths of hundreds of nanometers. These values correspond to resistances of <6 and $\sim 10\,k\Omega/\mu m$ for metallic and semiconducting SWNTs, respectively. The resistivities of nanometer-diameter metallic and semiconducting SWNTs are on order $10^{-6}\,\Omega\,cm$, comparable to good metals.

Given the dispersionless nature of the metallic SWNT bands, and some uncertainty in how to define the carrier density (relative to the sub-band bottom or band bottom), it is difficult to assign a mobility to the metallic SWNT devices. For semiconducting nanotubes, there is a well-defined effective mass and carrier density is readily defined relative to the bandgap edge; thus the mobility can be defined. Furthermore, a fairly linear dependence of the conductance on gate voltage (and hence the carrier density n) also argues for the usefulness of the mobility in describing the properties of semiconducting nanotubes; over a significant range, the carrier mobility is constant and the conductivity is Drude-like, increasing linearly with n. Semiconducting nanotubes have extraordinary room temperature mobilities, reaching values up to $20\,000\,cm^2/V\cdot s$, exceeding high-mobility Si MOSFETs.

V. Nanotube Nanoelectronic Devices

This section will discuss efforts to fabricate useful electronic devices based on carbon nanotubes. The fundamental motivation for this work is the impending end of Moore's law scaling in silicon CMOS electronics. This has motivated a great deal of interest in the search for alternative technologies that may replace or augment CMOS. Such technologies generally come under the heading "nanoelectronics", since the relevant length scale at which they will become competitive with CMOS is thought to be significantly less than 100 nm.

Many nanoelectronics schemes have in common the idea of "bottom-up" assembly. In contrast to "top-down" assembly, where circuits are laid out precisely and then shrunk to sub-micrometer dimensions using, for example, lithography, "bottom-up" assembly starts with individual nanometer-scale components, and uses weak forces, for example, molecular recognition, to "self-assemble" the components into some useful larger structure. The hope is that the significant drawbacks of self assembly (high defect density, simplicity of structure) will be overcome by its advantages (small component size, low cost).

At first glance, nanotubes seem an ideal material for bottom-up fabrication: both metals and semiconductors are present, nanotubes are extremely robust to chemical and physical treatment, and carbon chemistry is well known. However, significant hurdles exist which have yet to be addressed: no method exists for separating nanotubes according to electronic property; solubility of nanotubes is poor; and while organic carbon chemistry is well known, graphene chemistry is not. Efforts to overcome these difficulties will be discussed in the next section.

Nevertheless, work on nanoelectronic nanotube devices progresses. Out of necessity, the first model devices have been constructed using a top-down approach. These include nanotube FETs and SETs, the basics of which were discussed in Section IV. This section will continue the discussion with greater focus on tailoring the characteristics of these devices to produce useful behavior.

One of the simplest geometries which may be considered in a bottom-up fabrication scheme is a grid of perpendicular wires, often called a crossbar array. The fabrication of such a grid only requires the fabrication of an array of aligned, evenly spaced long wires (twice, at right angles to each other). Devices in the crossbar array may be formed naturally at the junctions between wires (the wires are the devices), or may be placed at the junctions through further self-assembly (the wires are the interconnects). In either case, electronic programmability of devices is a useful characteristic, since this allows the use of the use of the very simple geometry to create more complicated devices. Nanotube–nanotube junctions have been explored as both electronic and programmable electromechanical devices as well as robust nanoscale interconnects.

1.   FIELD-EFFECT TRANSISTORS

As discussed in Section IV.3, the semiconducting SWNT channel acts like an FET; positive gate voltage shuts off conduction, while the conduction is enhanced by negative gate voltage. Because the field is rapidly developing, it is worthwhile to briefly review the history of the attempts to understand the device behavior of SWNT FETs before continuing to discuss efforts to modify that device behavior.

Tans et al. (1998b) reported the first electrical transport measurements on an individual semiconducting SWNT, as discussed in Section IV.3.a. Their devices were made using laser ablation derived SWNT material using the tube-on-top method. Tans et al. considered that the semiconducting SWNT acts as an intrinsic *ballistic* channel between the source and drain electrodes. Because of difference in work function ϕ between the metal electrodes (platinum, with a work function $\phi = 5.7\,eV$) and the nanotube (with $\phi = 4.5\,eV$), the Fermi level in the nanotube is pinned in the valence band at the contacts. Far from the electrodes, the nanotube is intrinsic. The bands bend towards the intrinsic position over the electrostatic screening length, which may be very long in one dimension. This creates a barrier for hole transport through the nanotube, which may be lowered by applying a negative voltage to the gate electrode. The analogous semiconductor device is termed a barrier impact transit time (BARITT) diode. Since the barrier height depends strongly on the ratio of the device length to the electronic screening length, the BARITT model predicts a large sensitivity of the device characteristics to device length, which was not observed in subsequent research.

Martel *et al.* (1998) also fabricated semiconducting SWNT devices from laser ablation derived material using the tube-on-top technique. They observed similar device characteristics as reported by Tans *et al.*, but explained the transistor behavior with a qualitatively different model: conduction through the semiconducting SWNT is *diffusive*. The semiconducting SWNT acts as the channel of a MOSFET, apparently intrinsically doped slightly p-type, and the device operates in a p-channel depletion mode (positive gate voltage shuts off the conduction, and negative gate voltage enhances the conduction). The conductance saturates at large negative V_g due to the series resistance of the contacts. The linear region in the conductance as a function of gate voltage was used to extract a hole mobility with a rather low value of $20 \, \mathrm{cm^2/V \cdot s}$.

As discussed above, scanned probe experiments (see Section IV.3) shed light on this problem. Electronic transport through laser ablation derived semiconducting SWNTs is dominated by a number of large conductance barriers (Bachtold *et al.* 2000), which explains the low mobility of the devices measured by Martel. *et al.* In CVD-grown semiconducting SWNTs, the intrinsic resistance is much lower; the mean-free path is several hundred nanometers, and the hole mobility may be as high as $20\,000 \, \mathrm{cm^2/V \cdot s}$. However, the transport is still diffusive, as seen from EFM experiements (Fuhrer *et al.* 2001), and semiconducting SWNT devices have a well-defined resistance per length.

a. Transconductance

As discussed in Section IV.3.a, the conductance of an FET is given by $G = c_g(V_t - V_g)\mu/L$. The transconductance $\mathrm{d}I_{SD}/\mathrm{d}V_g$ is then maximized by increasing the gate capacitance and the carrier mobility. The transconductances of the first SWNT FETs were very low, simply because most experiments to date have used gate oxide thicknesses of hundreds of nanometers. More recently, researchers have investigated a number of ways to increase the gate coupling. Bachtold *et al.* (2000) deposited SWNTs on top of oxidized Al wires which acted as gate electrodes. In this way FETs were constructed with dielectric thicknesses of a few nanometers. Although low-mobility laser ablation derived SWNTs were used, transconductances of up to 0.3 μS were obtained, and transistor–resistor logic devices showing gain were constructed.

An electrolytic solution may be used to gate CVD-grown semiconducting nanotube devices (Rosenblatt *et al.*, unpublished results) (see Fig. 9). In this way the effective dielectric thickness could be made very small, on the order of a few nanometers. A maximum transconductance of 20 μS was observed. Normalizing this to the device width of ~2 nm, this gives a transconductance per unit width of ~10 mS/μm; significantly better than current-generation MOSFETs, but reasonable in light of the high hole mobility of semi-conducting SWNTs.

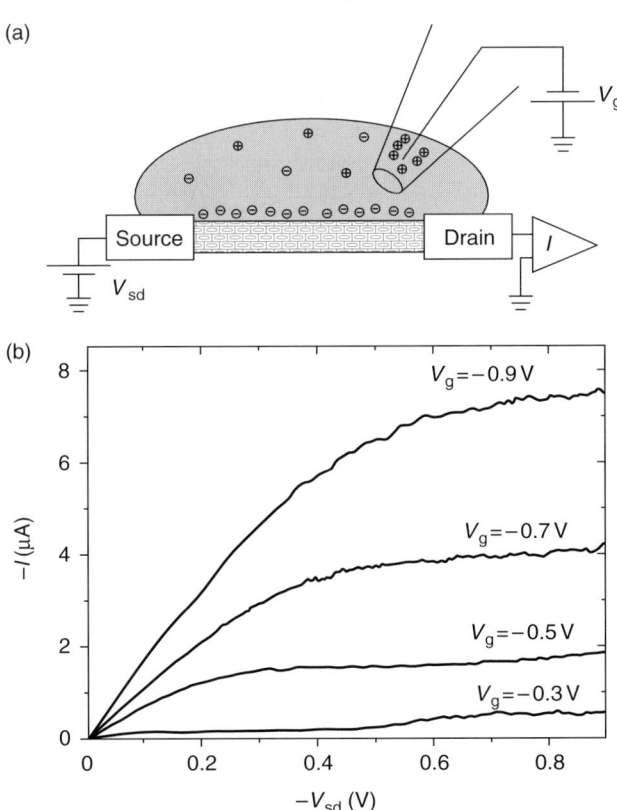

FIG. 9.   Gating an SWNT FET using an ionic solution (images courtesy P. L. McEuen). A schematic of the device is shown in (a). A pipette containing a NaCl solution (1–100 mM) is used to place a drop of solution over the device. An electrode inside the pipette acts as a gate terminal. The device chartacteristics are shown in (b). The maximum transconductance of the device is 20 μS.

b. Doping

As-prepared semiconducting SWNT FETs are normally-on (depletion mode) p-channel FETs. The p-channel conduction results from pinning of the Fermi level of the SWNT at the electrode at or near the valence band edge, and normally-on behavior results from doping by ambient species such as oxygen. Control of the nanotube-electrode interface, and control of the chemical doping of the nanotube channel may be used to alter these characteristics.

n-Type doping of SWNTs was first accomplished using alkalai metals that donate electrons to the tube. This has been used to create n-type transistors (Bockrath *et al.* 2000; Kong *et al.* 2000a; Derycke *et al.* 2001), p–n junctions (Zhou *et al.* 2000), and p–n–p devices (Kong *et al.* 2002). Alkalai metals are not air-stable, however, so other techniques are under development, such as using polymers for charge-transfer doping (Kong and Dai 2001).

Martel *et al.* (2001) found that annealing of nanotubes with titanium electrodes forms a covalently bonded TiC–nanotube contact, with the Fermi level in the SWNT at the interface pinned near the center of the gap. The hole doping due to oxygen could be removed by passivating the device with SiO_2, followed by treatment at high temperature. After this process, ambipolar transistor behavior is observed, with efficient injection of both electrons and holes. This is intriguing, because typically the sum of the Schottky barrier heights for electrons and holes should equal the band gap ($\sim$600 meV for the laser ablation derived nanotubes used in this study). Apparently once electrons or holes are induced into the bulk of the SWNT, the Schottky barrier width in the 1D nanotube becomes extremely small, so tunneling through the barrier dominates the conduction.

c. Summary

The high mobility of semiconducting SWNTs leads to excellent device properties in SWNT FETs. While the first devices had low transconductances due to poor quality nanotubes and extremely thick gate dielectrics, recent experiments indicate that, with a suitably thin gate dielectric, the transconductance of SWNT FETs may exceed that of the best silicon devices when scaled by the device dimensions. Work on doping the nanotube channel and controlling the nanotube–metal interface has progressed significantly. n-Type doping has been demonstrated, and ambipolar transistor action has been achieved by using covalently bonded TiC contacts to the nanotube.

Several unique aspects of the nanotube FET have been pointed out as possible advantages for applications. Unlike conventional MOSFETs, the conducting channel of the nanotube FET is exposed at the surface. High sensitivity of the nanotube FET to chemical environment has already been demonstrated, and it is natural to think of the applicability of nanotube FETs as chemical sensors. SWNT FETs can operate in water, including in ionic solutions, suggesting possible sensing applications in a biological or chemical "lab on a chip". The small size of the nanotube FET suggests memory applications; the use of SWNT FETs in floating-gate memories could reduce the amount of stored charge, and greatly increase speed and density.

2. SINGLE-ELECTRON TRANSISTORS

Since the first SETs were fabricated by Fulton and Dolan (1987) in 1987, there have been efforts to increase the operating temperature for single-electron devices. There now exist a handful of reports on SETs operating at room temperature. Room temperature SETs have been fabricated by nano-oxidation of a Ti metal film using a scanned-probe tip (Matsumoto *et al.* 1996), random oxidation to form constrictions in a thin silicon wire (Zhuang *et al.* 1998), and even using the original technique of Fulton and Dolan (electron-beam lithography and shadow evaporation to form $Al/AlO_x/Al$ structures) by modifying the electron-beam lithography to achieve exceedingly fine resolution (Pashkin *et al.* 2000). A charging energy of over 150 meV was reported for an SET using a single C_{60} molecule as an island, but the device was not shown to be operable at room temperature (Park *et al.* 2000).

Since the capacitance of SWNT SETs scales inversely with length (see Section IV.2.b) one may predict that a SWNT device with length shorter than $\sim$50 nm will have a charging energy greater than the thermal energy at room temperature. Lefebvre *et al.* (2000) fabricated very short SWNT SETs via a metal-on-top technique, with the gap between electrodes defined by a nanotube used as a shadow mask. They found an electron addition energy in their devices on order 100 meV, but no Coulomb oscillations were observed at room temperature, possibly due to thermal activation of electrons over the tunnel barrier formed at the nanotube–electrode interface.

Postma *et al.* (2000) had shown that tunnel junctions could be produced in metallic SWNTs through AFM manipulation to produce a kink (see also Section IV.2.c). Bozovic *et al.* (2001) took advantage of this technique to produce pairs of tunnel junctions in SWNTs to form SETs with island sizes less than 100 nm. These small islands acted as SETs. One island with length $\sim$50 nm had an addition energy of $\sim$70 meV, and showed Coulomb oscillations up to a temperature of about 165 K. Postma *et al.* (2001) used the double-kink technique to produce a device with a length of $\sim$25 nm, which showed Coulomb oscillations at room temperature (see Fig. 10). This device had an addition energy of $\sim$120 meV.

The AFM manipulation technique used to fabricate double-kink SWNT SETs is rather uncontrollable, and difficult to envision as a method of mass production. However, these devices show in general some of the advantages

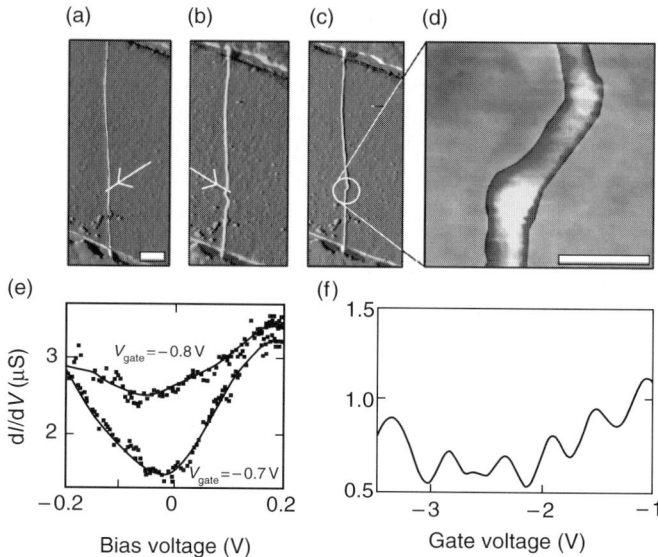

FIG. 10. Room temperature nanotube SET (Postma *et al.* 2001) (images courtesy Cees Dekker). An AFM tip was used to modify a nanotube device by dragging over the device along the lines shown in (a) and (b). This formed two kinks in the nanotube shown in (c) and (d). The scale bar in (a)–(c) is 200 nm, and the scale bar in (d) is 20 nm. The differential conductance as a function of bias voltage (e) at 300 K shows significant variation with gate voltage. The differential conductance as a function of gate voltage (f) at 260 K shows periodic oscillations due to Coulomb blockade.

of making SETs from SWNTs, which in principle may be generalized to other fabrication techniques.

The reason that scaling of the SWNT SETs to small lengths produces high charging energy devices is that the capacitance is dominated by the length of the nanotube, not the area of the contacts. In SETs fabricated by, for example, the $Al/AlO_x/Al$ process the total capacitance is typically dominated by the capacitance of the tunnel junctions. In order to produce smaller capacitance devices, one must make smaller area tunnel junctions (Pashkin *et al.* 2000). The problems with this approach are that fabricating small tunnel junctions by brute-force lithography is extremely difficult, and the resistance of the tunnel junction increases as the area decreases. With SWNT SETs, a small-area tunnel junction is self-assembled; the area of the SWNT controls the area of the junction. Lithography need only define the long dimension of the SET, on the order of 25 nm for room temperature operation.

This advantage may also be seen in the ratio of gate capacitance to total capacitance C_g/C_Σ for the SWNT SETs. Both double-kink devices discussed above have ratios of C_g/C_Σ of around 0.3, while the room temperature SET described by Pashkin *et al.* has a ratio C_g/C_Σ of about 3×10^{-3}, two orders of magnitude smaller. That this ratio determines the maximum transconductance of an SET can be shown by a rough calculation as follows: The transconductance is the change in current through the device divided by the change in gate voltage. The maximum current modulation will be produced if the device operates at a source-drain bias $V_{sd} \approx E_c = e/2C_\Sigma$. The maximum possible conductance through a single-electron charging device is $G_{max} \approx e^2/h$. Thus, the maximum possible current is $I_{max} = G_{max}V_{sd} = e^3/2hC_\Sigma$. The current varies from minimum to maximum over a change in gate voltage $\Delta V_g/2 = e/2C_g$. Therefore, the transconductance is $2I_{max}/\Delta V_g \approx (e^2/h)$ (C_g/C_Σ). This simple derivation assumes that the device is classical, that is, $\Delta E \ll k_B T$, which is not true for the SWNT devices. However, the general result holds, increasing C_g/C_Σ increases the transconductance of the device.

The result above suggests that not only would SWNT islands make good SETs, but metal islands with SWNT leads would also make good SETs, since they too would have small junction capacitances.

3.   NANOTUBE HETEROJUNCTION DEVICES

a.   Linear Nanotube Heterojunctions

Iijima observed discrete transitions in the curvature of the shells making up MWNTs, and posited that the insertion of a pentagon or heptagon in the graphene lattice was responsible (Iijima 1993). It was pointed out that nanotubes of different diameter and/or chirality may be joined seamlessly by the introduction of pairs of pentagons and heptagons into the graphene lattice (Dunlap 1994). All carbon metal–semiconductor diodes were envisioned by joining a metallic and a semiconducting SWNT by a single pentagon–heptagon pair (Charlier et al. 1996; Chico et al. 1996; Saito et al. 1996). Chico et al. (1998) explored the possibility of an on-tube quantum dot produced by a pair of pentagon–heptagon defects which connected a metallic (5,5) SWNT to two semiconducting (6,4) SWNT leads.

Collins et al. (1997) provided the first experimental hints at the existence of on-nanotube electronic devices. They probed a tangled mat of laser ablation produced SWNT bundles with an STM tip. The tip was brought into electrical contact with the mat, and it was found that the tip could often be retracted for several micro meters before electrical contact was lost. They proposed a model of a sliding contact between tip and nanotube bundle as the tip was retracted. During retraction of the tip, current–voltage (I–V) characteristics of the tip–mat junction were acquired at various retraction distances. It was found that the I–V characteristics would sometimes abruptly change from symmetric to strongly rectifying. These events were interpreted as the tip sliding past a defect in a nanotube which acted as an on-tube device. The experiments of Collins et al. were performed essentially blind; the STM was not used in imaging mode. Hence, there was no direct evidence that the rectifying behavior resulted from an on-tube device, and not, for example, a junction between two different SWNTs in a bundle, or a shift of the STM tip from one SWNT to another in a bundle.

In fabricating tube-on-top electrical devices (see Section III.2), Yao et al. (2000) occasionally observed individual SWNTs with sharp kinks along their length. Out of 500 devices, four were found with single kinks, and one with two kinks. These kinks were interpreted as individual pentagon–heptagon pair defects causing an abrupt change of chirality of the SWNT at the defect. Electrical measurements on two kinked SWNTs were reported in detail. The first device unfortunately spanned only three electrodes, so independent two-terminal measurements could only be performed on one side of the kink. The portion of the nanotube on this side of the kink junction had a two-probe conductance of $110\,k\Omega$, with no gate voltage dependence, indicating a metallic SWNT. Two-probe measurements across the kink, however, gave an immeasurably low conductance ($<4\,pS$) at zero bias, but a strong non-linear onset of conduction when around $+1–2$ V was applied to the metallic SWNT. The onset of conduction shifted significantly with the application of a gate voltage, becoming more conducting at positive bias for negative gate voltages. Because of the gate voltage dependent conductivity of the kink segment, the side of the kink which contact only a single electrode was presumed to be a semiconducting SWNT. The rectifying behavior was then attributed to the formation of a metal–semiconductor Schottky barrier at the kink, although the very high threshold voltage ($1–2$ V) (Odintsov 2000) and the incorrect sign of the rectification (a Schottky barrier between a metal and p-doped semiconductor should conduct for negative bias applied to the metal) indicated that the device may be somewhat more complicated. (A second device was found to be a junction between two metallic SWNTs; the conductance of this device was consistent for an end-to-end Luttinger liquid junction as discussed in Section IV.2.c.)

b.   Nanotube T and Y Junctions

More complex all-carbon covalently bonded nanotube structures have also been envisioned (Hamada 1993; Menon and Srivastava 1997). Hamada studied the band structure of 1D, 2D, and 3D superstructures formed by the joining of SWNTs with pentagons and heptagons, and envisioned an all-carbon circuit network. Menon and Srivastava (1997) studied two T junctions connecting metallic and semiconducting SWNTs. The junctions were shown to be stable under molecular dynamics relaxation. Andriotis et al. (2001) found that Y junctions between SWNTs may rectify currents.

Such ideal junctions between SWNT have been difficult to realize experimentally. However, a number of authors have reported branching structures of larger tubular carbon nanofibers, synthesized by templated pyrolysis of acetylene in nanochannel alumina (Li *et al.* 1999), pyrolysis of nickelocene and thiophene (Satishkumar *et al.* 2000), pyrolysis of methane over cobalt supported on MgO (Li *et al.* 2001), and pyrolysis of methane over iron particles on roughened silicon (Ting and Chang 2002). These structures are all tens of nanometers in diameter and thus not electronically one dimensional at room temperature. Moreover, those structures for which TEM images exist show poor graphitization (Li *et al.* 1999, 2001; Satishkumar *et al.* 2000), and thus do not fit the strict definition of nanotube used in this chapter. However, electrical measurements of individual examples as well as ensembles of Y junction fibers synthesized inside nanochannel alumina templates show interesting rectifying behavior, though it is difficult to connect these structures with the rectifying Y junctions considered theoretically (Andriotis *et al.* 2001).

c. Crossed-Nanotube Junctions

Devices consisting of single junctions between individually electrically contacted nanotubes have been fabricated using a metal-on-top method (Fuhrer *et al.* 2000). Each crossed-SWNT device consisted of two crossed individual SWNTs or small bundles (diameter < 3 nm) of SWNTs with four electrical contacts, one on each end of each SWNT or bundle. The inset of Fig. 11 shows an AFM image of such a crossed nanotube device; two crossed SWNTs interconnect Cr/Au contacts.

This configuration allows each SWNT in the junction to be measured independently to determine its properties. The room temperature two-terminal conductance as a function of gate voltage $G(V_g)$ characteristics measured across the individual SWNTs allowed for the assignment of each as metallic or semiconducting. Each crossed-SWNT device can be composed of two metallic SWNTs (MM), one metallic and one semiconducting SWNT (MS), or two semiconducting SWNTs (SS).

Figure 11 shows the four-terminal current–voltage (I–V) characteristic of an MM junction at 200 K (filled squares). The slope of the I–V curve corresponds to a resistance of 200 kΩ, or a conductance of $0.13e^2/h$. Similar measurements of three other MM junctions gave conductances of $0.086e^2/h$, $0.12e^2/h$, and $0.26e^2/h$. The measurements of SS junctions are often complicated by the presence of high resistance barriers in the laser-ablation synthesized semiconducting SWNT (see Section IV.3). Nevertheless, two-terminal conductances of SS junctions as high as $0.011e^2/h$ and $0.06e^2/h$ (the higher conductance curve is represented by the filled circles in Fig. 11) were observed.

The measured conductances of MM junctions correspond to a transmission probability for the junction $T_j = G/(4e^2/h) \approx 0.02$–$0.06$. Thus, an

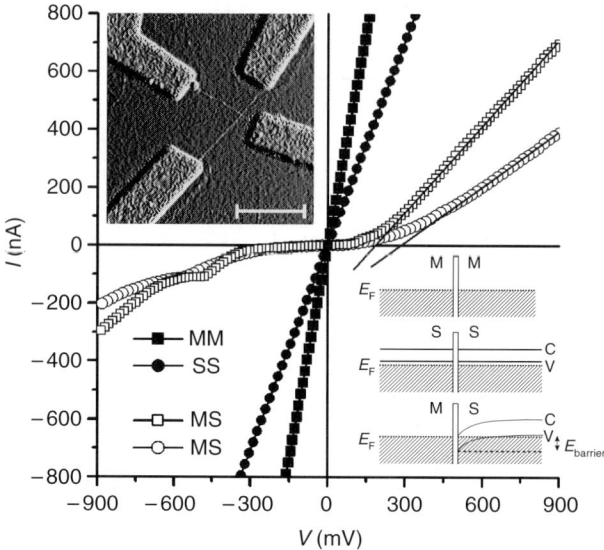

FIG. 11. Current–voltage characteristics of crossed carbon nanotube junctions (Fuhrer *et al.* 2000). The *I–V* curves of a metal–metal junction (filled squares), a semiconductor–semiconductor junction (filled circles), and two metal–semiconductor junctions (open squares and circles) are shown. The lower right inset shows the expected band profiles on either side of the junction for each case. The non-linear behavior of the metal–semiconductor junctions is due to the formation of a Schottky barrier at the junction, as shown in the inset. The solid lines in the main panel extrapolate the forward bias region to zero current, giving an estimate of the barrier height of 190–290 meV. The upper left inset show an AFM image of a typical crossed nanotube device. Two narrow crossing lines are SWNTs, the larger blocks are Cr/Au electrodes. The scale bar is 1 μm.

electron arriving at the junction in one SWNT has a few percent chance of tunneling into the other SWNT. MM junctions make surprisingly good tunnel contacts, despite the extremely small junction area on the order of $1\,\text{nm}^2$. In order to understand these results, first-principles density functional calculations of the conductance of MM junctions were performed. For two (5,5) SWNTs separated by the van der Waals distance of 0.34 nm, a transmission $T_j \approx 2 \times 10^{-4}$ was found. However, when the contact force between the nanotubes due to interaction with the SiO_2 substrate was included (Hertel *et al.* 1998), the nanotubes deformed significantly at the junction. In this case, it was found $T_j \approx 0.04$, in excellent agreement with the experimental result. The high conductance of MM junctions indicates that metallic SWNT junctions may be useful for branching interconnects.

The MS case is qualitatively different from the MM and SS cases. Charge transfer at the junction between a doped semiconducting SWNT and a metallic SWNT is expected to form a Schottky barrier (Odintsov 2000) with the Fermi level E_F of the metallic SWNT aligned with the center of the band gap of the semiconducting SWNT at the junction (see lower inset, Fig. 11). The total barrier transmission probability T_{MS} is then given by $T_{MS} \approx T_j \, T_d$, where T_d is the transmission probability for tunneling through the depletion region to the location of the metal SWNT and T_j is the probability of tunneling between the SWNTs. $T_{MS} \approx 2 \times 10^{-4}$ for both MS devices in Fig. 11. If we assume that $T_j \approx 0.04$ for the MS junctions (comparable to the MM value), then $T_d \approx 5 \times 10^{-3}$, in excellent agreement with a calculation by Odintsov (2000), who found $T_d \approx 5 \times 10^{-3}$ and a corresponding depletion width of 7 nm for a doping level similar to the experimental case.

A Scottky barrier should also show asymmetric transport; indeed this is the case for the MS junction barrier (see Fig. 11). The V-intercept of the linear positive-bias region gives a gross measure of the barrier height: $E_{barrier} = 190$ and 290 meV for the two devices. This agrees reasonably well with the expected barrier height $E_{barrier} = E_g/2 \sim 250$–$350$ meV for 1–1.5-nm semiconducting SWNTs ($E_g \sim 500$–700 meV).

Poor rectification is obtained because the MS Schottky barrier is rather leaky. Better rectification was achieved with a three-terminal device which takes advantage of the double-width depletion barrier in the semiconducting nanotube (Fig. 12). A voltage was applied to one end of the semiconducting SWNT in an MS device, while the other end was grounded through a current-measuring amplifier. If the metal SWNT as well as one end of the semiconducting SWNT are grounded, the barrier to holes will remain intact

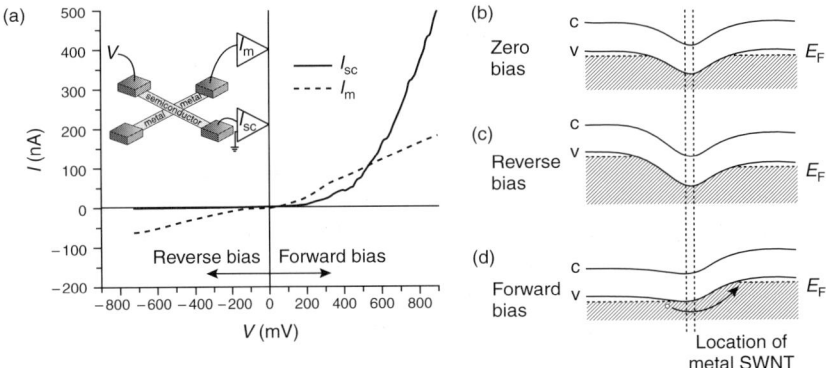

FIG. 12. Three-terminal characteristics of a metal–semiconductor crossed-nanotube junction (Fuhrer *et al.* 2000). In (a), the bias is applied to one end of the semiconducting nanotube, the current flowing to ground is measured at the other end (see inset). The metallic nanotube is also grounded through a current amplifier. The solid curve in (a) shows the current in the semiconducting nanotube as a function of bias voltage. The dotted curve shows the current flowing from the metallic nanotube. The schematic operation of the device is shown in (b)–(d). At zero bias (b), a depletion barrier exist in the semiconducting nanotube on either side of the metallic nanotube. On reverse bias (c), the barrier on the biased side increases, and no current flows. On forward bias (d), the barrier on the biased side is decreased, and holes may pass the barrier.

when a negative voltage is applied, because the metallic SWNT remains at roughly the same potential as the grounded end of the semiconducting SWNT [Fig. 12(c)]. However, when a positive bias is applied such that the potential difference between the metallic and semiconducting SWNT is greater than the barrier height, holes may pass the barrier, and current will flow through the semiconducting SWNT [Fig. 12(d)]. Such a response is observed in the measured current leaving the semiconducting SWNT [I_S in Fig. 12(a)].

The current from the semiconducting SWNT is approximately 100 times greater for a bias of $+700\,\text{mV}$ than for $-700\,\text{mV}$. It has been noted that the ineffective screening inherent to 1D systems poses problems for nanotube Schottky devices (Leonard and Tersoff 1999): nanometer-scale depletion regions are likely to be leaky barriers to tunneling. This three-terminal device points out at least one solution to this problem; a good rectifier is constructed from narrow Schottky barriers. The active length of this device is on the order of 15 nm, demonstrating that useful devices consisting of only a few thousands of atoms can be constructed from SWNTs.

4. MECHANICAL DEVICES

Carbon nanotubes are exceedingly light and stiff. This suggests that nanomechanical devices based on carbon nanotubes could operate at very high speeds. Indeed, simple oscillators made from sub-micro meter nanotube beams have characteristic mechanical frequencies in the Gigahertz range (Reulet *et al.* 1999). In addition, nanotubes are exceptionally tough; the elastic limit (elongation) for SWNTs is ~6%. Nanotubes may be bent through large angles, and even buckled, and return to their original shape elastically (Yakobson and Avouris 2001). This suggests that nanotube mechanical devices could operate reversible for large numbers of cycles with no fatigue.

Rueckes *et al.* have proposed a mechanically bistable nanotube device which would act as a memory or programmable logic element (Rueckes *et al.* 2000). The device consists of two nanotubes; one lies on a substrate, while the other nanotube crosses it, suspended above it at a distance of a few nanometers (see Fig. 13). The total energy of the device was analyzed as a function of the nanotube separation for a range of initial separations and nanotube lengths, and bistability of the device was found for a broad range of parameters. The two stable states correspond to the nanotubes in close contact (held by van der Waals forces), and relaxed. In order to switch between the two states, opposite polarity voltages would be applied between the two nanotubes to cause them attract each other while like polarity voltage (relative to a third gate electrode) would cause the nanotubes to separate.

Rueckes *et al.* fabricated a demonstration device using bundles of SWNT deposited onto pre-fabricated gold electrodes with a height of 150 nm and a

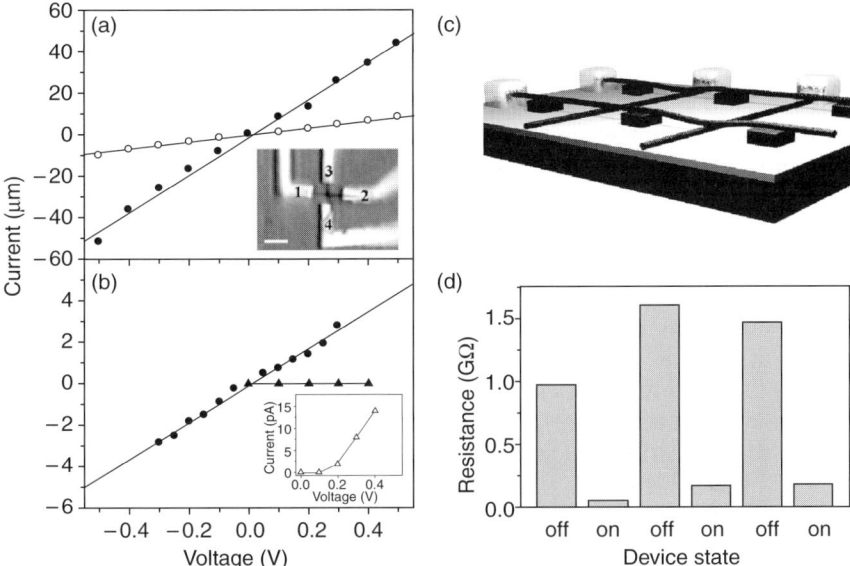

FIG. 13. Carbon nanotube mechanical memory (Rueckes *et al.* 2000) (images courtesy C. M. Lieber). The inset of (a) shows an optical micrograph (scale bar 4 μm) of a device consisting of two crossing bundles of SWNTs (dark lines) deposited on 150 nm high Cr/Au electrodes (white). The *I–V* characteristics measured from contacts 1 to 2 (solid circles) and 3 and 4 (open circles) are shown in (a). The *I–V* curves of the junction in the OFF state (open triangles) and ON state (solid circles) are shown in (b). The junction was switched to the ON state by the application of ±2.5 V. The schematic of a possible nanotube mechanical memory array is shown in (c). Panel (d) shows the ON/OFF resistance for a device similar to that shown in (a), but which showed reversible switching. This device could be switched ON with application of ±5 V, and OFF at 40 V.

separation of ∼4 μm (see Fig. 13). They were able to demonstrate irreversible switching of one device from a separated state ($>$10 GΩ) to a contacted state (112 kΩ) at a voltage of 2.5 V. Another device switched reversibly from separated (1.36 GΩ) to contacted (140 MΩ) with the application of ±5 V to contact and 40 V to separate the SWNT bundles. The authors calculate that in principle, bistable devices could be made as small as 5–20 nm, with switching speeds of 100–200 GHz.

5. NANOTUBE INTERCONNECTS

Nanometer-scale integrated electronic devices will require enormous demands on interconnects. The power density dissipated by such devices will be tremendous; in ballistic devices this power will be dissipated in the interconnects. Interconnects must be low resistance not only to reduce power consumption, but to increase device speed.

Nanotubes are excellent candidates for interconnects given their low electrical and thermal resistance, and stability at high temperature and current. As shown in Section IV.2.a, the resistivity of metallic single-walled carbon nanotubes is on order $10^{-6}\,\Omega$ cm, comparable or superior to copper. Furthermore, the conduction of electrons is quasi-ballistic; the energy of hot electrons is dissipated on a length scale exceeding many micro meters. Nanotubes may be used as exceedingly fine interconnects to nanoscale devices, for example in a crossbar geometry, in which molecular-sized devices are located at the junctions between crossing wires in an array. The ballistic nature of electron transport in nanotubes would allow the power to be dissipated far from the devices in macroscopic metal wires.

Even if longer nanotube interconnects are employed, such that the electron energy relaxes within the nanotube, the very high thermal conductivity and excellent thermal stability of nanotubes suggests that they would be superior to conventional metals. The phonon thermal conductivity of carbon nanotubes has been measured to be approximately 2000 W/m/K at room temperature, comparable to diamond, the best-known room temperature thermal conductor (Kim *et al.* 2001).

The current-carrying capacity of a single SWNT is limited by optical phonon emission to about $2.5 \times 10^9\,\mathrm{A/cm^2}$ for a 1 nm diameter tube—orders of magnitude larger than current densities found in present-day interconnects. Moreover, metal interconnects typically fail via electromigration. Electromigration should be enormously reduced in covalently bonded carbon nanotubes—there are no low-energy defects or dislocations that can lead to migration of atoms. Nanotubes have been observed to carry current densities of $10^9\,\mathrm{A/cm^2}$ for periods of weeks without failure (Wei *et al.* 2001).

VI. Outlook for Nanotubes in Nanoelectronics

As evidenced above, nanotube electronic devices have demonstrated some very attractive properties: high-speed, high-sensitivity FETs for memories, sensors, or logic circuits; high-transconductance SETs operating at room temperature; densely packed junction devices; robust, high-speed mechanical devices; and low-loss interconnects. However, significant roadblocks stand in the way of developing any of these devices for use in a technology.

In order for nanotube devices to find a home in an electronic technology, two major obstacles must be overcome: nanotubes must be prepared with uniform electronic properties, and methods must be found to place individual nanotubes precisely on a substrate. Any nanoelectronics technology faces similar versions of these challenges; however it is worth discussing some of the problems that will be specific to nanotubes, and some of the efforts that have been made toward solving these problems.

There are two approaches to nanotube synthesis for electronic devices. In the first, large quantities of nanotubes are synthesized "off-chip", processed, then precisely placed on substrates using self-assembly techniques. In the second technique, nanotubes are grown "on chip" from specific sites on a substrate, and guided in specific directions through the use of, for example, electric fields. Both approaches face hurdles, but the nature of the hurdles is somewhat different in each case.

In the off-chip synthesis, the major challenge is to produce a physically and electronically uniform product that is soluble, so that self-assembly techniques such as chemical recognition can be used to precisely place the nanotubes on suitably prepared substrates. The holy grail of off-chip synthesis would be two vials of nanotubes; one containing a stable solution of metallic SWNTs of fixed length and diameter, another semiconducting SWNTs of fixed length and diameter. At this point, nanotubes would be analogous to any other component of "molecular electronics" and could be treated the same way; the hope being that the chemical self-assembly techniques will be developed to place molecule-sized objects with precision on substrates.

On-chip synthesis attempts to combine the self-assembly and synthesis in order to remove some of the intermediate processing. Here, the major challenge is to produce a suitable catalyst which under the proper conditions will produce a nanotube with uniform electronic properties. All that remains, then, is to self-assemble these catalyst particles into the desired positions, and then perform the growth procedure.

1. UNIFORMITY OF PROPERTIES

The unique "size effect" in carbon nanotubes—that the band structure changes enormously, from metal to semiconductor, with tiny changes in diameter and chirality—presents both a significant opportunity and a significant challenge. The opportunity is to realize a complete electronic system using only covalently bonded carbon networks. The challenge is that such structures must be constructed perfectly—small changes in the components will have drastic changes in the electronic properties.

There are several possible solutions to this challenge. One possibility is to enormously overbuild devices, and then remove unwanted components. For example, working transistors have been made from bundles containing both metallic and semiconducting SWNTs by first depleting the semiconducting nanotubes of carriers using a gate voltage, and then destroying the metallic nanotubes with a large voltage pulse (Collins *et al.* 2001). Another solution to the challenge would be the synthesis or extraction of a nanotube material that is electronically uniform. This section will address some of the large body of research that has focused on this solution. (A third possibility—to find a new material that has some of the advantageous properties of carbon nanotubes, but is electronically uniform—will be discussed in Section VII.)

Smalley *et al.* synthesized SWNT with a narrow diameter distribution via laser ablation and concluded that the majority of the SWNT were metallic (10,10) nanotubes (Thess *et al.* 1996); the consensus is now that this was not the case. However, research continues toward finding a synthesis route to produce nanotubes of a single structure. Recently, small bundles of nanotube were synthesized via a novel method; layers of nickel and C_{60} were alternately evaporated onto a molybdenum surface through small apertures, followed by annealing at 950°C in vacuum and a magnetic field of 1.5 T (Schlittler *et al.* 2001). Small rods protruding from the substrate were found to be *crystals* of hundreds of nanotubes, each with identical diameter and helicity, although these parameter varied from crystal to crystal. It remains to be seen whether this technique can be scaled to produce large quantities of uniform nanotubes.

Controlling the uniformity of nanotubes has also been a challenge in chemical vapor deposition growth. Typically CVD-grown nanotubes are more heterogeneous than laser ablation derived nanotubes; presumably because of difficulties in making the catalyst particles monodisperse. Various research groups have attacked this issue. The iron-storage protein ferritin was used to produce uniform-size catalyst particles consisting of ~200 or ~1100 Fe atoms (Li *et al.* 2001). These catalyst particles were then used to grow nanotubes with diameters of 1.5 ± 0.4 and 3.0 ± 0.9 nm, respectively. Similar results were obtained with discrete-sized iron nanoparticles synthesized in solution; particles with average diameters of 3, 9, and 13 nm were used to grow nanotubes of average diameter 3, 7, and 12 nm (Cheung *et al.* 2002).

Barring the synthesis of uniform nanotubes, the extraction of nanotubes of a particular type from bulk sample is a desirable goal. The first step in such a scheme is solubilization of the nanotubes, then hopefully differences in their electrical, magnetic, structural, or chemical properties may be used to separate them via chromatography, electrophoresis, or selective chemical functionalization.

Solubilization of carbon nanotubes has, however, remained a major challenge, although a number of important steps in this direction have been taken. Nanotubes have been systematically shortened and suspended in surfactant solutions (Liu *et al.* 1998), where they may be purified via chromatography (Niyogi *et al.* 2001), although this procedure causes damage to the nanotube sidewalls. The damage, however, can allow functionalization of the sidewalls which can aid in solubility (Chen *et al.* 1998, 2001; Riggs *et al.* 2000). Nanotubes have also been rendered soluble by fluorination (Mickelson *et al.* 1998).

2. SELF-ASSEMBLY

Once nanotubes are solubilized, self-assembly schemes can be utilized to place them with precision on substrates. Such schemes rely on localized forces, such as chemical recognition or local electric fields, to guide the nanotubes into place. Shortened SWNTs exhibit sidewall carboxyl groups that have been used as the basis of functionalization, and the science of nanotube sidewall chemistry is proceeding rapidly.

Liu *et al.* were able to take advantage of the sidewall functionalization to place nanotubes precisely on substrates (Liu *et al.* 1999). They prepared SiO_2 substrates with a methyl-terminated self-assembled monolayer (SAM), to which it was observed that shortened SWNTs had poor adhesion. This monolayer was locally removed via electron beam radiation, and subsequently replaced with an amine-terminated SAM. When the substrate was exposed to a suspension of shortened SWNTs, the nanotubes stuck preferentially on the exposed areas. However, a few nanotubes were observed on areas outside the exposure. Electronic devices made by this method also showed poor electrical properties, probably due to the presence of sidewall defects.

VII. Beyond Carbon Nanotubes

An alternative solution to the problem of non-uniform electronic properties in carbon nanotubes is to find a material that is more electronically uniform, but still retains some of the advantageous properties of carbon nanotubes. Immediately after the discovery of carbon nanotubes, the possibility of synthesizing tubular structures of other layered materials was realized, and now a number of examples exist; the most well characterized being the $B_xC_yN_z$ system and the transition-metal dichalcogenides, both discussed below. Crystalline nanowires are another possible system. The vapor–liquid–solid reactions used to produce nanotubes of carbon may be generalized to a broad range of materials. Nanowires, however, lack the unique property of nanotubes that the surface states are eliminated.

1. INORGANIC NANOTUBES

Following the analogy of constructing nanotubes from single sheets of graphite, one may consider nanotubes of other layered compounds. Indeed, the synthesis of nanotubes from several other layered compounds has been reported.

The single sheet of hexagonal boron nitride (h-BN) is probably the closest analog to graphene; h-BN may be thought of as the III–V analog of graphite.

The inversion symmetry that allows a crossing of the π and π^* bands at the Fermi level is removed in h-BN, and consequently h-BN is a wide gap semiconductor, with $E_g = 5.8$ eV (Zunger et al. 1976). Nanotubes of h-BN were predicted to be stable and be semiconducting with large band gaps (Rubio et al. 1994); for tube larger than ~ 1 nm the band gap is fairly constant (on order 5.5 eV), tending towards the bulk value for large diameters (Blase et al. 1994).

Soon after the prediction of their stability, multiwalled h-BN nanotubes were indeed synthesized, by arc discharge between a BN-packed tungsten rod and a copper electrode (Chopra et al. 1995). Following this discovery, both single and multi-walled h-BN nanotube have been produced by a variety of methods (Golberg et al. 1996; Loiseau et al. 1996; Weiqiang et al. 1998; Yu et al. 1998; Cummings and Zettl 2000). The synthesis techniques have been refined, and now highly pure monodisperse single-wall (Lee et al. 2001) and, intriguingly, double-wall (Cummings and Zettl 2000) h-BN nanotubes have been produced. Still, electronic conduction properties of h-BN nanotubes have not been reported; it remains unclear whether these large band gap semiconducting nanotubes will be useful for electronics.

Perhaps more intriguing electronically are other nanotubes from the $B_x C_y N_z$ family. Besides BN, there exist other $B_x C_y N_z$ analogs of graphite that form stoichiometric structures; for example BC_2N and BC_3 (Krishnan et al. 1988). Nanotubes of BC_2N were considered theoretically (Miyamoto et al. 1994a). Two possible structures for the BC_2N sheet were considered, with one structure leading to metallic or semiconducting behavior dependent on chirality (similar to the carbon nanotube case) and the other giving only semiconducting behavior, with a moderate gap of ~ 1.3 eV. The second case is particularly interesting; since these nanotubes were formed from a sheet which possesses a strongly anisotropic structure, chiral nanotubes could be envisioned in which the current-carrying states may have an angular momentum about the tube axis (Miyamoto et al. 1996, 1999).

Nanotubes of BC_3 have also been considered theoretically (Miyamoto et al. 1994b), and found to have a lower formation energy compared to the sheet than similar-diameter carbon nanotubes, suggesting their stability. Band structure calculations show metallic or small gap (~ 0.2 eV) semiconducting character, tending to uniformly small gap semiconducting at larger diameters.

Arcing a composite rod of BN and graphite indeed produced multiwalled nanotubes (Weng-Sieh et al. 1995); within the same sample individual nanotubes were identified by electron energy-loss spectroscopy (EELS) performed in a TEM as being stoichiometrically BC_2N and BC_3. Arcing a hafnium diboride rod together with a graphite rod in a nitrogen atmosphere produced intriguing structures of B, C, and N (Suenaga et al. 1997). Multi-walled nanotubes and onion-like nanoparticles were observed with adjacent walls of pure C and pure BN. One particular MWNT studied by EELS in a

TEM had 14 walls; the inner three were carbon, the next six BN, and the outer five were again carbon. Such structure have led to fanciful description of nano-coaxial cables, however, no electrical transport results have been reported.

Layered transition-metal dichalcogenides have also been explored as a basis for constructing nanotubes. These materials consist of layers three atoms thick; a single layer of transition metal capped on either side by layers of sulfur, selenium, or tellurium. These three atom thick layers are weakly bound together through van der Waals interactions. Molybdenum disulfide is perhaps the most familiar example, finding use as a dry lubricant. Both WS_2 and MoS_2 nanotubes have been synthesized in significant quantities (Tenne and Zettl 2001). Typically, the results are large MWNTs. However, it was recently reported that crystals of monodisperse SWNTs of MoS_2 with diameter of 0.96 nm could be produced in large quantities, raising interest in these nanotubes as electronic nanowires with uniform properties (Remskar *et al.* 2001). Similar to these MoS_2 nanotubes, crystals of chains of Mo_3Se_3 with width of $\sim$0.6 nm have been prepared (Hornbostel *et al.* 1995). These fibers are soluble in polar solvents, and electrical measurements on single fibers show that they are conducting at room temperature (Venkataraman and Lieber 1999).

2. SEMICONDUCTOR NANOWIRES

It has been known for over 30 years that narrow crystalline wires of silicon may be prepared through a vapor–liquid–solid reaction with a metal catalyst, for example gold (Wagner and Ellis 1964); however, the technique was limited by the size of the catalyst particles to produce wires of >100 nm diameter (Wagner 1970). By analogy to the laser ablation production of nanotubes, silicon and other semiconductor nanowires may be produced by laser ablation of a catalyst-containing target; small particles of catalyst are formed in the ablation plume, and nanowires grow via a VLS reaction (Morales and Lieber 1998; Zhang *et al.* 1998). Crystalline semiconductor nanowires may also be grown from solid-supported nanoparticle catalysts (Yazawa *et al.* 1992; Westwater *et al.* 1997) and via solution–liquid–solid reactions (Trentler *et al.* 1995; Holmes *et al.* 2000). Crystalline nanowires of elemental semiconductors (e.g., Si, Ge), and compounds (e.g., GaAs, GaN, InP) have been prepared; given the universality of the Vapor–Liquid–Solid reaction, the possible list of nanowires which may be prepared is very large.

Many of the advances in fabricating electronic devices from semiconductor nanowires have come from the Lieber group at Harvard. Their approach has been to form devices at the crossing junctions of nanowires. Aligned nanowire arrays have been self-assembled by deposition from a suspension flowing through a microchannel over a substrate (Huang *et al.* 2001a).

Arrays of crossed nanowires may be obtained by sequential deposition of more nanowires at an angle to the first array. In this way, a number of crossed nanowire devices may be formed. Silicon nanowires doped p- and n-type with boron or phosphorus were found to make rectifying junctions. Junctions between p-type silicon and n-type GaN nanowires could act as either diodes, or, after oxidation of the silicon nanowire, FETs (Huang *et al.* 2001b).

Crystalline semiconductor nanowires have a number of possible advantages over carbon nanotubes; the electronic properties as well as processing techniques of inorganic semiconductors are already well understood, their electronic properties may be controlled through doping (Cui *et al.* 2000), and they may be oxidized controllably to form barriers (Cui and Lieber 2001). However, because of the high doping levels and large numbers of surface states, it is unlikely the nanowires will achieve the high conductivities or high mobilities demonstrated in nanotube FETs. The current state of self-assembly and integration of nanowire devices is, however, more advanced compared to nanotube devices. Given the similarity to existing semiconductor technology in materials processing and device behavior, semiconductor nanowires may be a model system to study nanoscale devices produced by self-assembly.

VIII. Conclusions

Interest in carbon nanotubes for nanoelectronics was sparked by the theoretical prediction of a 1D conductor that was impervious to the Peierls distortion, and whose electronic properties could be tuned through changes in atomic structure. Nanometer-diameter SWNTs have since been mass produced, and are indeed metallic or semiconducting depending sensitively on their diameter and helicity. Experiments have since revealed a material that has exceeded all expectations: metallic SWNTs rival the best metals in conductivity at room temperature, and semiconducting SWNTs have room temperature mobilities comparable to the best semiconductors.

A range of electronic devices have been demonstrated which exploit the unique properties of carbon nanotubes. Nanotube FETs have transconductances per width that rival the best silicon FETs. Nanotube SETs operate at room temperature due to natural nanometer-size low-capacitance junctions. Junctions between nanotubes demonstrate the ultimate in small: working two- and three-terminal devices containing a few thousand atoms that can rectify microampere currents. Nanotube interconnects carry current densities of greater than 10^9 A/cm^2 for extended periods of time without failure.

These devices are proofs of the principle that useful nanoelectronic devices can be constructed from nanotubes. However, they certainly do not represent a technology; each was painstakingly constructed one by one using slow serial fabrication techniques. The challenge now lies in engineering. No techniques yet exist for the mass production of electronically uniform nanotube materials, or the placement of those nanotubes precisely into electronic circuits. However, the advances in the past 5 years since the first SWNT devices were constructed have been enormous. There is every reason to expect that the next 5 years will be equally productive.

REFERENCES

Anantram, M. P., and T. R. Govindan, Conductance of carbon nanotubes with disorder: a numerical study, *Phys. Rev. B (Condens. Matter)* **58**, 4882 (1998).

Andriotis, A. N., M. Menon, D. Srivastava, and L. Chernozatonskii, Rectification properties of carbon nanotube "Y-junctions", *Phys. Rev. Lett.* **87**, 066802 (2001).

Bachtold, A., M. S. Fuhrer, S. Plyasunov, M. Forero, E. H. Anderson, A. Zettl, and P. L. McEuen, Scanned probe microscopy of electronic transport in carbon nanotubes, *Phys. Rev. Lett.* **84**, 6082 (2000).

Bandow, S., A. M. Rao, K. A. Williams, A. Thess, R. E. Smalley, and P. C. Eklund, Purification of single-wall carbon nanotubes by microfiltration, *J. Phys. Chem. B* **101**, 8839 (1997).

Bethune, D. S., C. H. Kiang, M. S. Devries, G. Gorman, R. Savoy, J. Vazquez, and R. Beyers, Cobalt-catalysed growth of carbon nanotubes with single-atomic-layer walls, *Nature* **363**, 605 (1993).

Bezryadin, A., A. R. M. Verschueren, S. J. Tans, and C. Dekker, Multiprobe transport experiments on individual single-wall carbon nanotubes, *Phys. Rev. Lett.* **80**, 4036 (1998).

Blase, X., A. Rubio, S. G. Louie, and M. L. Cohen, Stability and band gap constancy of boron nitride nanotubes, *Europhys. Lett.* **28**, 335 (1994).

Bockrath, M., D. H. Cobden, P. L. McEuen, N. G. Chopra, A. Zettl, A. Thess, and R. E. Smalley, Single-electron transport in ropes of carbon nanotubes, *Science* **275**, 1922 (1997).

Bockrath, M., D. H. Cobden, L. Jia, A. G. Rinzler, R. E. Smalley, L. Balents, and P. L. McEuen, Luttinger-liquid behaviour in carbon nanotubes, *Nature* **397**, 598 (1999).

Bockrath, M., J. Hone, A. Zettl, P. L. McEuen, A. G. Rinzler, and R. E. Smalley, Chemical doping of individual semiconducting carbon-nanotube ropes, *Phys. Rev. B (Condens. Matter)* **61**, R10606 (2000).

Bozovic, D., M. Bockrath, J. H. Hafner, C. M. Lieber, P. Hongkun, and M. Tinkham, Electronic properties of mechanically induced kinks in single-walled carbon nanotubes, *Appl. Phys. Lett.* **78**, 3693 (2001).

Charlier, J. C., P. Lambin, and T. W. Ebbesen, Electronic properties of carbon nanotubes with polygonized cross sections, *Phys. Rev. B (Condens. Matter)* **54**, R8377 (1996).

Chen, J., M. A. Hamon, H. Hu, Y. Chen, A. M. Rao, P. C. Eklund, and R. C. Haddon, Solution properties of single-walled carbon nanotubes, *Science* **282**, 95 (1998).

Chen, J., A. M. Rao, S. Lyuksyutov, M. E. Itkis, M. A. Hamon, H. Hu, R. W. Cohn, P. C. Eklund, D. T. Dolbert, R. E. Smalley, and R. C. Haddon, Dissolution of full-length single-walled carbon nanotubes, *J. Phys. Chem. B* **105**, 2525 (2001).

Cheung, C. L., A. Kurtz, H. Park, and C. M. Lieber, Diameter-controlled synthesis of carbon nanotubes, *J. Phys. Chem. B* **106**, 2429 (2002).

Chico, L., V. H. Crespi, L. X. Benedict, S. G. Louie, and M. L. Cohen, Pure carbon nanoscale devices: nanotube heterojunctions, *Phys. Rev. Lett.* **76**, 971 (1996).

Chico, L., M. P. Lopez Sancho, and M. C. Munoz, Carbon-nanotube-based quantum dot, *Phys. Rev. Lett.* **81**, 1278 (1998).

Chopra, N. G., R. J. Luyken, K. Cherrey, V. H. Crespi, M. L. Cohen, S. G. Louie, and A. Zettl, Boron nitride nanotubes, *Science* **269**, 966 (1995).

Cobden, D. H., M. Bockrath, P. L. McEuen, A. G. Rinzler, and R. E. Smalley, Spin splitting and even-odd effects in carbon nanotubes, *Phys. Rev. Lett.* **81**, 681 (1998).

Cobden, D. H., M. Bockrath, N. G. Chopra, A. Zettl, P. L. McEuen, A. Rinzler, A. Thess, and R. E. Smalley, Transport spectroscopy of single-walled carbon nanotubes, *Physica B* **251**, 132 (1998a).

Collins, P., A. Zettl, H. Bando, A. Thess, and R. Smalley, Nanotube nanodevice, *Science* **278**, 100 (1997).

Collins, P. G., M. S. Arnold, and P. Avouris, Engineering carbon nanotubes and nanotube circuits using electrical breakdown, *Science* **292**, 706 (2001).

Cui, Y., and C. M. Lieber, Functional nanoscale electronic devices assembled using silicon nanowire building blocks, *Science* **291**, 851 (2001).

Cui, Y., X. Duan, J. Hu, and C. M. Lieber, Doping and electrical transport in silicon nanowires, *J. Phys. Chem. B* **104**, 5213 (2000).

Cumings, J., and A. Zettl, Mass-production of boron nitride double-wall nanotubes and nanococoons, *Chem. Phys. Lett.* **316**, 211 (2000).

Dai, H., A. G. Rinzler, P. Nikolaev, A. Thess, D. T. Colbert, and R. E. Smalley, Single-wall nanotubes produced by metal-catalyzed disproportionation of carbon monoxide, *Chem. Phys. Lett.* **260**, 471 (1996).

Datta, S., *Electronic Transport in Mesoscopic Systems*, Cambridge University Press, Cambridge (1995).

Dekker, C., Carbon nanotubes as molecular quantum wires, *Phys. Today* **52**, 22 (1999).

Derycke, V., R. Martel, J. Appenzeller, and P. Avouris, Carbon nanotube inter- and intramolecular logic gates, *Nanoletters* **1**, 453 (2001).

Dunlap, B. I., Relating carbon tubules, *Phys. Rev. B (Condens. Matter)* **49**, 5643 (1994).

Egger, R., and A. O. Gogolin, Effective low-energy theory for correlated carbon nanotubes, *Phys. Rev. Lett.* **79**, 5082 (1997).

Eriksson, M. A., R. G. Beck, M. Topinka, J. A. Katine, R. M. Westervelt, K. L. Campman, and A. C. Gossard, Cryogenic scanning probe characterization of semiconductor nanostructures, *Appl. Phys. Lett.* **69**, 671 (1996).

Fisher, M. P. A., and A. Glazman, *Mesoscopic Electron Transport*, Kluwer Academic, Boston, MA (1997).

Forró, L., and C. Schönenberger, Physical properties of multi-wall nanotubes, in *Topics in Applied Physics*, edited by M. S. Dresselhaus, G. Dresselhaus, and P. Avouris, Vol. 80, Springer, Berlin, p. 329 (2001).

Fuhrer, M. S., J. Nygård, L. Shih, M. Forero, Y.-G. Yoon, M. S. C. Mazzoni, H. J. Choi, J. Ihm, S. G. Louie, A. Zettl, and P. L. McEuen, Crossed nanotube junctions, *Science* **288**, 494 (2000).

Fuhrer, M. S., M. Forero, A. Zettl, and P. L. McEuen, Ballistic transport in semiconducting carbon nanotubes, in *Electronic Properties of Molecular Nanostructures*, edited by H. Kuzmany, J. Fink, M. Mehring, and S. Roth, AIP Conference Proceedings, New York, p. 401 (2001).

Fulton, T. A., and G. J. Dolan, Observation of single-electron charging effects in small tunnel junctions, *Phys. Rev. Lett.* **59**, 109 (1987).

Golberg, D., Y. Bando, M. Eremets, K. Takemura, K. Kurashima, and H. Yusa, Nanotubes in boron nitride laser heated at high pressure, *Appl. Phys. Lett.* **69**, 2045 (1996).

Hafner, J. H., M. J. Bronikowski, B. R. Azamian, P. Nikolaev, A. G. Rinzler, D. T. Colbert, K. A. Smith, and R. E. Smalley, Catalytic growth of single-wall carbon nanotubes from metal particles, *Chem. Phys. Lett.* **296**, 195 (1998).

Hafner, J. H., C. Chin-Li, T. H. Oosterkamp, and C. M. Lieber, High-yield assembly of individual single-walled carbon nanotube tips for scanning probe microscopies, *J. Phys. Chem. B* **105**, 743 (2001).

Hamada, N., Electronic band structure of carbon nanotubes: toward the three-dimensional system, *Materials Science & Engineering B* **19**, 181 (1993).

Hamada, N., S. Sawada, and A. Oshiyama, New one-dimensional conductors—graphitic microtubules, *Phys. Rev. Lett.* **68**, 1579 (1992).

Hertel, T., R. E. Walkup, and P. Avouris, Deformation of carbon nanotubes by surface van der Waals forces, *Phys. Rev. B (Condens. Matter)* **58**, 13870 (1998).

Holmes, J. D., K. P. Johnston, R. C. Doty, and B. A. Korgel, Control of thickness and orientation of solution-grown silicon nanowires, *Science* **287**, 1471 (2000).

Hornbostel, M. D., S. Hillyard, J. Silcox, and F. J. DiSalvo, Nanometer width molybdenum selenide fibers, *Nanotechnology* **6**, 87 (1995).

Huang, Y., M. Okada, K. Tanaka, and T. Yamabe, Estimation of Peierls-transition temperature in metallic carbon nanotube, *Solid State Commun.* **97**, 303 (1996).

Huang, Y., X. Duan, Q. Wei, and C. M. Lieber, Directed assembly of one-dimensional nanostructures into functional networks, *Science* **291**, 630 (2001a).

Huang, Y., X. Duan, Y. Cui, L. J. Lauhon, K. H. Kim, and C. M. Lieber, Logic gates and computation from assembled nanowire building blocks, *Science* **294**, 1313 (2001b).

Iijima, S. Helical microtubules of graphitic carbon, *Nature* **354**, 56 (1991).

Iijima, S., Growth of carbon nanotubes, *Materials Science & Engineering B* **19**, 172 (1993).

Iijima, S., and T. Ichihashi, Single-shell carbon nanotubes of 1-nm diameter, *Nature* **363**, 603 (1993).

Javey, A., M. Shim, and H. Dai, Electrical properties and devices of large-diameter single-walled carbon nanotubes, *Appl. Phys. Lett.* **80**, 1064 (2002).

Journet, C., W. Maser, P. Bernier, A. Loiseau, M. de la Chapelle, S. Lefrant, P. Deniard, R. Lee, and J. Fischer, Large-scale production of single-walled carbon nanotubes by the electric-arc technique, *Nature* **388**, 756 (1997).

Kane, C. L., and E. J. Mele, Size, shape, and low energy electronic structure of carbon nanotubes, *Phys. Rev. Lett.* **78**, 1932 (1997).

Kane, C., L. Balents, and M. P. A. Fisher, Coulomb interactions and mesoscopic effects in carbon nanotubes, *Phys. Rev. Lett.* **79**, 5086 (1997).

Kim, P., L. Shi, A. Majumdar, and P. L. McEuen, Thermal transport measurements of individual multiwalled nanotubes, *Phys. Rev. Lett.* **87**, 215502 (2001).

Kong, J., and H. Dai, Full and modulated chemical gating of individual carbon nanotubes by organic amine compounds, *J. Phys. Chem. B* **105**, 2890 (2001).

Kong, J., H. T. Soh, A. Cassell, C. F. Quate, and H. Dai, Synthesis of single single-walled carbon nanotubes on patterned silicon wafers, *Nature* **395**, 878 (1998).

Kong, J., C. Zhou, E. Yenilmez, and H. Dai, Alkaline metal-doped n-type semiconducting nanotubes as quantum dots. *Appl. Phys. Lett.* **77**, 3977 (2000a).

Kong, J., N. R. Franklin, C. Zhou, M. G. Chapline, S. Peng, K. Cho, and H. Dai, Nanotube molecular wires as chemical sensors, *Science* **287**, 622 (2000b).

Kong, J., E. Yenilmez, T. W. Tombler, W. Kim, H. Dai, R. B. Laughlin, L. Liu, C. S. Jayanthi, and S. Y. Wu, Quantum interference and ballistic transmission in nanotube electron waveguides, *Phys. Rev. Lett.* **87**, 106801 (2001).

Kong, J., J. Cao, and H. Dai, Chemical profiling of single nanotubes: intramolecular p–n–p junctions and on-tube single-electron transistors, *Appl. Phys. Lett.* **80**, 73 (2002).

Kouwenhoven, L. R., C. M. Marcus, P. L. McEuen, S. Tarucha, R. Westervelt, and N. S. Wingreen, Electron transport in quantum dots, in *Mesoscopic Electron Transport* edited by L.L. Sohn, L. P. Kouwenhoven, and G. Schon Kluwer, Boston, MA (1997).

Krishnan, K. M., J. Kouvetakis, T. Sasaki, and N. Bartlett, Characterization of newly synthesized novel graphite films, in *Better Ceramics Through Chemistry III. Symposium* edited by C. J. Brinker, D. E. Clark, and D. R. Ulrich, Material Research Society, Pittsburgh, PA, p. 527 (1988).

Landauer, R., *IBM J. Res. Dev.* **1**, 223 (1958).

Lee, R. S., J. Gavillet, M. L. de la Chapelle, A. Loiseau, J. L. Cochon, D. Pigache, J. Thibault, and F. Willaime, Catalyst-free synthesis of boron nitride single-wall nanotubes with a preferred zig-zag configuration, *Phys. Rev. B (Condens. Matter Mater. Phys.)* **64**, 121405 (2001).

Lefebvre, J., M. Radosavljevic, and A. T. Johnson, Fabrication of nanometer size gaps in a metallic wire, *Appl. Phys. Lett.* **76**, 3828 (2000).

Leonard, J., and F. Tersoff, Novel length scales in nanotube devices, *Phys. Rev. Lett.* **83**, 5174 (1999).

Li, J., C. Papadopoulos, and J. Xu, Nanoelectronics: growing Y-junction carbon nanotubes, *Nature* **402**, 253 (1999).

Li, W. Z., J. G. Wen, and Z. F. Ren, Straight carbon nanotube Y junctions, *Appl. Phys. Lett.* **79**, 1879 (2001).

Li, Y., W. Kim, Y. Zhang, M. Rolandi, D. Wang, and H. Dai, Growth of single-walled carbon nanotubes from discrete catalytic nanoparticles of various sizes, *J. Phys. Chem. B* **105**, 11424 (2001).

Liang, W., M. Bockrath, D. Bozovic, J. H. Hafner, M. Tinkham, and H. Park, Fabry–Perot interference in a nanotube electron waveguide, *Nature* **411**, 665 (2001).

Liu, J., A. G. Rinzler, H. Dai, J. H. Hafner, R. K. Bradley, P. J. Boul, A. Lu, T. Iverson, K. Shelimov, C. B. Huffman, F. Rodriguez-Macias, Y.-S. Shon, T. R. Lee, D. T. Colbert, and R. E. Smalley, Fullerene pipes, *Science* **280**, 1253 (1998).

Liu, J., M. J. Casavant, M. Cox, D. A. Walters, P. Boul, L. Wei, A. J. Rimberg, K. A. Smith, D. T. Colbert, and R. E. Smalley, Controlled deposition of individual single-walled carbon nanotubes on chemically functionalized templates, *Chem. Phys. Lett.* **303**, 125 (1999).

Loiseau, A., F. Willaime, N. Demoncy, G. Hug, and H. Pascard, Boron nitride nanotubes with reduced numbers of layers synthesized by arc discharge, *Phys. Rev. Lett.* **76**, 4737 (1996).

Louie, S. G., Electronic properties, junctions, and defects of carbon nanotubes, in *Topics in Applied Physics* edited by M. S. Dresselhaus, G. Dresselhaus, and P. Avouris, Vol. 80, Springer, Berlin, p. 113 (2001).

Martel, R., T. Schmidt, H. R. Shea, T. Hertel, and P. Avouris, Single- and multi-wall carbon nanotube field-effect transistors, *Appl. Phys. Lett.* **73**, 2447 (1998).

Martel, R., V. Derycke, C. Lavoie, J. Appenzeller, K. K. Chan, J. Tersoff, and P. Avouris, Ambipolar electrical transport in semiconducting single-wall carbon nanotubes, *Phys. Rev. Lett.* **87**, 256805 (2001).

Matsumoto, K., M. Ishii, K. Segawa, Y. Oka, B. J. Vartanian, and J. S. Harris, Room temperature operation of a single electron transistor made by the scanning tunneling microscope nanooxidation process for the TiO_x/Ti system, *Appl. Phys. Lett.* **68**, 34 (1996).

McEuen, P. L., Single-wall carbon nanotubes, *Phys. World* (2000).

McEuen, P. L., M. Bockrath, D. H. Cobden, Y.-G. Yoon, and S. G. Louie, Disorder, pseudospins, and backscattering in carbon nanotubes, *Phys. Rev. Lett.* **83**, 5098 (1999).

Menon, M., and D. Srivastava, Carbon nanotube "T junctions": nanoscale metal–semiconductor–metal contact devices, *Phys. Rev. Lett.* **79**, 4453 (1997).

Mickelson, E. T., C. B. Huffman, A. G. Rinzler, R. E. Smalley, R. H. Hauge, and J. L. Margrave, Fluorination of single-wall carbon nanotubes, *Chem. Phys. Lett.* **296**, 188 (1998).

Mintmire, J. W., and C. T. White, Universal density of states for carbon nanotubes, *Phys. Rev. Lett.* **81**, 2506 (1998).

Mintmire, J. W., B. I. Dunlap, and C. T. White, Are fullerene tubules metallic? *Phys. Rev. Lett.* **68**, 631 (1992).

Miyamoto, Y., A. Rubio, M. L. Cohen, and S. G. Louie, Chiral tubules of hexagonal BC_2N, *Phys. Rev. B (Condens. Matter)* **50**, 4976 (1994a).

Miyamoto, Y., A. Rubio, S. G. Louie, and M. L. Cohen, Electronic properties of tubule forms of hexagonal BC_3, *Phys. Rev. B (Condens. Matter)* **50**, 18360 (1994b).

Miyamoto, Y., S. G. Louie, and M. L. Cohen, Chiral conductivities of nanotubes, *Phys. Rev. Lett.* **76**, 2121 (1996).

Miyamoto, Y., A. Rubio, S. G. Louie, and M. L. Cohen, Self-inductance of chiral conducting nanotubes, *Phys. Rev. B (Condens. Matter)* **60**, 13885 (1999).

Morales, A. M., and C. M. Lieber, A laser ablation method for the synthesis of crystalline semiconductor nanowires, *Science* **279**, 208 (1998).

Nakanishi, T., and T. Ando, Numerical study of impurity scattering in carbon nanotubes, *J. Phys. Soc. Jpn.* **68**, 561 (1999).

Niyogi, S., H. Hu, M. A. Hamon, P. Bhowmik, B. Zhao, S. M. Rozenzhak, J. Chen, M. E. Itkis, M. S. Meier, and R. C. Haddon, Chromatographic purification of soluble single-walled carbon nanotubes (s-SWNTs), *J. Am. Chem. Soc.* **123**, 733 (2001).

Nygard, J., D. H. Cobden, M. Bockrath, P. L. McEuen, and P. E. Lindelof, Electrical transport measurements on single-walled carbon nanotubes, *Appl. Phys. A (Mater. Sci. Process.)* **A69**, 297 (1999).

Nygard, J., D. H. Cobden, and P. E. Lindelof, Kondo physics in carbon nanotubes, *Nature* **408**, 342 (2000).

Odintsov, A. A. Schottky barriers in carbon nanotube heterojunctions, *Phys. Rev. Lett.* **85**, 150 (2000).

Odom, T. W., H. Jin-Lin, P. Kim, and C. M. Lieber, Atomic structure and electronic properties of single-walled carbon nanotubes, *Nature* **391**, 62 (1998).

Park, H., J. Park, A. K. L. Lim, E. H. Anderson, A. P. Alivisatos, and P. L. McEuen, Nano-mechanical oscillations in a single-C_{60} transistor, *Nature* **407**, 57 (2000).

Park, J., and P. L. McEuen, Formation of a p-type quantum dot at the end of an n-type carbon nanotube, *Appl. Phys. Lett.* **79**, 1363 (2001).

Pashkin, Y. A., Y. Nakamura, and J. S. Tsai, Room-temperature Al single-electron transistor made by electron-beam lithography, *Appl. Phys. Lett.* **76**, 2256 (2000).

Peierls, R. E. *Quantum Theory of Solids*, Oxford University Press, New York (1955).

Poncharal, P., Z. L. Wang, D. Ugarte, and W. A. De Heer, Electrostatic deflections and electromechanical resonances of carbon nanotubes, *Science* **283**, 1513 (1999).

Postma, H. W. C., M. de Jonge, and C. Dekker, Electrical transport through carbon nanotube junctions created by mechanical manipulation, *Phys. Rev. B (Condens. Matter)* **62**, R10653 (2000).

Postma, H. W. C., T. Teepen, Y. Zhen, M. Grifoni, and G. Dekker, Carbon nanotube single-electron transistors at room temperature, *Science* **293**, 76 (2001).

Rao, A. M., E. Richter, S. Bandow, B. Chase, P. C. Eklund, K. A. Williams, S. Fang, K. R. Subbaswamy, M. Menon, A. Thess, R. E. Smalley, G. Dresselhaus, and M. S. Dresselhaus, Diameter-selective Raman scattering from vibrational modes in carbon nanotubes, *Science* **275**, 187 (1997).

Remskar, M., A. Mrzel, Z. Skraba, A. Jesih, M. Ceh, J. Demsar, P. Stadelmann, F. Levy, and D. Mihailovic, Self-assembly of subnanometer-diameter single-wall MoS_2 nanotubes, *Science* **292**, 479 (2001).

Reulet, B., A. Y. Kasumov, M. Kociak, R. Deblock, I. I. Khodos, Y. B. Gorbatov, V. T. Volkov, C. Journet, and H. Bouchiat, Bolometric detection of mechanical bending waves in suspended carbon nanotubes, *Condens. Matter Arch.* (xxx.lanl.gov) (1999).

Riggs, J. E., Z. Guo, D. L. Carroll, and Y.-P. Sun, Strong luminescence of solubilized carbon nanotubes, *J. Am. Chem. So.* **122**, 5879 (2000).

Rinzler, A. G., J. Liu, H. Dai, P. Nikolaev, C. B. Huffman, F. J. Rodriguez-Macias, P. J. Boul, A. H. Lu, D. Heymann, D. T. Colbert, R. S. Lee, J. E. Fischer, A. M. Rao, and R. E. Smalley, Large-scale purification of single-wall carbon nanotubes: process, product, and characterization, *Appl. Phys. A (Mater. Sci. Process.)* **67**, 29 (1998).

Rubio, A., J. L. Corkill, and M. L. Cohen, Theory of graphitic boron nitride nanotubes, *Phys. Rev. B (Condens. Matter)* **49**, 5081 (1994).

Rueckes, T., K. Kim, E. Joselevich, G. Y. Tseng, C. L. Cheung, and C. M. Lieber, Carbon nanotube-based nonvolatile random access memory for molecular computing, *Science* **289**, 94 (2000).

Saito, R., M. Fujita, G. Dresselhaus, and M. S. Dresselhaus, Electronic structure of graphene tubules based on C_{60}, *Phys. Rev. B (Condens. Matter)* **46**, 1804 (1992a).

Saito, R., M. Fujita, G. Dresselhaus, and M. S. Dresselhaus, Electronic structure of chiral graphene tubules, *Appl. Phys. Lett.* **60**, 2204 (1992b).

Saito, R., G. Dresselhaus, and M. S. Dresselhaus, Tunneling conductance of connected carbon nanotubes, *Phys. Rev. B (Condens. Matter)* **53**, 2044 (1966).

Satishkumar, B. C., P. J. Thomas, A. Govindaraj, and C. N. R. Rao, Y-junction carbon nanotubes, *Appl. Phys. Lett.* **77**, 2530 (2000).

Schlittler, R. R., J. W. Seo, J. K. Gimzewski, C. Durkan, M. S. M. Saifullah, and M. E. Welland, Single crystals of single-walled carbon nanotubes formed by self-assembly, *Science* **292**, 1136 (2001).

Schonenberger, C., and L. Forro, Multiwall carbon nanotubes, *Phys. World* (2000).

Schonenberger, C., A. Bachtold, C. Strunk, J. P. Salvetat, and L. Forro, Interference and interaction in multi-wall carbon nanotubes, *Appl. Phys. A (Mater. Sci. Process.)* **A69**, 283 (1999).

Sedeki, A., L. G. Caron, and C. Bourbonnais, Electron–phonon coupling and Peierls transition in metallic carbon nanotubes, *Phys. Rev. B (Condens. Matter)* **62**, 6975 (2000).

Su, W. P., J. R. Schrieffer, and A. J. Heeger, Soliton excitations in polyacetylene, *Phys. Rev. B (Condens. Matter Mater. Phys.)* **22**, 2099 (1980).

Suenaga, K., C. Carbon, N. Demoncy, A. Loiseau, H. Pascard, and F. Williame, Synthesis of nanoparticles and nanotubes with well-separated layers of boron nitride and carbon, *Science* **278**, 653 (1997).

Tans, S. J., M. H. Devoret, H. Dai, A. Thess, R. E. Smalley, L. J. Georliga, and C. Dekker, Individual single-wall carbon nanotubes as quantum wires, *Nature* **386**, 474 (1997).

Tans, S. J., M. H. Devoret, R. J. A. Groeneveld, and C. Dekker, Electron–electron correlations in carbon nanotubes, *Nature* **394**, 761 (1998a).

Tans, S. J., R. M. Verschueren, and C. Dekker, Room temperature transistor based on a single carbon nanotube, *Nature* **393**, 49 (1998b).

Tenne, R., and A. K. Zettl, Nanotubes from inorganic materials, in *Topics in Applied Physics*, edited by M. S. Dresselhaus, G. Dresselhaus, and P. Avouris, Vol. 80, Springer, Berlin, p. 81 (2001).

Thess, A., R. Lee, P. Nikolaev, H. Dai, P. Petit, J. Robert, C. Xu, Y. H. Lee, S. G. Kim, A. G. Rinzler, D. T. Colbert, G. E. Scuseria, D. Tomanek, J. E. Fischer, and R. E. Smalley, Crystalline ropes of metallic carbon nanotubes, *Science* **273**, 483 (1996).

Ting, J.-M., and C.-C. Chang, Multijunction carbon nanotube network, *Appl. Phys. Lett.* **80**, 324 (2002).

Treacy, M. M. J., T. W. Ebbesen, and J. M. Gibson, Exceptionally high Young's modulus observed for individual carbon nanotubes, *Nature* **381**, 678 (1996).

Trentler, T. J., K. M. Hickman, S. C. Goel, A. M. Viano, P. C. Gibbons, and W. E. Buhro, Solution–liquid–solid growth of crystalline III–V semiconductors: an analogy to vapor–liquid–solid growth, *Science* **270**, 1791 (1995).

Venkataraman, L., and C. M. Lieber, Molybdenum selenide molecular wires as one-dimensional conductors, *Phys. Rev. Lett.* **83**, 5334 (1999).

Wagner, R. S, in *Whisker Technology*, edited by A. P. Levitt Wiley-Interscience, New York, p. 47 (1970).

Wagner, R. S., and W. C. Ellis, *Appl. Phys. Lett.* **4**, 89 (1964).

Wei, B. Q., R. Vajtai, and P. M. Ajayan, Reliability and current carrying capacity of carbon nanotubes, *Appl. Phys. Lett.* **79**, 1172 (2001).

Weiqiang, H., Y. Bando, K. Kurashima, and T. Sato, Synthesis of boron nitride nanotubes from carbon nanotubes by a substitution reaction, *Appl. Phys. Lett.* **73**, 3085 (1998).

Weng-Sieh, Z., K. Cherrey, N. G. Chopra, X. Blase, Y. Miyamato, A. Rubio, M. L. Cohen, S. G. Louie, A. Zettl, and R. Gronsky, Synthesis of $B_xC_yN_z$ nanotubules, *Phys. Rev. B (Condens. Matter)* **51**, 11229 (1995).

Westwater, J., D. P. Gosain, S. Tomiya, S. Usui, and H. Ruda, Growth of silicon nanowires via gold/silane vapor–liquid–solid reaction, *J. Vac. Sci. Technol. B (Microelectron. Nanometer Struct.)* **15**, 554 (1997).

White, C. T., and T. N. Tudorov, Carbon nanotubes as long ballistic conductors, *Nature* **393**, 240 (1998).

Wildoer, J. W. G., L. C. Venema, A. G. Rinzler, R. E. Smalley, and C. Dekker, Electronic structure of atomically resolved carbon nanotubes, *Nature* **391**, 59 (1998).

Wong, E. W., P. E. Sheehan, and C. M. Lieber, Nanobeam mechanics: elasticity, strength and toughness of nanorods and nanotubes, *Science* **277**, 1971 (1997).

Woodside, M. T., C. Vale, P. L. McEuen, C. Kadow, K. D. Maranowski, and A. C. Gossard, Imaging interedge-state scattering centers in the quantum Hall regime, *Phys. Rev. B (Condens. Matter Mater. Phys.)* **64**, 041310 (2001).

Yakobson, B. I., and P. Avouris, Mechanical properties of carbon nanotubes, in *Topics in Applied Physics*, edited by M. S. Dresselhaus, G. Dresselhaus, and P. Avouris, Vol. 80, Springer, Berlin, p. 287 (2001).

Yao, Z., C. L. Kane, and C. Dekker, High-field electrical transport in single-wall carbon nanotubes, *Phys. Rev. Lett.* **84**, 2941 (2000).

Yao, Z., C. Dekker, and P. Avouris, Electrical transport through single-wall carbon nanotubes, in *Topics in Applied Physics*, edited by M. S. Dresselhaus, G. Dresselhaus, and P. Avouris, Vol. 80, Springer, Berlin, p. 147 (2001).

Yazawa, M., M. Koguchi, A. Muto, M. Ozawa, and K. Hiruma, Effect of one monolayer of surface gold atoms on the epitaxial growth of InAs nanowhiskers, *Appl. Phys. Lett.* **61**, 2051 (1992).

Yu, D. P., X. S. Sun, C. S. Lee, I. Bello, S. T. Lee, H. D. Gu, K. M. Leung, G. W. Zhou, Z. F. Dong, and Z. Zhang, Synthesis of boron nitride nanotubes by means of excimer laser ablation at high temperature, *Appl. Phys. Lett.* **72**, 1966 (1998).

Zhang, Y. F., Y. H. Tang, N. Wang, D. P. Yu, C. S. Lee, I. Bello, and S. T. Lee, Silicon nanowires prepared by laser ablation at high temperature, *Appl. Phys. Lett.* **72**, 1835 (1998).

Zhou, C., J. Kong, E. Yenilmez, and H. Dai, Modulated chemical doping of individual carbon nanotubes, *Science* **290**, 1552 (2000).

Zhuang, L., L. Guo, and S. Y. Chou, Silicon single-electron quantum-dot transistor switch operating at room temperature, *Appl. Phys. Lett.* **72**, 1205 (1998).

Zunger, A., A. Katzir, and A. Halperin, Optical properties of hexagonal boron nitride, *Phys. Rev. B (Solid State)* **13**, 5560 (1976).

Advanced Semiconductor and Organic Nano-Techniques (Part II)
H. Morkoç (Ed.)

CHAPTER 7

Short Wavelength III-Nitride Lasers

A. V. Nurmikko

BROWN UNIVERSITY, PROVIDENCE, RHODE ISLAND

I. Introduction

1. HISTORICAL OVERVIEW

While wide bandgap semiconductors were recognized as prime candidates for efficient compact blue/green light sources in the early 1960s, nearly three decades passed before the promise of blue/green LEDs and diode lasers finally began to see light at the end of the tunnel in the early 1990s. New epitaxial techniques proved to be crucial for the synthesis of wideband gap quantum well (QW) heterostructures while also playing a seminally profound role in solving longstanding problems of doping in both GaN- and ZnSe-based semiconductors.

The first diode laser demonstrations in the blue–green spectrum were made in ZnSe-based QW heterostructures in 1991 (Haase *et al.* 1991; Jeon *et al.* 1991). Rapid developments shortly thereafter in III-nitrides compounds brought LEDs based on these materials to technological viability by mid-1990s (Nakamura *et al.* 1993). After the first pulsed violet InGaN QW diode lasers were demonstrated by Nakamura (1996) and Akasaki (1996), major improvements were made in the performance and device durability, especially at the Nichia Chemical Company, and the first commercially available devices appeared in 2000. As one example of the impact of this development, it now appears increasingly likely that the 405–415 nm InGaN laser, or perhaps its even shorter wavelength AlGaInN cousins, will form the basis of the next generation of optical disk technology, to supercede the current DVD format. There have been recent feasibility demonstrations using the 405-nm laser in conjunction with a rewritable dual-layer phase-change standard 12-cm optical disk, yielding a storage capacity of 27 Gbytes per side with a user data transfer rate of 33 Mbps (Akiyama *et al.* 2001). This compares with approximately 5 Gbytes per side for the present DVD using a 670-nm red laser. For reasons described later, the approximately 405–415 nm wavelength range is best suited to achieving high performance in the InGaN QW system for a diode laser, in contrast with the LEDs which thrive throughout the blue and blue-green regime. It appears difficult to reach the longer blue and green wavelengths with the InGaN QW system, at least from the present vantage point. The extension into the ultraviolet (UV) is another challenge to the III-nitrides, where progress in the near term can be expected.

The rapid success of the short wavelength III-nitride laser devices, in particular, has been somewhat of a surprise, because the underlying heterostructure is typically quite defected, specifically in terms of a high dislocation density. The first InGaN QW diode lasers were demonstrated in a material where a dislocation density on the order of $10^{10}\,\mathrm{cm}^{-2}$ was present, underscoring the extraordinary robustness of the nitride materials. Under typical current densities, $J < 1\,\mathrm{kA\,cm}^{-2}$, such a high concentration of extended state defects would be unthinkable in a continuous-wave GaAs-type diode laser. Nonetheless, innovative growth techniques have been required for long-lived diode lasers ($>1000\,\mathrm{h}$ in CW operation), particularly through the use of lateral epitaxial regrowth to reduce the density of threading dislocations. By contrast, history now shows that ZnSe-based II–VI QW diode lasers were quite vulnerable to the presence of defects (Gunshor and Nurmikko 1997). Although the dislocation-related defect density was subsequently reduced to less than $10^3\,\mathrm{cm}^{-2}$, extending the lifetime of a CW green laser (500–520 nm) to nearly 1000 h, it is still an open question how much further reduction in the defect density is required for a technologically viable device (a real application push, since it is unclear if InGaN lasers will reach into the green).

In this section, we provide an introduction to the basic physical properties of III-nitride semiconductors relevant for high performance light emitters. Following, a discussion of contemporary device science blue/violet diode lasers, including some examination of the physics of stimulated emission in wide-gap semiconductor light emitters is covered in Section II. Next is an overview of current efforts at fabricating advanced blue/violet emitters such as vertical cavity devices, and multiwavelength emitters presented in Section III. Section IV, briefly highlights the prospects of extending the III-nitride emitters into the UV spectrum. As a counterpoint, Section V provides an abbreviated overview of the impressive recent gains that have made the InGaAsN system, a competitive candidate for *infrared* lasers at communications wavelengths ($\sim 1.3\,\mu m$). While this subject is not the main focus of this chapter, we survey the progress from the viewpoint of the potential impact that these infrared sources may have on future optical telecommunication systems.

2. MATERIAL AND PHYSICAL PROPERTIES OF WIDE BANDGAP SEMICONDUCTORS

We begin by summarizing generic physical properties of wide-gap III-nitride semiconductors, which were responsible for the frustrations associated with early attempts to develop practical light emitter devices. As with challenges to create p–n junctions in any wide-gap material, the AlGaInN system might perhaps be better termed "semi-insulator", with physical features quite distinct from, say, GaAs. On the other hand, its optical properties offer new opportunities for optical devices, even reaching to the fundamental study of light–matter interaction.

a. Lattice Matching and Heteroepitaxy

With a view towards heteroepitaxy for blue/green, violet and (future) UV emitters, Fig. 1 shows the bandgap of a the AlGaInN material system as a function of the in-plane lattice constant (basal plane of the wurtzite structure), with emphasis on the AlGaN and InGaN ternaries, relative to SiC, ZnO, and sapphire substrates. The challenge of lattice matching to the substrate is much more complex than in conventional III–V and II–VI semiconductor heteroepitaxy, and still not fully resolved in terms of a device-optimal choice for the III-nitrides. Note how the bandgap of the AlGaInN system covers a wide range, from approximately 1.9 eV to about 6.2 eV. For a review of the epitaxial crystal growth on the group-III nitrides by the

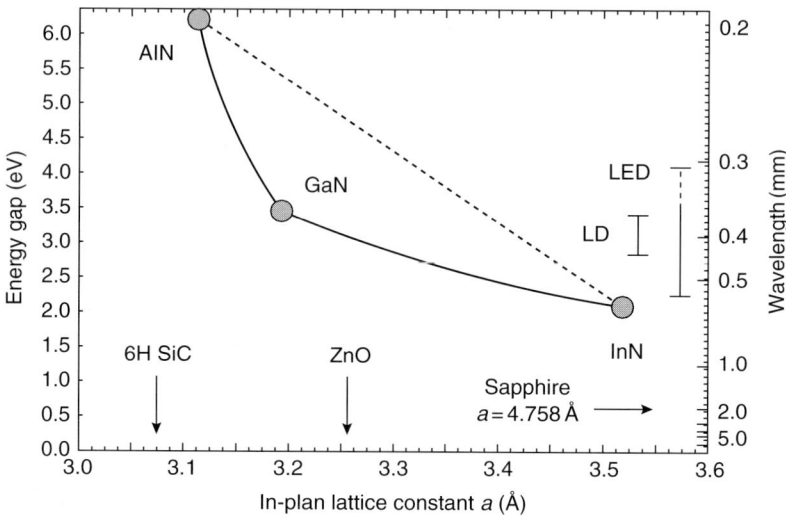

FIG. 1. Energy bandgap vs lattice constant for heteroepitaxy of the AlGaInN system and selected substrates. The approximate operating wavelength range of LEDs and diode lasers (LD) is indicated.

MOCVD approach, the reader is referred to an article by DenBaars and Keller (1999).

A satisfactory substrate solution for commercial high-brightness LEDs (e.g., those produced by Nichia Co. of Japan and Lumileds Lighting in the US), as well as for the first CW demonstrations of violet lasers, is the growth of a thick (up to several micrometers) GaN buffer layer atop of the vastly lattice mismatched sapphire substrate (13.5% mismatch). The use of GaN or AlN nucleation and/or stress control layers in MOCVD epitaxy is viewed as essential for the subsequent layered growth. Nonetheless, high densities of threading dislocations permeate the active light emitting regions. As an alternative, blue LEDs grown on SiC substrates were introduced some time ago (by Cree Lighting), followed by the first violet diode laser demonstration of an InGaN QW device grown on a SiC substrate (Bulman *et al.* 1997; Edmond *et al.* 1997). SiC is attractive as a substrate for the high performance nitride emitters because of its good thermal and electrical conductivity (vertical injection possible through a bottom device contact). For edge emitters, SiC permits ready cleaving of end facets. Other "substrate solutions" include efforts to create a natural single crystal GaN substrate, either by increasing the buffer layer thickness to hundreds of microns by hydride vapor phase epitaxy (Molnar *et al.* 1997) or by the growth a sufficently high quality and large area GaN bulk crystal substrates (Porowski *et al.* 1997). Finally, we note that bulk AlN substrates are becoming commercially available at the time of this writing (Rojo *et al.* 2001).

b. *Epitaxial Lateral Overgrowth (ELOG)*

The InGaN QW diode lasers operate at high current densities ($>1 \, \mathrm{kA/cm^2}$) even with the presence of the extremely high densities of dislocations, but their device lifetime is nonetheless compromised such that technologically viable device performance is not obtained. By contrast, these extended defects do not appear to effect the device lifetime in the LED regime ($<100 \, \mathrm{A/cm^2}$), although they have a finite effect on electrical-to-optical conversion efficiency. One effective growth strategy, which has been developed to reduce the density of the extended defects arising from the use of lattice mismatched substrates, utilizes lateral epitaxial overgrowth (ELOG). For III-nitrides, the techniques were introduced Nam *et al.* (1997), Usui *et al.* (1997), and others and involve interrupting the initial growth of the GaN layer to deposit and pattern a dielectric film before resuming growth. As shown in Fig. 2, a stripe-patterned amorphous dielectric film (SiO_2) defines windows of exposed GaN for the crystalline regrowth. Proper orientation of the stripes relative to crystallographic axes of the underlying GaN promotes rapid lateral growth over the dielectric and planarization of the growing GaN layer. The benefit of this process is truncation at the dielectric of the threading dislocations that originate at the sapphire–GaN interface and the near absence of these extended defects in the laterally grown GaN directly above the dielectric stripes. Above the dielectric, the two laterally propagating growth fronts from the opposite sides of a stripe coalesce to form an interfacial layer (the "suture" in Fig. 2), the structural quality of which

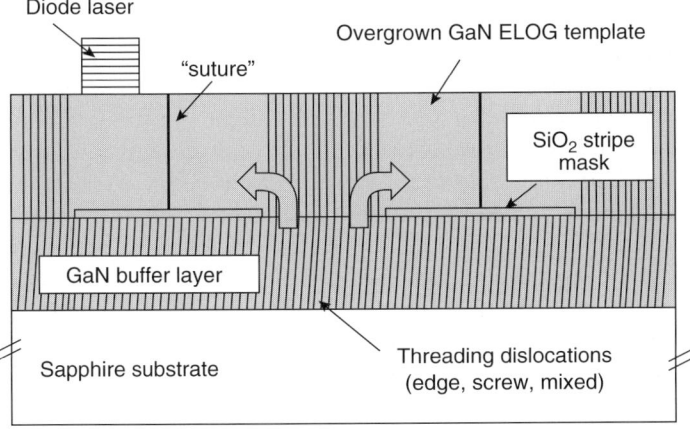

FIG. 2.   Schematic of epitaxial lateral overgrowth (upper panel). The lateral growth fronts, emanating from spaces between the SiO_2 stripes, coalesce near the center of each mesa to for a "suture" region which commonly still contains defects. The electron micrograph images show how the "wing" regions between the seed areas and the sutures have significantly fewer dislocations (L. Romano, private communication).

depends on the degree of tilt misalignment between the coalescing fronts. The region above the dielectric between the coalescence interface and the edge of the window defines the "wing" of the ELOG material where the density of threading dislocations is reduced by several orders of magnitude, on either sapphire or SiC substrates, perhaps to values as low as $10^5 \, \mathrm{cm}^{-2}$ and below. One measure of the impact of reduced dislocation density by ELOG is the greatly reduced reverse leakage current obtained in GaN p–n junctions. In connection with ultraviolet AlGaN photodetectors, a reduction of the reverse-bias leakage current by up to six orders of magnitude has been reported (Parish et al. 1999).

As discussed in Section II, the use of ELOG "substrate template" has substantially improved the performance of laser diodes, while significantly extending the device lifetime. Groups at Nichia, Sony and others have demonstrated lasers with threshold current densities below $J \sim 3 \, \mathrm{kA \, cm}^{-2}$, fabricated from the portions of the ELOG wafer where dislocations are minimized. The characteristics of lasers fabricated from such ELOG wafers have been systematically investigated, specifically in terms of variations in threshold current density for diodes fabricated above the wing, window, and coalescence regions. Yet, the threshold current density even in the best lasers remains somewhat high, raising questions about the microscopic mechanisms involved in the formation of optical gain and stimulated emission in the InGaN QW system, as discussed in Section II.

3. SOME PHYSICAL PARAMETERS: A COMPARISON

Some of the key physical parameters of the wide gap GaN, ZnSe, and GaAs are compared in Table I. Note the large polar energy for the chemical bond for the wide gap GaN and ZnSe, as well as the large covalent energy for

TABLE I

COMPARISON OF SELECTED MATERIAL PARAMETERS FOR GaAs, GaN, AND ZnSe

	Covalent energy (eV)	Polar energy (eV)	Young's modulus (GPa)	Thermal conductivity (W/cm/K)	T_{growth} (°C)	Optical phonon (meV)	Frohlich coupling (meV)^a	Exciton binding (meV)^b
GaAs	5.55	1.88	120	0.45	~ 650	36	5	~ 10
ZnSe	5.55	3.80	80	0.19	~ 300	30	36	-40
GaN	8.85	3.92	150	1.3	> 900	90	?	22^+

$^a T = \Gamma_0 + \Gamma_{\mathrm{LO}} / [\exp(\hbar\omega_{\mathrm{LO}}/kT) - 1]$.
^{b}A-exciton for bulk GaN.

the former. The strength of the underlying chemical bonds in GaN makes the material robust, and able to withstand high current densities (tens of kA/cm^2) even in the presence of a large number of threading dislocations and related microstructural defects. The epitaxial growth temperatures for the nitride semiconductors are considerably higher than those used for the growth of GaAs. However, since the optimal growth temperatures for AlGaN, GaN, and InGaN can cover a range as much as two hundred degrees (in descending order), the MOCVD epitaxial growth of quality AlGaInN heterostructures is nontrivial. Table I also shows the impact of the strong electron–hole Coulomb interaction (quantified here as the exciton binding energy) which leads to large enhancements in the interband optical cross-sections for the GaN and ZnSe. This "built-in" property, generic for both the III-nitrides and the II–VI compounds, has already been studied in much detail in ZnSe quantum wells, including diode lasers, demonstrating the impact of such a many-body enhancement to the optical gain and stimulated emission. We will return to this subject in Section II. Table I also includes a measure of the strength of the coupling of electrons to optical phonons, a parameter of importance in electronic scattering and hot electron relaxation. The carrier mobility in wide-gap semiconductors at room temperature (and above) is strongly dominated by the LO-phonon scattering, leading, in particular, to low values of the hole mobility. While the exact values for the Frohlich coupling constant are largely unavailable for GaN, these are expected to exceed those in ZnSe (which are large in their own right). Of course, at high carrier injection such as encountered in a diode laser, partial screening of the Frohlich interaction is expected. The actual details of the dynamical screening process are so far unexplored in the nitride diode lasers, and are further complicated by the presence of large piezoelectrically induced electric fields in the wurtzite crystals.

a. p-Doping of the AlGaInN Wide-gap Semiconductor

We will only survey the vital but vast topic of p-doping in wide bandgap III-nitrides in this chapter. Many scientific issues, especially at the microscopic level, remain open concerning the practice and optimization of p-doping in the AlGaInN system, particularly in regard to current efforts to extend the range of light emitters into the UV. The reader is referred to representative review articles (Gotz and Johnson 1999). In the next section we link the p-doping up with the issues of ohmic contact formation, which represents another specific challenge in the blue–violet diode emitters.

The microscopic issue concerning p-doping in the AlGaInN system is the nature of the local microscopic environment for the intentional Mg-dopants and their correlation with other impurities, point defects, and the extended defect microstructure. In the early work by Amano *et al.* (1989), practical levels of p-type doping of GaN with Mg was achieved by low energy electron beam irradiation (LEEBI) "activation" treatment of this acceptor. Nakamura and coworkers at Nichia Chemical Company built on this fundamental breakthrough to achieve even higher p-doping (perhaps up to $10^{18}\,cm^{-3}$ range) while demonstrating low resistivity in GaN epilayers (Nakamura *et al.* 1992). Their approach relied on high temperature thermal annealing activation of Mg. Subsequently, Nakamura and coworkers have employed short period modulation-doped (MD) p-GaN/AlGaN superlattices to achieve high p-doping levels $10^{18}\,cm^{-3}$ range; enhanced doping efficiency in such heterostructures with a *lateral* resistivity as low as $0.2\,\Omega\,cm$ has been reported (Kozodoy *et al.* 1999). Short period superlattices, both in terms of enhanced doping, improved contacts, as well as playing a role as "dislocation filters" are now included in many current violet laser designs, and may play a central role in the efforts to extend these devices into the UV. Recent activity is focusing on the study of the *vertical* transport in the MD superlattices within the AlGaInN system, as part of efforts to extend the emission wavelengths deeper into the UV.

Precise microscopic description of the impurity states in GaN:Mg is hampered by the scarce experimental information about local chemical structure associated with the doping sites, including the impact of the defect microstructure on their electrical activation. It is generally acknowledged that typical Mg concentrations on the order of $10^{20}\,cm^{-3}$ are required for p–n junction devices. Hence, on a statistical basis, less than 1% of the Mg dopant provides free holes. We now understand that acceptors in GaN form acceptor–hydrogen complexes during epitaxial growth that removes the acceptor energy level from the bandgap, a mechanism previously established for acceptors in Si and GaAs. In MOCVD growth there are numerous sources of hydrogen (e.g., the primary source for nitrogen, NH_3, as well as several metalorganic compounds). Interstitial hydrogen is incorporated into the GaN during growth and forms Mg–H complexes during cooling. Subsequently, these complexes can be thermally dissociated, with the H migrating out of the material or to other stable, electrically inactive sites, with creation of p-type conductivity. Fundamental understanding of the local electronic structure of these complexes has been advanced through computational studies of Neugebauer and Van de Walle at Xerox (Van de Walle and Johnson 1999), and the existence of the Mg–H complex was experimentally demonstrated by combined vibrational mode spectroscopy and Hall-effect measurements by Götz *et al.* (1996) at Xerox. On the other hand, there has been no experimental determination of the physical descriptors for the kinetics of the hydrogenation process such as the thermal dissociation energy or the diffusivities of monatomic, interstitial hydrogen in its different charge states.

A fundamental impediment to achieving high hole concentrations at room temperature is the high activation energy for thermal ionization of Mg acceptors in GaN, measured by several investigators to be approximately $\sim$200 meV at low densities. In the one-electron approximation, the binding energy depends on the dielectric constant and effective mass of the material. The nitride system possesses low dielectric constants ($\varepsilon \sim 9.5$ for GaN vs 13 for GaAs) and large effective masses (e.g., the heavy hole mass in GaN is $\sim$0.75m_0), which result in large binding energies and only partial ionization of the available acceptors at room temperature under thermal equilibrium. The problem can be further exacerbated by the presence of significant concentrations of n-type background contaminants that partially compensate the p-type dopants. The low p-type doping (typically $10^{17}\,\mathrm{cm}^{-3}$) leads to high contact resistance with almost any metal of choice. Ongoing research is focusing on means to increase the p-doping concentration and identify a shallower new p-dopant to improve the operating voltage and power efficiency of present lasers and LEDs. Increasing the bandgap for future ultraviolet light emitters by employing AlGaN at elevated Al concentrations ($>$30%) raises another fundamental issue due to possible coupling of the acceptor states with phonons. In an analogous case studied in the II–VI semiconductor alloy p-ZnMgSSe, such "polaron" (or self-trapping) formation was found to result in very deep acceptor levels, creating a fundamental barrier for p-doping (Han *et al.* 1994). The unintended incorporation of oxygen makes the p-doping of AlGaN a practical challenge, due to its high level of reactivity with Al.

The role of dislocations and other extended defects on p-type doping is less clear. Underlining the impacts of the defect microstructure are recent experimental observations that dopants may concentrate in spatial regions of high dislocation density (Bertram *et al.* 1999). To put the challenges into perspective, we note that device quality pn-junctions for an InGaN blue–violet lasers have been realized in only about a dozen laboratories worldwide at the time of this writing, due at least in part with the difficulty in implementing p-doping in device practice.

b.   Ohmic Contacts for p-AlGaInN

A technical challenge related to p-doping is the fabrication of low-resistance ohmic contacts to p-type materials, suitable for the high injection case encountered in a diode laser or high power LED. At the simplest level, the problem exists due to the lack of metals with suitably large work functions, which are needed to reduce the (potentially very large) potential energy barrier for hole injection. While reasonably low resistance contacts could be formed more easily on n-type materials, with specific resistances of $10^{-4}\,\Omega\,\mathrm{cm}^2$ and better, the poor contacts to p-type p-GaN caused the first LEDs and diode lasers to operate at unacceptably high voltages. For example, the first pulsed

InGaN QW diode lasers in 1996 had a voltage threshold exceeding 20 V. To obtain a low resistance contact between a metal and a wide bandgap semiconductor (p- or n-type) presents a fundamental dilemma of electrically bridging two materials with vastly different electronic band structure. The difficulty of the problem scales qualitatively with the bandgap of the semiconductor. In the case of GaN and its alloys, the relative ease of achieving heavy n-type doping, together with the position of the surface Fermi level, presents a rather low barrier for electron injection with many common metals ($\ll 1$ eV).

In this section, we outline the current state of affairs for the p-contacts to the p-AlInGaN system. While the best InGaN QW diode lasers have been operated very recently at threshold voltages well below 5 V, there seem to be little guidance to low contact resistance for p-GaN based on an understanding of the underlying microscopic physics. We note that in case of the wide-gap II–VI semiconductors, an approach using superlattice band-structure engineering was employed to achieve "electronic state impedance matching" at the contact interface (Fan *et al.* 1992; Hiei *et al.* 1993). This in turn led to the demonstration in 1993 of the continuous wave SQW ZnCdSe green diode laser at voltages as low as 4 V.

c.   Ohmic Contacts to p-AlGaInN

A somewhat bewildering host of p-contacting schemes have been reported in the literature, including metals and their combinations that involve Au, Ni, Ti, Pd, Pt, W, WSi, Ni/Au, Pt/Au, Cr/Au, Pd/Au, Au/Mg/Au, Pd/Pt/Au, Ni/Cr/Au, Ni/Pt/Au, Pt/Ni/Au, Ni/Au–Zn, Ni/Mg/Ni/Si, etc. In addition to its purely electronic function, a metal such as Pd may be helpful as a gathering agent for the extraction of residual hydrogen from p-GaN. Because much of the information related to contact fabrication is proprietary, it is difficult to gauge the state of the art in an open review. It appears that an acceptable p-contact resistance in a typical commercial blue LED is probably on the order of $R_c \sim 10^{-2}$–$10^{-3}\ \Omega\,\text{cm}^2$, frequently implemented with Ni/Au-based contacts. The contacts require a thermal annealing step, typically in nitrogen ambient. This range of specific contact resistance, is quite unacceptable for a high current density device as the diode laser, and should be compared, for example, with $R_c \sim 10^{-5}\ \Omega\,\text{cm}^2$ and below found in a long wavelength III–V laser. From the best reported results for the InGaN MQW diode laser, one can surmise that a contact resistance on the order of $10^{-3}\ \Omega\,\text{cm}^2$ has been now achieved routinely, and with some effort the values in practical devices probably reaching into the range $R_c \sim 10^{-5}\ \Omega\,\text{cm}^2$ with great care in wafer cleaning, etc. prior to metallization. These values are still high and contribute to the electrical and thermal budget of present blue/violet lasers, possibly impacting the device lifetime.

It is difficult to gauge the published reports on contact resistance to p-GaN, because of the sensitivity of contact resistance to both the quality of the underlying p-GaN material and to the detailed history of the process used to fabricate the contacts (recall that p-GaN requires its own post-growth thermal activation in the first place). Invariably, some form of thermal annealing (typically at temperatures around 500°C) is associated with the p-contact formation, sometimes in conjunction with the postgrowth activation of the free holes, often separate of this step. p-Type contact schemes based on Ni/Au and on Pd/Pt/Au are presently the most popular choice because they are easy to fabricate and have been extensively used in blue LEDs. The Ni/Au contact provided the starting point for the diode lasers, judged, for example, by the current-voltage characteristics of InGaN MQW laser diodes by Nakamura *et al.* (1998). However, even in this case, the microscopic characteristics of the contact remain largely unknown and subject to empirical design/process parameters. To illustrate the point, there are several reports in recent literature of a low contact resistance scheme with Ni/Au, where the thermal annealing in oxygen rich ambience might have resulted in the formation of NiO (Ho *et al.* 1999; Sheu *et al.* 1999). We note that NiO is a wide gap, usually p-type semiconductor so that an "accidental" band structure engineered GaN/NiO/Ni–Au contact scheme might occur in this case. However, other alternatives have been suggested, including one where the incorporation of O_2 into the activation ambient lowers the sheet resistivity of p-GaN by enhancement of outdiffusion of H from the p-GaN layer (Hull *et al.* 2000) The subject requires continued basic research, particularly for the AlGaInN devices under development for UV emitters.

d. Band Structure Engineering Approaches to p-Contacts

Recent approaches have employed "band structure engineering" approaches to the III-nitride p-contact problem. The first uses the already mentioned short period superlattices, which can yield high hole concentrations ($>10^{18}$ cm^{-3}), for *vertical* transport. For example, Zhou and coworkers used a p-AlGaN/GaN superlattice to obtain low contact resistance (Zhou *et al.* 2000). Kumakura *et al.* (2000) extended the concept to InGaN/GaN superlattices, recognizing further that the smaller bandgap of InGaN compared to GaN should reduce the metal–semiconductor contact barrier and hence the contact resistance. These authors also showed the p-InGaN contact was less susceptible to dry etching damage than a comparable p-GaN case, a feature of importance in device fabrication.

We highlight next an entirely different recent approach, where a (Esaki) tunnel junction (TJ) has been employed to provide the effective contact to p-GaN. A tunnel junction is composed of a heavily doped p^{++}/n^{++} diode where interband tunneling across the very narrow depletion layer provides a low resistance that is constant for small bias voltages across the junction. The use of the tunnel junction in conventional III–V semiconductors can be found in certain specific edge-emitting and vertical-cavity lasers (Wierer *et al.* 1999). For example, this approach has been recently employed to minimize the portion of p-type material in an InP-based VCSEL (Ortsiefer *et al.* 2000) since the InP-based p-layer has relatively high resistivity, which leads to serious heating problems. Following recent demonstrations of Mg-doping of InGaN layers with very high carrier concentrations ($p \sim 1 \times 10^{19} \, \text{cm}^{-3}$ at 300 K), a workable tunnel junction based on the successful growth of heavily doped p^{++}/n^{++} thin InGaN/GaN junction has been achieved (Jeon *et al.* 2001; Takeuchi *et al.* 2001), in spite of the concern for the Mg doping associated memory effects in MOCVD in the growth of thin, heavily doped layers. In the nitride case, a TJ becomes especially useful as a contacting scheme with good lateral current spreading, given that the conductivity of n-type GaN is nearly 100 times higher than that of p-GaN. To achieve hole injection with the TJ embedded in the GaN-based light emitting diode structure, Takeuchi and colleagues grew 30 nm GaN:Si^{++} and 15 nm InGaN:Mg^{++} atop a GaN–InGaN QW p–n LED heterostructure. In particular, the TJ was grown directly on the p-GaN layer that normally completes the LED device structure and resides on top of the pn-junction light emitter device. As shown in the schematic of Fig. 3 for the entire device, the top contact can now be made to the n$^+$ GaN:Si layer with a low contact resistance. Figure 4 shows the forward voltage of the LED, including the voltage drop across the reverse-biased TJ, was measured to be ~ 4.1 V at 50 A/cm^2 injection current density, while that of a standard LED with a conventional contact structure is 3.5 V. The TJ can eliminate the need for a highly resistive p-AlGaN cladding layer in short-wavelength laser diodes and the semi-transparent electrode required for current spreading in conventional

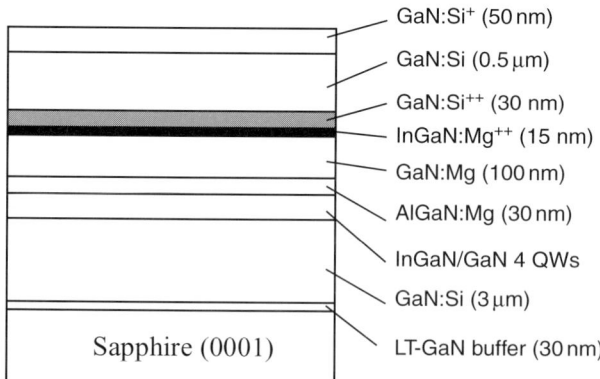

GaN:Si$^+$ (50 nm)

GaN:Si (0.5 μm)

GaN:Si^{++} (30 nm)

InGaN:Mg^{++} (15 nm)

GaN:Mg (100 nm)

AlGaN:Mg (30 nm)

InGaN/GaN 4 QWs

GaN:Si (3 μm)

Sapphire (0001)

LT-GaN buffer (30 nm)

FIG. 3.   Schematic of a blue LED structure incorporating a p^{++}/n^{++} InGaN:Mg/GaN:Si tunnel junction.

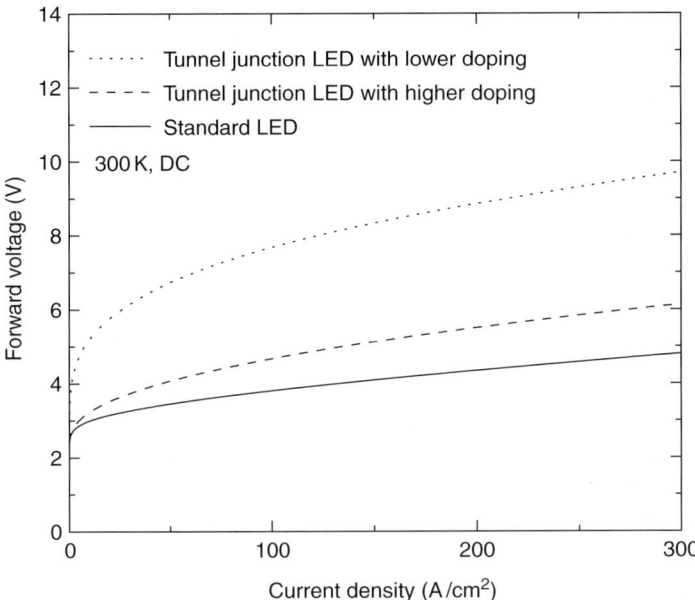

FIG. 4. *I–V* characteristics of the LED with the lower doping levels at the tunnel junction, the LED with the higher doping levels at the tunnel junction, and our standard LED with p-contact (Takeuchi *et al.* 2001).

GaN-based LEDs. We will return to this subject when discussing vertical cavity devices in Section III.

II.   Device Design, Performance, and Physics of Optical Gain of the InGaN Quantum Well Blue/Violet Diode Lasers

1.   INTRODUCTION TO DEVICE ISSUES

In this section, we review the central device science and optical engineering features of the blue/violet edge-emitting InGaN MQW diode lasers. The extraordinary progress made with these devices within the past three years now seems to have assured them a permanent place in optoelectronics applications (Nakamura 1999). At the time of this writing, about half a dozen research groups around the world have reported CW room temperature operation, with the best devices reaching hundreds to 1000 h in lifetime, although the extrapolated lifetime of 15 000 h at Nichia still stands by itself. Among other groups to mention are those at the laboratories of Sony, NEC, Toyoda Gosei, and Sharp in Japan, Samsung in Korea, Xerox PARC and Cree Lighting in the US, and OSRAM in Germany.

Below, we address representative InGaN QW heterostructures for diode lasers, comment on some fabrication techniques, and point out continuing device challenges. The latter include questions concerning the threshold current density and the continued efforts to create substrate templates for reducing the misfit (threading) dislocation density for improved device performance and lifetime. At a fundamental level, there is evidence that the InGaN alloy, which forms the optically active MQW medium, exhibits compositional and/or quantum well layer thickness disorder that impacts on the gain spectrum of the laser. This aspect has so far restricted the operation of the InGaN-based devices at practical threshold current densities to the narrow ~405–415 nm range in the violet, leaving the longer blue and green regions to await future developments, perhaps involving complementary material approaches. The extension of laser operation into the UV represents, on the other hand, a contemporary and different set of major opportunities and challenges, exploiting the full range of the quaternary system, discussed in Section IV.

While the prospects for developing the green ($\lambda \sim 500$ nm) ZnCdSe QW diode laser into a practical technology are hampered by the serious, unresolved degradation problems, a brief examination of its basic device science offers a useful point of reference for the III-nitride blue/violet laser. The best threshold current densities that have been achieved in the ZnCdSe SQW lasers (<200 A/cm^2) (Katayama *et al.* 1998), are about an order of magnitude lower than those achieved in the best InGaN QW lasers today ($\sim$2–3 kA/cm^2). At the level of simple effective mass theory, the electronic structure and optical properties of the II–VI and III-nitrides are rather comparable, so that this comparison in the threshold current density suggests a significant improvement in the future operation of the nitride diode lasers. In particular, electron–hole Coulomb correlations are predicted to offer a significant enhancement to the peak gain in the III-nitrides, a piece of physics at a fundamental level that has been already identified in the operation of the II–VI green diode laser devices.

2. DESIGN AND PERFORMANCE OF THE BLUE/VIOLET InGaN DIODE LASER

a. Device Heterostructure Design

The active region of present violet edge-emitting nitride diode lasers and high power blue LEDs are based on the InGaN/GaN/AlGaN MQW separate confinement heterostructure (SCH), grown either on the (nonconducting) sapphire or conducting SiC substrate. In Fig. 5, the layering details (exclusive of the substrate template) are shown for a contemporary InGaN SCH/MQW diode laser, which combines the generic features of a separate

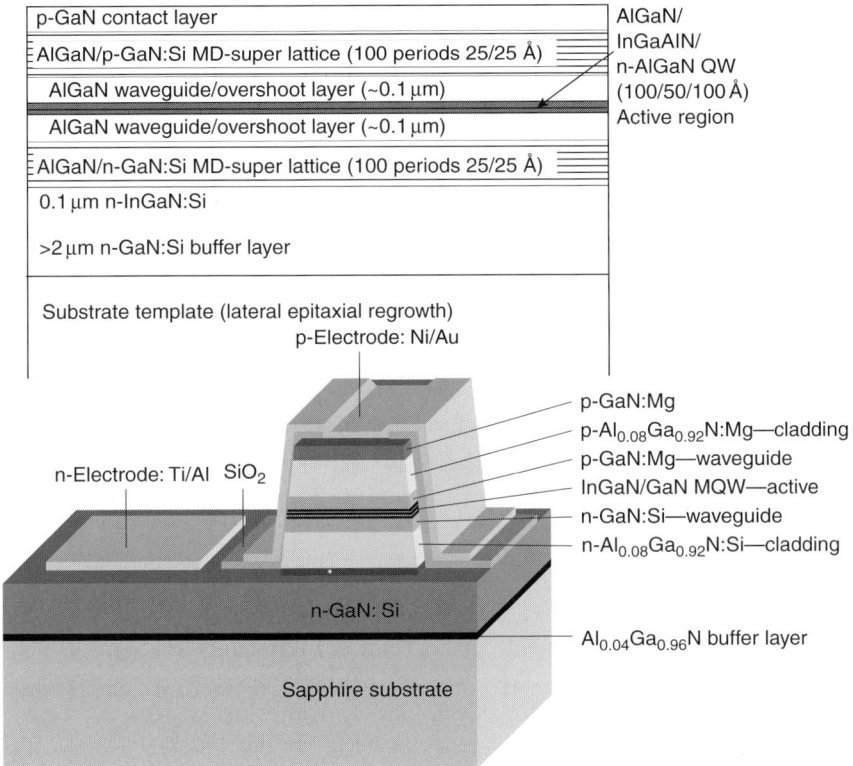

FIG. 5.   Schematic of layer structure for a SCH MQW violet laser AlGaInN heterostructure.

confinement edge-emitting diode laser within the constraints imposed by the growth of the AlGaN, GaN, and InGaN cladding, waveguide, and QW active layers, respectively. A wide range of temperatures is required when growing the three different materials, anywhere from about $T = 800°C$ for InGaN to nearly $T = 1100°C$ for AlGaN.

In the heterostructure diode laser structural schematic of Fig. 5 (upper trace), one might find a low temperature grown GaN nucleation layer ($\sim$250 Å, not shown), a pair of modulation doped superlattices on both n- and p-side of the junction, and the active MQW region sandwiched between AlGaN waveguide layers. The superlattice approach to modulation doping is particularly useful for the p-doping in terms of lateral and vertical transport, and has been shown to have a role as a "dislocation" as well. The AlGaN waveguide layers may house additional thin AlGaN layers (not shown) for blocking the electron/hole overshoot of the active QW region, strategically positioned near the gain medium. The role of the blocking layer also involves other considerations, such as prevention of the

InGaN decomposition under the high temperature growth of the structure above the MQW section. While the range of In concentrations for LEDs ranges from a about $x \sim 0.01$–0.4, the range for InGaN diode lasers at practical current threshold has so far been rather limited ($x \sim 0.10$–0.15), with typical wavelength of emission near $\lambda = 410$ nm. The threshold current density increases rapidly with the indium concentration, by more than a factor of 5 over the range from about 410 to 450 nm (Nagahama *et al.* 2001a). The optimum number of quantum wells has been determined empirically to be on the order of 3, dictated by the modal gain required to offset the cavity losses. One pragmatic challenge, is the ease with which cracks form in AlGaN-rich heterostructures, induced by stresses introduced into the AlGaN layers due to lattice mismatch and due to the differences in the coefficients of thermal expansion with respect to the thick GaN layers. As already mentioned in Section I, the initial buffer layers can be synthesized essentially on a different substrate template, prepared by epitaxial lateral overgrowth that significantly reduces the density of threading dislocations (of screw, mixed, and edge type) in the active region.

Many of the details in optimizing the SCH design and the active QW region have been determined empirically, with a limited amount of the kind of device modeling which is routinely applied, for example, to the infrared (IR) communication lasers. For example, precise information about the band offsets at the InGaN/GaN heterojunction is not available from spectroscopic work, owing largely to the very soft bandedge of InGaN. We will discuss the impact of this problem on the optical gain below. The common anion supposition implies that the electronic confinement in valence band is weak, perhaps on the order of 50 meV or less, while the conduction band offset might be expected to be on the order of 250–300 meV for the typical average In composition ($x \sim 0.10$–0.15) in the laser devices. However, both the In compositional and QW thickness fluctuations, coupled to the presence of large (fluctuating) strain and piezoelectric fields makes it very difficult to obtain an operationally meaningful measure of the QW confinement. There is empirical agreement that QW thicknesses on the order of 40 Å are optimal for the diode laser performance. Unlike the case of GaAs- and ZnSe-based lasers, only a limited amount of systematic work exists concerning the designs of the SCH/MQW nitride lasers and the phenomenon of current overshoot. Yet, the reports suggest that a finite leakage current competes with the radiative recombination current in both the diode lasers as well as LEDs at high injection, indicating that the finite QW confinement is a factor.

The first InGaN QW diode lasers were devices grown on thick GaN buffer layers on the (0001) *c*-face of sapphire. This meant that dislocation densities in the range of 10^8–10^{10} cm^{-2} were threading into the active region (Ponce *et al.* 1994; Lester *et al.* 1995). Later, we show examples of high spatial resolution microscopies that strongly suggest how these types of extended state defects in the nitrides do not overwhelm radiative recombination processes. Also remarkable, Nakamura reported a CW InGaN MQW diode laser operating for several hundred hours with a threshold current density of $3.6\,\text{kA/cm}^2$ in this type of high defect density environment. A first order answer to this tolerance of GaN and its related compounds to high injection and temperature can be found in the large covalent bond energy that makes this family of semiconductors exceptionally stable mechanically.

For truly long-lived blue and UV diode lasers (lifetime $>10^4\,\text{h}$), the high dislocation density is not acceptable. Furthermore, the thermal budget even in today's very best InGaN diode lasers is still very high, due in part to the high threshold current density in the edge-emitting devices. Innovative approaches are now beginning to be implemented to address these issues. The ELOG technique outlined earlier, is increasingly becoming a standard building block for laser design/fabrication. Recall that spatially patterned GaN growth is implemented by employing a few micrometer-scale apertured SiO_2 mask on the GaN buffer layer. The epitaxial growth nucleates on the exposed GaN within the stripes of the SiO_2 template and, under a sufficiently high temperature ($T > 1050°C$) and constituent flow in the MOCVD reactor, acquires a large lateral growth rate, once the aperture area is filled with GaN. By the standards of the nitrides, there can be very few ($<10^5\,\text{cm}^{-2}$) dislocations in the overgrown areas above the SiO_2 mask, specifically "wing area". These tend to be of the edge dislocation type, parallel to the (0001) plane after a $90°$ bend in the regrown region. After a thick (say $>20\,\mu\text{m}$ GaN) layer has been overgrown, this new "substrate" is ready for the subsequent deposition of the device heterostructure. Sakai and colleagues, among others, made detailed studies of the dislocation propagation during the ELOG process (Sakai *et al.* 1998).

Nakamura *et al.* (1998) first adapted the ELOG approach for their InGaN MQW laser diodes. A significant improvement in the room temperature cw device lifetime to beyond 100 hours was achieved, apparently as a direct consequence of the reduced dislocation density. Recently, the Nichia group applied an SiO_2-free ELO technique (where SiO_2 is used as an etch mask to define GaN window region and then etched away prior to the continued epitaxy), originally developed by Nam *et al.* (1997), to further extend the device lifetime longer than 1000 hours under high power ($\sim$50 mW) CW operation (Nakamura *et al.* 1999). It is worth noting that the ELOG techniques, while being very desirable for the advancement of the violet lasers, are likely to be essential for nitride-based short wavelength photodiode detectors. This follows from the necessity of reducing the reverse bias leakage current, dark current, which is otherwise greatly increased by contributions from the dislocations as well as related extended state and point defects.

Among contemporary examples of the impact of the ELOG-based substrates on the violet diode lasers, the group at Sony Laboratories has studied the correlation between photoluminescence efficiency, surface profile of cleaved laser facets, and laser performance on ELOG-based substrates

(Tojyo *et al.* 2001). The relationship between the dislocation density (measured as etch pit density) and PL efficiency of the approximately 5-μm thick GaN substrate is shown in Fig. 6, whereas Fig. 7 shows the spatially resolved PL emission, indicating that the "best quality" material is found under the wing region. Only the mixed dislocation type was found in this region, at concentrations less than $10^5 \, \text{cm}^{-2}$. To prepare the optical resonator required for the complete diode laser structure, the sapphire substrate was thinned to about 100 μm and the laser diode facets were subsequently fabricated by cleaving (as an alternative to their fabrication by dry etching). Interestingly and usefully, the surface flatness and morphology were superior for those portions of the cleaved facets that coincided with the wing region of the ELOG substrate. Figure 8 shows the results of AFM profiling where the smoothness of the laser facet can reach values better than 1 nm, in contrast with roughness measured on cleaved laser devices growth directly on sapphire (without the ELOG process) beyond 10 nm. In case of the ELOG devices by Tojyo and others, lifetime tests under 30 mW CW output power indicate good performance beyond 1000 h.

The diode laser device degradation and lifetime by various ELOG approaches has been systematically studied by Nagagama *et al.* (2000, 2001). Figure 9 shows a schematic for comparing three different substrate/template cases where the ELOG idea is implemented at levels of successively increasing complexity and physical thickness. The first case corresponds to

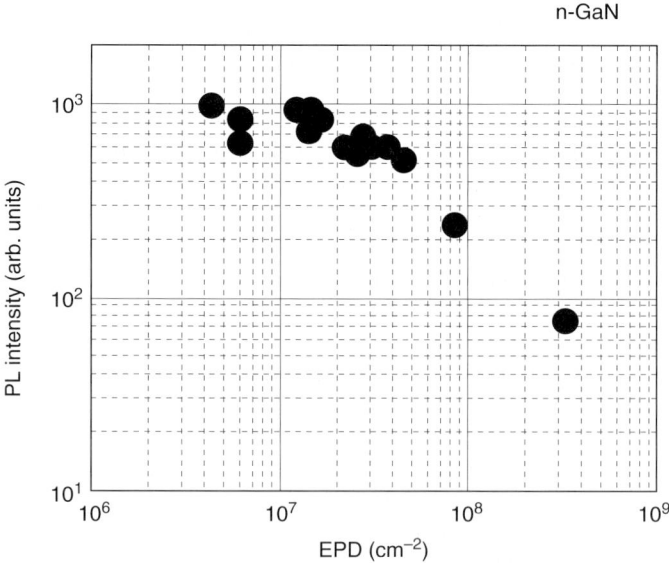

FIG. 6. Correlation between photoluminescence intensity and etch pit density (dislocation density) in GaN substrate, demonstrating the impact of ELOG techniques (Tojyo *et al.* 2001).

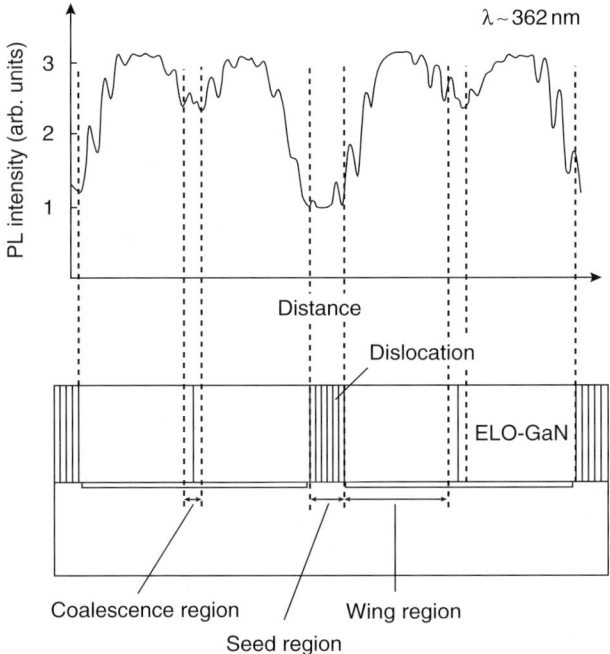

FIG. 7.   Spatially resolved photoluminescence intensity in ELOG-GaN (Tojyo *et al.* 2001).

the usual ELOG already outlined. The second involves the use of hydride vapor phase epitaxy (HVPE) (Molnar *et al.* 1997) to synthesize a very thick (200 μm) GaN atop the ELOG undercarriage; subsequently most of the substrate and HVPE-grown thick GaN are removed by polishing leaving an about 150 μm thick free-standing GaN substrate. In this case, the HVPE layer is assumed to direct residual dislocations towards the edge of the wafer. In the third instance, the substrate/template growth continues atop the HVPE GaN with another ELOG step, which is then ready for the hetero-epitaxy of the active device. The backside of the device is polished to remove the substrate, including some of the HVPE GaN, to form a free-standing device film so that cleaving techniques can be used to form a laser cavity with facets in the {1100} direction. Cathodoluminescence images show how the dislocations are reduced with the increased complexity of the ELOG-based substrate/templates, specifically in terms of their density in GaN above the SiO$_2$ mask window region. With the laser devices grown on the three substrate/templates, the lifetimes at 30 mW average power were reported to be 700, 300, and 15 000 h, respectively, at a device case temperature of 60°C (Nagahama *et al.* 2000). This comparison dramatically illustrates how important it is to reduce dislocations in order to achieve truly long-lived violet diode lasers. Even so, the ELOG intensive approach probably still

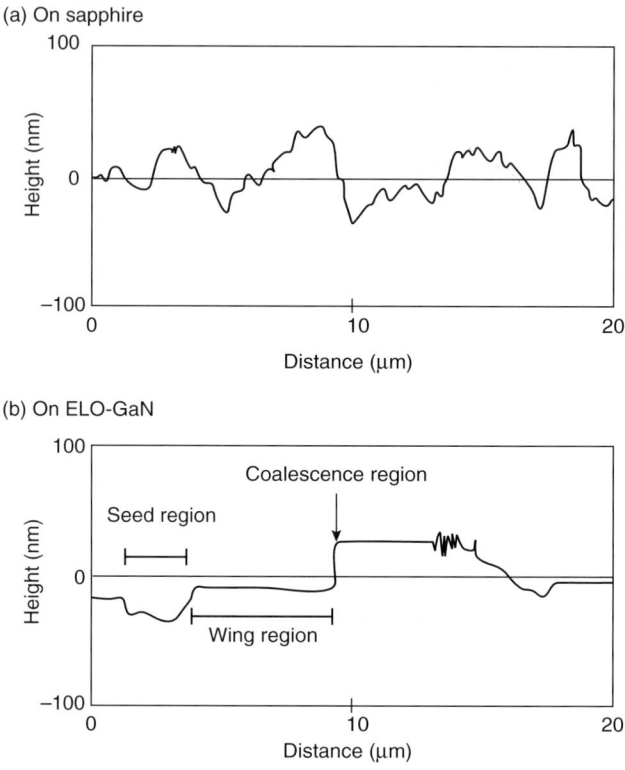

FIG. 8.   Effect of GaN ELOG substrate/template on facet roughness of cleaved diode laser (Tojyo *et al.* 2001).

leaves a residual dislocation density which would be considered unrealistic for a viable conventional III–V diode laser.

   Given the substantial heat that is generated is the present diode lasers (typically in excess of 500 mW), the presence of the sapphire substrate presents another hindrance in the development and flexible application of the InGaN MQW violet laser. The thermal conductivity of sapphire at room temperature is about 50 W/m/K, compared with about 130 W/m/K for GaN and 490 W/m/K for SiC. Moreover, the nonconducting sapphire prevents the present edge emitting diode laser structures from being flip-chip mounted for heatsinking since both the n- and p-type contact reside on the same side of the device (see ahead). Hence, the introduction of the HVPE growth technique for synthesis of very thick (>100 μm) GaN buffer layers offers a solution to the management of both the thermal budget and conductivity problems. Nakamura and coworkers first adapted the approach to separate InGaN QW heterostructures with their thick GaN buffer, to transfer laser devices onto another "artificial" host substrate. Recently, other techniques have been

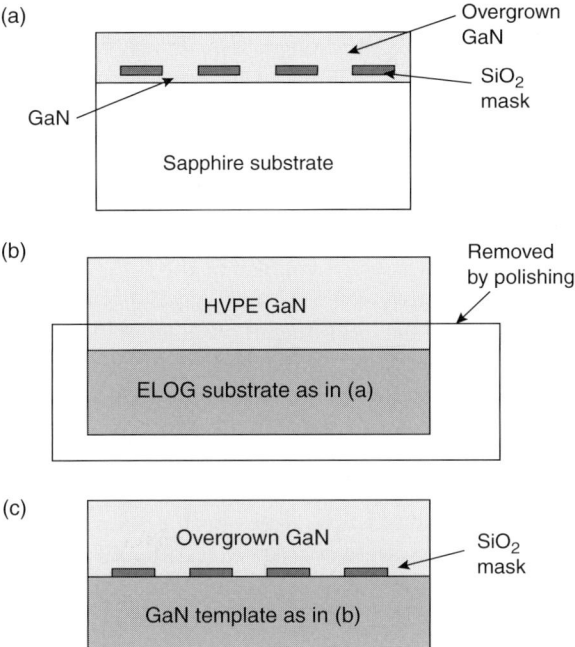

FIG. 9. Schematic of ELOG substrate/templates for InGaN MQW diode laser, at varying levels of complexity (Nagahama *et al.* 2000).

developed for separating the nitride heterostructure from its sapphire sub-strate, for instance by employing a UV laser ablation technique, which has been applied by the Xerox group to transfer the violet diode laser on a copper substrate (Kneissl 2000).

b. Edge Emitter Diode Laser Performance

While not detailing the fabrication of edge emitting lasers, we note briefly that the SCH/MQW InGaN/AlGaN heterostructure material is fabricated into index-guided mesa structures (bottom trace of Fig. 5) by employing dry-etching techniques. The electron cyclotron resonance plasma (ECR), chemi-cally assisted ion beam etching (CAIBE), or standard reactive-ion etching, all using chlorine chemistry, is known to produce acceptable etch quality for the vertical walls of the few micrometer wide lateral (ridge) waveguide as well the resonator end facets. The etched n-GaN surface onto which metallization is applied for the n-type contact likewise has an acceptable electronic quality following properly applied dry etching. As already noted, in case of the HVPE grown thick buffer layers, or the SiC substrate, the optical resonator can been formed also by cleaving of free standing ~100 μm films, following the (laborious) removal of the substrate/template.

There appears to be significant variation in the few reported values for cavity losses in the InGaN MQW diode lasers, in part due to the uneven pace of device development between various laboratories. Extracting the loss coefficient from threshold current density variation as a function of resonator length and/or facet reflectivity suggests, roughly, a typical value of 20–50 cm^{-1} for the optical losses. Some of this should be apportioned to the quality of the end mirror facets for the devices grown on sapphire, where dry etching is required for mirror fabrication.

As an example of the device performance of the InGaN MQW violet laser, operating in the 405–415 nm wavelength range, Fig. 10 shows the voltage–current characteristics and the light output per uncoated facet of a cw InGaN diode laser at room temperature (Nakamura *et al.* 1998), grown on a sapphire substrate. The devices were ridge waveguide structures with dimensions of $2 \times 600\,\mu$m, and could be augmented by reflective facet coatings. The threshold current density was approximately 4 kA/cm^2, at a voltage of 5.5 V (under the uncertain assumption of a uniform current flow). A slope efficiency (differential) as high as 1.0 W/A has been obtained from such devices by the Nichia group. Output powers well exceeding 50 mW can be reached, observed under high injection (>100 mA), the level required for the writing process in most current optical disk media. Due to the small ridge width, a rather stable fundamental transverse mode operation is possible, however, in order to avoid the dependence of the far field pattern on the injection level (so-called "kinks") the ridge waveguide design requires high accuracy in the index of refraction data as well as high precision in the ridge etch depth. These constraints have recently led Kijima *et al.* (2001) to explore a new approach to stabilizing of the transverse mode by using a spin-on-glass

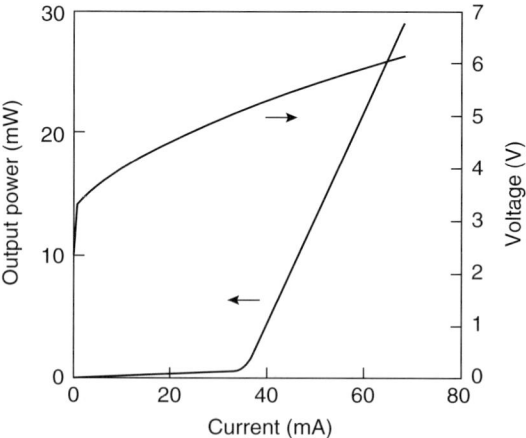

FIG. 10.   *I–V* characteristics and the optical power output per uncoated facet of a CW InGaN diode laser at room temperature (Nakamura *et al.* 1998).

(SOG) insulator in place of SiO2 as the dielectric insulator at the mesa edges (bottom trace of Fig. 5). The light absorptive characteristics of this useful process material constrain and stabilize the lowest transverse mode. Figure 11 shows the *L–I* characteristics and the "kink-free" far-field pattern, the latter showing FWHM angles parallel and perpendicular to the junction plane of 9.1 and 22.4°, respectively.

An example of the emission spectrum below and above lasing threshold is shown in Fig. 12 for an InGaN MQW diode laser grown on an SiC substrate (Song *et al.* 1998) and displaying a single longitudinal mode. Early on, the InGaN diode laser spectral output on sapphire substrates showed a complex longitudinal mode pattern where the presence of subcavities is apparent (i.e., with length less than the device physical length). Evidently, specific crystalline defects, such as cracks that are frequently induced within the growth of the AlGaN waveguide layers, formed "accidental" resonators that provide additional optical feedback, augmenting the external resonator. However, good progress is being made to create stable, single longitudinal mode blue and violet lasers, even if the single longitudinal mode aspect is not necessary for optical storage applications (in contrast to the single transverse mode requirement).

In terms of InGaN MQW diode laser efficiency, by employing the ELOG-based substrate/template approaches, some of the above cited laser performance parameters have improved, though the principal advantage of this approach is to reduce the device degradation. Threshold current densities have been reduced to the range of 2–3 kA/cm^2 in leading laboratories (Nichia, Sony, Xerox) and the slope efficiency increased up to about 1.5 W/A in best cases. The threshold voltages, likewise have dropped to the 4–5 V range, still dictated in part by the quality of the p-contacts. We also note recent work in which linear arrays of the violet InGaN MQW have been fabricated, with total output power exceeding 3 W (Goto *et al.* 2001).

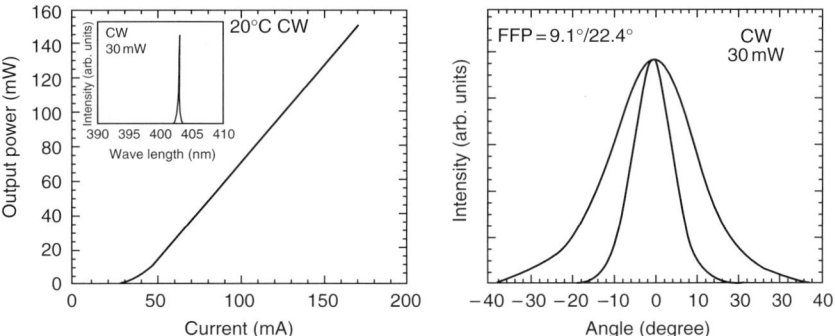

FIG. 11. *L–I* characteristics of a diode laser using SOG current blocking layer (left panel); far-field emission pattern from the transverse-mode stabilized device at 30 mW output power (Kijima *et al.* 2001).

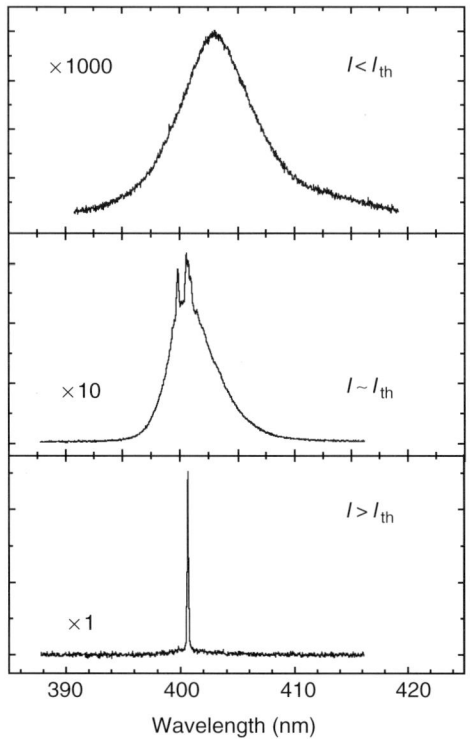

Fig. 12.   Emission spectrum below and above lasing threshold for an InGaN MQW diode laser grown on SiC substrate (Song *et al.* 1998b).

3.   PHYSICS OF OPTICAL GAIN IN THE InGaN MQW DIODE LASER

A still somewhat puzzling aspect about the operation of present violet InGaN QW lasers concerns the high electron–hole pair densities that are induced from the electrical injection conditions. For an estimated carrier lifetime of several nanoseconds, the bulk equivalent electron–hole (e–h) pair density at threshold in excess of 10^{19} cm^{-3} is obtained. In fact, Nakamura and coworkers quoted a density of 2×10^{20} cm^{-3} for the first Nichia CW laser, a very high figure in a semiconductor laser. This high e–h pair density is only partly explained by the somewhat higher effective masses in terms of formation of population inversion, when compared with GaAs. On the other hand, since the effective masses in ZnSe and GaN are of comparable magnitude and the threshold current density as well as the e–h pair density in a ZnCdSe QW laser are more than one order of magnitude lower, an important question arises concerning other factors that influence the optical gain of the InGaN device. In strong contrast with the optical properties of the binary GaN, the characteristics of the InGaN QW system at the band-edge include a spectrally very broad luminescence emission, even at cryogenic temperatures (linewidth on the order of 100 meV for In concentration $x > 0.20$). Up to injection levels of at least 10^{18} cm^{-3}, luminescence spectroscopy (including time resolved) shows evidence for localized e–h pair states that reflect the large energy range available for such localization (Chichibu *et al.* 1997a; Narukawa *et al.* 1997; Lefebvre *et al.* 2001). The degree of localization increases more rapidly with In concentration than expected for a simple random alloy, suggesting lattice disorder for the thin QWs, which is both structural and compositional. How this can influence the optical gain and the necessary level of electron-hole pair injection in an InGaN MQW diode laser is a subject of the following discussion.

The optical fingerprints of thin InGaN QWs show strong departures from a usual random alloy. A predominant view, at least qualitatively established via many microprobe studies, is that a propensity for In clustering may originate from a finite degree of thermodynamic immiscibility of the InN and GaN constituents in a normal solid solution at the near 800°C growth temperature. If such "compositional anomalies" occur, they would profoundly affect the bandedge electronic states which form the "electronic power supply" for optical emission, both for LEDs and diode lasers. Here we adopt the view that the presence of such deviations from a random alloy lead into a competition for electronic excitations between localized and extended electronic states. We also note that another possibility exists for explaining the spectroscopic observations in InGaN QWs, namely disorder due to quantum well thickness fluctuations. In particular, recent low-temperature photo- and cathodoluminescence spectroscopy by Monemar *et al.* (2001) has led these authors to argue that quantum well thickness or related disorder, not phase separation is responsible for the optical properties near the bandedges. It is, of course, possible that the reality involves a mixture of the two sets of disorder effects.

A key issue is the nature of those band-edge electronic states that supply the requisite optical gain. Typically, the mean values of x_{In} in the laser devices are in the range of $x_{In} \sim 0.1$–0.2. High-resolution electron microscopy studies have argued that "clustering"-type disorder takes place in the InGaN QW and the term "quantum dot" was applied to such clusters (Narukawa *et al.* 1997). When this type of nanoscale heterogeneous semiconductor forms the active laser material, a question that arises concerns the competition between localization and many-electron correlations within the available e–h pair states, given the very high pair densities required even in best present devices to reach lasing threshold ($>10^{19}$ cm^{-3}).

Early on, Nakamura and Fasol (1997) used the Hakki–Paoli method to acquire gain spectra on diode laser devices and an electrical injection-optical probe method was applied to study the formation of optical gain in InGaN QW p–n junction heterostructures (Kuball *et al.* 1997). Significant insight to the optical gain spectra of the InGaN laser active medium has been acquired subsequently, based on the analysis of the spontaneous emission spectra of the diode laser, in conjunction with its threshold characteristics. This approach makes use of the fundamental relationships between spontaneous emission, stimulated emission, and absorption (Henry *et al.* 1980). For a gain medium possessing such idiosyncrasies as the InGaN QW, the "Henry" approach is advantageous, given the central question of the fundamental gain characteristics and their relationship with radiative processes.

First gain spectroscopic measurements on actual InGaN QW diode lasers ($x = 0.15$) were made by Song *et al.* (1998), to show the pronounced extension of the gain spectra associated with the $n = 1$ QW transition into the low energy region. At threshold, finite gain was found as much as 200 meV below its peak position, indicating a degree of broadening that is most uncharacteristic of common semiconductor lasers. A peak gain coefficient of approximately 3200 cm^{-1} was measured. Subsequently, Kneissl *et al.* (2001) measured the gain spectra over a range of indium compositions for the practical operating range of the violet InGaN diode laser, shown in Fig. 13, where the reader's attention is directed to the dependence of the gain bandwidth on the composition. As detailed by Song *et al.* (1998), with increasing injection one reaches the transparency condition relatively easily, at levels of injection that were not very different from that of the conventional LED regime. With increasing current, gain emerges over the large spectral range, indicative of the participation of a corresponding range of electronic states. The position of the peak gain somewhat blue shifts at higher injection levels, but considerably less than anticipated from a one electron state-filling picture, possibly due to many-body bandgap renormalization effects. Qualitatively, we may now understand one reason for the high e–h pair density required for laser operation, apart from extrinsic reasons such as unwanted optical losses. That is, while the spectrally integrated gain is, in fact, quite large, its peak value (determining the lasing threshold) is much diluted at the expense of the remarkably large broadening.

These observations provide an extension to arguments (Chichibu *et al.* 1996, 1997a; Kuball *et al.* 1997) that radiative recombination processes at the lowest interband transition in the InGaN QW are profoundly influenced by localized e–h pair states at room temperature, within an energy range which is nearly an order of magnitude larger than estimated for a simple random alloy. Whether this aspect of InGaN is intrinsic or subject to specific MOCVD growth conditions has yet to be established; suffice it to say that all available optical data on InGaN QWs and thin films to date display the striking extension of the bandedge states so that, for example, excitonic features in absorption at the $n = 1$ QW states have not been unambiguously identified. By contrast, gain spectroscopy performed on widegap ZnCdSe QW diode lasers by the same method described below shows very clearly the characteristic influence of the strong excitonic enhancement of the peak gain and an overall optical response at the $n = 1$ HH exciton with the pronounced

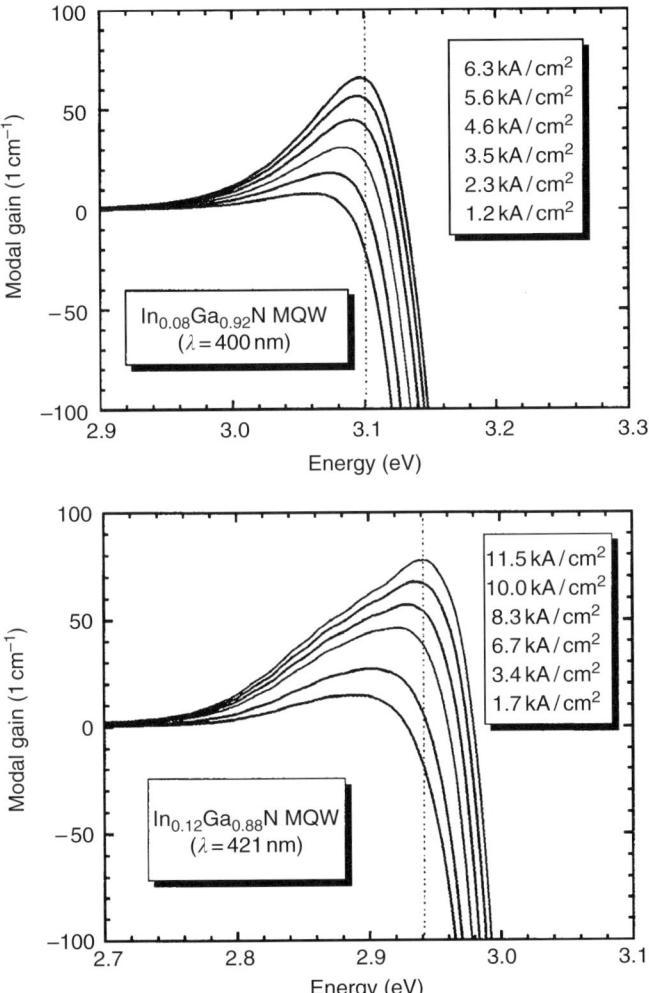

FIG. 13. Gain/absorption spectrum of a InGaN MQW diode laser of two different In composition, at varying injection levels at room temperature (Kneissl *et al.* 2001).

Coulomb correlations in evidence. Such effects are clearly masked by the disorder contributions in the InGaN QW, making it difficult to isolate predicted many-body interactions in the dense e–h system within the active region of the blue diode laser.

To summarize, experiments show how filling of the localized states is a necessary prerequisite prior to the buildup of a sufficient population inversion for threshold gain in the present devices. On the other hand, since transparency is reached at a rather low injection level ($n = p \sim 10^{18} \, \text{cm}^{-3}$) it may be possible to reduce the threshold current by designing a laser resonator with very low optical losses. The near "clamping" of the Fermi level E_F at higher injection may be due to a significant increase in the effective density of states. However, the question that we will defer to the next section is whether these states are still localized or extended. We wish to emphasize that the issue of the In compositional anomalies increases in severity very rapidly as the In concentration reached about $x_{In} = 0.1$ and beyond; in fact for $x_{In} \ll 0.1$, InGaN QWs behave very nearly as a random alloy. The work at Xerox PARC laboratories (Kneissl 2000) and in the author's group has shown how the gain spectra does indeed significantly narrow as the In concentration is reduced, for lasers operating in the violet ($\sim$395–405 nm). On the other hand, to maintain adequate electronic/optical confinement, one then needs to increase the Al concentration in the cladding layers, adding a different type of materials science challenge.

a.   Electronic Microstructure and Carrier Localization InGaN Quantum Wells

In case of the blue and green QW LEDs, there is general agreement that the pronounced compositional anomalies encountered in InGaN enhance the overall light emission efficiency. This is so because of the strong localization that accompanies the compositional fluctuations on the spatial scale of the electron and hole Bohr radii. However, the presence of the large average strain and local fluctuations in the strain, coupled with the large piezoelectric coefficients in GaN/AlGaN heterostructures (Im et al. 1998), make it difficult to determine the specific contributions of these effects to the nature of near band-edge states existing in InGaN laser devices. The piezofields (estimated up to MV/cm) not only influence the energies of the conduction and valence band edges sizably ($\sim$100 meV) but can have a direct impact on the dilution of the e–h overlap and interaction. The former is of relevance to the one-electron matrix element while the latter influence many-body electronic phenomena, specifically the excitonic effects described at the end of this section. Kollmer et al. (1999) have studied the optical transitions in undoped InGaN QWs by time-resolved photoluminescence to argue that the observed spectral shifts are consistent with the built-in piezoelectric fields. In an InGaN MQW diode laser, the high e–h pair density will, of course, significantly screen the piezoelectric contributions, but so far there is no detailed experimental studies to address this question in the actual devices.

There have been a number of experimental reports, employing microscope and microprobe techniques that have provided real images of the In aggregation in InGaN epitaxial thin films and quantum wells. At one end of the spatial spectrum, Kisielowski et al. (1997) have employed atomic resolution transmission electron microscopy scale to show evidence of clustering on the scale of a lattice constant in connection with studies of strain in the

InGaN/GaN QWs, although recent studies by Monemar *et al.* (2001) appear to dispute this. At the other end of the apparent size distribution, both cathodoluminescence (CL) (Chichibu *et al.* 1997b) and near-field optical microscopy (NSOM) (Crowell *et al.* 1998; Vertikov *et al.* 1998a; Bertram *et al.* 2001) have been used to acquire luminescence-based images on a spatial scale down to ~100 nm. Such a strikingly wide distribution of the cation aggregates (augmented by, perhaps, QW roughness) is in strong contrast with the random alloy behavior, where spatial compositional (and crystal potential energy) fluctuations are confined to the atomic scale.

Among the reports on high-resolution microscopy of InGaN QWs and epilayers, Chichibu *et al.* (1997b) studied spectral variations of luminescence from InGaN/GaN QWs by CL at low temperatures. In that work, the mean In-cluster size was estimated to be less than 60 nm, though the CL images revealed In-rich (deficient) areas up to half a micron in diameter. The NSOM is an attractive alternative to CL, allowing sub-wavelength resolution in photoluminescence (PL) measurements with energetically direct and accurately measurable carrier injection into the QW, together with simultaneous topographic imaging. Here we show an example of the use of *collection*-mode NSOM, where issues of carrier diffusion can be minimized when studying the local spectral variations in the PL emission on a sub-100 nm scale under high e–h pair injection such as encountered in a diode laser (Vertikov *et al.* 1998b).

Typical near-field PL images taken on four undoped 30 Å thick $In_xGa_{1-x}N$ QWs of with 90 Å thick GaN barriers are shown in Fig. 14 for three different wavelengths of emission ($x \approx 0.2$). The injection of e–h pairs ($\sim 10^{19}$ cm^{-3}) was resonant into the QWs by choice of the excitation wavelength. Figure 14(a) shows the NSOM image recorded on the higher-energy side of the far-field PL spectrum at $\lambda = 450$ nm ($\hbar\omega = 2.75$ eV), Fig. 14(b) was recorded at the center of the spectrum at $\lambda = 460$ nm ($\hbar\omega = 2.70$ eV), while Fig. 14(c) is the NSOM image taken at $\lambda = 470$ nm ($\hbar\omega = 2.64$ eV) on the lower-energy side of the PL spectrum (Vertikov *et al.* 1998b). The images reveal darker and brighter regions several hundred nanometers in extent. They strongly suggest that the light emission in the room temperature InGaN MQW lasers occurs from the extended states, once localized states are filled. The excitation and position dependence of the NSOM spectra in Fig. 14(a) and (b) gives another view of this state-filling process. A significant part of the PL spectra in each particular point of the sample is broadened *homogeneously* (on the scale of the instrument resolution), and one can argue that all localized states and a significant amount of extended states, perhaps "mixed" with higher-energy localized states, contribute to the local radiative recombination under high injection levels. Also, sites with larger local band gap (e.g., In-deficient regions) can easily become interconnected through the carrier diffusion.

A different application of the NSOM imaging technique has been recently used to measure the e–h (ambipolar) diffusion in InGaN MQWs under

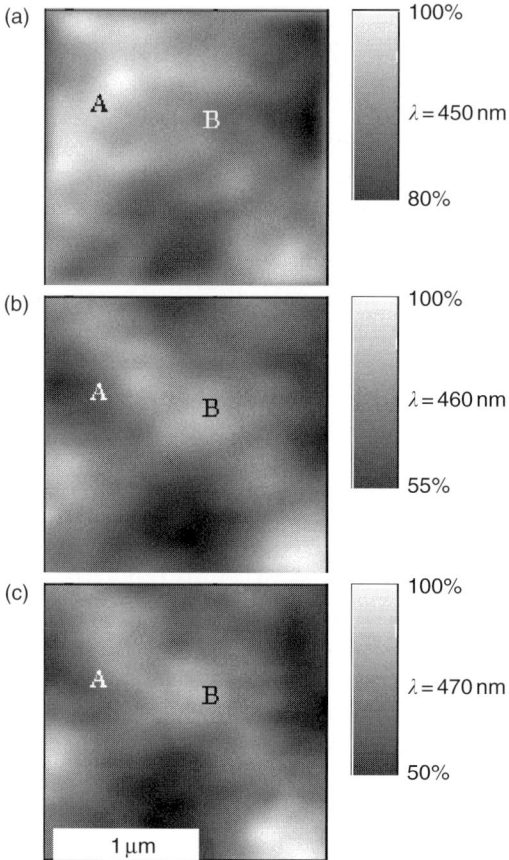

FIG. 14. Collection-mode NSOM PL from an undoped InGaN/GaN MQW for three different wavelengths. Traces (a), (b), and (c) correspond to PL NSOM images at $\lambda = 450$ nm (2.75 eV), $\lambda = 460$ nm (2.70 eV), and at $\lambda = 470$ nm (2.64 eV), respectively. Markers "A" and "B" highlight regions of complimentary optical contrast at different emission wavelengths (Vertikov *et al.* 1999).

optical injection (Vertikov *et al.* 1999). The technique involves the setting up of an excitation interference grating with two blue laser beams, on a spatial scale of about 200 nm, and the direct imaging of the PL intensity variations with the NSOM fiber tip as the high spatial resolution light collector. In these types of experiments, one studies the grating contrast in the PL profiles, diminished due to carrier diffusion. Numerical fitting of such PL profiles with the solutions of the diffusion equation for e–h densities

$$\Delta n(x, t) = \Delta p(x, t) \propto [e^{-t/\tau} + \gamma_0 \, e^{-(t/\tau)(1+(KL_{\mathrm{D}})^2)} \cos(2\pi x/\Lambda)]$$

yields the gives diffusion length L_D. Here, τ is the (carrier density independent) recombination lifetime, γ_0 is the contrast parameter of the excitation
grating, and is the grating period. Vertikov and coworkers studied a number
of InGaN MQW samples grown at different laboratories to find that the
diffusion lengths at room temperature could vary very widely under low e–h
injection, presumably reflecting the individual microstructure of a particular
MQW. On the other hand, in the high injection regime typical of a blue diode
laser ($>10^{19}\,\mathrm{cm}^{-3}$) the values of diffusivity converge to those roughly
expected for a "free" e–h gas.

b.   Excitonic Contributions in Green–Blue ZnSe-based
   QW Diode Lasers

In terms of basic optical properties, the wide bandgap semiconductors are
exceptional in the strength of the excitonic effects that dominate the linear
optical response. Exciton binding energies exceeding $E_x > 40\,\mathrm{meV}$ have been
measured for ZnCdSe QWs and the nitride compounds are expected to
possess considerably larger values. Noting that the exciton oscillator strength
scales roughly as the square of E_x, the robustness of 2D excitons (against
screening and thermal dissociation) implies a significant contribution to the
optical gain even at room temperature. Whereas the compositional/struc-
tural roughness in InGaN QWs, coupled by piezofield Stark effects has made
it difficult to isolate excitonic features in their optical spectra to date, the
heteroepitaxial quality of II–VI structures has allowed a good deal of such
spectroscopy to be performed. In particular, both excitonic and biexcitonic
spectrally distinct features have been clearly identified in ZnSe-based QWs
up into a high-density regime at cryogenic temperatures where stimulated
emission and laser action commences. We illustrate the point next, for the
room temperature ZnCdSe QW diode laser device that is demonstrably
subject to pronounced e–h Coulomb effects.

In a high-density e–h system (2D), the direct e–h Coulomb interaction is
accompanied by exchange and correlation effects which profoundly increase
in importance with the particle density. Physically, the two extreme limits are
easily identified, for example, in a GaAs bulk crystal or QW: the low-density
bound state (exciton) regime and the high-density e–h plasma. For the latter,
the exchange and correlation effects are seen mainly through the bandgap
renormalization effect, which shifts the emission wavelength of such a
semiconductor laser to longer wavelengths (by up to several tens of meV). In
terms of the actual gain spectrum (lineshape or magnitude), however, the
many-body effects induce only rather small corrections to calculations made
in the one electron picture that usually include some phenomenological
damping rates. Sophisticated theoretical approaches, converging to the
semiconductor Bloch equations, have been applied to describe the gain
spectra of various III–V lasers with impressive success (Haug and Koch 1993)

and these methods are presently also being tested with the widegap semi-conductors (and Koch 1995; C. Ell and H. Hang, private communication). This the approach might be qualitatively described as having its starting point in the e–h plasma limit to which the e–h Coulomb interactions are added. One can also adopt a viewpoint that takes the opposite limit as a starting point, that is a gas of non-interacting (or weakly interacting) excitons subject to many-body interactions. This pathway is natural for an experimentalist when considering, for example, the ZnSe QW or "disorder-free" future GaN QW laser in which the temperature is gradually increased from liquid helium to room temperature and where the role of bound e–h pair states at low temperatures is beyond dispute. Approaching the problem from this limit presents, however, significant challenges to the semiconductor Bloch equation approach, for example, due to the difficulty of including the screening by the bound states into the formalism. Yet, evidence in the ZnCdSe QWs strongly suggest that the Coulomb correlations remain very potent, when compared, for example, with a GaAs laser.

Chronologically, localized excitons were first invoked by the authors to define optical gain in the ZnCdSe QW (Ding *et al.* 1992), with supporting and complementary evidence reported by other groups (Alferov *et al.* 1994; Kawakami *et al.* 1994), especially by Cingolani (1996). Figure 15 compares gain/absorption spectra for a ZnCdSe (Ding *et al.* 1994) and GaAs SQW diode laser in the vicinity of the $n = 1$ QW HH exciton transition (Kesler and Harder 1990) obtained in each case by the same experimental technique of correlating edge stimulated emission with top spontaneous emission. Note how, as a function of increasing current injection, the HH and LH exciton features are rapidly bleached in the GaAs QW so that at laser threshold any semblance to excitonic resonances is absent. This is in striking contrast with the case for the II–VI laser, where a partially bleached HH exciton resonance and a nearly intact LH resonance are clearly present at laser threshold, with gain appearing some 60 meV below the HH exciton resonance. For some phenomenological ideas and interpretations, as well as experimental details leading to Fig. 15, the reader is referred to Ding *et al.* (1994). The role of the Coulomb interactions remains practically relevant, for example, in the context of present efforts to extend the III-nitride lasers into the UV, the exciton states might be expected to be increasingly stable against dissociation and screening. The typical range of enhancement of the *peak* gain for room temperature ZnCdSe QW lasers is estimated to be about a factor of 2–3, clearly a device-enhancing feature for designing low threshold devices. In considering the laser threshold condition, we remark that while the excitonic enhancements also imply an accelerated spontaneous (radiative) decay rate, the concentration of oscillator strength in the gain spectrum near the $n = 1$ HH exciton resonance specifically enhances the peak gain coefficient, whereas the spontaneous emission rate is proportional to the integral of the corresponding spontaneous emission spectrum.

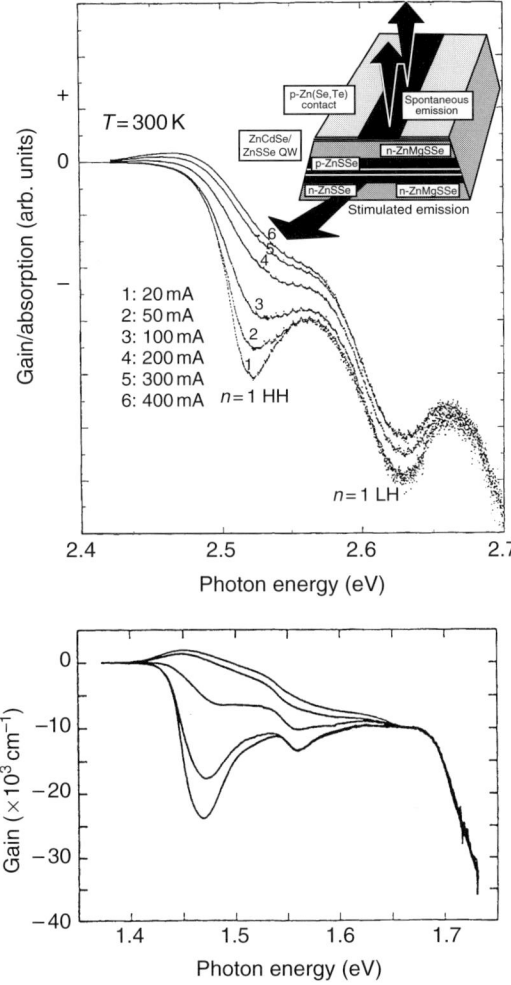

FIG. 15. Comparison of the gain/absorption spectra for SQW ZnCdSe and GaAs SCH diode lasers at room temperature, as a function of injection current (upper and lower traces, respectively), in the vicinity of the $n = 1$ HH and LH transition (Kresler *et al.* 1991; Ding *et al.* 1994).

As another experimental illustration of the impact of the e–h Coulomb correlation at room temperature, we note results where ZnCdSe QW diode lasers have been studied in a high magnetic field, applied perpendicular to the QW layer plane (Song *et al.* 1997). The choice of QW design parameters was made in order to approach the quasi-two dimensional limit, that is, to enhance the exciton binding energy, which for this concentration range of Cd is estimated to be at least 40 meV, based on earlier studies on the ZnCdSe/ZnSSe QW (Pelekanos *et al.* 1992). Figure 16 shows the dependence of the

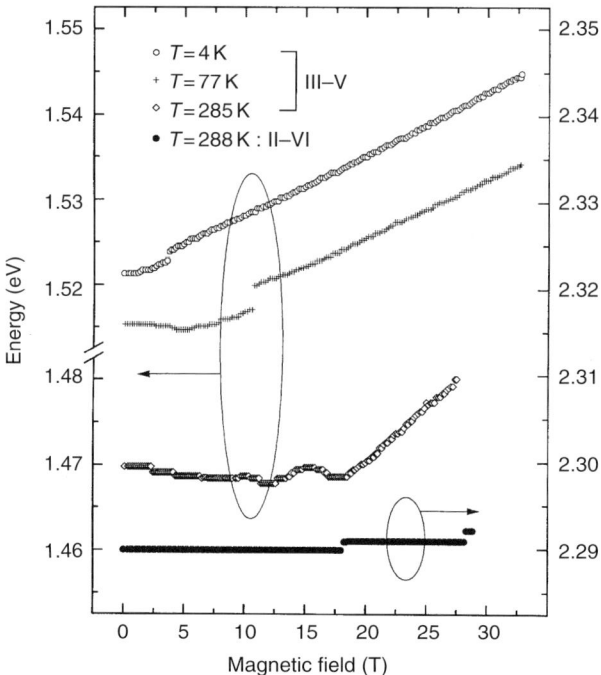

Fig. 16. The emission photon energy for a GaAs QW diode laser (upper traces) and ZnCdSe QW diode laser (lowest trace) as a function of magnetic field, oriented perpendicular to the layer plane (Song *et al.* 1997).

room temperature laser photon energy of the II–VI diode laser as a function of the magnetic field, compared data obtained over the same field range for a GaAs SQW diode laser at three different temperatures. The laser injection condition corresponds to an e–h pair density of $1.9 \times 10^{12}\,\mathrm{cm}^{-2}$ in the ZnCdSe single QW. In a free carrier picture, a Landau shift, linear in the magnetic field and of the order of $\Delta E_{\mathrm{L}} \sim 0.5\,\mathrm{meV/T}$, would be expected (either due to the shift of the lowest conduction band Landau level or the jumps of the Fermi level between such occupied levels). Clearly, the experimental result is about one order of magnitude smaller and incompatible with free e–h behavior (including longitudinal mode mode hops as the gain spectrum shifts). On the other hand, the agreement is much better when compared with the diamagnetic shifts measured for the $n = 1$ HH exciton transition in ZnCdSe QWs, typically, $\Delta E_{\mathrm{dia}} \sim 2\,\mu\mathrm{eV/T}^2$, with the reduced exciton mass of $\mu^* \sim 0.1 m_{\mathrm{o}}$. Given the exciton binding energy of about 40 meV, this implies that at $B = 30\,\mathrm{T}$, we are still in the regime where $E_{\mathrm{x}} > \hbar \omega_{\mathrm{c}}$, the electron cyclotron energy. Note the strong contrast with the GaAs QW diode laser which exhibits temperature dependent "Landau-level jumps" (oscillations) of the Fermi level at low B-fields (in terms of the Fermi level), but where the the spectral shifts in the emission evolve into a linear cyclotron-like shift (with some influence of the many-body interactions on the effective masses evident, however).

III.   Approaches to Advanced Heterostructure Blue/Violet Light Emitters

1.   VERTICAL CAVITY EMITTERS

As the examples in preceding chapters illustrate, the nitride optoelectronic materials frequently stand apart from other III–V semiconductors in terms of their device-related properties. This fact extends to the device science and engineering of the vertical cavity blue and near-UV emitters. From a device engineering point of view, the primary challenge presented by the nitride vertical cavity emitters is two-fold. First, the fabrication of a high-Q optical resonator is non-trivial. Secondly, dictated mainly by the presently low conductivity of the p-GaN and its alloys, the electrical injection schemes must incorporate lateral current spreading schemes. In the following, we consider these issues separately, while showing examples of solutions from the authors' laboratories. We note that important parallel work is being pursued, for example, at the University of California, Santa Barbara, and wish to acknowledge important ideas put forth by Prof. K. Iga and his collaborators (Honda *et al.* 1995). Elsewhere, for example, Mackowiak *et al.* (2001) have put forth innovative theoretical ideas concerning current injection into a nitride-based VCSEL.

2.   OPTICAL RESONATOR DESIGN AND FABRICATION: DEMONSTRATION OF OPTICALLY PUMPED VCSEL OPERATION IN 380–410 nm RANGE

As the choice for high reflectivity mirror materials in the fabrication of nitride microcavity resonators, *in situ* as-grown AlGaN/GaN distributed Bragg reflectors (DBR) are feasible in principle, but, due to the small index of refraction contrast, require a large number of layer pairs. Nonetheless, several groups have demonstrated the epitaxial growth of such DBRs (Someya and Arakawa 1998; Krestnikov *et al.* 1999; Larger *et al.* 1999; Ng *et al.* 2000). Furthermore, there are reports that argue for the observation of vertical cavity, or "surface" lasing under optical intense pumping from GaN or InGaN MQW or thin-film heterostructures that are encased by *in situ* grown AlGaN/GaN DBR reflectors (Redwing *et al.* 1996; Krestnikov *et al.* 1999). Someya *et al.* (1999) have shown stimulated emission employing a "hybrid" structure composed of one *in situ* grown AlGaN/GaN DBR and one dielectric DBR.

a. All-Dielectric DBR Resonator

A useful testbed for vertical cavity nitride light emitters has been the all-dielectric mirror cavity. One fabrication technique has been shown to produce microcavity resonators with quality factor Q approaching 1000 (Song et al. 1999). In this approach, the sapphire substrate was separated by pulsed excimer laser ablation (Kelly et al. 1997; Wong et al. 1998) and a specific process sequence was used to sandwich the InGaN/GaN/AlGaN heterostructure between two DBR stacks of SiO_2/HfO_2. Good morphology is crucial to the realization of true VCSEL operation with a mean roughness of 2–3 nm over areas on the order of several hundred square micrometers. In the absence of good morphology, lasing can be obtained but is readily dominated by in-plane stimulated emission. Optically pumped quasi-CW VCSEL operation was achieved by photo-excitation at 355 nm, outside the reflectance band of the DBRs and slightly below the bandgap of the AlGaN cladding layer, ensuring a predominant creation of e–h pairs directly into the InGaN QWs. The upper trace of Fig. 17 shows the spontaneous emission

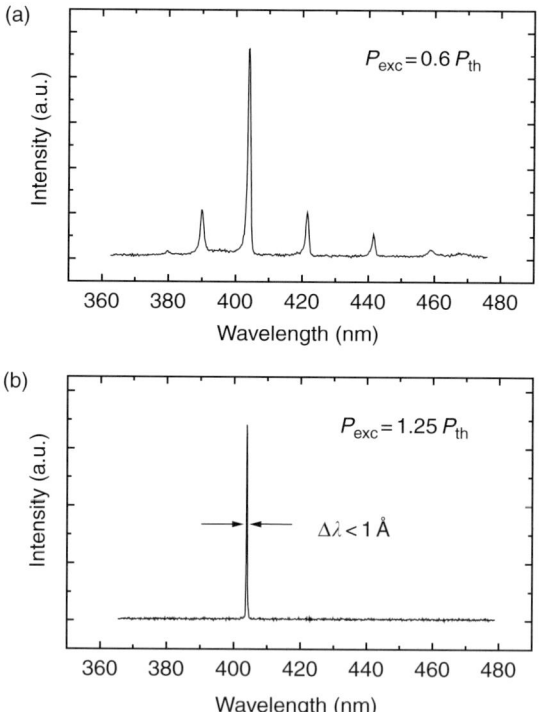

FIG. 17.   Upper trace: spontaneous emission spectra of the optically pumped InGaN MQW VCSEL below threshold. Lower trace: stimulated emission spectra above threshold under quasi-CW pumping conditions at $T = 258$ K (Song et al. 2000a).

spectrum at temperature $T = 258$ K at an average incident power of approximately 17 mW (Song *et al.* 2000b). Several well-defined cavity modes are seen, with a typical modal linewidth of approximately 0.6 nm, limited by scattering from residual morphological roughness. The bottom trace of Fig. 17 shows the onset of stimulated emission (at threshold input power of $P_{th} = 32$ mW). A well-defined far-field pattern was acquired with a FWHM radiation angle of approximately 5°, for the *linearly polarized* nearly Gaussian beam emerging from an 20-μm diameter aperture.

From the conditions of excitation, one could estimate that the threshold conditions of this type of optically pumped VCSEL correspond to approximately those in the CW edge-emitting diode lasers in terms of equivalent electrical injection (assuming an approximately 1 ns carrier lifetime).

b. Stress Engineering of AlGaN/GaN DBRs

The large number of AlGaN/GaN layer pairs (>50) to achieve the high reflectivities present sizable difficulties in the control of cracks and the DBR morphology. It has been discovered recently (Han *et al.* 2001) that the use of AlGaN interlayers is effective in controlling mismatch-induced stress and suppressing the formation of cracks otherwise occurred during growth of AlGaN directly upon GaN epilayers. This idea has been applied in conjunction with *in situ* monitoring to control of stress evolution during growth of AlGaN/GaN DBR mirrors by metal–organic vapor phase epitaxy. The employment of an AlN interlayer at the beginning of a thick (~5 μm) DBR growth leads to a substantial modification of the initial stress evolution. Tensile growth stress can be brought under control and nearly eliminated through multiple insertions of AlN interlayers. Using this technique, crack-free growth of 60 pairs of $Al_{0.20}Ga_{0.20}N$/GaN DBR mirrors has been achieved over the entire 2 in. wafer with a maximum reflectivity of at least 99%. Even so, the associated spectral bandwidths are relatively small (10–20 nm).

Such DBRs, with high-quality morphology and a peak reflectivity $R \sim 0.991$, have been applied to demonstrate quasi-CW operation, at room temperature, of an optically pumped $In_xGa_{1-x}N$ ($x \sim 0.03$) MQW VCSEL at near $\lambda = 383$ nm (H. Zhou *et al.* 2000). The vertical cavity scheme combined a high reflectivity *in situ* grown multilayer GaN/$Al_{0.25}Ga_{0.75}N$ and post-growth dielectric SiO_2/HfO_2 DBR. A photograph of the side profile of the low divergence beam of circular cross section is shown in the inset of Fig. 18, which also displays the input/output power characteristics of a device with lasing threshold at pump power of 30 mW (average VCSEL output powers up to 3 mW were measured). However, finite thickness variation across the wafer led to spectral shifts of the cavity modes (relative to InGaN MQW gain spectrum) so that significant increases in threshold were encountered for devices fabricated from near the edge of the wafer. When accounting for the

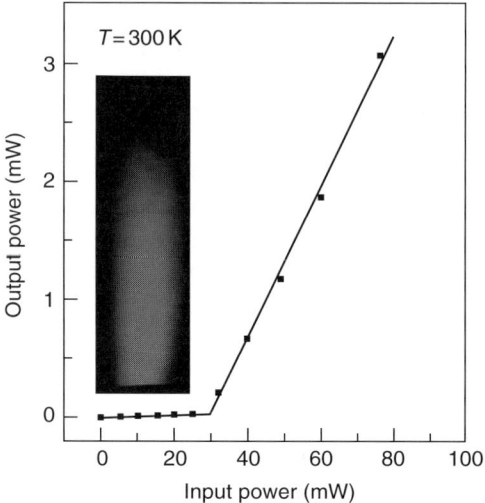

FIG. 18.   Average input vs output power of a violet optically pumped VCSEL device. The inset shows the beam far-field (side) profile captured on a screen (Zhou *et al.* 2000).

optical excitation volume, the fractional absorption of the pump, and using an e–h recombination time of approximately 0.5 ns, we estimate that the device is about 25% efficient and that the threshold corresponds roughly to a carrier density of approximately 10^{19} cm^{-3}.

3.   ELECTRICAL INJECTION: RESONANT CAVITY LEDS

The need for lateral injection of holes from the p-side of the junction to the optically active volume poses a major challenge for a III-nitride vertical cavity diode emitters. The low conductivity of typical MOCVD-grown p-GaN (on the order of $1\,\Omega^{-1}\,$cm^{-1}) dictates that a high conductivity *intra-cavity* layer be inserted between the top p-GaN layer and the adjacent DBR. This layer must not contribute to optical losses so that both its intrinsic absorption as well as morphology (in terms of optical scattering) must be considered carefully. Two approaches have been taken in this direction. First, appropriately processed indium-tin oxide (ITO) is a useful thin film material both from the standpoint of low optical losses ($\sim$1% single pass absorption in 1000 Å thick film for wavelengths >400 nm) and in terms of forming an electrical contact to p-GaN (resistivity of magnetron-sputtered and thermally annealed ITO approximately $4 \times 10^{-4}\,\Omega\,$cm). Useful demonstrations of blue and vertical cavity LEDs have been made by employing ITO as an intracavity element for both approaches to optical resonator fabrication outlined above (all-dielectric DBRs and the hybrid configuration with one *in situ* grown AlGaN/GaN DBR, respectively) (Song *et al.* 1999). Second, recently developed nitride-based (Esaki) TJ, outlined in Section I (Jeon *et al.* 2001; Takeuchi *et al.* 2001) provides a heterostructure building block which generally extends the design versatility for nitride optoelectronic devices.

The schematic diagram of a vertical cavity violet LED is shown in Fig. 19 which combines the hybrid resonators structure with a hole injection scheme that is based on the use of a p^{++}/n^{++} InGaN/GaN tunnel junction (Diagne *et al.* 2001). AlN strain-relief layers were used in the deposition of a 60-layer-pair quarter-wave GaN/$Al_{0.25}Ga_{0.75}N$ bottom DBR stack. Those growths that yielded a root-mean-square average roughness no worse than 4 nm over an area of 1×1 mm^2 were deemed suitable for continuation of the epitaxy. The active p–n junction region was grown directly atop the GaN/(Al,GaN) DBR, composed typically of seven $In_{0.08}Ga_{0.92}N$ QWs ($L_w = 40$ Å) with GaN barriers ($L_B = 60$ Å), and surrounded by approximately 1000 Å thick $Al_{0.07}Ga_{0.93}N$ current blocking/carrier confinement layers. The tunnel junction was included in the superstructure of the nitride segment and capped by a n-GaN layer. Note that in this case the positive bias to the device was applied through a contact to this n-layer. Lateral current spreading on the scale of ~ 100 μm was obtained, given the nearly 100 times higher n-type conductivity of GaN. The TJ itself was grown atop the 1000 Å thick p-GaN layer as a p^{++}/n^{++} InGaN/GaN bilayer with the thicknesses of the layers approximately 150 and 300 Å, respectively. The doping levels were approximately 1×10^{20} cm^{-3} Mg and 6×10^{19} cm^{-3} Si for the junction.

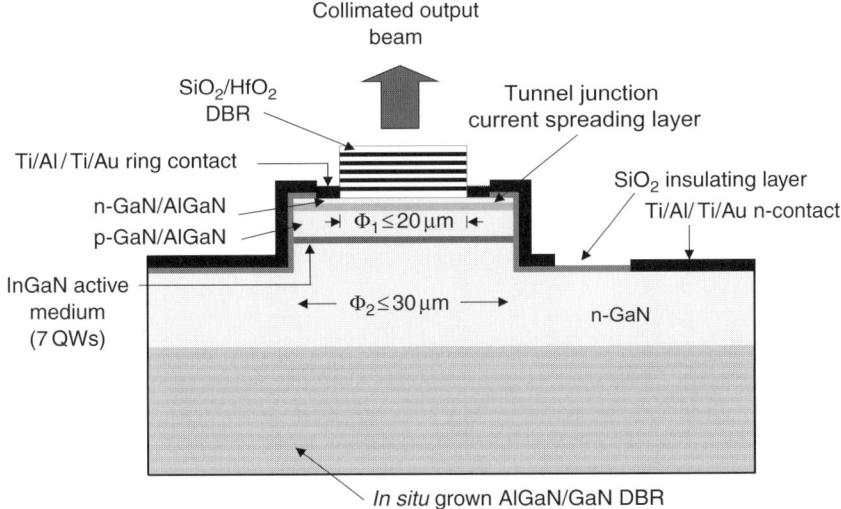

FIG. 19.   Schematic of an RCLED device with one as-grown GaN/AlGaN and one dielectric DBR mirrors, incorporating a nitride tunnel junction as the top current spreading layer.

The vertical cavity was completed by capping the structure with a multilayer $\lambda/4$ stack of SiO_2/HfO_2 ($R > 0.995$), deposited by reactive ion beam sputtering. The top dielectric DBR was patterned so that the device had an effective optical aperture varying from 10 to 30 μm.

The current density vs voltage of a typical device is shown in Fig. 20 up to a high continuous injection level (~ 1 kA/cm^2), which shows evidence of series resistance, assigned in part to the presence of the TJ. Nonetheless, lateral current spreading was clearly accomplished so that the far-field light emission from the devices was uniform in its average intensity across the emitting aperture. Figure 21 shows the output spectrum of a typical device at operating current density of approximately 0.2 kA/cm^2. The emission was observed in the direction normal to the planar device, within an angular view of

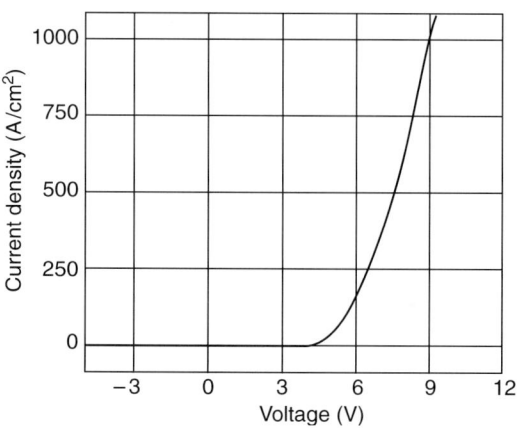

FIG. 20.   Current density vs voltage of a hybrid RCLED device with a 20 μm diameter mesa defining the vertical current path.

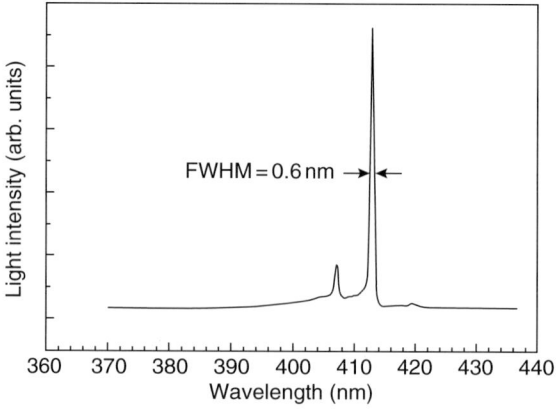

FIG. 21.   Emission spectrum of the RCLED hybrid device (Diagne *et al.* 2001).

approximately 10°. While the optical resonator is rather thick ($>10\lambda$), only two vertical cavity modes are seen, demonstrating the restrictive spectral bandwidth of the AlGaN DBR. The dominant mode at $\lambda = 413$ nm, which coincides with the high reflectivity region of the DBR and the peak of the QW PL emission, has a spectral linewidth of approximately 0.6 nm.

Recently, strong evidence for stimulated emission has been obtained from these types of vertical cavity emitters, at high current injection levels (>10 kA/cm^2) (He *et al.* 2002). The top panel of Fig. 22 shows the *L–I* curve of a seven-QW device with ITO hole current spreading layer at two different temperatures. At 15 kA/cm^2, threshold-like *L–I* curve was observed,

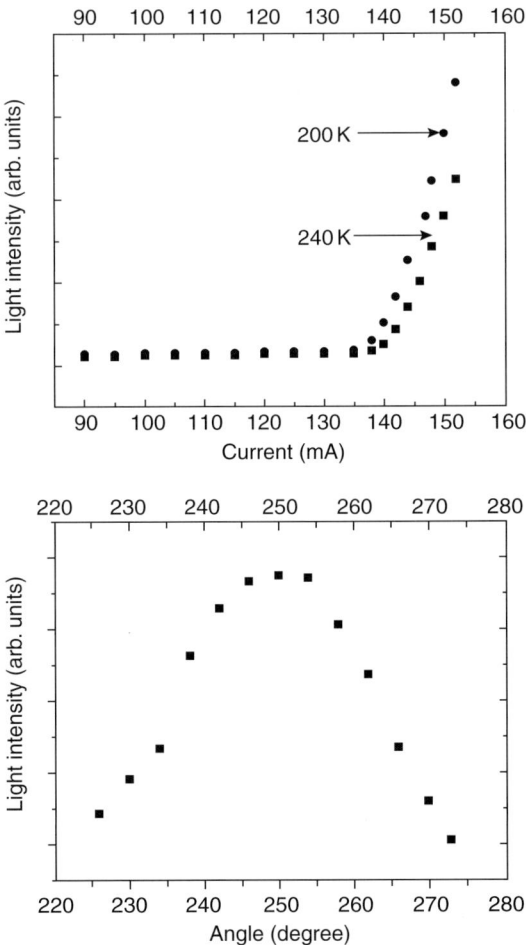

FIG. 22.   Top panel: light output vs current for vertical cavity device in high injection, showing a stimulated emission threshold. Bottom panel: corresponding far-field pattern for the dominant cavity mode.

suggesting the possible existence of stimulated emission. Pulsed output powers in excess of 10 mW have been measured. The far-field pattern of the dominant cavity mode is shown in the bottom panel of Fig. 22, showing a half angle of approximately 12°. This angle refers to the measured external angle of divergence after passing through the sapphire substrate (as the device output was extracted through the AlGaN DBR). Spectral narrowing to less than 0.4 nm is observed. Although stimulated emission has been achieved in these structures, several puzzling features remain unresolved at this writing. For example, the directionality of the emission is neither as narrow nor as well polarized as with the optically pumped (albeit much simpler) structures. Furthermore there is a finite, nearly isotropic background emission present as well. The presence of intracavity losses (from ITO or the TJ) is a factor in raising the onset of threshold current, given the rather small total thickness of the InGaN MQW region (~300 A).

The examples shown above suggest that the recent implementation of vertical cavity LEDs has provided the important building blocks for the further development of viable diode VCSELs in the violet. The current device challenge is two-fold: high-quality epitaxy and relatively complex device processing. For example, layer thickness and composition (Al, In) control over a 2–3 in. wafer area is still quite difficult, yet required so that spectral overlap, for example, between the high reflectivity band of the AlGaN DBR and the gain maximum of the InGaN MQW gain remain in spectral synchrony.

We also remark that the application of the lateral epitaxial overgrowth techniques discussed in Sections I and II may aid to create flexibility for designing and implementing blue/violet RCLEDs and VCSELs at least in two different ways. First, the patterned growth can be adapted for creating a buried bottom dielectric DBR mirror. An illustration is shown in Fig. 23

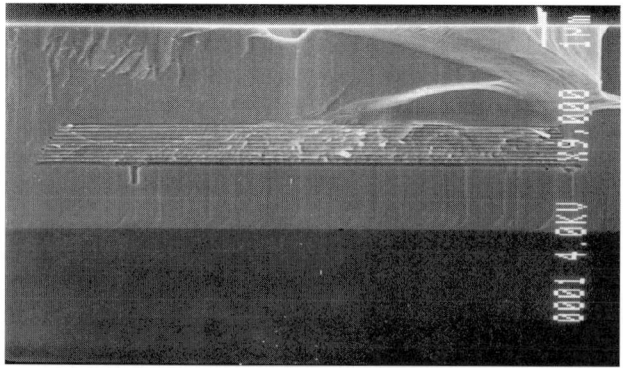

Fig. 23. Use of lateral epitaxial GaN overgrowth to "bury" a dielectric DBR. The imperfections are due to breaking of the sample for access to an edge view in this cross-sectional SEM image.

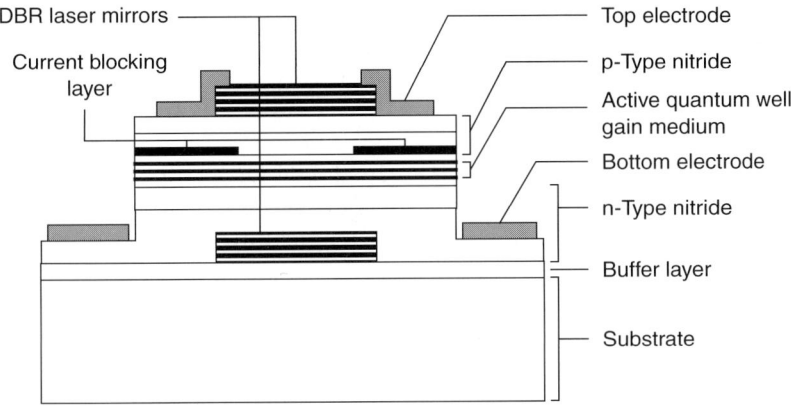

FIG. 24. Schematic drawing illustrating a possible blue VCSEL device structure which features a buried dielectric DBR and current confining aperture.

where a patterned HfO_2/SiO_2 multilayer dielectric stack was deposited on a GaN buffer layer prior to subsequent regrowth of GaN. The particular dielectric stack was terminated with an SiO_2 layer so that the ELOG process would occur normally. The second application of the lateral epitaxy pertains to building in a current aperturing scheme in order to alleviate the problem that arises from the competition between vertical and lateral transport in the nitride devices (especially on the p-side). Figure 24 shows the schematic of a possible arrangement where an SiO_2 defined current aperture is implemented in conjunction with lateral epitaxial growth.

a. Monolithic Dual-Wavelength LED

The TJ employed above for lateral conduction purposes is also an enabler for monolithic, electrically segmented multiwavelength emitters. This feature is attractive for lasers and LEDs at short visible wavelengths and the UV, for applications in displays, solid state lighting, and spectroscopy of biochemical/biological molecules. A three-terminal, dual wavelength blue/green LED consisting of two electrically independent InGaN QWs of different indium composition within a single vertical heterostructure has been demonstrated in the InGaN QW system (Ozden *et al.* 2001a), and we describe it briefly here as an example of the kind of optoelectronic device versatility in the III-nitride family, which is only beginning at this writing. The device incorporated the p^{++}/n^{++} InGaN/GaN TJ so as to operate a time-multiplexed two-color blue/green LED source (here at 470 and 535 nm). The nitride heterostructure used in first experiments is shown schematically in Fig. 25. The TJ segment was inserted between the two active $In_xGa_{1-x}N$ QW emitter segments

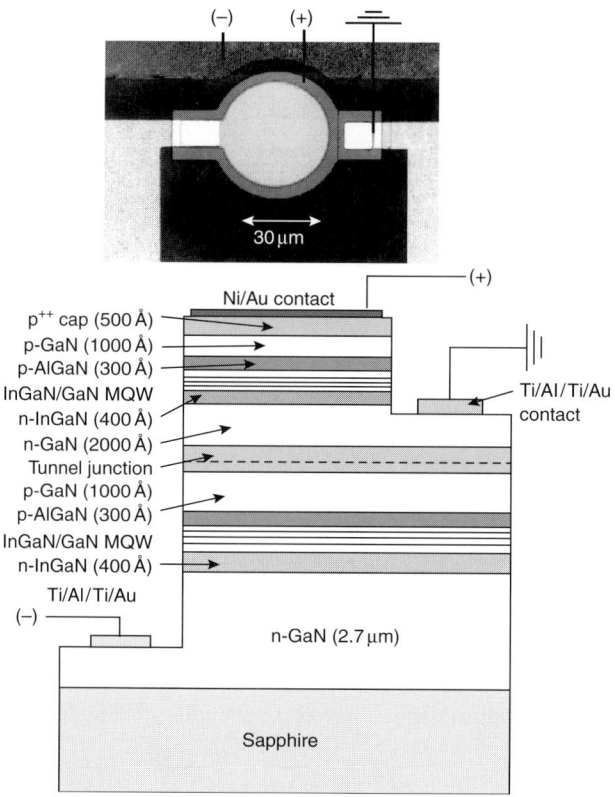

FIG. 25. Schematic view of the two-wavelength blue/green LED, indicating the active regions, the TJ, and the bias arrangement. A plan view photograph of a device is shown at the top (Ozden *et al.* 2001a).

($L_w = 30$ Å). The TJ accomplished two tasks: a means for: (a) electrically sectioning the nitride heterostructure into two independent LEDs; and (b) lateral current spreading for the bottom device.

The fabrication of the three-terminal device required two etch steps, for the bottom LEDs n-GaN contact and the common ground immediately atop the TJ segment. The diameters of the top and the bottom LEDs were 60 and 80 µm, respectively. In the LED injection regime (~ 100 A/cm²), the TJ in the bottom device typically added about 1 V to the forward "turn-on" characteristics. Upper trace in Fig. 26 shows its spectral characteristics (dashed curve corresponds to each LED being switched on in a time sequential manner, solid line is for the LED operated as a simple two-terminal device with a constant voltage applied across the top p-GaN and lowest n-GaN). The electrical independence of the blue and green segments allowed us to "program" the 470 and 535 nm LEDs for any time sequence, up to speeds

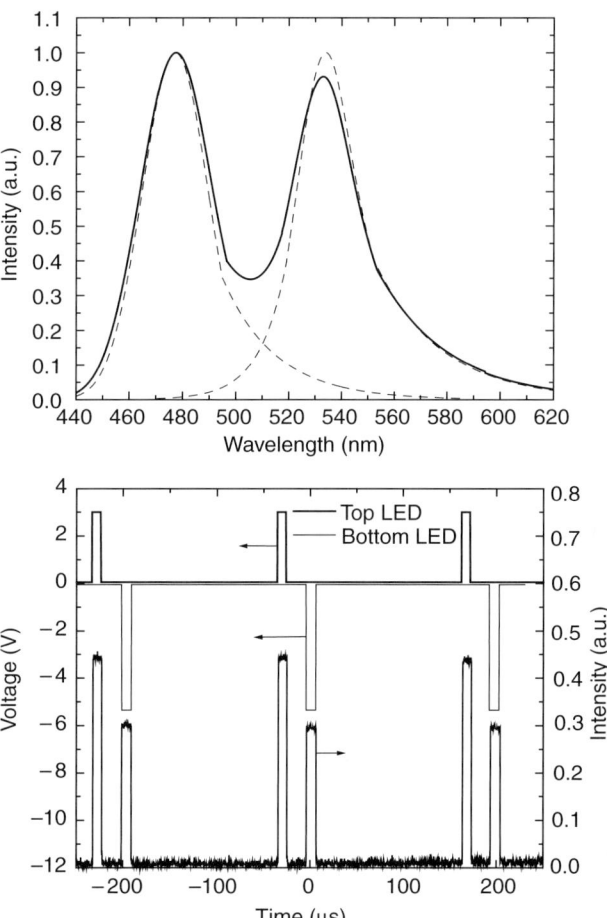

FIG. 26. Top: comparison of the emission spectrum from the blue and green LEDs when the devices are activated independently in a three-terminal case (weak solid trace), and their simultaneous activation as a two-terminal LED (bold trace). Bottom: time-sequenced activation of the two LEDs (amplitudes and duty cycle chosen arbitrarily).

of 100 MHz. The bottom trace in Fig. 26 shows such time-sequenced two-wavelength emission in the three-terminal configuration. However, as the current–voltage characteristics as well as their effective current apertures of the two LEDs were different, the applied voltage for each LED needed to be adjusted to obtain a comparable optical output from the devices for now. The device has been applied to time-gated "spectrometer-free" fluorescence fingerprinting of molecular dye labels such as those used in DNA sequencing. This application idea is based on the fact that the fluorescent dyes are known in terms of their absorption and emission spectra, and the emission wavelengths of the two LEDs are chosen to match their absorption spectra, respectively.

Finally, we note that for the types of multiwavelength emitters envisioned above, as well as single color devices, programmable 2D arrays are needed. For example, in prospective schemes for fluorescence-based chipscale detection of chemical and biochemical species, such arrays add significantly to device throughput. Likewise, specialized display application can benefit from high density arrays of monolithic LEDs and vertical lasers. Several examples of single wavelength LED arrays have appeared recently, including a matrix addressable 1024 element array scheme (Ozden *et al.* 2001b).

IV.   Extension of the Wavelength into the UV

1.   CURRENT STATUS OF NEAR-UV DIODE LASERS

The extension of the III-nitride lasers into the ultraviolet is attractive from both applications and device science point of view. Very high density optical storage (including holographic storage) and compact biochemical-labs-on-chip are two examples of potential technology insertion. Basic device science is on interest as the physics of optical gain and electrical character-istics of very wide bandgap semiconductors ($E_g \sim 4\,eV$) are largely unex-plored subjects to date. In this section, we briefly survey recent developments to extend the diode lasers from the violet into the near-UV spectral range. Following a specific example of an approach to edge-emitting lasers, we will focus on current materials and device efforts which are concentrated on UV LEDs. (A note for the reader: this is a contemporary subject which possesses a resemblance to a moving target.)

As noted earlier in this chapter, the threshold current density for the ternary InGaN QW diode laser increases rapidly for wavelengths *longer* that approximately 420 nm. This is due to a combination of factors, but domi-nated by the compositional and QW thickness disorder in the InGaN/GaN gain medium. By contrast, at wavelengths *shorter* than about 400 nm, the observed increase in the threshold current has been interpreted to be a combination of weakened electronic confinement and the increasing role by nonradiative point defects, as the In concentration is reduced towards zero. It is also well documented that the efficiency of violet LEDs drops significantly as the In content is reduced, interpreted as being due to loss of e–h locali-zation in the InGaN QWs and corollary susceptibility to nonradiative defects in the background crystal material. While a diode laser operates in a higher injection regime, where the carrier localization appears not to benefit the formation of optical gain directly (Section II), sizable non-radiative recom-bination currents are, of course, also a most unwelcome factor.

The basic design rules for a separate confinement, quantum well laser (SCH/QW) in the III-nitride system flow from the heterostructure diagram of Fig. 1 for the AlGaInN system. In making the transition from the violet into the UV, choices for the active medium when reaching the approximately 360 nm wavelength range are basically down to a binary GaN, ternary AlGaN, or a quaternary AlGaInN QW. Each presents its material challenge, both in terms of point defects in the active medium and extended defects emanating from lattice mismatch with the substrate and the cladding layer of the SCH structure, etc. For example, while photopumped laser action from GaN thin films and GaN/AlGaN QWs has been reported under intense excitation, obtaining a reasonable PL or EL efficiency at room temperature has remained problematic. The quaternary involves the issue of simultaneous incorporation of both In and Al, but offers a the added virtue of a "tunable" materials, both in terms of the optical emission wavelength and lattice constant. Of course, issues associated with the control of electrical properties with the p–n junction encasing the active medium, are of paramount importance, and increase rapidly in difficulty with the increasing bandgap of associated materials.

As an illustration of the present status of diode laser research in the near-UV, we cite the work of Nagahama *et al.* with both GaN QW and AlInGaN QWs, which has led to demonstrations of CW edge emitting lasers at room temperature near 370 nm (Nagahama *et al.* 2001b,c). In each instance, the substrate was formed by a thick ($\sim$20 μm) GaN lateral-epitaxially overgrown template on a sapphire substrate, aimed at dislocation reduction (see Sections I and II). Likewise, in both cases a modulation-doped short period superlattice was incorporated in the "cladding" layer region for enhanced doping ($\sim$100 periods of 25/25 Å n-$Al_{0.05}Ga_{0.95}N/Al_{0.10}Ga_{0.90}N$ with Si and correspondingly for the p-doped SL with Mg-dopant; see Fig. 2). Finally, in contrast to the common violet diode laser active medium, each near-UV diode laser employed only a single QW as the gain medium: a 100 Å thick GaN QW with $Al_{0.15}Ga_{0.85}N$ barriers and a 75 Å thick $Al_xIn_yGa_{1-x-y}N$ QW with $Al_{0.15}In_{0.01}Ga_{0.84}N$ barriers for the binary and quaternary cases, respectively. As an example of a quaternary QW device, Fig. 27 shows the voltage–current and light output–current performance of a device ($Al_{0.03}In_{0.03}Ga_{0.94}N$ QW) under CW operation at room temperature, with maximum output power reaching several milliwatts. The corresponding emission spectrum above threshold is shown in Fig. 28, centered near 371.7 nm, including a red-shift due to heating (under pulsed operation, the lasing is centered around 366.4 nm). Lastly, Fig. 29 demonstrates the wavelength "tunability" of the lasers for different Al and In compositions in the quaternary well, and, equally important, the corresponding variations of the threshold current density J_{th}. Note particularly the increase in J_{th} with increasing Al concentrations, up to about 12 kA/cm^2 for $X_{Al} = 0.08$. Nagahama *et al.* suggest that this increase is mainly the result of the crystal

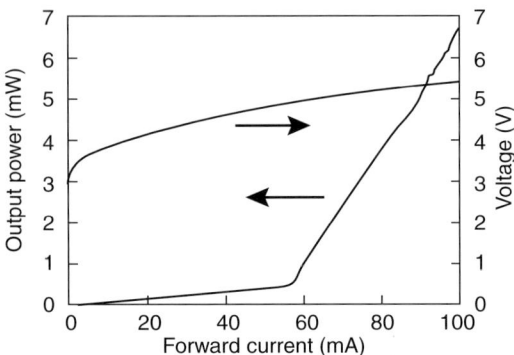

FIG. 27. *L–I* and *V–I* characteristics of $Al_{0.03}In_{0.03}Ga_{0.94}N$ SQW laser diode under 25°C CW operation (Nagahama *et al.* 2001b).

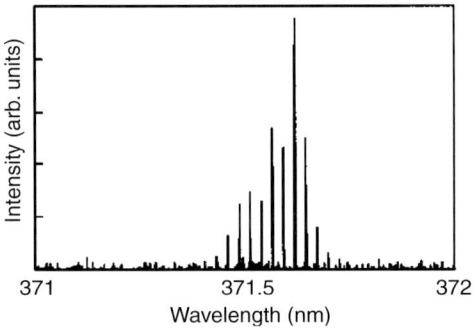

FIG. 28. Emission spectrum of CW $Al_{0.03}In_{0.03}Ga_{0.94}N$ SQW laser diode operation at output power of 2 mW (Nagahama *et al.* 2001b).

quality of the quaternary $Al_xIn_yGa_{1-x-y}N$, both in terms of general morphology and defects. Overall, these results broaden the initial basis for development efforts of III-nitride materials for light emitters into the deeper UV, a subject we approach next.

2. CHALLENGES FOR LIGHT EMITTERS FOR THE 350–280 NM UV RANGE

As already noted, the UV range of approximately 350 nm down to 280 nm represents a pioneering territory for semiconductor-based light emitters, with significant potential for a new optoelectronics technology base. For example, a wide range of photochemistry and photobiology applications require sources in this region of the spectrum. Since the extension of III-nitride diode

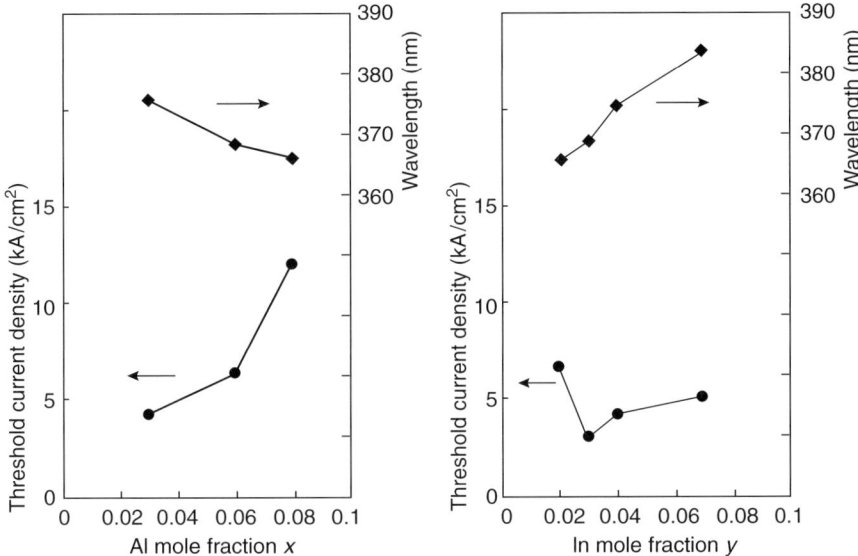

FIG. 29.   Threshold current density and emission wavelength of $Al_xIn_{0.04}Ga_{(0.96-x)}N$ SQW diode lasers as a function of the Al mole fraction x (left panel); $Al_{0.03}In_yGa_{(0.97-y)}N$ SQW diode lasers as a function of the In mole fraction y (Nagahama *et al.* 2001b).

lasers to this UV range is a truly major endeavor, requiring significant research and development efforts, it is useful to consider the situation initially from the viewpoint of UV light emitting diodes, which themselves would find additional applications such as for solid state (white) lighting. Below we summarize the present understanding of the problems regarding the light emitting medium and key issues of electronic transport for UV LEDs, which are subject to significant contemporary interest.

Within the past 2 years, there has been a nearly continuous path of progress, beginning with very modest initial LED efficiencies, of both the wavelength and device performance approaching the 300 nm spectral regime. Han *et al.* (2000) employed the AlGaN ternary in $Al_{0.03}Ga_{0.97}N/$ $Al_{0.17}Ga_{0.83}N$ MQW LED structures that emit at 337 nm (well width $\sim$50 Å). Kinoshita *et al.* (2000) employed $Al_{0.25}Ga_{0.75}N/Al_{0.03}Ga_{0.97}N$ MQW structures with narrower well widths (10–30 Å) to shift the emission to 333 nm. Very recently, Wong *et al.* (2001) reported an $Al_{0.49}Ga_{0.51}N/Al_{0.23}Ga_{0.77}N$ double heterostructure LED emitting at 321 nm, while Stampfl *et al.* demonstrated quaternary AlGaInN MQW devices that emitted at wavelengths as short as 305 nm. Figure 30 shows the LED emission spectra obtained from the quaternary devices, together with PL from these heterostructures, that featured GaN/AlGaN modulation-doped superlattices within the p–n junction, the entire structure grown by a pulsed atomic layer epitaxial approach.

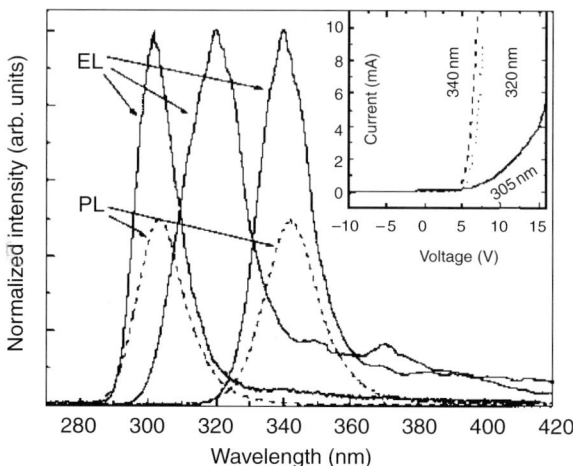

FIG. 30. Room temperature dc EL spectra of LEDs with three different MQW configurations. Dotted line shows RT PL spectra at 305 and 340 nm. The *I–V* curves for LEDs are depicted in the inset (Khan *et al.* 2001).

These proof-of-concept ternary AlGaN and quaternary AlGaInN emitters, aimed chiefly at demonstrating the feasibility of the wavelength extension into the UV, have much reduced optical output compared to the high efficiency blue and green InGaN QW emitters. We have already referred to the observation of the monotonic and rapid decrease of optical efficiency as indium fraction decreases in InGaN epilayers violet and near-UV LEDs. For example, the choice of ternary AlGaN as an active medium has met with some questions as to its device viability in deeper UV, even if understanding of the reduced light emission mechanism from AlGaN, especially at higher ($x_{Al} > 0.30$) Al compositions, is still at a primitive stage from a microscopic point of view. In fact, early evidence has suggested the presence of a high density of point defects in AlGaN ($x_{Al} > 30\%$), with luminescence properties reminiscent of other highly defected semiconductor systems and not unlike low-temperature grown GaAs and amorphous Si, where the excited or injected carriers nonradiatively recombine at the deep levels. The existence of these defects can be purely intrinsic due to local stoichiometric deviations, or extrinsic via the unintentional incorporation by Al-reactive chemical impurities, notably oxygen. Nonetheless, encouraging recent results in the 350 nm spectrum range have been achieved by Nishida and Kobayashi (2001) in $Al_{0.06}Ga_{0.94}N/Al_{0.12}Ga_{0.88}N$ SQW structures (20 Å well thickness), grown on GaN substrates. These authors reported the result of 10 mW optical output with an estimated internal quantum efficiency of about 80% and an external emission efficiency of about 1% in devices without any light extraction enhancement schemes.

Returning now to the AlGaInN quaternary, the primary challenge associated with the growth of AlGaInN for extension into the deeper UV is the interplay between two surface science processes that impose opposite constraints on the thermodynamic growth parameters: the need to maintain sufficient diffusion length of Al and Ga species during epitaxial growth, and the need to suppress the desorption and to enhance the physical incorporation of indium. Han *et al.* (2000) explored the bandgap and lattice constant versus both indium and aluminum compositions of MOCVD grown AlGaInN (Al < 30%, In < 10%) by a combination of PL, high-resolution X-ray diffraction, and Rutherford backscattering spectroscopy. For example, UV luminescence efficiency at 350 nm higher than that from binary GaN (at 364 nm) was reported from a quaternary epilayers (Al = 0.14, In = 0.04). To improve crystallinity and compositional controllability, Zhang *et al.* (2001) devised a novel technique of flow modulation during deposition of AlGaInN (the "pulsed atomic layer epitaxy"). However, the utility of quaternary AlGaInN as an *efficient* UV active medium below 350 nm awaits further proof in a diode setting. Preliminary finding by Hirayama *et al.* (2001) appear to affirm its benefit in improving LED efficiency around 345 nm. In addition to the empirical material optimization of AlGaInN or AlGaN for UV emission, current research is focusing on attaining an understanding on a microscopic scale of the nature of nonradiative recombination mechanisms in AlGaInN and assess the roles of extended and point defects.

Control of the electrical properties within the widegap III-nitrides is, likewise, under intense scrutiny. For example, conductivity control in the ternary AlGaN is quite complex as many material issues become tangled to make systematic investigation and interpretation less than straightforward. Limited information on doping of AlGaN (both n- and p-types) up to 30% is available (Katsuragawa *et al.* 1998). Stampfl *et al.* (1999) used first principle calculations to predict the qualitative behavior of n- and p-type dopants in AlGaN. Mechanisms such as local lattice relaxation (shallow-to-deep transition), compensation, and deepening of ionization energies are expected to occur with several common dopants. Thus the task of maintaining the Fermi level in AlGaN in proximity to either the conduction or valence band will be a nontrivial exercise as Al-concentration exceeds about 30%. There have been reports of AlGaN devices (including emitters (Wong *et al.* 2001) and detectors (Brown *et al.* 2001; Monroy *et al.* 2001) that utilize "p-type AlGaN" with Al concentrations as high as 60% but precise information concerning transport properties is unavailable. While other acceptor candidates such as beryllium and carbon have been suggested, the use of Mg remains to date the pragmatic path towards p-type conduction. We note that the superlattice doping approach, given the fundamental material constraints of dopant solubility and ionization energy, nonetheless suggests an enhancement of free hole concentration through field-assisted ionization of acceptors (Schubert *et al.* 1996). Illustrating an "extra" degree of freedom, utilization of piezoelectric and spontaneous polarization in AlGaN/GaN ($X_{Al} < 20\%$) superlattice structures was reported effective in boosting up the free hole concentration (Kozodoy et al. 1999; Saxler et al. 1999). One consequence of such tailoring of band profiles is the occurrence of space-charge region that tends to impede the vertical transport across the heterojunction superlattice planes. To reach beyond the band profile engineering along growth direction, it may be possible to exploit the field-assisted activation concept through band-edge fluctuation along the transverse (in-plane) directions. The compositional fluctuation in InGaN, embodied in the form of nano-clusters or quantum dots with bandgap differences of 50–300 meV, might be sufficient to create in-plane electric fields and further enhance the activation of Mg in p-type GaN:In or AlGaN:In (Kumakura et al. 2000b).

Finally, the evolution of the deeper III-nitride UV light emitter technology will also be dependent on advances in substrate development. The absence of high-quality homoepitaxial (GaN) substrates with large area remains a serious inhibiting factor to the development of III-nitrides devices and technologies. Both sapphire and silicon carbide substrates have been demonstrated viable to support heteroepitaxy of GaN with acceptable structural quality (dislocation density $< 10^9 \, \text{cm}^{-2}$) through innovative schemes of intermediate buffer layers. In the context of blue and green LEDs, as well as violet and near-UV edge-emitting lasers, in which the active regions are routinely grown on thick ($>3 \, \mu\text{m}$) GaN buffer layers or ELOG templates, the selection of these two substrates does not result in significant difference in device performance. The scenario is expected to change substantially in deeper UV emitters, however, due to a combination of structural and electrical issues. Structural quality of thick AlGaN template on sapphire, prepared through the two-step procedure and required to provide lattice matching to the AlGaN-containing active region, is found to deteriorate quickly as the Al content is increased (Amano et al. 1999). It is also speculated that the lateral overgrowth approach, which is so effective for defect reduction in GaN, cannot be effectively applied to AlGaN due to a limited surface diffusion coefficient of Al. In addition to the microstructural consideration, injection and horizontal transport of electrons in n-type AlGaN (Al $> 30\%$) will likely make sapphire (or other insulating wafers) a less favorable choice. The doping and metal contacting of p-type AlGaN is granted a challenge independent of wafer selection. It is expected, however, that the effect of an increasing resistivity in n-type AlGaN will be amplified due to the geometric constraints ($R = \rho l/A$) of dual front-contact scheme for standard geometry LEDs or edge-emitting lasers on insulating wafers. Metal contact to the mesa-etched n-type AlGaN surface adds yet another uncertainty to device processing. These adverse effects in electrical transport are largely removed with the use of conducting wafers such as n-type 6H SiC in a vertical configuration. Elimination of an extra (n-type) electrode would free up additional surface area for p-type transparent electrode, which should partially offset the loss of light extraction due to an opaque substrate. At this moment bulk GaN, AlN substrates, and thick GaN layers by hydride vapor phase epitaxy are still regarded as laboratory products; affordable mass production with consistent quality will inject new variables and undoubtedly alter the landscape again.

V. The GaInNAs Quaternary IR Lasers

While the frontier at blue and near-UV regime of the spectrum has been the main focus of this chapter for III-nitride based semiconductor sources, we briefly survey progress in the contrasting end of the spectrum, namely the 1.3–1.5 μm regime where the quaternary GaInNAs is emerging as a significant competitor to the more established GaInAsP material.

Semiconductor lasers in the 1.3–1.5 μm range form the optical source backbone of today's fiber communication systems. The main workhorse is based on the GaInAsP/InP quantum well distributed feedback Bragg-grating (DFB) edge-emitters, but there is intense development work in the areas of VCSELs as well. Gain switched DFB lasers operate in commercial systems at data rates of 10 Gb/s and are expected to dominate, at least initially, the expected upcoming 40 Gb/s transmission links as well. One of the (very few) weaknesses of the GaInAsP QW DFB laser concerns their temperature stability, attributed at least in part to the small conduction bad offsets in the GaInAsP/InP heterostructures. Ingenious schemes are being developed to address this issue. In case of VCSELs, an additional challenge concerns the small index of refraction contrast for GaInAsP/InP based DBR mirrors which require a very large number of layer pairs, a requirement which among other things introduces excess electrical resistance to the laser' s internal circuit.

It was recognized a few years ago that the quaternary $Ga_xIn_{1-x}N_{1-y}As_y$ could offer a number of advantages over the established IR lasers (Kondow et al. 1996). Elementary arguments for bandoffsets, inputting the large electronegativity of nitrogen, suggested that the conduction band offsets on the order of 500 meV could be achieved for $y \sim 0.01$ and $x \sim 0.7$. Moreover, pseudomorphic growth of GaInNAs on GaAs has been shown to be possible in this composition range with critical thickness on the order of 10 nm for the compressively strained QWs, thereby providing for an active medium for a $\lambda \sim 1.3$ μm laser with strong carrier confinement that can be maintained even at high temperatures.

In contrast with the III-nitride lasers at blue and NUV wavelengths, the GaInNAs ternary infrared lasers have been subject to successful growth by both molecular beam (MBE) and metallo-organic chemical vapor deposition (MOCVD) epitaxies on GaAs substrates, in that chronological order. Initial growth studies showed how the wavelength of spontaneous emission could be controlled by changing the concentration of In and N (or both). However, with increasing In concentration the compressive QW strain increases and thus the critical thickness of the QW decreases, creating a special challenge for reaching the 1.5 μm communication wavelengths. The strategy of compensating for the increase in compressive strain by increasing the nitrogen content has encountered the problem that the GaInNAs/GaAs QW luminescence intensity decreases and its spectral linewidth increases dramatically, when compared, for example, with GaInAsP, leading to large increases in the laser threshold current density due to broadening of the gain spectrum. There is strong evidence that the incorporation of nitrogen into InGaAs leads to pronounced alloy fluctuations, beyond the random alloy regime, and, depending on growth conditions, may be severe enough to cause pronounced segregation into In- and N-rich regions. This behavior is, or course, well documented in the widegap InGaN and AlInGaN blue and near-UV light emitter materials where it plays a key role in LED and laser device characteristics, as discussed extensively in Section III of this chapter. In the GaInNAs system, however, postgrowth thermal annealing can significantly aid in improving the In–Ga interdiffusion and are commonly employed (Xin *et al.* 1999).

Both GaInNAs/GaAs QW edge-emitting lasers and VCSELs have been demonstrated in the past several years, with continuous improvement in the threshold current density. Among the many groups that have reported on room temperature CW laser studies, we mention the low threshold current density $J_{th} = 270\,A/cm^2$ achieved by solid source MBE growth (Livshits *et al.* 2000) and the $J_{th} = 450\,A/cm^2$ for approximately 1.3 μm devices grown by MOCVD (Kawaguchi *et al.* 2001). A large variation in the threshold current and output powers can be found in recent literature, depending on the exact wavelength and device configuration (number of QWs, optical confinement scheme, and desired output power range). Edge-emitting lasers up to 1.52 μm have been recently demonstrated, although at high threshold current density (Fischer *et al.* 2000).

In comparison with the GaInAsP-based VCSELs, GaInNAs offers an immediate fabrication advantage for vertical cavity emitters, since the well-established, high index contrast GaAs/AlGaAs multilayer DBR reflectors can be readily incorporated during *in situ* growth. (Likewise, resonant cavity photodiodes are being actively pursued.) Following initial work with optically pumped structures, electrical injection of a 1.3 μm GaInAsP/GaAs MQW VCSEL was achieved (Choquette *et al.* 2000). Very recently, relatively low threshold >1.3 μm VCSELs have been grown by MOCVD (Takeuchi *et al.* 2002). The InGaAsN 3QW VCSEL structures were grown with a Si-doped, 40 pair bottom DBR, mirco-cavity, and a top 28 pair C-doped DBR including an oxidation layer. Room temperature CW operation of 1.305 μm VCSELs, with $J_{th} = 5.3\,kA/cm^2$ and maximum power over 0.7 mW at room temperature. Room temperature CW lasing was achieved at wavelengths as long as 1312 nm, with somewhat lower output power.

VI. Summary

The aim of this chapter has been to give a review of recent research activity, which has shaped a new generation of short wavelength III-nitride semiconductor sources. As shown by the many examples, seminal breakthroughs only a few years ago have led to these sources poised to impact a number of areas of optoelectronics and other technologies. For example, the question of using violet diode lasers in a future generation optical disk technology is most likely no longer a question of if but when, and largely independent of technical issues. On the other hand, while this progress would have been difficult to anticipate a decade ago, many exciting challenges and opportunities exists in the field, which in many ways has barely squeezed itself from the research laboratories. Among these are further developments in vertical cavity emitters and high brightness directional LEDs and, perhaps, most important, the further penetration of the III-nitride emitters into the UV for applications as varied as biomedical diagnostics and solid state lighting. For these, and related technological challenges, close coupling of basic material and wide bandgap semiconductor science to device design and fabrication is required.

REFERENCES

Akasaki, I., S. Sota, H. Sakai, T. Tanaka, M. Koike, and H. Amano, Shortest wavelength semiconductor laser diode, *Electron. Lett.* **32** (12), 1105 (1996).

Akiyama, T., M. Uno, H. Kitaura, K. Narumi, R. Kojima, K. Nishiuchi, and N. Yamada, Rewritable dual-layer phase-change optical disk utilizing a blue–violet laser, *Jpn. J. Appl. Phys.* **40** (3B), 1598 (2001).

Alferov, Zh., S. Ivanov, P. Kopeev, A. Lebedev, N. Ledentsov, M. Maximov, I. Sedova, T. Shubina, and A. Toropov, Exciton-induced enhancement of optical waveguide confinement in (Zn,Cd)(Se,S) quantum well laser heterostructures, *Superlat. Microstruct.*, **15**, 65 (1994).

Amano, H., M. Kito, K. Hiramatsu, and I. Akasaki, p-type conduction in Mg-doped GaN treated with low-energy electron beam irradiation (LEEBI), *Jpn. J. Appl. Phys.* **28** (12), L2112 (1989).

Amano, H., M. Iwaya, N. Hayashi, T. Kashima, S. Nitta, C. Wetzel, and I. Akasaki, Control of dislocations and stress in AlGaN on sapphire using a low temperature interlayer, *Phys. Stat. Sol. (b)* **216**, 683 (1999).

Bertram, F., T. Riemann, J. Christen, A. Keschner, A. Hoffmann, C. Thomsen, T. Shibata, and N. Sawaki, Strain relaxation and strong impurity incorporation in epitaxial laterally overgrown GaN: direct imaging of different growth domains by cathodoluminescence microscopy and micro-Raman spectroscopy, *Appl. Phys. Lett.* **74**, 359 (1999).

Bertram, F., S. Srinivasan, L. Geng, F. A. Ponce, T. Riemann, J. Christen, S. Tanaka, H. Omiya, and Y. Nakagawa, Spatial variation of luminescence of InGaN alloys measured by highly-spatially-resolved scanning catholuminescence, *Phys. Stat. Sol. (b)* **228**, 35 (2001).

Brown, J. D., J. Li, P. Srinivasan, J. Matthews, J. F. Schetzina, Solar-blind AlGaN heterostructure photodiodes, *MRS Int. J. Nitride Semicond. Res.* **5S1** (2000).

Bulman, G. E., K. Doverspike, S. T. Sheppard, T. W. Weeks, H. S. Kong, H. M. Dieringer, J. A. Edmond, J. D. Brown, J. T. Swindell, and J. F. Schetzina, Pulsed operation lasing in a cleaved-facet InGaN/GaN MQW SCH laser grown on 6H-SiC, *Electron. Lett.* **33** (18), 1556 (1997).

Chichibu, S., T. Azuhata, T. Sota, and S. Nakamura, Spontaneous emission of localized excitons in InGaN single and multiquantum well structures, *Appl. Phys. Lett.* **69**, 4188 (1996).

Chichibu, S., T. Azuhata, T. Sota, and S. Nakamura, Luminescences from localized states in InGaN epilayers, *Appl. Phys. Lett.* **70**, 2822 (1997a).

Chichibu, S., K. Wada, and S. Nakamura, Spatially resolved cathodoluminescence spectra of InGaN quantum well, *Appl. Phys. Lett.* **71**, 2346 (1997b).

, W. W., and S. W. Koch, Many-body Coulomb effects in room-temperature II–VI quantum well semiconductor lasers, *Appl. Phys. Lett.* **66**, 3004 (1995).

Choquette, K. D., J. F. Klem, A. J. Fischer, O. Blum, A. A. Allerman, I. J. Fritz, S. R. Kurtz, W. G. Breiland, R. Sieg, K. M. Geib, J. W. Scott, and R. L. Naone, Room temperature continuous wave inGaAsN quantum well vertical-cavity lasers emitting at 1.3 μm, *Electron. Lett.* **36**, 1388 (2000).

Cingolani, R., II–VI blue/green light emitters: device physics and epitaxial growth, in *Semiconductors and Semimetals*, Vol. 44, edited by R. L. Gunshor, and A. V. Nurmikko, Academic Press, New York, p. 163 (1996).

Crowell, P. A., D. K. Young, S. Keller, E. L. Hu, and D. D. Awschalom, Near-field scanning optical spectroscopy of an InGaN quantum well, *Appl. Phys. Lett.* **72**, 927 (1998).

DenBaars, S. P. and S. Keller, in *Gallium Nitride*, edited by J. Pankove, and T. Moustakas, Chapter 5, Academic Press, San Diego, p. 11 (1999).

Diagne, M., Y. He, H. Zhou, E. Makarona, A. V. Nurmikko, J. Han, K. E. Waldrip, J. J. Figiel, T. Takeuchi, and M. Krames, A vertical cavity violet light emitting diode incorporating an AlGaN distributed bragg mirror and a tunnel junction, *Appl. Phys. Lett.* **79**, 3720 (2001).

Ding, J., H. Jeon, T. Ishihara, M. Hagerott, A. V. Nurmikko, N. Samarth, H. Luo, and J. Furdyna, Excitonic gain and laser emission in ZnSe-based quantum wells, *Phys. Rev. Lett.* **60**, 1707 (1992).

Ding, J., M. Hagerott, P. Kelkar, A. V. Nurmikko, D. C. Grillo, L. He, J. Han, and R. L. Gunshor, The role of coulomb correlated electron–hole pairs in in ZnSe-based quantum well diode lasers, *Phys. Rev. B* **50**, 5787 (1994).

Edmond, J., G. Bulman, H. S. Kong, M. Leonard, K. Doverspike W. Weeks, J. Niccum, S. Sheppard, G. Negley, J. D. Brown, J. T. Swindell, T. Overocker, J. F. Schetzina, Y-K. Song, M. Kuball, and A. Nurmikko, *Proceedings of the Second International Conference on Nitride Semiconductors ICSNS'97*, Tokushima, Japan, 448 (1997).

Fan, Y., J. Han, L. He, J. Saraie, R. L. Gunshor, M. Hagerott, H. Jeon, A. V. Nurmikko, G. C. Hua, and N. Otsuka, Graded band gap ohmic contact to p-ZnSe, *Appl. Phys. Lett.* **61** (26), 3160 (1992).

Fischer, M., M. Reinhardt, and A. Forchel, GaInNAs/GaAs laser diodes operating at 1.52 μm, *Electron. Lett.* **36**, 1208 (2000).

Goto, S., T. Tojyo, Y. Yabuki, T. Hino, S. Ansai, H. Yamanaka, Y. Moriya, Y. Ito, Y. Hamaguchi, S. Uchida, and M. Ikeda, Super high output power over 3 W in blue–violet AlGaInN laser diode array, *International Symposium on Compound Semiconductors*, Tokyo (2001).

Gotz, W., and N. M. Johnson, in *Gallium Nitride*, edited by J. Pankove, and T. Moustakas, Chapter 5, Academic Press, San Diego, p. 185 (1999).

Gotz, W., N. M. Johnson, J. Walker, D. P. Bour, and R. A. Street, Activation of acceptors in Mg-doped GaN grown by metalorganic chemical vapor deposition, *Appl. Phys. Lett.* **68** (5), 667 (1996).

Gunshor, R. L., and A. V. Nurmikko, editors, *II–VI Blue–Green Light Emitters: Device Physics and Epitaxial Growth, Semiconductors and Semimetals*, Vol. 44, Academic Press, New York (1997).

Haase, M. A., J. Qiu, J. M. DePuydt, and H. Cheng, Blue–green diode lasers, *Appl. Phys. Lett.* **59**, 1272 (1991).

Han, J., M. D. Ringle, Y. Fan, R. L. Gunshor, and A. V. Nurmikko, D (donor) X center behavior for holes implied from observation of metastable acceptor states, *Appl. Phys. Lett.* **65** (25), 3230 (1994).

Han, J., J. J. Figiel, G. A. Petersen, S. M. Myers, M. H. Crawford, and M. A. Banas, Metalorganic vapor-phase epitaxial growth and characterization of quaternary AlGaInN, *Jpn. J. Appl. Phys.* **39**, 2372 (2000).

Han, J., K. E. Waldrip, S. R. Lee, J. J. Figiel, S. J. Hearne, G. A. Petersen, and S. M. Myers, Control and elimination of cracking of AlGaN using low-temperature AlGaN interlayers, *Appl. Phys. Lett.* **78**, 67 (2001).

Haug, H., and S. W. Koch, *Quantum Theory of the Optical and Electronic Properties of Semiconductors*, World Scientific, Singapore (1993).

He, Y., Q. Zhang, A. V. Nurmikko, J. Han, T. Takeuchi, and M. Krames (in press) (2002).

Henry, C. H., R. A. Logan, and F. R. Merritt, Measurement of gain and absorption in AlGaAs lasers, *J. Appl. Phys.* **51**, 3042 (1980).

Hiei, F., M. Ikeda, M. Ozawa, T. Miyajima, A. Ishibashi, and K. Akimoto, Ohmic contacts to p-type ZnSe using ZnTe/ZnSe multiquantum wells, *Electron. Lett.* **29** (10), 878 (1993).

Hirayama, H., A. Kinoshita, A. Hirata, and Y. Aoyagi, Growth and optical properties of quaternary InAlGaN for 300 nm band UV-emitting devices, *Phys. Stat. Sol. (a)* **188**, 83 (2001).

Ho, J.-K., C.-S. Jong, C.-N. Huang, C.-Y. Chen, C. C. Chien, K.-K. Shih, Low-resistance ohmic contacts to p-type GaN, *Appl. Phys. Lett.* **74** (9) 1275 (1999).

Honda, T., A. Katsube, T. Sakaguchi, F. Koyama, and K. Iga, Threshold estimation of GaN-based surface emitting lasers operating in ultraviolet spectral region, *Jpn. J. Appl. Phys.* **34** (7a), 3527 (1995).

Hull, B. A., S. E. Mohney, J. C. Ramer, and H. S. Venugopalan, Influence of oxygen on the activation of p-type GaN, *Appl. Phys. Lett.* **76** (16), 2271 (2000).

Im, J. S., H. Kollmer, J. Off, A. Sohmer, F. Scholz, and A. Hangleiter, Reduction of oscillator strength due to piezoelectric fields in GaN/Al$_x$Ga$_{1-x}$N quantum wells, *Phys. Rev. B* **57**, R9435 (1998).

Jeon, H., J. Ding, W. Patterson, A. V. Nurmikko, W. Xie, D. C. Grillo, M. Kobayashi, and R. L. Gunshor, Blue–green injection laser diodes in (Zn,Cd)Se/ZnSe quantum wells, *Appl. Phys. Lett.* **59**, 3619 (1991).

Jeon, S.-R., Y-H. Song, H-J. Jang, G. M. Yang, S. W. Hwang, and S. J. Son, Lateral current spreading in GaN-based light-emitting diodes utilizing tunnel contact junctions, *Appl. Phys. Lett.* **78**, 3265 (2001).

Katayama, K, H. Yao, F. Nakanishi, H. Doi, A. Saegusa, N. Okuda, T. Yamada, H. Matsubara, M. Irikura, T. Matsuoka, T. Takebe, S. Nishine, and T. Sirakawa, Lasing characteristics of low threshold ZnSe-based blue/green laser diodes grown on conductive ZnSe substrates, *Appl. Phys. Lett.* **73**, 102 (1998).

Katsuragawa, M., S. Sota, M. Komori, C. Anbe, T. Takeuchi, H. Sakai, H. Amano, and I. Akasaki, Thermal ionization energy of Si and Mg in AlGaN, *J. Crystal. Growth* **189–190**, 528 (1998).

Kawakami, K., I. Hauksson, J. Simpson, H. Stewart, I. Gailbraith, K. A. Prior, and B. C. Cavenett, Photoluminescence excitation spectroscopy of the lasing transition in ZnSe-(Zn,Cd)Se quantum wells, *J. Crystal. Growth* **138**, 759 (1994).

Kawaguchi, M., T. Miyamoto, E. Gouardes, D. Schlenker, T. Kondo, F. Koyama, and K. Iga, Lasing characteristics of low-threshold GaInNAs lasers grown by metalorganic chemical vapor deposition, *Jpn. J. Appl. Phys.* **40**, L744 (2001).

Kelly, M. K., O. Ambacher, R. Dimitrov, R. Handschuh, and M. Stutzmann, Optical process for liftoff of Group III-nitride films, *Phys. Stat. Sol.* A **159**, R3 (1997).

Kesler, M. P., and C. Harder, Excitonic effects in gain and index in GaAlAs quantum well lasers, *Appl. Phys. Lett.* **57**, 123 (1990).

Khan, M. A., V. Adivarahan, J. Zhang, C. Chen, E. Kuokstis, A. Chitnis, M. Sahatalov, J. Yang, and G. Simin, Stripe geometry ultraviolet light emitting diodes at 305 nanometers using quaternary AlInGaN multiple quantum wells, *Jpn. J. Appl. Phys.* **40**, L1308 (2001).

Kijima, S., T. Tojyo, S. Goto, M. Takeya, T. Asano, T. Hino, S. Uchida, and M. Ikeda, Novel techniques for stabilizing transverse mode in AlGaInN-based laser diodes, *Phys. Stat. Sol. (a)* **188**, 55 (2001).

Kinoshita, A., H. Hirayama, M. Ainoya, Y. Aoyagi, and A. Hirata, Room-temperature operation at 333 nm of AlGaN/AlGaN quantum-well light-emitting diodes with Mg-doped superlattice layers, *Appl. Phys. Lett.* **77**, 175 (2000).

Kisielowski, C., Z. Lilienthal-Weber, and S. Nakamura, Atomic scale indium distribution in a GaN/In$_{0.43}$Ga$_{0.57}$N/Al$_{0.1}$Ga$_{0.9}$N quantum well structure, *Jpn. J. Appl. Phys.* **36**, 6932 (1997).

Kneissl, M., C. VandeWalle, D. P. Bour, L. T. Romano, C. P. Master, J. E. Northrup, and N. M. Johnson, Performance and optical gain characteristics of InGaN QW laser diodes, *J. Lumin.* **135**, 97 (2000).

Kneissl, M., W. S. Wong, D. W. Treat, M. Teepe, N. Myiashita, and N. M. Johnson, Continuous-wave operation of InGaN multiple-quantum-well laser diodes on copper substrates obtained by laser lift-off, *IEEE J. Select. Top. Quantum Electron.* **7** (2), 188 (2001).

Kollmer, H., J. S. Im, S. Heppel, J. Off, F. Scholz, and A. Hangleiter, Intra- and interwell transitions in GaInN/GaN multiple quantum wells with built-in piezoelectric fields, *Appl. Phys. Lett.* **74**, 82 (1999).

Kondow, M., K. Uomi, A. Niwa, T. Kitatani, S. Watahiki, and Y. Yazawa, GaInAsN: A novel material for long wavelength laser diodes with excellent high temperature performance, *Jpn. J. Appl. Phys.* **35**, 1273 (1996).

Kozodoy, P., M. Hansen, S. P. DenBaars, and U. K. Mishra, Enhanced Mg doping efficiency in Al$_{0.2}$Ga$_{0.8}$N/GaN superlattices, *Appl. Phys. Lett.* **74** (24), 3681 (1999).

Krestnikov, I., W. Lundin, A. V. Sakharov, V. Semenov, A. Usikov, A. F. Tsatsulnikov, Zh. Alferov, N. Ledentsov, A. Hoffmann, and D. Bimberg, Room-temperature photopumped InGaN/GaN/AlGaN vertical-cavity surface-emitting laser, *Appl. Phys. Lett.* **75**, 1192 (1999).

Kuball M., E.-S. Jeon, Y.-K. Song, A. V. Nurmikko, P. Kozodoy, A. Abare, S. Keller, L. A. Coldren, U. K. Mishra, S. P. DenBaars, and D. A. Steigerwald, Gain spectroscopy on InGaN/GaN quantum well diodes, *Appl. Phys. Lett.* **70**, 2580 (1997).

Kumakura, K., T. Makimoto, and Kobayashi, N. Activation energy and electrical activity of Mg in Mg-doped In$_x$Ga$_{1-x}$N (X 0.2), *Jpn. J. Appl. Phys.* **39**, L337 (2000a).

Kumakura, K., T. Makimoto, and N. Kobayashi, High hole concentrations in Mg-doped InGaN grown by MOVPE, *J. Crystal. Growth* **221**, 267 (2000b).

Langer, R., A. Barski, J. Simon, N. Pelekanos, O. Konovalov, R. Andre, and L. S. Dang, High-reflectivity GaN/GaAlN Bragg mirrors at blue/green wavelengths grown by molecular beam epitaxy, *Appl. Phys. Lett.* **74**, 3610 (1999).

Lefebvre, P., T. Taliercio, S. Kalliakos, A. Morel, X. B. Zhang, M. Gallart, T. Bretagnon, B. Gil, N. Grandjean, B. Damilano, and J. Massies, Carrier dynamics in group-III nitride low-dimensional systems: localization versus quantum-confined stark effect, *Phys. Stat. Sol. (b)* **228**, 65 (2001).

Lester, S. D., F. A. Ponce, M. G. Craford, and D. A. Steigerwald, High dislocation densities in high efficiency GaN-based light-emitting diodes, *Appl. Phys. Lett.* **66**, 1249 (1995).

Livshits, D. A., A. Yu, Egorov, and H. Riechert, *Electron. Lett.* **36**, 1381 (2000).

Mackowiak, P., R. P. Sarzala, and W. Nakwaski, Novel design for nitride VCSELs, *Proc. Int. Workshop on Nitride Semiconductors, IPAP Conf. Series* **1**, 889 (2001).

Molnar, R. J., W. Goetz, L. T. Romano, N. M. Johnson, Growth of gallium nitride by hydride vapor-phase epitaxy, *J. Crystal Growth* **178** (1–2),147 (1997).

Monemar, B., P. P. Paskov, G. Pozina, T. Paskova, J. P. Bergman, M. Iwaya, S. Nitta, H. Amano, and I. Akasaki, Optical characterization of InGaN/GaN MQW structures without in phase separation, *Phys. Stat. Sol. (b)* **288**, 157 (2001).

Monroy, E., F. Calle, J. L. Pau, E. Munoz, F. Omnes, B. Beaumont, and P. Gibart, Application and performance of GaN based UV detectors, *Phys. Stat. Sol. (a)* **185**, 91 (2001).

Nagahama, S., N. Iwasa, M. Senoh, T. Matsushita, Y. Sugimoto, H. Kiyoku, T. Kozaki, M. Sano, H. Matsumura, H. Umemoto, K. Chocho, and T. Mukai, High-power and long-lifetime InGaN multi-quantum-well laser diodes grown on low-dislocation-density GaN substrates, *Jpn. J. Appl. Phys.* **39**, L647 (2000).

Nagahama, S., N. Iwasa, M. Senoh, T. Matsushita, Y. Sugimoto, H. Kiyoku, T. Kozaki, M. Sano, H. Matsumura, H. Umemoto, K. Chocho, and T. Mukai, *Phys. Stat. Sol. (a)* **188**, 1 (2001).

Nagahama *et al.* (2001a).

Nagahama, S., T. Yanamoto, M. Sano, and T. Mukai, Ultraviolet GaN single quantum well laser diodes, *Jpn. J. Appl. Phys.* **40**, L785 (2001b).

Nagahama, S., T. Yanamoto, M. Sano, and T. Mukai, Characteristics of ultraviolet laser diodes composed of quaternary $Al_xIn_yGa_{(1-x-y)}N$, *Jpn. J. Appl. Phys.* **40**, L785 (2001c).

Nakamura, S., InGaN-based violet laser diodes, *Semicond. Sci. Technol.* **14**, R27 (1999).

Nakamura, S., and G. Fasol, *The Blue Laser Diode*, Springer, Berlin (1997).

Nakamura, S., N. Iwasa, M. Senoh, and T. Mukai, Hole compensation mechanism of P-type GaN films, *Jpn. J. Appl. Phys.* **31** (5A), 1258 (1992).

Nakamura, S., M. Senoh, and T. Mukai, p-GaN/N-InGaN/N-GaN double heterostructure blue-light-emitting diodes, *Jpn. J. Appl. Phys.* **32** (1A-B), L8 (1993).

Nakamura, S., M. Senoh, S. Nagahama, N. Iwasa, T. Yamada, T. Matsushita, H. Kiyoku, and Y. Sugimoto, InGaN-based multi-quantum-well-structure laser diodes, *Jpn. J. Appl. Phys.* **35** (1B), L74 (1996).

Nakamura, S., M. Senoh, S.-I. Nagahama, N. Iwasa, T. Yamada, T. Matsushita, H. Kiyoku, Y. Sugimoto, T. Kozaki, H. Umemoto, M. Sano, and K. Chocho, InGaN/GaN/AlGaN-based laser diodes with modulation-doped strained-layer superlattices grown on an epitaxially laterally overgrown GaN substrate, *Appl. Phys. Lett.* **72** (2), 211 (1998).

Nakamura, S., M. Senoh, S. Nagahama, T. Masushita, H. Kiyoku, Y. Sugimoto, T. Kozaki, H. Umemoto, M. Sano, and T. Mukai, Violet InGaN/GaN/AlGaN-based laser diodes operable at 50 degrees C with a fundamental transverse mode, *Jpn. J. Appl. Phys.* **38**, L226 (1999).

Nam, O.-H., M. D. Bremser, T. S. Zheleva, and R. F. Davis, Lateral epitaxy of low defect density GaN layers via organometallic vapor phase epitaxy, *Appl. Phys. Lett.* **71** (18), 2638 (1997).

Narukawa, Y., Y. Kawakami, Sz. Fujita, Sg. Fujita, and S. Nakamura, Recombination dynamics of localized excitons in $In_{0.20}Ga_{0.80}N$–$In_{0.05}Ga_{0.95}N$ multiple quantum wells, *Phys. Rev. B* **55**, R1938 (1997).

Nishida T., and N. Kobayashi, Ten-milliwatt operation of an AlGaN-based light emitting diode grown on GaN substrate, *Phys. Stat. Sol. (a)* **188**, 113 (2001).

Ng, H. M., T. D. Moustakas, and S. N. G. Chu, High reflectivity and broad bandwidth AlN/GaN distributed Bragg reflectors grown by molecular-beam epitaxy, *Appl. Phys. Lett.* **76**, 2818 (2000).

Ortsiefer, M., R. Shau, G. Bohm, F. Kohler, G. Abstreiter, and M.-C. Amann, Low-resistance InGa(Al)As tunnel junctions for long wavelength vertical-cavity surface-emitting lasers, *Jpn. J. Appl. Phys.* **39** (4A), 1727 (2000).

Ozden, I., M. Diagne, A. V. Nurmikko, J. Han, and T. Takeuchi, A matrix addressable 1024 element blue light emitting InGaN QW diode array, *Phys. Stat. Sol (b)*, (in press) (2001a).

Ozden, I., E. Makarona, A. V. Nurmikko, T. Takeuchi, and M. Krames, A dual-wavelength indium gallium nitride quantum well light emitting diode, *Appl. Phys. Lett.* **79**, 2532 (2001b).

Parish, G. S. Keller, P. Kozodoy, J. P. Ibbetson, H. Marchand, P. T. Fini, S. B. Fleischer, S. P. DenBaars, and U. K. Mishra, and E. J. Tarsa, Effect of growth termination conditions on the performance of AlGaN/GaN high electron mobility transistors, *Appl. Phys. Lett.* **75**, 247 (1999).

Pelekanos, N. T., J. Ding, M. Hagerott, A. V. Nurmikko, N. Samarth, and J. Furdyna, Quasi-two dimensional excitons in (Zn,Cd)Se/ZnSe quantum wells: reduced exciton–LO phonon coupling due to confinement effects, *Phys. Rev. B* **45**, 6037 (1992).

Ponce, F. A., J. S. Major, Jr., W. E. Plano, and D. F. Welch, Crystalline structure of AlGaN epitaxy on sapphire using AlN buffer layers, *Appl. Phys. Lett.* **65**, 2302 (1994).

Porowski, S., M. Bockowski, B. Lucznik, M. Wroblewski, S. Krukowski, I. Grzegory, M. Leszczynski, G. Nowak, K. Pakula, and J. Baranowski, GaN crystals grown in the increased volume high pressure reactors, *Mat Res. Symp. Proc.* **449**, 35 (1997).

Redwing, J. M., D. A. S. Loeber, N. G. Anderson, M. A. Tischler, and J. S. Flynn, An optically pumped GaN–AlGaN vertical cavity surface emitting laser, *Appl. Phys. Lett.* **69**, 1 (1996).

Rojo, J. C., G. A. Slack, K. Morgan, B. Raghothamachar, M. Dudley, and L. J. Schowalter, Report on the growth of bulk aluminum nitride and subsequent substrate preparation, *J. Crystal. Growth* **231**, 317 (2001).

Sakai, A., H. Sunakawa, and A. Usui, Transmission electron microscopy of defects in GaN films formed by epitaxial lateral overgrowth, *Appl. Phys. Lett.* **73**, 481 (1998).

Saxler, A., W. C. Mitchel, P. Kung, and M. Razeghi, Aluminum gallium nitride short-period superlattice doped with magnesium, *Appl. Phys. Lett.* **74**, 2023 (1999).

Schubert, E. F., W. Grieshaber, and I. D. Goepfert, Enhancement of deep acceptor activation in semiconductors by superlattice doping, *Appl. Phys. Lett.* **69**, 3737 (1996).

Sheu, J. K., Y. K. Su, G. C. Chia, P. L. Koh, M. J. Jou, C. M. Chang, C. C. Liu, and W. C. Hung, High-transparency Ni/Au ohmic contact to p-type GaN, *Appl. Phys. Lett.* **74** (16), 2340 (1999).

Someya, T., and Y. Arakawa, Highly reflective $GaN/Al_{0.34}Ga_{0.66}N$ quarter-wave reflectors grown by metal organic chemical vapor deposition, *Appl. Phys. Lett.* **73**, 3653 (1998).

Someya, T., R. Werner, A. Forchel, M. Catalano, R. Cingolani, and Y. Arakawa, Room temperature lasing at blue wavelengths in gallium nitride microcavities, *Science* **285**, 1905 (1999).

Song, Y.-K., A. V. Nurmikko, T. Schmiedel, C.-C. Chu, J. Han, W.-L. Chen, and R. L. Gunshor, Spectroscopy of a ZnCdSe/ZnSSe quantum well diode laser in high magnetic fields, *Appl. Phys. Lett.* **71**, 2874 (1997).

Song, Y.-K., M. Kuball, A. V. Nurmikko, G. E. Bulman, K. Doverspike, S. T. Sheppard, T. W. Weeks, M. Leonard, H. S. Kong H. Dieringer, and J. Edmond, Gain characteristics of InGaN/GaN quantum well diode lasers, *Appl. Phys. Lett.* **72**, 1418 (1998).

Song, Y.-K., Private Communication (1998a).

Song, Y.-K., H. Zhou, M. Diagne, I. Odzen, A. Vertikov, A. V. Nurmikko, C. Carter-Coman, S. Kern, F. A. Kish, and M. R. Krames, A vertical cavity light emitting InGaN QW heterostructure, *Appl. Phys. Lett.* **74**, 3441 (1999).

Song, Y.-K., M. Diagne, H. Zhou, A. V. Nurmikko, R. P. Schneider Jr., and T. Takeuchi, Resonant cavity InGaN quantum well blue light emitting diodes, *Appl. Phys. Lett.* **77**, 1744 (2000).

Song, Y.-K, A. V. Nurmikko, R. P. Schneider, C. P. Kuo, M. R. Krames, R. S. Kern, C. Carter-Coman, and F. A. Kish, A quasicontinuous wave, optically pumped violet vertical cavity surface emitting laser, *Appl. Phys. Lett.* **76**, 1662 (2000b).

Stampfl, C., J. Neugebauer, and C. G. Van de Walle, Doping of AlGaN alloys, *Mater. Sci. Eng. B* **59**, 253 (1999).

Takeuchi, T., G. Hasnain, M. Hueschen, C. Kocot, M. Blomqvist, Y-.L., Chang, D. Lefforge, R. Schneider, M. R. Krames, L. W. Cook, and S. A. Stockman, GaN-based light emitting diodes with tunnel junctions, *Jpn. J. Appl. Phys.* **40**, 861 (2001).

Takeuchi, T., Y.-L. Chang, M. Leary, A. Tandon, H.-C. Luan, D. Bour, S. Corzine, R. Twist, and M. Tan, Low threshold 1.3 μm InGaAsN vertical cavity surface emitting lasers grown by metalorganic chemical vapor deposition, *SPIE Conference*, San Jose, CA (2002).

Tojyo, T., T. Asano, M. Takeya, T. Hino, S. Kijima, S. Goto, S. Uchida, and M. Ikeda, GaN-based high power blue-violet laser diodes, *Jpn. J. Appl. Phys.* **40**, 3206 (2001).

Usui, A., H. Sunakawa, A. Sakai, and A. A. Yamaguchi, Thick GaN epitaxial growth with low dislocation density by hydride vapor phase epitaxy, *Jpn. J. Appl. Phys.* **36** (7B), L899 (1997).

Van de Walle, C., and N. M. Johnson, Hydrogen in III–V nitrides, in *Semiconductors and Semimetals*, Vol. 57, Academic Press, New York, p. 157 (1999).

Vertikov, A., A. V. Nurmikko, K. Doverspike, G. Bulman, and J. Edmond, Role of localized and extended electronic states in InGaN/GaN multiple quantum wells under high injection, inferred from near-field optical microscopy, *Appl. Phys. Lett.* **73**, 493 (1998a).

Vertikov, A., M. Kuball, A. V. Nurmikko, Y. Chen, and S.-Y. Wang, Near-field optical study of InGaN/GaN quantum wells, *Appl. Phys. Lett.* **72**, 2645 (1998b).

Vertikov A., I. Ozden, and A. V, Nurmikko, Diffusion and relaxation of excess carriers in InGaN quantum wells in localized versus extended states, *J. Appl. Phys.* **86**, 4697 (1999).

Wierer, J. J., D. A. Kellogg, and N. Holonyak, Jr. Tunnel contact junction native-oxide aperture and mirror vertical-cavity surface-emitting lasers and resonant-cavity light-emitting diodes, *Appl. Phys. Lett.* **74** (7), 926 (1999).

Wong, W. S., T. Sands, and N. W. Cheung, Damage-free separation of GaN thin films from sapphire substrates, *Appl. Phys. Lett.* **72**, 599 (1998).

Wong, M. M., J. C. Denyszyn, C. J. Collins, U. dhury, T. G. Zhu, K. S. Kim, and R. D. Dupuis, AlGaN/AlGaN double-heterojunction ultraviolet light-emitting diodes grown by metal organic chemical capour deposition, *Electron. Lett.* **37**, 1188 (2001).

Xin, X. P., K. L. Kavanagh, M. Kondow, and C. W. Tu, Effects of rapid thermal annealing on GaInAsN/GaAs multiple quantum wells, *J. Crystal. Growth* **201**, 419 (1999).

Zhang, J. P., E. Kuokstis, Q. Fareed, H. N. Wang, J. W. Yang, G. Simin, M. Asof Khan, G. Tamulaitis, G. Kurilcik, S. Jursenas, A. Zukauskas, R. Gaska, and M. Shur, Pulsed atomic layer epitaxy of quaternary AlInGaN layers for ultraviolet light emitters, *Phys. Stat. Sol. (a)* **188**, 95 (2001).

Zhou, H., M. Diagne, E. Makarona, A. V. Nurmikko, J. Han, K. E. Waldrip, and J. J. Figiel, Near ultraviolet optically pumped vertical cavity laser, *Electron. Lett.* **36**, 1777 (2000).

Zhou, L., A. T. Ping, F. Khan, A. Osinsky, and I. Adesida, Ti/Pt/Au ohmic contacts on p-type GaN/Al$_x$Ga$_{1-x}$N superlattices, *Electron. Lett.* **36** (1), 91 (2000).

Index

407

ISBN 0-12-507062-4